MANUFACTURING AUTOMATION

METAL CUTTING MECHANICS, MACHINE TOOL VIBRATIONS, AND CNC DESIGN

Second Edition

YUSUF ALTINTAS

University of British Columbia

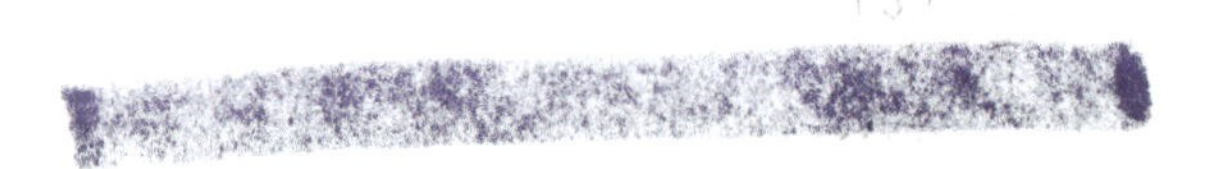

CAMBRIDGE UNIVERSITY PRESS
Cambridge, New York, Melbourne, Madrid, Cape Town,
Singapore, São Paulo, Delhi, Tokyo, Mexico City

Cambridge University Press
32 Avenue of the Americas, New York, NY 10013-2473, USA

www.cambridge.org
Information on this title: www.cambridge.org/9780521172479

First published 2012

Printed in the United States of America

A catalog record for this publication is available from the British Library.

Library of Congress Cataloging in Publication data

Altintas, Yusuf, 1954–
Manufacturing automation : metal cutting mechanics, machine tool vibrations, and CNC design /
Yusuf Altintas. – 2nd ed.
p. cm.
Includes bibliographical references and index.
ISBN 978-1-107-00148-0 (hardback) – ISBN 978-0-521-17247-9 (paperback)
1. Machining – Automation. 2. Machine-tools – Vibration. 3. Machine-tools – Numerical
control. I. Title.
TJ1185.5.A48 2011
670–dc23 2011041033

ISBN 978-1-107-00148-0 Hardback
ISBN 978-0-521-17247-9 Paperback

MANUFACTURING AUTOMATION

Second Edition

Metal cutting is one of the most widely used methods of producing the final shape of manufactured products. The technology involved in metal cutting operations has advanced considerably in recent years, along with developments in materials, computers, and sensors.

This book treats the scientific principles of metal cutting and their practical application to solving problems encountered in manufacturing. The subjects of mathematics, physics, computers, and instrumentation are discussed as integration tools in analyzing or designing machine tools and manufacturing processes.

The book begins with the fundamentals of metal cutting mechanics. The basic principles of vibration and experimental modal analysis are applied to solving problems on the shop floor. A special feature is the in-depth coverage of chatter vibrations, a problem experienced daily by practicing manufacturing engineers. The essential topics of programming, design, and automation of computer numerically controlled (CNC) machine tools, numerically controlled (NC) programming, and computer-aided design/computer-aided manufacturing (CAD/CAM) technology are discussed. The text also covers the selection of drive actuators, feedback sensors, the modeling and control of feed drives, the design of real-time trajectory generation and interpolation algorithms, and CNC-oriented error analysis in detail. Each chapter includes examples drawn from industry, design projects, and homework problems.

Advanced undergraduate and graduate students, and practicing engineers, as well, will find in this book a clear and thorough way to learn the engineering principles of metal cutting mechanics, machine tool vibrations, CNC system design, sensor-assisted machining, and CAD/CAM technology.

Yusuf Altintas is a Fellow of the Royal Society of Canada and NSERC Pratt & Whitney Canada Research Chair Professor of Mechanical Engineering and Director of the Manufacturing Automation Laboratory at the University of British Columbia.

CONTENTS

PREFACE

Metal cutting is one of the most widely used manufacturing processes to produce the final shape of products, and its technology continues to advance in parallel with developments in materials, computers, sensors, and actuators. A blank is converted into a final product by cutting extra material away by turning, drilling, milling, broaching, boring, and grinding operations conducted on computer numerically controlled (CNC) machine tools. The second edition of this book helps students and engineers understand the scientific principles of metal cutting technology and the practical application of engineering principles to solving problems encountered in manufacturing shops. The book reflects the author's industrial and research experience, and his manufacturing engineering philosophy as well.

Engineers can learn best by being shown how to apply the fundamentals of physics to actual machines and processes that they can feel and visualize. Mathematics, physics, computers, algorithms, and instrumentation then become useful integration tools in analyzing or designing machine tools and machining processes.

Metal cutting operations take place between a cutting tool and workpiece material mounted on a machine tool. The motion of the machine tool is controlled by its CNC unit, and the numerically controlled (NC) commands to CNC are generated on computer-aided design/computer-aided manufacturing (CAD/CAM) systems. The productivity and accuracy of the metal removal operation depend on the preparation of NC programs, planning of machining process parameters and cutting conditions, cutter geometry, work and tool materials, machine tool rigidity, and performance of the CNC unit. Manufacturing engineers who are involved in machining and machine tool technology must be familiar with each of these topics. It is equally important to link them and to be able to apply them in an interdisciplinary fashion to solve machining problems.

The beginning chapters of this book provide detailed mathematical models of metal cutting, milling, turning, and drilling operations. The macromechanics of cutting, which is applicable to solving problems on the shop floor and in machine tool design, is emphasized. Although required in work and tool material design – the micromechanics of cutting – basic principles of machinability, tool wear mechanisms, and chipping are briefly introduced to provide a complete picture. The design of machine tools requires knowledge of structures, mechanics of solids, vibrations, and kinematics, subjects that are covered in dedicated mechanical engineering texts. This text builds on that knowledge, applying the principles of vibration and experimental modal analysis to machine tools and metal cutting. Mathematical methods are simplified so that they can be easily used to solve machining vibration problems. Chatter vibrations in machining are treated in depth in this text because the problem is experienced daily by practicing manufacturing engineers.

The last three chapters of the book are dedicated to programming, design, and automation of CNC machine tools. Numerically controlled programming and CAD/CAM technology are briefly covered, but with sufficient explanation so that the reader can start programming and using CNC machine tools. The selection of drive actuators, feedback sensors, modeling and analysis of feed drives, the design of real-time trajectory generation and interpolation algorithms, and CNC-oriented error analysis are presented in more detail than can be found in other texts. Open CNC design philosophy and improvement of accuracy and productivity by adding sensors and algorithms to CNC machine tools are also covered.

Students learn best by dealing with real manufacturing problems. The contents of this book are based on experimentally proven engineering principles that are widely used in applied research laboratories and industry. The examples and problems presented in each chapter originate from the research and industrial problems solved by the author and his graduate students. Interdisciplinary problems are posed as industrial projects so that readers can apply all the necessary techniques simultaneously. They solve the basic metal cutting mechanics problem first, followed by milling mechanics, static deflection of end mills and corresponding surface-form error modeling, and vibration model of the end mill and chatter stability. For example, the chain of knowledge is exercised in solving problems associated with milling of an aircraft structure, a project that originated from industry. Similarly, in another project, the reader is guided step by step through the programming, real-time modeling, and control of a CNC machine tool. Because all the projects were tried in the author's laboratory, a number of teaching and research setups are provided in the book to aid instructors.

The book is intended as a text for senior undergraduate and graduate students and practicing manufacturing engineers who wish to learn the engineering principles of metal cutting, machine tool vibration, experimental modal analysis, NC programming and CAD/CAM technology, CNC system design, and sensor-based machining. The book can also be used by researchers who wish to study metal cutting mechanics, machine tool vibrations, feed drive design and control, and CNC and sensor-based machining.

Acknowledgments

The contents of each chapter mostly originated from the author's own engineering, research, and teaching experience. Each chapter is based on a number of graduate student theses supervised at the Manufacturing Automation

Laboratory founded and directed by the author at the University of British Columbia.

The studies of former graduate students and associates E. Budak, A. Spence, E. Shamoto, I. Lazoglu, Haikun Ren, and F. Atabey contributed to Chapter Two, which deals with metal cutting. The theses of E. Budak, S. Engin, P. Lee, S. Park, M. Namazi, D. Merdol, J. Roukema, Z. Dombavari, and M. Eynian were very helpful in writing Chapters Three and Four, where machine tool vibrations are presented. The theses of K. Erkorkmaz, B. Sencer, and C. Okwudire were helpful in writing Chapters Five and Six, which presents the principles of CNC design. The graduate research theses of N. A. Erol and K. Munasinghe were instrumental in forming the final chapter, which deals with sensor-fused machining. The author acknowledges the contributions of all of his former and present graduate students to the accumulation and dissemination of knowledge in machining, machine tools, machine tool vibrations, and control.

Several machinists, engineers, and professors contributed to the author's manufacturing engineering experience. The author received significant practical training from the machine tool design engineers, technologists, and machinists at M.K.E. Top Otomotiv Factory in Kirikkale, Turkey; the machinists and the process planners at Pratt & Whitney Canada in Montreal; and those at Canadian Institute of Metal Working in Hamilton. The author's basic engineering education with rich machine design and analytical content from Istanbul Technical University, CAD/CAM education from the University of New Brunswick, and machine tool engineering background from McMaster University were most valuable in overall development of his manufacturing engineering and research skills. Professors G. Pritschow, U. Heisel, T. Moriwaki, F. Klocke, M. Weck, H. van Brussel, G. Bryne, G. Stepan, T. Altan, and A. G. Ulsoy and industrial colleagues Dr. M. Zatarin (Ideko), Dr. M. Fujishima (Mori Seiki), M. Lundblad (Sandvik), and D. McIntosh (Pratt & Whitney Canada) provided strong personal friendship and research partnership. Machine Tool Technology Research Foundation (MTTRF) and Mori Seiki loaned the experimental machine tools, and Sandvik Coromant and Mitsubishi Materials donated the cutting tools for research. The author was most influenced by the research and machine tool engineering style and philosophy of his mentor the late Professor J. Tlusty.

I acknowledge the valuable editorial support provided by the Cambridge University Press editor, Peter Gordon. I thank those at Cambridge University Press and Aptara for their cooperation and assistance.

Machine tool and metal cutting engineering is a multidisciplinary area that demands knowledge from various fields if one is to be an effective manufacturing engineer and researcher. This requires significant work and effort that cannot be accomplished without the sacrifices of close family members. The author's wife Nesrin, daughter Cagla, and son Hasan bore with the author, who missed an endless number of family days in establishing the Manufacturing Automation Research Laboratory at the University of British Columbia, where the book has been developed and improved for twenty-five years. The author's mother Hatice and late father Hasan, who were hard-working, honest, and warm wine makers living in the little town Bekili (Denizli, Turkey), were more than role models and a source of support for the author. The author's brother Asim and sister Ummuhan and her husband Ibrahim have been more than family members; they have been very close friends. The author owes the completion of the book to those who supported him throughout his career and life.

MANUFACTURING AUTOMATION

CHAPTER ONE

INTRODUCTION

The areas of machine tools, metal cutting, computer numerically controlled (CNC), computer-aided manufacturing (CAM), and sensor-assisted machining are quite wide, and each requires the academic and engineering experience to appreciate a manufacturing operation that uses all of them in an integrated fashion.

Although it is impossible to be an expert in all these subjects, a manufacturing engineer must be familiar with the engineering fundamentals for the precision and economical manufacturing of a part. This book emphasizes only the fundamentals of metal cutting mechanics, machine tool vibrations, feed drive design and control, CNC design principles, sensor-assisted machining, and the technology of programming CNC machines. The book is based on more than 120 journal articles and more than 60 research theses that reflect the engineering, research, and teaching experience of the author.

The book is organized as follows.

Chapter Two covers the fundamentals of metal cutting mechanics. The mechanics of two-dimensional orthogonal cutting is introduced first. The laws of fundamental chip formation and friction between the rake and flank faces of a tool during cutting are explained. The relationships among the workpiece material properties, tool geometry, and cutting conditions are presented. Identification of the shear angle, the average friction coefficient between the tool's rake face and moving chip, and the yield shear stress during machining is explained. The oblique geometry of practical cutting tools used in machining is introduced. The mechanics of oblique cutting for three-dimensional practical tools are explained, and methods in predicting the cutting forces in all directions are presented with the use of the laws of oblique cutting mechanics. The mechanics of turning, milling, and drilling, which constitute the majority of machining operations in the manufacturing industry, are presented. Algorithms for predicting the milling forces in three Cartesian coordinates are derived and illustrated with sample experimental results. Efficient force prediction algorithms for widely used helical end mills are presented. The chapter also briefly discusses the modes and causes of tool wear and breakage, that are important in evaluating the machinability of parts.

Chapter Three deals with static deformations and vibrations during machining. The static deformations occur because of the elastic deflections of both parts and machine during machining. When the static deformation is beyond the tolerance limit, the part is scrapped. Sample formulations are provided to predict the magnitude and location of static deformations in bar turning and end milling. The methods can be extrapolated to other machining operations such as grinding and drilling. One of the most common problems in machining originates from dynamic deformations (i.e., relative structural vibrations between the tool and workpiece). The most common vibrations are due to self-excited chatter vibrations, which grow until the tool jumps out of the cutting zone or breaks because of the exponentially growing dynamic displacements between the tool and workpiece. To understand the machine tool vibrations, the fundamental principles of single – and multi–degree-of-freedom vibration theory are summarized first. Because the machine tool chatter is mainly investigated by analyzing experimental data, the fundamentals of the experimental modal analysis techniques are presented. The modal analysis technique allows the engineer to represent a complex machine tool or workpiece structure by a set of commonly used mathematical expressions that engineers can understand. The technique not only allows one to analyze the chatter vibrations, but it gives a clear message to the machine tool engineer about the structural source of the vibrations during machining, which leads to the improved design.

Chapter Four presents the theory of chatter vibrations in both orthogonal and oblique machining operations both in the frequency and discrete time domain. The mathematical model of regenerative vibrations, which occur in subsequent tool passes during machining, is presented. The methods of determining chatter vibration–free axial depths of cuts and spindle speeds in orthogonal cutting operations are presented with and without process damping. Mathematical models of predicting chatter stability in turning, drilling, and milling operations are introduced. The techniques are explained with the aid of results obtained from simulation and machining tests. The engineer is presented with methods that increase the machining productivity by avoiding chatter vibrations.

Chapter Five introduces the CNC technology and its principles of operation and programming. First, standard NC commands accepted by all CNC machine tools are summarized. These include the format of an NC code accepted by the CNC of the machine tool, motion commands such as linear and circular contouring along a tool path, miscellaneous commands such as spindle and coolant control, and automatic cycles. Later, the introductory principles of CAM are presented. NC programs generated by CAM systems are processed by CNC units that generate position commands to each drive based on trajectory generation and real-time interpolation algorithms. The mathematical details of generating smooth trajectory with velocity, acceleration, and jerk limits of the machine are covered. Real-time interpolation of linear, circular, and splined paths are presented with examples.

Engineers who know how to use and program CNC machine tools must familiarize themselves with the design and internal operational principles of CNC. Chapter Six describes the fundamentals of CNC design, starting with the selection of drive motors and servoamplifiers. Mathematical modeling of feed servodrives is presented in detail. The transfer functions of mechanical drive inertia and friction, servomotor, amplifier, and velocity and position feedback sensors are explained with their practical interpretations. The transformation of continuous-time domain models of the physical system into discrete computer time domain models is explained. Design and tuning procedures for the digital control of feed drives are presented with real-life examples. Advanced control techniques for precision tracking and active vibration damping of feed drives are presented. The chapter is complemented with the design of electrohydraulic machine tool drives to show that the CNC design principles are general and can be applied to any mechanical system regardless of the actuators.

The recent trend in machining is to add intelligence to the machine tools and CNC, as discussed in Chapter Seven. Sensors that can measure the forces, vibrations, temperature, and sound during machining are installed on the machine tools. Mathematical models that correlate the relationship between the measured sensor signals and the state of machining are formed. The mathematical models are coded into real-time algorithms that monitor the machining process and send commands to CNC for corrective actions. The chapter includes simple but fundamental machining process control algorithms along with their theoretical foundations. Adaptive control of cutting forces, in-process monitoring of tool failure, and chatter detection algorithms are presented with their experimental validation and engineering application.

Sample problem sets are included at the end of each chapter. The problems mostly originated from the actual design, application, and experiments conducted at the author's manufacturing automation research laboratory; hence, they are designed to give a realistic feeling for engineering students. Because the book contains multiple engineering disciplines applied to machine tool engineering problems in an integrated fashion, most of the basic mechanical engineering concepts are assumed to be understood by the readers. However, the basic principles of Laplace and z transforms, as well as least squares – based identification techniques, are provided in the appendix.

The advanced mathematical models developed in the author's laboratory are simplified to teach the basic principles of metal cutting mechanics, machine tool vibrations, and control in this second edition of the book. The details of the full mathematical models are published in the research theses of graduate students and journal articles supervised by the author. The advanced algorithms are also packaged in CUTPRO © Advanced Machining Process Simulation Software [66], which is licensed to research centers and machining industry worldwide.

CHAPTER TWO

MECHANICS OF METAL CUTTING

2.1 INTRODUCTION

The final shapes of most mechanical parts are obtained by machining operations. Bulk deformation processes, such as forging and rolling, and casting processes are mostly followed by a series of metal-removing operations to achieve parts with desired shapes, dimensions, and surface finish quality. The machining operations can be classified under two major categories: cutting and grinding processes. The cutting operations are used to remove material from the blank. The subsequent grinding operations provide a good surface finish and precision dimensions to the part. The most common cutting operations are *turning*, *milling*, and *drilling* followed by special operations such as *boring, broaching, hobing, shaping*, and *form cutting*. However, all metal cutting operations share the same principles of mechanics, but their geometry and kinematics may differ from each other. The mechanics of cutting and the specific analysis for a variety of machining operations and tool geometries are not widely covered in this text. Instead, a brief introduction to the fundamentals of cutting mechanics and a comprehensive discussion of the mechanics of milling operations are presented. Readers are referred to established metal cutting texts authored by Armarego and Brown [25], Shaw [96], and Oxley [83] for detailed treatment of the machining processes.

2.2 MECHANICS OF ORTHOGONAL CUTTING

Although the most common cutting operations are three-dimensional and geometrically complex, the simple case of two-dimensional orthogonal cutting is used to explain the general mechanics of metal removal. In orthogonal cutting, the material is removed by a cutting edge that is perpendicular to the direction of relative tool–workpiece motion. The mechanics of more complex three-dimensional oblique cutting operations are usually evaluated by geometrical and kinematic transformation models applied to the orthogonal cutting process. Schematic representations of orthogonal and oblique cutting processes are shown in Figure 2.1. The orthogonal cutting resembles a shaping process with a straight tool whose cutting edge is perpendicular to the cutting velocity (V). A metal chip with a width of cut (b) and depth of cut (h) is sheared away from the workpiece. In orthogonal cutting, the cutting is assumed to be uniform along the cutting edge; therefore, it is a two-dimensional plane strain deformation process without side spreading of the material. Hence, the cutting forces are exerted only in the directions of velocity and uncut chip thickness, which are called tangential (F_t) and feed forces (F_f). However, in oblique cutting, the cutting edge is oriented with an inclination angle (i) and the additional third force acts in the radial direction (F_r).

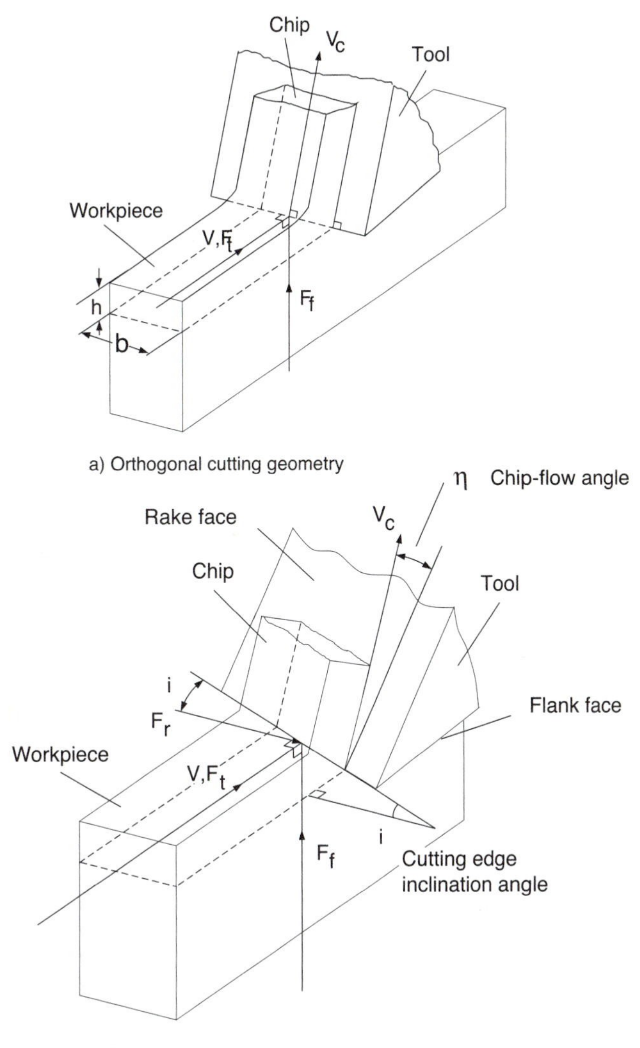

Figure 2.1: Geometries of orthogonal and oblique cutting processes.

There are three deformation zones in the cutting process as shown in the cross-sectional view of the orthogonal cutting (see Fig. 2.2). As the edge of the tool penetrates into the workpiece, the material ahead of the tool is sheared over the primary shear zone to form a chip. The sheared material, the chip, partially deforms and moves along the rake face of the tool, which is called the secondary deformation zone. The friction area, where the flank of the tool rubs the newly machined surface, is called the tertiary zone. The chip initially sticks to the rake face of the tool, which is called the *sticking region*. The friction stress is approximately equal to the yield shear stress of the material at the sticking zone where the chip moves over a material stuck on the rake face of the tool. The chip stops sticking and starts sliding over the rake face with a constant *sliding friction* coefficient. The chip leaves the tool, losing contact with the rake face of the tool. The length of the contact zone depends on the cutting speed, tool geometry, and material properties. There are basically two types of assumptions in the analysis of the primary shear zone. Merchant [75] developed an orthogonal cutting model by assuming that the shear zone is a thin plane. Others, such as Lee and Shaffer [67] and Palmer and Oxley [84], based their analysis on a thick shear deformation zone, proposing "shear angle

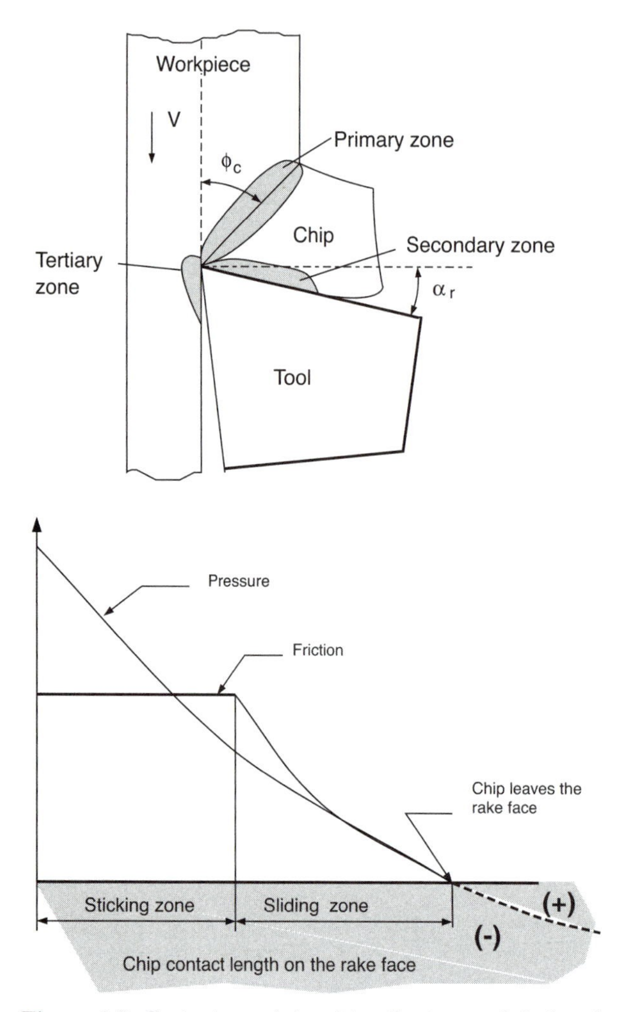

Figure 2.2: Contact zone (−) and tensile stresses (+) after the chip

prediction" models in accordance with the laws of plasticity. In this text, the primary shear deformation zone is assumed to be a thin zone for simplification.

The deformation geometry and the cutting forces are shown on the cross-sectional view of the orthogonal cutting process (see Fig. 2.3). It is assumed that the cutting edge is sharp without a chamfer or radius and that the deformation takes place at the infinitely thin shear plane. The shear angle ϕ_c is defined as the angle between the direction of the cutting speed (V) and the shear plane. It is further assumed that the shear stress (τ_s) and the normal stress (σ_s) on the shear plane are constant; the resultant force (F_c) on the chip, applied at the shear plane, is in equilibrium to the force (F_c) applied to the tool over the chip–tool contact zone on the rake face; an average constant friction is assumed over the chip–rake face contact zone. The contact forces originating from tertiary zone are assumed to be zero, and all cutting forces are due to shearing process or chip–rake face contact. From the force equilibrium, the resultant force (F_c) is formed from the feed (F_{fc}) and tangential (F_{tc}) cutting forces:

$$F_c = \sqrt{F_{tc}^2 + F_{fc}^2}. \tag{2.1}$$

The feed force (or thrust force) is in the direction of uncut chip thickness, and the tangential cutting force (or power force) is in the direction of cutting velocity. The cutting forces acting on the tool will have equal amplitude but opposite directions with respect to the forces acting on the chip. The mechanics of orthogonal cutting for two deformation zones are shown as follows.

Primary Shear Zone

The shear force (F_s) acting on the shear plane is derived from the geometry as follows:

$$F_s = F_c \cos(\phi_c + \beta_a - \alpha_r), \tag{2.2}$$

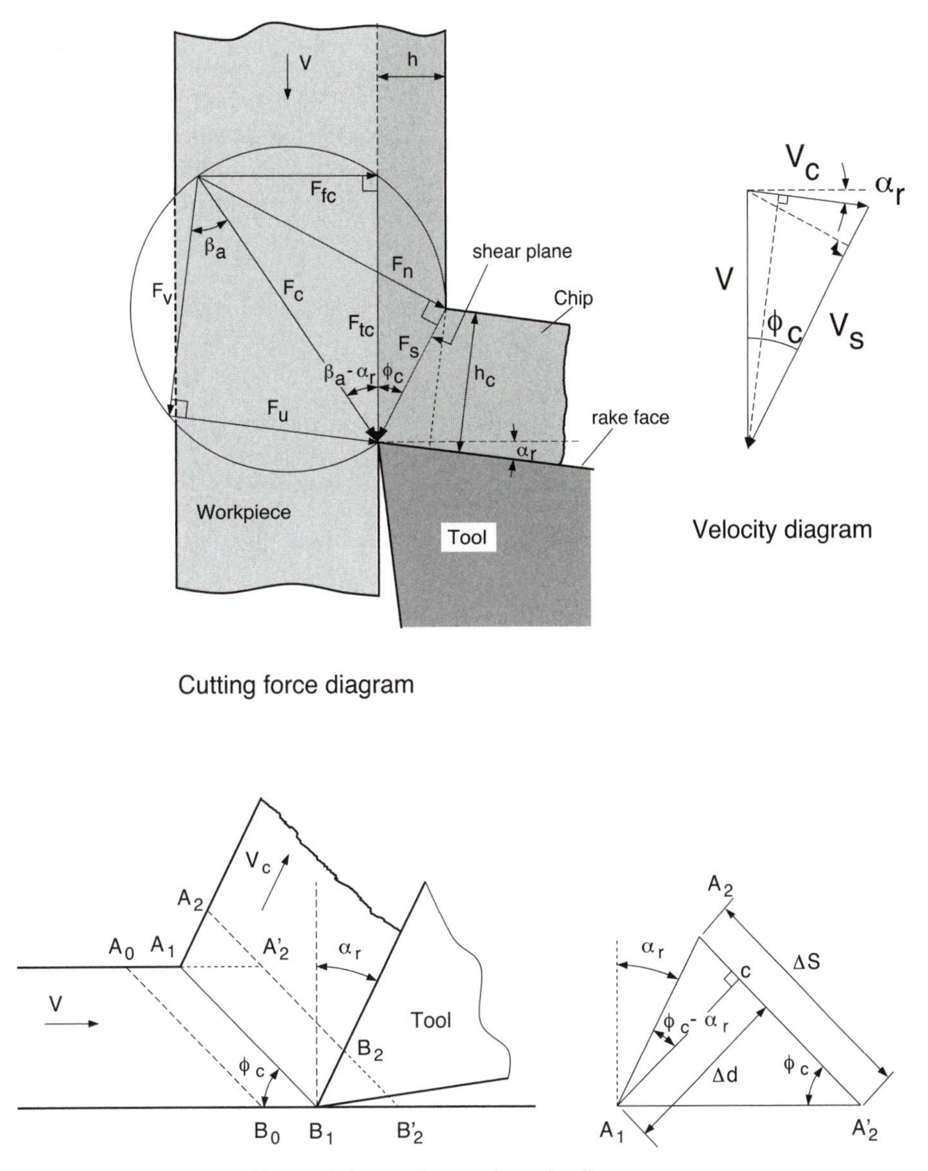

Figure 2.3: Mechanics of orthogonal cutting.

where β_a is the average friction angle between the tool's rake face and the moving chip, and α_r is the rake angle of the tool. The shear force can also be expressed as a function of the feed and tangential cutting forces as follows:

$$F_s = F_{tc} \cos \phi_c - F_{fc} \sin \phi_c. \tag{2.3}$$

Similarly, the normal force acting on the shear plane is found to be

$$F_n = F_c \sin(\phi_c + \beta_a - \alpha_r) \tag{2.4}$$

or

$$F_n = F_{tc} \sin \phi_c + F_{fc} \cos \phi_c. \tag{2.5}$$

With the assumption of uniform stress distribution on the shear plane, the shear stress (τ_s) is found to be

$$\tau_s = \frac{F_s}{A_s}, \tag{2.6}$$

where the shear plane area (A_s) is

$$A_s = b \frac{h}{\sin \phi_c}, \tag{2.7}$$

where b is the width of cut (or the depth of cut in turning), h is the uncut chip thickness, and (ϕ_c) is the shear angle between the direction of cutting speed (V) and the shear plane. The normal stress on the shear plane (σ_s) is

$$\sigma_s = \frac{F_n}{A_s}. \tag{2.8}$$

The cutting velocity (V) is resolved into two components (see the velocity diagram shown in Fig. 2.3). The material is sheared away from the workpiece with the shear velocity (V_s). From the velocity hodograph shown, we have

$$V_s = V \frac{\cos \alpha_r}{\cos(\phi_c - \alpha_r)}. \tag{2.9}$$

The shear power spent in the shear plane is

$$P_s = F_s \cdot V_s, \tag{2.10}$$

which is converted into heat. The corresponding temperature rise on the shear plane (T_s) is

$$P_s = m_c c_s (T_s - T_r), \tag{2.11}$$

where m_c is the metal removal rate [kg/s], c_s is the specific coefficient of heat for the workpiece material [Nm/kg°C], and T_r is the shop temperature. The metal removal rate is found from the cutting process conditions,

$$\left.\begin{aligned} m_c &= Q_c \rho, \\ Q_c &= bhV \quad [\mathrm{m}^3/\mathrm{s}], \end{aligned}\right\} \tag{2.12}$$

where ρ[kg/m^3] is the specific density of the workpiece material. The shear plane temperature (T_s) can be calculated from Eqs. (2.9) to (2.12):

$$T_s = T_r + \frac{P_s}{m_c c_s}. \tag{2.13}$$

The formulation given above considers that the entire plastic deformation takes place only at the shear plane and that all the heat is also consumed at the shear plane. This assumption is shown to overestimate the temperature prediction proposed by Boothroyd [30], who considered that some of the plastic deformation takes place over a shear zone of finite thickness and that some of the heat is dissipated to the work material and the chip, away from the thin shear plane. Oxley [83] used the following modified temperature prediction:

$$T_s = T_r + \lambda_h(1 - \lambda_s)\frac{P_s}{m_c c_s}, \tag{2.14}$$

where λ_h ($0 < \lambda_h \leq 1$) is a factor that considers the plastic work done outside the thin shear zone, and λ_s is the proportion of the heat conducted into the work material. For a plain carbon steel, an average value for $\lambda_h \approx 0.7$ can be assumed [107]. The heat conducted into the work material is evaluated with the following experimentally evaluated empirical equation [83]:

$$\begin{aligned} \lambda_s &= 0.5 - 0.35\log(R_T \tan\phi_c), \quad \text{for } 0.04 \leq R_T \tan\phi_c \leq 10, \\ \lambda_s &= 0.3 - 0.15\log(R_T \tan\phi_c), \quad \text{for } R_T \tan\phi_c \geq 10, \end{aligned} \tag{2.15}$$

where ϕ_c is the shear angle and R_T is a nondimensional thermal number given by

$$R_T = \frac{\rho c_s V h}{c_t}, \tag{2.16}$$

where c_t is the thermal conductivity of the work material with units [W/(m°C)]. Note also that the heat transmitted to the work material can not be more than the total energy generated, and a negative influx of the heat into the shear plane is not possible ($0 \leq \lambda_s \leq 1$).

The shear plane length L_c is found from the chip deformation geometry as follows:

$$L_c = \frac{h}{\sin\phi_c} = \frac{h_c}{\cos(\phi_c - \alpha_r)}. \tag{2.17}$$

The chip compression ratio (r_c) is the ratio of the uncut chip thickness over the deformed (h_c) one as follows:

$$r_c = \frac{h}{h_c}. \tag{2.18}$$

The shear angle is found from the geometry as a function of rake angle and the chip compression ratio as follows:

$$\phi_c = \tan^{-1}\frac{r_c \cos\alpha_r}{1 - r_c \sin\alpha_r}. \tag{2.19}$$

The shear strains and strain rates in metal cutting are significantly higher than those found from standard tensile tests and metal-forming operations. The geometry of a deformed chip is shown in Figure 2.3. Assume that an undeformed chip section $A_0B_0A_1B_1$ is moving with workpiece velocity V. The workpiece material is deformed plastically at the shear plane (B_1A_1), and the cut chip slides over the rake face with a chip velocity V_c. After Δt shearing time, the uncut metal strip $A_0B_0B_1A_1$ becomes a chip with a geometry of $A_1B_1B_2A_2$. Hence, the chip is shifted from the expected position $B_2'A_2'$ to the deformed position B_2A_2 because of shearing in the shear plane with a shear angle of

ϕ_c. Because of plane strain deformation, $A_2'A_2 = B_2'B_2$. The shear strain (γ_s) is defined as the ratio of deformation ($\Delta s = A_2'A_2$) over the nominal distance between the deformed and undeformed planes ($\Delta d = A_1C$) as follows:

$$\gamma_s = \frac{\Delta s}{\Delta d} = \frac{\overline{A_2A_2'}}{\overline{A_1C}} = \frac{\overline{A_2'C}}{\overline{A_1C}} + \frac{\overline{CA_2}}{\overline{A_1C}} = \cot\phi_c + \tan(\phi_c - \alpha_r).$$

By rearranging, the shear strain can be expressed as

$$\gamma_s = \frac{\cos\alpha_r}{\sin\phi_c \cos(\phi_c - \alpha_r)}. \tag{2.20}$$

The shear strain rate is

$$\gamma_s' = \frac{\gamma_s}{\Delta t}.$$

Assuming that the shear zone increment is Δs and that the thickness of shear deformation zone is Δd, the shear strain and shear velocity can be defined as $\gamma_s = \Delta s/\Delta d$ and $V_s = \Delta s/\Delta t$, respectively. The shear strain rate is then defined as

$$\gamma_s' = \frac{V_s}{\Delta d} = \frac{V\cos\alpha_r}{\Delta d\cos(\phi_c - \alpha_r)}. \tag{2.21}$$

Because the shear zone thickness Δd is extremely small in cutting, Eq. (2.21) indicates the presence of very high shear strain rates. Especially when the shear zone is assumed to be a plane with zero thickness, the strain rate becomes infinite, which can not be true. However, the thin shear plane approximation is useful for the macromechanics analysis of metal cutting. For practical and approximate predictions, the thickness of the shear zone can be approximated as a fraction of the shear plane length (i.e., $\Delta d \approx 0.15$–$0.2\, L_c$). For more accurate analysis, the shear zone thickness must be evaluated by freezing the machining process with a quick stop test and measuring the zone thickness with a scanning electron microscope (SEM).

Secondary Shear Zone

Two components of the cutting force are acting on the rake face of the tool (Fig. 2.3): the normal force F_v,

$$F_v = F_{tc}\cos\alpha_r - F_{fc}\sin\alpha_r, \tag{2.22}$$

and the friction force F_u on the rake face,

$$F_u = F_{tc}\sin\alpha_r + F_{fc}\cos\alpha_r. \tag{2.23}$$

In the orthogonal cutting analysis shown here, it is assumed that the chip is sliding on the tool with an average and constant friction coefficient of μ_a. In reality, the chip sticks to the rake face for a short period and then slides over the rake face with a constant friction coefficient [119]. The average friction coefficient on the rake face is given as

$$\mu_a = \tan\beta_a = \frac{F_u}{F_v}. \tag{2.24}$$

The friction angle β_a can alternatively be found from the tangential and feed forces as follows:

$$\tan(\beta_a - \alpha_r) = \frac{F_{fc}}{F_{tc}}. \quad \rightarrow \quad \beta_a = \alpha_r + \tan^{-1}\frac{F_{fc}}{F_{tc}}. \tag{2.25}$$

The deformed chip slides on the rake face of the tool with the velocity of

$$V_c = r_c V = \frac{\sin\phi_c}{\cos(\phi_c - \alpha_r)} V. \tag{2.26}$$

The friction power spent on the tool chip contact face is

$$P_u = F_u V_c. \tag{2.27}$$

The total power consumed in cutting is the sum of energy spent in the shear and friction zones as follows:

$$P_{tc} = P_s + P_u. \tag{2.28}$$

From the equilibrium of cutting forces and the velocities, the total power is also equal to the cutting power drawn from the spindle motor as follows:

$$P_{tc} = F_{tc} V. \tag{2.29}$$

The friction power increases the temperature of tool and chip. As can be seen from Eq. (2.27), if the velocity is increased, the friction power and thus the temperature of the tool increase. Excessive heat will cause an undesirable high temperature in the tool, which leads to the softening of the tool material and its accelerated wear and breakage. However, the production engineer desires an increased cutting velocity to obtain a high metal removal rate (Eq. 2.12) for productivity gains. The manufacturing researchers' challenge has been to decrease the cutting force F_u and move the heat toward the chip with better tool geometry design, and to develop heat-resistant tool materials that can preserve their hardness at elevated temperatures. Although the prediction of the temperature distribution at the tool–chip interface is rather complex, the following simplified analysis is still useful for metal cutting engineers.

The friction power consumed at the tool–chip interface (Eq. 2.27) is converted into heat via

$$P_u = m_c c_s \Delta T_c, \tag{2.30}$$

where ΔT_c is the average temperature rise in the chip. Boothroyd [30] and Stephenson [104] assumed a constant sticking friction load with a constant rectangular plastic zone at the tool–chip interface. The experimental temperature measurement and assumed plastic deformation zone led to the following empirical temperature relationship [83]:

$$\log\left(\frac{\Delta T_{\mathrm{m}}}{\Delta T_{\mathrm{c}}}\right) = 0.06 - 0.195\delta\sqrt{\frac{R_{\mathrm{T}}h_{\mathrm{c}}}{l_{\mathrm{t}}}} + 0.5\log\left(\frac{R_{\mathrm{T}}h_{\mathrm{c}}}{l_{\mathrm{t}}}\right), \tag{2.31}$$

where ΔT_m is the maximum temperature rise of the chip at the rake face–chip interface, which has a total contact length of l_{t}. The nondimensional number δ is the ratio of the plastic layer thickness over the deformed chip thickness (h_{c}) on the tool rake face–chip interface. The average temperature rise (T_{int}) at the rake face–chip interface is given by

$$T_{\mathrm{int}} = T_{\mathrm{s}} + \lambda_{\mathrm{int}}\Delta T_m, \tag{2.32}$$

where T_{s} is the average shear plane temperature and λ_{int} (i.e., ≈ 0.7) is an empirical correction factor that accounts for temperature variations along the chip–tool contact zone. For an accurate analysis, both the plastic layer thickness (δh_{c}) and l_{t} must be measured with a microscope that has a large magnification (such as a SEM). Our experiments indicated that the thickness of the plastic layer on the rake face is observed to be between 5 and 10 percent of the deformed chip thickness ($\delta/h_{\mathrm{c}} \approx 0.05$–$0.1$). The contact length can be estimated approximately by assuming that the resultant cutting force acts in the middle of the contact length and parallel to the stress-free chip boundary. From the geometry of orthogonal cutting (Fig. 2.3), the chip–rake face contact length can be approximately predicted as

$$l_{\mathrm{t}} = \frac{h\sin(\phi_{\mathrm{c}} + \beta_{\mathrm{a}} - \alpha_{\mathrm{r}})}{\sin\phi_{\mathrm{c}}\cos\beta_{\mathrm{a}}}. \tag{2.33}$$

The prediction of temperature distribution at the tool–chip interface is very important in determining the maximum speed and feed rate that give the most optimal material removal rate without excessive tool wear. The binding materials within the cutting tools may be weakened or diffused to the moving chip material at their critical diffusion or melting temperature limits. The fundamental machinability study requires the identification of a maximum cutting speed and uncut chip values that correspond to the critical temperature limit where the tool wears rapidly. By using the approximate solutions summarized above, one can select a cutting speed and feed rate that would correspond to a tool–chip interface temperature (T_{int}) that lies just below the diffusion and melting limits of materials present in a specific cutting tool. The detailed and fundamental scientific and experimental treatment of the cutting process is covered in Oxley [83].

Tertiary Deformation Zone

The contact dimension and mechanics between the flank face of the tool and finished surface depend on the tool wear, preparation of the cutting edge, and the friction characteristics of the tool and work materials. Let us assume that the total friction force on the flank face is F_{ff}, and the force normal to the flank is F_{fn}. If the pressure (σ_f) on the flank face is uniform, which is an oversimplified assumption, the normal force on the flank can be expressed by $F_{fn} = \sigma_f VB \cdot b$, where VB is flank contact length and b is width of cut. An average friction (μ_f) coefficient between the flank face and the finished surface can be defined by $\mu_f = F_{ff}/F_{fn}$. The angle between the flank face and the finished surface is called *clearance or relief angle* (Cl_p). The total contact forces can be resolved into the tangential (F_{te}) and feed (F_{fe}) directions as

$$F_{te} = F_{fn}\sin(Cl_p) + F_{ff}\cos(Cl_p) \tag{2.34}$$
$$F_{fe} = F_{fn}\cos(Cl_p) + F_{ff}\sin(Cl_p).$$

TABLE 2.1. Orthogonal Cutting Database for Titanium Alloy Ti_6Al_4V

$\tau_s = 613$ (MPa)
$\beta_a = 19.1 + 0.29\alpha_r$ (deg)
$r_c = C_o h^{C_1}$
$C_o = 1.755 - 0.028\alpha_r$
$C_1 = 0.331 - 0.0082\alpha_r$
$K_{te} = 24$ (N/mm)
$K_{fe} = 43$ (N/mm)

It should be noted that the measured cutting forces may include both the forces due to shearing (F_{tc}, F_{fc}) and a tertiary deformation process "ploughing" or "rubbing" (F_{te}, F_{fe}) at the flank of the cutting edge. Thus the measured force components are expressed as a superposition of shearing and edge forces as follows:

$$\begin{aligned} F_t &= F_{tc} + F_{te}, \\ F_f &= F_{fc} + F_{fe}. \end{aligned} \tag{2.35}$$

Hence, the cutting force expressions (F_t, F_f) presented up to Eq. (2.35) represent only shearing forces (F_{tc}, F_{fc}). The edge forces (F_{te}, F_{fe}) must be subtracted from the measured tangential and feed forces before applying the laws of orthogonal cutting mechanics explained in this section.

It is difficult to predict the shear angle and stress in the shear plane and the average friction coefficient on the rake face by using the standard material properties obtained from tensile and friction tests. For accurate and realistic modeling, such fundamental parameters are identified from orthogonal cutting tests, where the deformed chip thickness and feed and tangential cutting forces are measured by using cutting tools with a range of rake angles. The influence of uncut chip thickness and cutting speed is also considered by conducting experiments over a wide range of feeds and cutting speeds.

The relationships shown in Table 2.1 are identified from statistical analysis of more than 180 orthogonal cutting tests conducted using tungsten carbide (WC) cutting tools and Ti_6Al_4V titanium alloy work material. A set of turning experiments resembling orthogonal cutting was conducted on titanium tubes (Ti_6Al_4V) with tools of different rake angles at different feeds and cutting speeds. The diameter of the tube was 100 mm, and the cutting speed range was 2.6 to 47 m/min. Cutting forces in the tangential (F_t) and feed (F_f) directions were measured with a force dynamometer. Two sample orthogonal cutting test results are shown in Figure 2.4. Small steps in cutting conditions were used to increase the reliability of the measured forces.

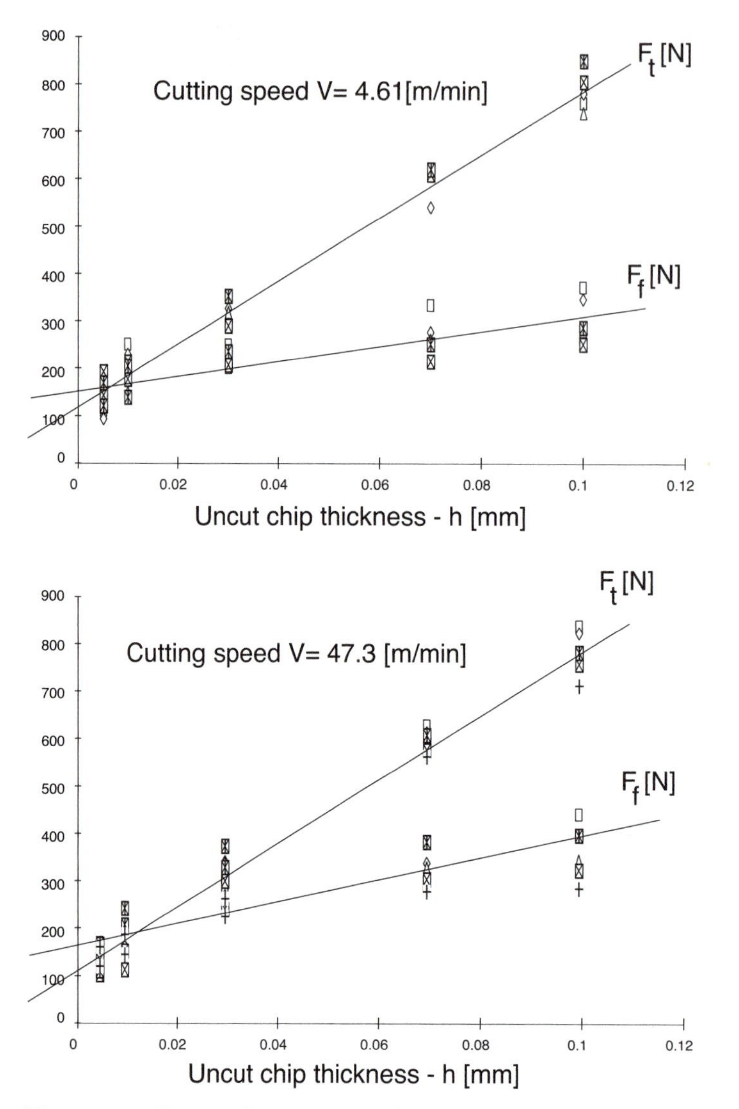

Figure 2.4: Cutting forces measured during orthogonal tuning of Ti_6Al_4V tubes.

The tests have been repeated a number of times at different feeds and cutting speeds to ensure the statistical reliability of measurements. The edge forces are obtained by extrapolating the measured forces to zero chip thickness. It can be seen that the edge forces do not vary significantly with cutting speeds for the particular titanium alloy used here. The average edge force coefficients K_{te} and K_{fe} represent the rubbing forces per unit width. The chip compression ratio (r_c), shear stress τ_s, shear angle ϕ_c, and friction angle β_a (Table 2.1) are calculated from the measured "cutting" component of the forces and the cutting ratio by applying the orthogonal cutting theory presented above.

2.3 MECHANISTIC MODELING OF CUTTING FORCES

Orthogonal cutting mechanics are not directly applicable to many practical cutting tools with corner radius, side cutting edge angle, and chip breaking grooves. It is more practical to carry out a few experiments to identify constant parameters of the tool geometry–workpiece material pair to model existing cutting tools. However, it must be noted that, for a tool design and analysis of a particular metal cutting process, oblique (i.e., three-dimensional) cutting mechanics and plasticity analysis are still necessary, and this is covered in Section 2.5.

As an example of mechanistic modeling, let us take the case of orthogonal cutting. We can extend the idea of model building to other cutting processes that are not orthogonal. In the previous section, the shear force is formulated as a function of measured feed and tangential cutting forces in the orthogonal cutting. The shear force can be expressed as a function of shear stress and shear angle (Eqs. 2.6 and 2.7) as follows:

$$F_s = \tau_s b \frac{h}{\sin\phi_c}. \tag{2.36}$$

From Eqs. (2.3) and (2.36), the resultant cutting force (F_c) can be expressed in terms of shear stress, friction and shear angles, width of cut, and feed rate as follows:

$$F_c = \frac{F_s}{\cos(\phi_c + \beta_a - \alpha_r)} = \tau_s bh \frac{1}{\sin\phi_c \cos(\phi_c + \beta_a - \alpha_r)}. \tag{2.37}$$

The tangential and feed forces can be expressed in terms of resultant force as follows:

$$\left.\begin{aligned} F_{tc} &= F_c \cos(\beta_a - \alpha_r), \\ F_{fc} &= F_c \sin(\beta_a - \alpha_r). \end{aligned}\right\} \tag{2.38}$$

Substituting Eq. (2.37) into Eq. (2.38), we can find the measured main cutting forces as functions of tool geometry and the cutting conditions (i.e., uncut chip thickness (h) and width of cut (a)) and process- and material-dependent terms ($\tau_s, \beta_a, \phi_c, \alpha_r$) as follows:

$$F_{tc} = bh \left[\tau_s \frac{\cos(\beta_a - \alpha_r)}{\sin\phi_c \cos(\phi_c + \beta_a - \alpha_r)} \right]. \tag{2.39}$$

Similarly the feed force is

$$F_{fc} = bh \left[\tau_s \frac{\sin(\beta_a - \alpha_r)}{\sin\phi_c \cos(\phi_c + \beta_a - \alpha_r)} \right]. \tag{2.40}$$

In metal cutting literature the cutting parameter called *specific cutting pressure* or *tangential cutting force coefficient* (K_{tc}) is defined as

$$K_{tc}\,[\text{N/mm}^2] = \tau_s \frac{\cos(\beta_a - \alpha_r)}{\sin\phi_c \cos(\phi_c + \beta_a - \alpha_r)} \tag{2.41}$$

and the feed force constant (K_{fc}) as

$$K_{fc}\,[\mathrm{N/mm^2}] = \tau_s \frac{\sin(\beta_a - \alpha_r)}{\sin\phi_c \cos(\phi_c + \beta_a - \alpha_r)}. \tag{2.42}$$

It is also customary to use another convention for cutting constants, where the feed force is assumed to be proportional to the tangential force with a ratio of

$$K_{fc} = \frac{F_{fc}}{F_{tc}} = \tan(\beta_a - \alpha_r), \tag{2.43}$$

where K_{fc} is dimensionless in this specific form. As can be seen from the definition (Eq. 2.41), the specific cutting pressure is a function of the yield shear stress of the workpiece (τ_s) material during cutting, the shear angle (ϕ_c), tool geometry (i.e., rake angle α_r), and the friction between the tool and the chip (β_a). In Eq. (2.41), only the tool geometry is known beforehand. The friction angle depends on the lubrication used, the tool–chip contact area, and the tool and workpiece materials. An accurate, analytical shear angle prediction remains the subject of continuing research. Previous research results are still insufficient to be used in predicting the shear angles accurately. The shear stress in the shear plane is also still in question with the present knowledge of the cutting process. If the shear plane is assumed to be a thick zone, which is more realistic than having a thin shear plane, there will be a work hardening, and the shear stress will be larger than the workpiece material's original yield shear stress measured from pure torsion or tensile tests. The temperature variation in the shear and the friction zones will also affect the hardness of the workpiece material; therefore, the shear stress in the primary deformation zone will vary. The shear yield stress varies as a function of chip thickness, and because of varying strain hardening of the material being machined, as well. Hence, it is customary to define the cutting forces mechanistically as a function of cutting conditions (i.e., b and h) and the cutting constants (K_{tc}) and (K_{fc}) as follows:

$$\left.\begin{aligned} F_t &= K_{tc}bh + K_{te}b, \\ F_f &= K_{fc}bh + K_{fe}b. \end{aligned}\right\} \tag{2.44}$$

The cutting constants (K_{tc}, K_{fc}), and the edge coefficients that do not contribute to the shearing (K_{te}, K_{fe}), are directly calibrated from metal cutting experiments for a tool–workpiece pair. Note that the edge coefficients change as the cutting tool wears or experiences chipping. It should be also noted that, to take the influence of the chip thickness on the friction and shear angles, and the yield shear stress, the cutting constants (K_t, K_f) are sometimes expressed as nonlinear functions of uncut chip thickness as follows:

$$\left.\begin{aligned} F_t &= K_t bh, \quad K_t = K_T h^{-p}, \\ F_f &= K_f bh, \quad K_f = K_F h^{-q}, \end{aligned}\right\} \tag{2.45}$$

where p and q are cutting force constants determined from the cutting experiments at different feed rates. Equation (2.45) represents basic nonlinearity in the cutting force expressions. This form is used when the edge forces are neglected in the mechanistic models. It must be noted that some work materials exhibit different yield stress and friction coefficients at different speeds, which lead to the speed dependence of cutting constants. The mechanistic cutting constant equation (2.45) can be extended to include cutting speed as a variable.

Example. The cutting conditions for turning an AISI-1045 steel workpiece are set as follows: depth of cut $b = 2.54$ mm; feed rate $c = 0.2$ mm/rev; spindle speed $n = 350$ rev/min; workpiece diameter $= 100$ mm; tool's rake angle $\alpha_r = +5^\circ$. Specific mass of the steel $\rho = 7{,}800$ kg/m^3; specific heat coefficient of steel $c_s = 470$ Nm/kg°C; thermal conductivity $c_t = 28.74$ [W/m°C]. The following measurements are observed from the experiment: deformed chip thickness $h_c = 0.44$ mm, feed force $F_f = 600$ N, tangential force $F_t = 1{,}200$ N. Assuming that the turning is an orthogonal metal cutting process, the following values are evaluated.

Resultant cutting force	$F = \sqrt{F_t^2 + F_f^2} = 1342.0$ N
Chip ratio	$r_c = \frac{h}{h_c} = 0.4545$
Shear angle	$\phi_c = \tan^{-1} \frac{r_c \cos\alpha_r}{1 - r_c \sin\alpha_r = 25^\circ}$
Friction angle	$\beta_a = \alpha_r + \tan^{-1} \frac{F_f}{F_t} = 31.6^\circ$
Friction coefficient	$\mu_a = \tan\beta_a = 0.6144$
Shearing force	$F_s = F\cos(\phi_c + \beta_a - \alpha_r) = 833.5$ N
Shear plane area	$A_s = b\frac{h}{\sin\phi_c} = 1.2$ mm^2
Shearing stress	$\tau_s = \frac{F_s}{A_s} = 693.4$ MPa
Normal force on the shear plane	$F_n = F\sin(\phi_c + \beta_a - \alpha_r) = 1051.7$ N
Normal stress on the shear plane	$\sigma_s = \frac{F_n}{A_s} = 876.43$ MPa
Cutting speed	$V = \pi D n = 110$ m/min
Shearing velocity	$V_s = V\frac{\cos\alpha_r}{\cos(\phi_c - \alpha_r)} = 116.6$ m/min $= 1.9436$ m/s
Shearing power	$P_s = F_s V_s = 1{,}620$ W
Metal removal rate	$m_c = Q_c\rho = bhV\rho = 7.2644 \cdot 10^{-3}$ kg/s
Nondimensional thermal number	$R_T = \frac{\rho c_s V h}{c_t} = 45.78, \quad R_T \tan\phi_c = 21.34 > 10$
Scale of heat conducted into work	$\lambda_s = 0.3 - 0.15\log(R_T \tan\phi_c) = 0.1$
Shear plane temperature	$T_s = T_r + \lambda_h(1 - \lambda_s)\frac{P_s}{m_c c_s} = 20 + 299 = 319^\circ$ C $(\lambda_h \approx 0.7)$
Friction force	$F_u = F\sin\beta_a = 703.2$ N
Normal force	$F_v = F\cos\beta_a = 1143$ N
Chip velocity	$V_c = r_c V = 50$ m/min $= 0.8333$ m/s
Friction power	$P_u = F_u V_c = 586$ W

Chip contact length	$l_t = \frac{h\sin(\phi_c+\beta_a-\alpha_r)}{\sin\phi_c\cos\beta_a} = 0.435$ mm
Total cutting power drawn	$P_t = P_u + P_s = 2{,}200$ W
Specific cutting pressure	$K_t = \frac{F_t}{bh} = 2{,}362.$ N/mm^2
Cutting force ratio	$K_f = \frac{F_f}{F_t} = 0.5$

2.4 THEORETICAL PREDICTION OF SHEAR ANGLE

The evaluation of shear angle, shear stress, and average friction coefficients from orthogonal metal cutting tests was summarized in the previous sections. There have been many attempts at predicting the shear angle theoretically, without relying on metal cutting experiments. Some of the most fundamental models, which assume a perfect rigid plastic workpiece material without any strain hardening, are briefly presented in this section. These models assume that the shear plane is thin, that the shear stress in the shear plane is equivalent to the yield shear stress of the material, and that the average friction is found from friction tests between the tool and workpiece materials, leaving only the shear angle as unknown. There have been two fundamental approaches to predict the shear angle as follows.

Maximum Shear Stress Principle

Krystof [64] proposed a shear angle relation based on the maximum shear stress principle (i.e., shear occurs in the direction of maximum shear stress). The resultant force makes an angle $(\phi_c + \beta_a - \alpha_r)$ with the shear plane (see Fig. 2.3), and the angle between the maximum shear stress and the principal stress (i.e., the resultant force) must be $\pi/4$. Therefore, the following shear angle relation is obtained:

$$\phi_c = \frac{\pi}{4} - (\beta_a - \alpha_r). \tag{2.46}$$

Later, Lee and Shaffer [67] derived the same shear angle relationship from a slip-line field model.

Minimum Energy Principle

Merchant [74] proposed applying the minimum energy principle in predicting the shear angle. By taking the partial derivative of the cutting power (Eqs. 2.29 and 2.38), one gets

$$\frac{dP_{tc}}{d\phi_c} = \frac{d(VF_{tc})}{d\phi_c} = \frac{-V\tau_s bh\cos(\beta_a-\alpha_r)\cos(2\phi_c+\beta_a-\alpha_r)}{\sin^2\phi_c\cos^2(\phi_c+\beta_a-\alpha_r)} = 0,$$

or, with $\cos(2\phi_c + \beta_a - \alpha_r) = 0$,

$$\phi_c = \frac{\pi}{4} - \frac{\beta_a - \alpha_r}{2}, \tag{2.47}$$

which predicts a larger shear angle than the maximum shear stress principle. Although the equations above proposed by Krystof, Merchant, and Lee and Shaffer do not yield qualitatively accurate shear angle predictions beause of oversimplified assumptions, they provided an important relationship among the shear angle (ϕ_c), the rake angle (α_r), which is the most fundamental for tool design, and the friction coefficient between the workpiece and cutting tool materials ($\tan\beta_a$). The forces and power consumed in cutting decrease with increasing shear angle. The expression indicates that the friction coefficient between the tool and the chip must be decreased by using lubricants or materials with a smaller friction coefficient, and the rake angle of the cutting tool must be increased as much as possible provided the weakened cutting edge is able to withstand the pressure and friction load exerted by the chip at the rake face contact zone.

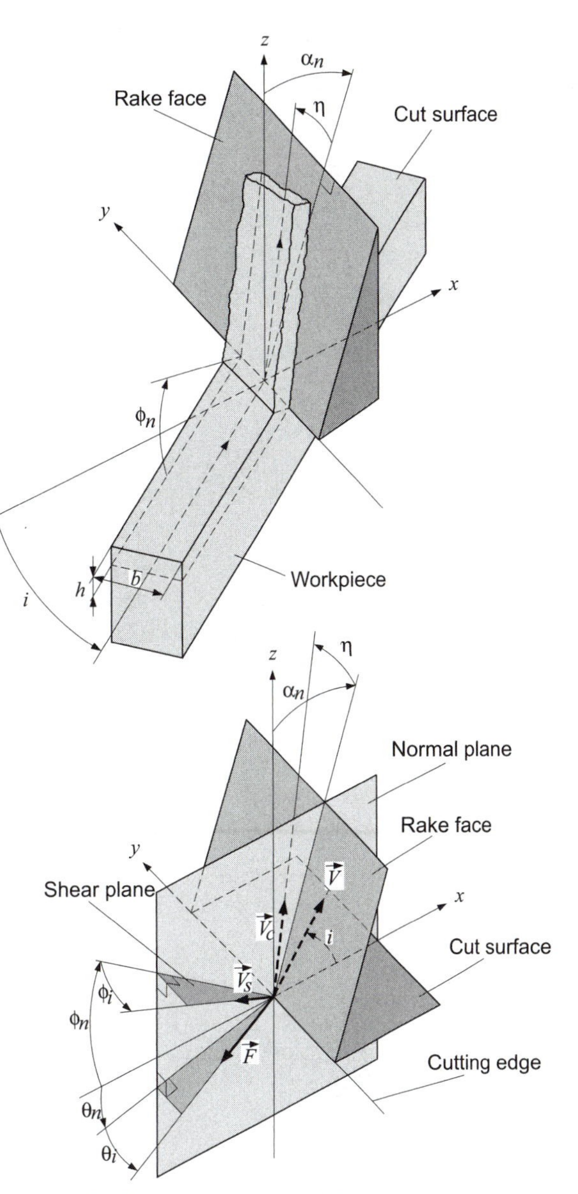

Figure 2.5: The geometry of the oblique cutting process.

2.5 MECHANICS OF OBLIQUE CUTTING

The geometry of oblique cutting is shown in Figure 2.5. The cutting velocity (V) is perpendicular to the cutting edge in orthogonal cutting, whereas, in oblique cutting, it is inclined at an acute angle i to the plane normal to the cutting edge.

2.5.1 Oblique Cutting Geometry

The difference in the geometry of two cutting mechanisms may be better explained by revisiting the orthogonal cutting geometry shown in Figure 2.1. A plane normal to the cutting edge, and inclined at an acute angle (i) with

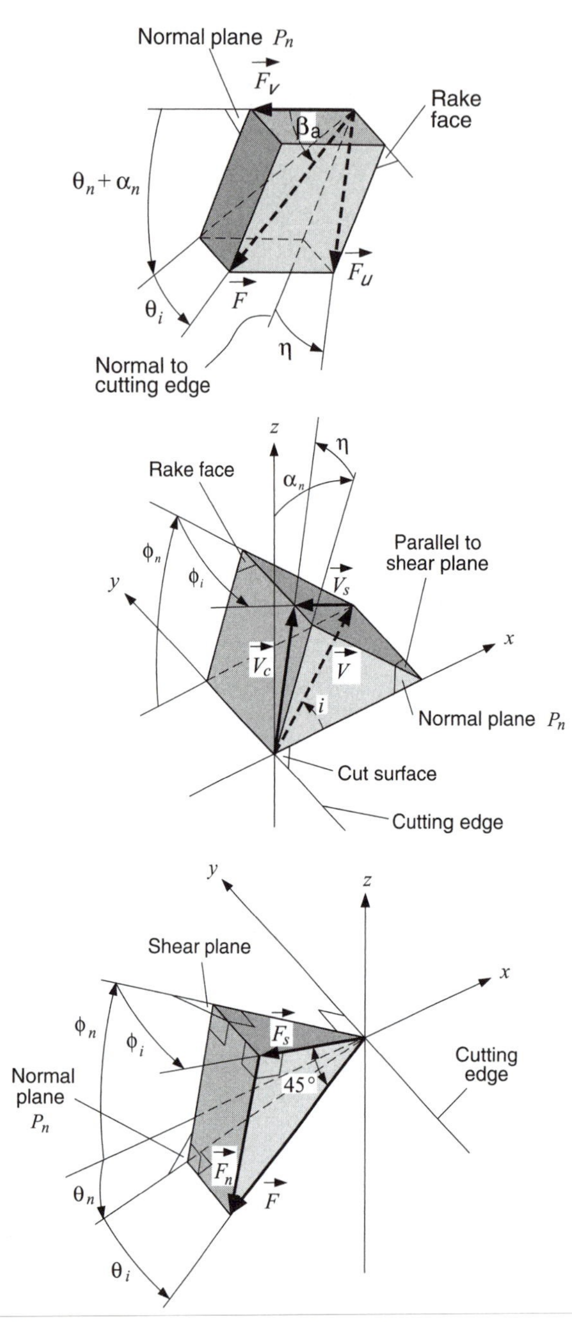

Figure 2.6: Force, velocity, and shear diagrams in oblique cutting.

the cutting velocity V is defined as the *normal plane* or P_n. Because the shear deformation is plane strain without side spreading, the shearing and chip motion are identical on all the normal planes parallel to the cutting speed V and perpendicular to the cutting edge. Hence, the velocities of cutting (V), shear (V_s), and chip (V_c) are all perpendicular to the cutting edge, and they lie in the velocity plane (P_v) parallel to or coincident with the normal plane (P_n). The resultant cutting force F_c, along with the other forces acting on the shear and chip–rake face contact zone, also lies on the same plane P_n in orthogonal cutting. There is no cutting force in the third direction (i.e., perpendicular to the normal plane), and the edge force at the tertiary zone is assumed to be zero. The cutting velocity has an *oblique* or inclination angle i in oblique cutting operations, and thus the directions of shear, friction, chip flow, and resultant cutting force vectors have components in all three Cartesian coordinates (x, y, z) (see Fig. 2.5). In Figure 2.5, the x axis is perpendicular to the cutting edge but lies on the cut surface, y is aligned with the cutting edge, and z is perpendicular to the xy plane.

The forces exist in all three directions in oblique cutting. The important planes in oblique cutting are the shear plane, the rake face, the cut surface xy, the *normal plane* xz or P_n, and the *velocity plane* P_v. Most analyses assume that the mechanics of oblique cutting in the normal plane are equivalent to that of orthogonal cutting; hence, all velocity and force vectors are projected on the normal plane. In Figure 2.6, the angle between the shear and xy planes is called the *normal shear angle* ϕ_n. The shear velocity lies on the shear plane but makes an *oblique shear angle* ϕ_i with the vector normal to the cutting edge on the normal plane. The sheared chip moves over the rake face plane with a *chip flow angle* η measured from a vector on the rake face but normal to the cutting edge. Note that this normal vector also lies on the normal plane P_n. The friction force between the chip and the rake face is collinear with the direction of chip flow. The angle between the z axis and normal vector on the rake face is defined as the *normal rake angle* α_n. The friction force on the rake face ($\vec{F_u}$) and normal force to the rake ($\vec{F_v}$) form the resultant cutting force $\vec{F_c}$ with a friction angle of β_a (see Fig. 2.6). The resultant force vector ($\vec{F_c}$) has an acute projection angle of θ_i with the normal plane P_n, which in turn has an in-plane angle of $\theta_n + \alpha_n$ with the normal force F_v. Here, θ_n is the angle between the x axis and the projection of $\vec{F_c}$ on P_n. The following geometric relations can be derived from Figure 2.6 as follows:

$$F_u = F_c \sin\beta_a = F\frac{\sin\theta_i}{\sin\eta} \to \sin\theta_i = \sin\beta_a \sin\eta, \tag{2.48}$$

$$F_u = F_v \tan\beta_a = F_v\frac{\tan(\theta_n + \alpha_n)}{\cos\eta} \to \tan(\theta_n + \alpha_n) = \tan\beta_a \cos\eta. \tag{2.49}$$

The velocities of chip ($\vec{V_c}$), shear ($\vec{V_s}$), and cutting ($\vec{V}$) form the velocity plane P_v as shown in Figure 2.6. Each velocity vector can be defined by its Cartesian components as follows:

$$\begin{aligned}
\vec{V} &= (V\cos i, & &V\sin i, & &0),\\
\vec{V_c} &= (V_c\cos\eta\sin\alpha_n, & &V_c\sin\eta, & &V_c\cos\eta\cos\alpha_n),\\
\vec{V_s} &= (-V_s\cos\phi_i\cos\phi_n, & &-V_s\sin\phi_i, & &V_s\cos\phi_i\sin\phi_n).
\end{aligned}$$

By eliminating V, V_c, and V_s from the velocity relation,

$$\vec{V_s} = \vec{V_c} - \vec{V},$$

the following geometric relation between the shear and the chip flow directions can be obtained [73]:

$$\tan\eta = \frac{\tan i\cos(\phi_n - \alpha_n) - \cos\alpha_n\tan\phi_i}{\sin\phi_n}. \tag{2.50}$$

The above relationships define the geometry of the oblique cutting process.

2.5.2 Solution of Oblique Cutting Parameters

There are five unknown oblique cutting parameters that define the directions of resultant force (θ_n, θ_i), shear velocity (ϕ_n, ϕ_i), and chip flow (η). In addition to the three equations (2.48–2.50) obtained from the oblique cutting geometry, two additional expressions are required to solve for the five unknown angles. There have been numerous proposed solutions based on the empirical chip flow direction [25, 102] and other empirical assumptions [70, 68, 35]. A theoretical shear angle prediction approach has been proposed by Shamoto and Altintas [95]; it parallels the maximum shear stress [64, 67] and the minimum energy [74] principles used in two-dimensional orthogonal cutting mechanics. The focus here is on the prediction of shear direction based on the laws of mechanics, not on the combined empirical predictions and geometric relations.

Maximum Shear Stress Principle

Krystof [64] applied the maximum shear stress criterion to predict the direction of shear angle in orthogonal cutting (i.e., $\phi_n = \pi/4 - \beta + \alpha_n$). Later, Lee and Shaffer [67] achieved the same orthogonal shear angle relationship using the slip-line field solution. Both assume that the shear occurs in the direction of maximum shear stress (where the angle between the shear velocity and resultant force directions is 45°; see Fig. 2.6). The same principle can be applied to oblique cutting, that is, the resultant force ($\vec{F_c}$) makes a 45° acute angle with the direction of shear as follows:

$$F_s = F_c(\cos\theta_i \cos(\theta_n + \phi_n)\cos\phi_i \\ + \sin\theta_i \sin\phi_i) = F_c \cos(45^\circ).$$

Furthermore, the same principle dictates that the projection of $\vec{F_c}$ to the shear plane coincides with the shear direction, that is, the component of the resultant force in the direction normal to the shear on the shear plane must be zero as follows:

$$F_c\ (\cos\theta_i \cos(\theta_n + \phi_n)\sin\phi_i - \sin\theta_i \cos\phi_i) = 0;$$

otherwise, the shear stress in the shear direction is not the maximum on the shear plane. The two expressions provide the necessary relationships between the shear and resultant force directions as follows:

$$\sin\phi_i = \sqrt{2}\sin\theta_i, \tag{2.51}$$

$$\cos(\phi_n + \theta_n) = \frac{\tan\theta_i}{\tan\phi_i}. \tag{2.52}$$

By solving the five equations (2.49) to (2.52), the five unknown angles (ϕ_n, ϕ_i, θ_n, θ_i, η) that describe the mechanics of oblique cutting can be obtained. However, direct analytical solutions of the equations are rather difficult; hence, they are solved by using an iterative numerical method. The numerical solution is obtained according to the block diagram shown in Figure 2.7. Note that the angles of friction (β_a), rake (α_n), and inclination (i) are known from the geometry and material tests and are considered as inputs to the system. The iterative solution is started by assuming an initial value for the chip flow angle (i.e., $\eta = i$) as proposed by Stabler [103]. The direction (θ_n, θ_i) of the resultant force vector $\vec{F_c}$ is obtained from Eqs. (2.48) and (2.49). Similarly, the shear direction angles (ϕ_n, ϕ_i) can be evaluated from Eqs. (2.51) and (2.52), followed by evaluating the new chip flow angle η_e from the velocity equation (2.50). The true chip flow angle is searched iteratively by using the following interpolation algorithm as follows:

$$\eta(k) = \nu\, \eta(k-1) + (1-\nu)\, \eta_e,$$

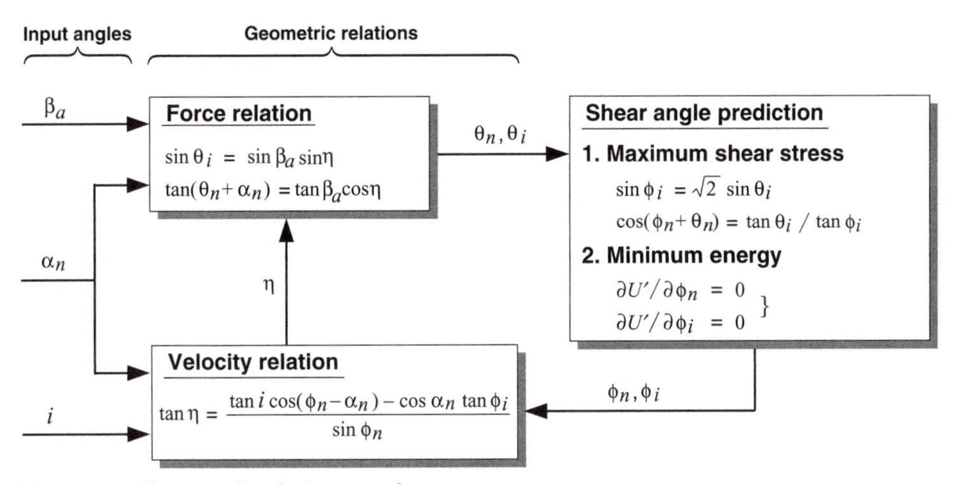

Figure 2.7: Shear angle solution procedure.

where k is the iteration counter, and the interpolation ratio ν is selected within the range of $0 < \nu < 1$. The ratio ν is dynamically updated for the rapid convergence, that is, decreased if $\eta(k)$ oscillates and increased if its value moves in the same direction. The iteration is continued until the chip flow angle converges within 10^{-12} percent. When the three-dimensional oblique cutting model introduced above is applied to two-dimensional orthogonal cutting, it yields the same shear angle expression proposed by Krystof [64] and by Lee and Shaffer [67] who used the maximum shear stress principle on orthogonal cutting (i.e., $i = \theta_i = \phi_i = 0 :\rightarrow: \phi_n = \pi/4 - (\beta_a - \alpha_n)$). Note that with identity (2.51), Eq. (2.52) reduces to $\cos(\phi_n + \theta_n) = 1/\sqrt{2}$ in orthogonal cutting.

Minimum Energy Principle

Merchant [74] proposed a shear angle prediction theory by applying the minimum energy principle to orthogonal cutting. The same principle is extended to oblique cutting here. From the geometry, the shear force is expressed as a projection of $\vec{F_c}$ in the direction of shear (Fig. 2.6) as follows:

$$F_s = F_c[\cos(\theta_n + \phi_n)\cos\theta_i \cos\phi_i + \sin\theta_i \sin\phi_i],$$

or as a product of shear stress and shear plane area (Fig. 2.5) as follows:

$$F_s = \tau_s A_s = \tau_s \left(\frac{b}{\cos i}\right)\left(\frac{h}{\sin\phi_n}\right),$$

where A_s, b, and h are the shear area, the width of cut, and the depth of cut (uncut chip thickness), respectively. By equating the two shear force expressions, the resultant force is derived as follows:

$$F_c = \frac{\tau_s bh}{[\cos(\theta_n + \phi_n)\cos\theta_i \cos\phi_i + \sin\theta_i \sin\phi_i]\cos i \sin\phi_n}. \tag{2.53}$$

The cutting power (P_{tc}) in oblique cutting can be expressed as a function of F_c (see Fig. 2.6) as follows:

$$P_{tc} = F_{tc}V = F_c(\cos\theta_i \cos\theta_n \cos i + \sin\theta_i \sin i)V,$$

and the nondimensional power (P_t') is obtained by substituting Eq. (2.53) for F_c. This gives

$$P_t' = \frac{P_{tc}}{V\tau_s bh} = \frac{\cos\theta_n + \tan\theta_i \tan i}{[\cos(\theta_n + \phi_n)\cos\phi_i + \tan\theta_i \sin\phi_i]\sin\phi_n}, \quad (2.54)$$

where the term $V\tau_s bh$ is constant. The minimum energy principle requires that the cutting power must be minimum for a unique shear angle solution. Because the direction of shear is characterized by the shear angles ϕ_n and ϕ_i, we have

$$\begin{aligned} \partial P_t'/\partial\phi_n &= 0, \\ \partial P_t'/\partial\phi_i &= 0, \end{aligned} \quad (2.55)$$

which provides two more equations in addition to the three geometric relations given by Eqs. (2.48), (2.49), and (2.50). Hence, the five unknown angles ($\phi_n, \phi_i, \theta_n, \theta_i, \eta$), that describe the mechanics of oblique cutting can be obtained. However, because finding an analytical solution of five equations is rather difficult, the solution is obtained by using a numerical iteration technique. The algorithm starts with an initial value of chip flow angle $\eta = i$ (Stabler [103]), followed by the evaluation of the remaining angles from Eqs. (2.48), (2.49), (2.51), and (2.52). After calculating the initial values of $\theta_n, \theta_i, \phi_n$, and ϕ_i, the cutting power (P_t') is obtained from Eq. (2.54). By changing the shear angles slightly (i.e., $\phi_n + \Delta\phi_n$ and $\phi_i + \Delta\phi_i$), the steepest descent direction ($\Delta P_t'/\Delta\phi_n, \Delta P_t'/\Delta\phi_i$) is evaluated. The shear angles are changed by a step ς in the steepest descent direction so that the cutting energy approaches the minimum value as follows:

$$\begin{Bmatrix} \phi_n(k) \\ \phi_i(k) \end{Bmatrix} = \begin{Bmatrix} \phi_n(k-1) \\ \phi_i(k-1) \end{Bmatrix} - \varsigma \begin{Bmatrix} \Delta P_t'/\Delta\phi_n \\ \Delta P_t'/\Delta\phi_i \end{Bmatrix}.$$

The numerical iteration is continued in accordance with Figure 2.7 until the nondimensional cutting power (P_t') converges to a minimum value. When the three-dimensional oblique cutting model introduced above is applied to two-dimensional orthogonal cutting, it yields the same shear angle expression proposed by Merchant [74] (i.e., $i = \theta_i = \phi_i = 0 :\rightarrow: \phi_n = \pi/4 - (\beta_a - \alpha_n)/2$).

Empirical Approach

There are a number of empirical models, and we introduce the one proposed by Armarego and Whitfield [28] here. In their model, two assumptions are used on the shear direction and chip length ratio: (1) the shear velocity is collinear with shear force, and (2) the chip length ratio in oblique cutting is the same as that in orthogonal cutting. The former assumption made by Stabler [103] is considered as one of the maximum shear stress criteria (Eq. 2.52), and by combining the three geometric equations derived in the previous section, it yields

$$\tan(\phi_n + \beta_n) = \frac{\cos\alpha_n \tan i}{\tan\eta - \sin\alpha_n \tan i}, \tag{2.56}$$

where $\beta_n = \theta_n + \alpha_n$. Thus, it is given by the following equation (see Eq. (2.49)):

$$\tan\beta_n = \tan\beta_a \cos\eta. \tag{2.57}$$

Armarego and Whitfield [28] made the latter assumption from their experiments, so that the normal shear angle ϕ_n can be obtained from the following equation of chip geometry:

$$\tan(\phi_n) = \frac{r_c(\cos\eta/\cos i)\cos\alpha_n}{1 - r_c(\cos\eta/\cos i)\sin\alpha_n}. \tag{2.58}$$

By solving these three equations numerically, the three unknown angles η, ϕ_n, and β_n are derived, or numerical iteration can be avoided if Stabler's [103] empirical chip flow rule (i.e., $\eta = i$) is applied to Eq. (2.58).

2.5.3 Prediction of Cutting Forces

The cutting force components are derived as projections of the resultant cutting force F_c that is found after subtracting the edge component (F_e) from the measured resultant force (F). The cutting force components are expressed as a function of shear yield stress τ_s, resultant force direction (θ_n, θ_i), oblique angle i, and obique shear angles (ϕ_i, ϕ_n) as given in Eq. (2.53) for F_c. The force components in the directions of cutting speed (F_{tc}), the thrust (F_{fc}), and the normal (F_{rc}) are given by the following equations (see Figs. 2.5 and 2.6):

$$\left.\begin{aligned}
F_{tc} &= F_c(\cos\theta_i\cos\theta_n\cos i + \sin\theta_i\sin i) \\
&= \frac{\tau_s bh(\cos\theta_n + \tan\theta_i\tan i)}{[\cos(\theta_n+\phi_n)\cos\phi_i + \tan\theta_i\sin\phi_i]\sin\phi_n}, \\
F_{fc} &= F_c\cos\theta_i\sin\theta_n \\
&= \frac{\tau_s bh\sin\theta_n}{[\cos(\theta_n+\phi_n)\cos\phi_i + \tan\theta_i\sin\phi_i]\cos i\sin\phi_n}, \\
F_{rc} &= F_c(\sin\theta_i\cos i - \cos\theta_i\cos\theta_n\sin i) \\
&= \frac{\tau_s bh(\tan\theta_i - \cos\theta_n\tan i)}{[\cos(\theta_n+\phi_n)\cos\phi_i + \tan\theta_i\sin\phi_i]\sin\phi_n}.
\end{aligned}\right\} \tag{2.59}$$

It is convenient to express the cutting forces in the following form:

$$\left.\begin{aligned}
F_t &= K_{tc}bh + K_{te}b, \\
F_f &= K_{fc}bh + K_{fe}b, \\
F_r &= K_{rc}bh + K_{re}b,
\end{aligned}\right\} \tag{2.60}$$

where the cutting constants contributed by the shear action correspond to

$$\left.\begin{aligned} K_{\rm tc} &= \frac{\tau_{\rm s}(\cos\theta_{\rm n} + \tan\theta_i \tan i)}{[\cos(\theta_{\rm n}+\phi_{\rm n})\cos\phi_i + \tan\theta_i \sin\phi_i]\sin\phi_{\rm n}}, \\ K_{\rm fc} &= \frac{\tau_{\rm s}\sin\theta_{\rm n}}{[\cos(\theta_{\rm n}+\phi_{\rm n})\cos\phi_i + \tan\theta_i \sin\phi_i]\cos i \sin\phi_{\rm n}}, \\ K_{\rm rc} &= \frac{\tau_{\rm s}(\tan\theta_i - \cos\theta_{\rm n}\tan i)}{[\cos(\theta_{\rm n}+\phi_{\rm n})\cos\phi_i + \tan\theta_i \sin\phi_i]\sin\phi_{\rm n}}. \end{aligned}\right\} \tag{2.61}$$

If Armarego's classical oblique model is used, the force expressions can be transformed into the following by using the geometric relations explained above:

$$\left.\begin{aligned} F_{\rm tc} &= bh\cdot\left[\frac{\tau_{\rm s}}{\sin\phi_{\rm n}}\frac{\cos(\beta_{\rm n}-\alpha_{\rm n}) + \tan i \tan\eta\sin\beta_{\rm n}}{\sqrt{\cos^2(\phi_{\rm n}+\beta_{\rm n}-\alpha_{\rm n}) + \tan^2\eta\sin^2\beta_{\rm n}}}\right], \\ F_{\rm fc} &= bh\cdot\left[\frac{\tau_{\rm s}}{\sin\phi_{\rm n}\cos i}\frac{\sin(\beta_{\rm n}-\alpha_{\rm n})}{\sqrt{\cos^2(\phi_{\rm n}+\beta_{\rm n}-\alpha_{\rm n}) + \tan^2\eta\sin^2\beta_{\rm n}}}\right], \\ F_{\rm rc} &= bh\cdot\left[\frac{\tau_{\rm s}}{\sin\phi_{\rm n}}\frac{\cos(\beta_{\rm n}-\alpha_{\rm n})\tan i - \tan\eta\sin\beta_{\rm n}}{\sqrt{\cos^2(\phi_{\rm n}+\beta_{\rm n}-\alpha_{\rm n}) + \tan^2\eta\sin^2\beta_{\rm n}}}\right]. \end{aligned}\right\} \tag{2.62}$$

Hence, the corresponding cutting constants are

$$\left.\begin{aligned} K_{\rm tc} &= \frac{\tau_{\rm s}}{\sin\phi_{\rm n}}\frac{\cos(\beta_{\rm n}-\alpha_{\rm n}) + \tan i \tan\eta\sin\beta_{\rm n}}{\sqrt{\cos^2(\phi_{\rm n}+\beta_{\rm n}-\alpha_{\rm n}) + \tan^2\eta\sin^2\beta_{\rm n}}}, \\ K_{\rm fc} &= \frac{\tau_{\rm s}}{\sin\phi_{\rm n}\cos i}\frac{\sin(\beta_{\rm n}-\alpha_{\rm n})}{\sqrt{\cos^2(\phi_{\rm n}+\beta_{\rm n}-\alpha_{\rm n}) + \tan^2\eta\sin^2\beta_{\rm n}}}, \\ K_{\rm rc} &= \frac{\tau_{\rm s}}{\sin\phi_{\rm n}}\frac{\cos(\beta_{\rm n}-\alpha_{\rm n})\tan i - \tan\eta\sin\beta_{\rm n}}{\sqrt{\cos^2(\phi_{\rm n}+\beta_{\rm n}-\alpha_{\rm n}) + \tan^2\eta\sin^2\beta_{\rm n}}}. \end{aligned}\right\} \tag{2.63}$$

The following practical approach can be used in predicting the oblique cutting forces from the orthogonal cutting database [35]:

- Evaluate the shear angle ($\phi_{\rm c}$), average friction angle ($\beta_{\rm a}$), and shear yield stress ($\tau_{\rm s}$) from orthogonal cutting tests (i.e., as given in Table 2.1).
- Assume that the orthogonal shear angle is equal to the normal shear angle in oblique cutting ($\phi_{\rm c} \equiv \phi_{\rm n}$); the normal rake angle is equal to the rake angle in orthogonal cutting ($\alpha_{\rm r} \equiv \alpha_{\rm n}$); the chip flow angle is equal to the oblique angle ($\eta \equiv i$) by adopting Stabler's chip flow rule [103]; the friction coefficient ($\beta_{\rm a}$) and shear stress ($\tau_{\rm s}$) are the same in both orthogonal and oblique cutting operations for a given speed, chip load, and tool–work material pair.
- Predict the cutting forces using the oblique cutting constants given in Eq. (2.63).

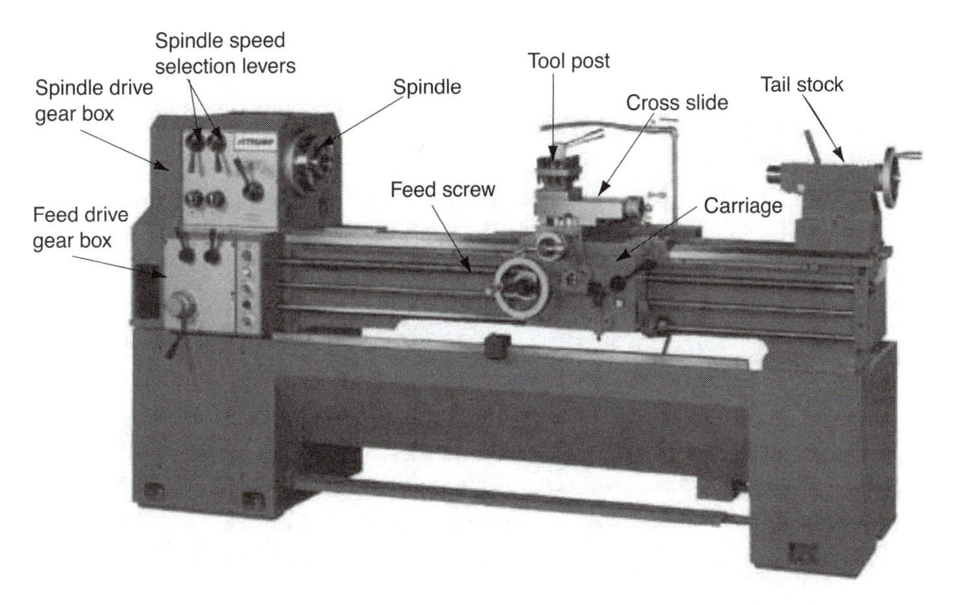

Figure 2.8: Conventional lathe.

Many practical oblique cutting operations, such as turning, drilling, and milling, can be evaluated by using the oblique cutting mechanics procedure outlined above.

2.6 MECHANICS OF TURNING PROCESSES

A typical conventional lathe is shown in Figure 2.8. The rotational workpiece is held in a chuck, which is bolted to the spindle, and a single-point cutting tool is attached to a tool post. A turning operation is used to machine cylindrical parts (see Fig. 2.9). When the workpiece is long and heavy, its two ends are held by the chuck and the center of the tail stock, respectively. The tool post is held on the top of a carriage, which has motions along the axis between the spindle and tail stock centers and is perpendicular to this axis. Conventional lathes have a single motor with constant speed. The speed is transmitted to spindle and feed drive gear boxes via belts. The speed is further reduced using gear combinations within the feed and spindle drive

Figure 2.9: A profile CNC turning operation with a carbide tool. Source: Mitsubishi Materials Corp.

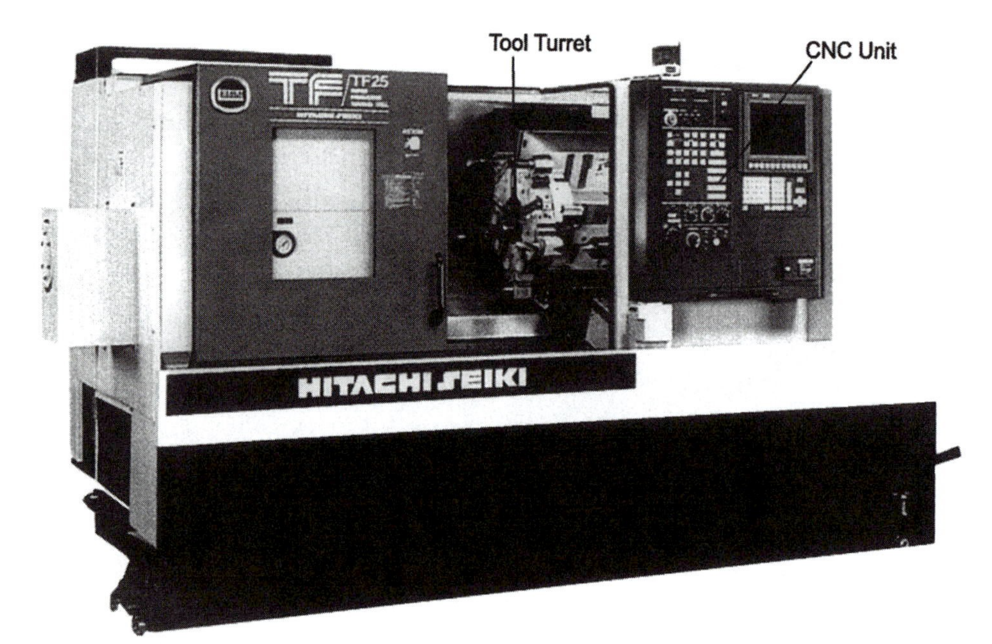

Figure 2.10: A CNC turning center. Source: Hitachi Seiki Co., Ltd.

boxes, which have shift arms with speed labels. However, in CNC lathes the spindle and feed speeds are directly programmed within NC programs, because they have computer-controlled stepless drives with a one or zero gear reduction level. A CNC turning center with a turret carrying multiple cutting tools is shown in Figure 2.10. The turret, which holds the tool, can be moved along or perpendicular to the spindle axis or both in CNC lathes.

If the tool moves along the main axis, the tool reduces the diameter of the cylindrical workpiece. If it moves perpendicular to the main axis, the tool removes material from the flat face of the workpiece, which is called a *facing* operation. The combination of cylindrical turning and facing can be used for chamfering and parting operations. The lathes allow synchronized linear motion of the carriage feed and the rotational motion of the spindle via gear boxes. The synchronized motion is used for threading operations.

The diagram of a typical cylindrical turning process is shown in Figure 2.11. The cutting tool moves parallel to the spindle and removes a skin from the blank, hence, reducing the diameter of the shaft. A standard turning tool geometry is shown in Figure 2.12. The important geometric parameters on the tool are tool nose radius, side rake, back rake, and side cutting edge angles. The chip lands and slides on the rake face of the tool. Side rake angle is the inclination of the rake face toward the cutting edge, whereas the back rake angle indicates the inclination toward the tip of the tool that is perpendicular to the surface of the workpiece in turning. In orthogonal cutting, there is zero back rake, and only side rake is considered. The tools are called positive, neutral (i.e., zero rake angle), or negative depending on the orientation of the rake angles.

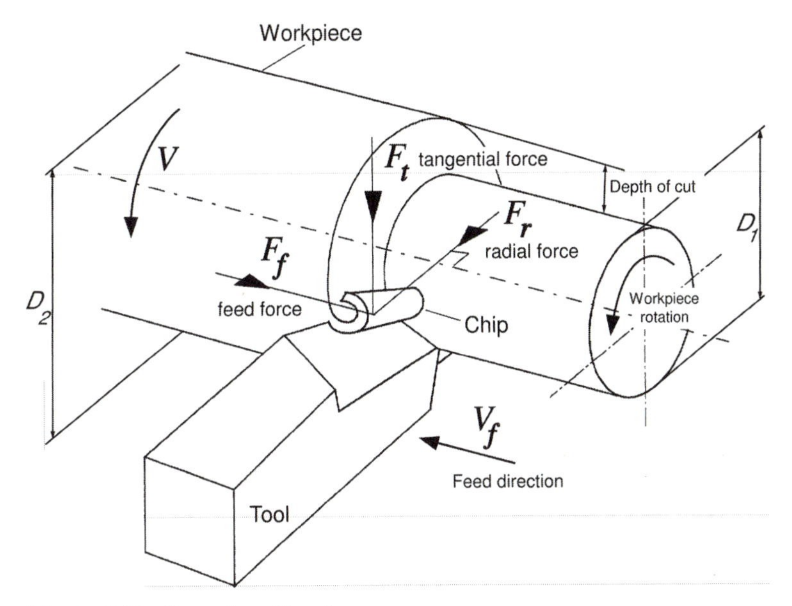

Figure 2.11: Geometry of turning process.

Positive rake angle produces higher shear angles; therefore, it helps to reduce the cutting forces. It also leaves a better surface finish because it assists the chip to flow away from the workpiece. Negative tools produce higher force than the positive tools for the same cutting conditions, because the negative rake angles decrease the shear angle. However, in interrupted cutting, where the tool periodically enters into and exits from the workpiece, the negative tools provide a greater shock resistance than positive tools because the initial material contact with the tool occurs away from the weak cutting edge. Cutting tool inserts made of carbide, ceramic, diamond, or cubic boron nitride (CBN) are usually clamped on tool holders that have a geometry like the one shown in Figure 2.12. Only one side of the positive tools can be used, because of the clearance angle at the edges. The negative carbide inserts have zero clearance angles; therefore, both sides of the insert can be used, which lowers tool costs. For example, eight edges of a negative but only four edges of a square positive insert can be used in machining. The tools have a small radius at their noses to minimize sharp feed marks on the finished surface. A large tool nose radius is not advisable because it makes the tool susceptible to self-excited vibrations, or chatter, in machining.

There are various definitions of tool angles. However, the following tool angle conversions based on back rake–side rake [25] are the most commonly accepted for analyzing the mechanics of oblique cutting:

$$\left.\begin{aligned} \tan\alpha_o &= \tan\alpha_f\cos\psi_r + \tan\alpha_p\sin\psi_r, \\ \tan i &= \tan\alpha_p\cos\psi_r + \tan\alpha_f\sin\psi_r, \\ \tan\alpha_n &= \tan\alpha_o\cos i, \end{aligned}\right\} \quad (2.64)$$

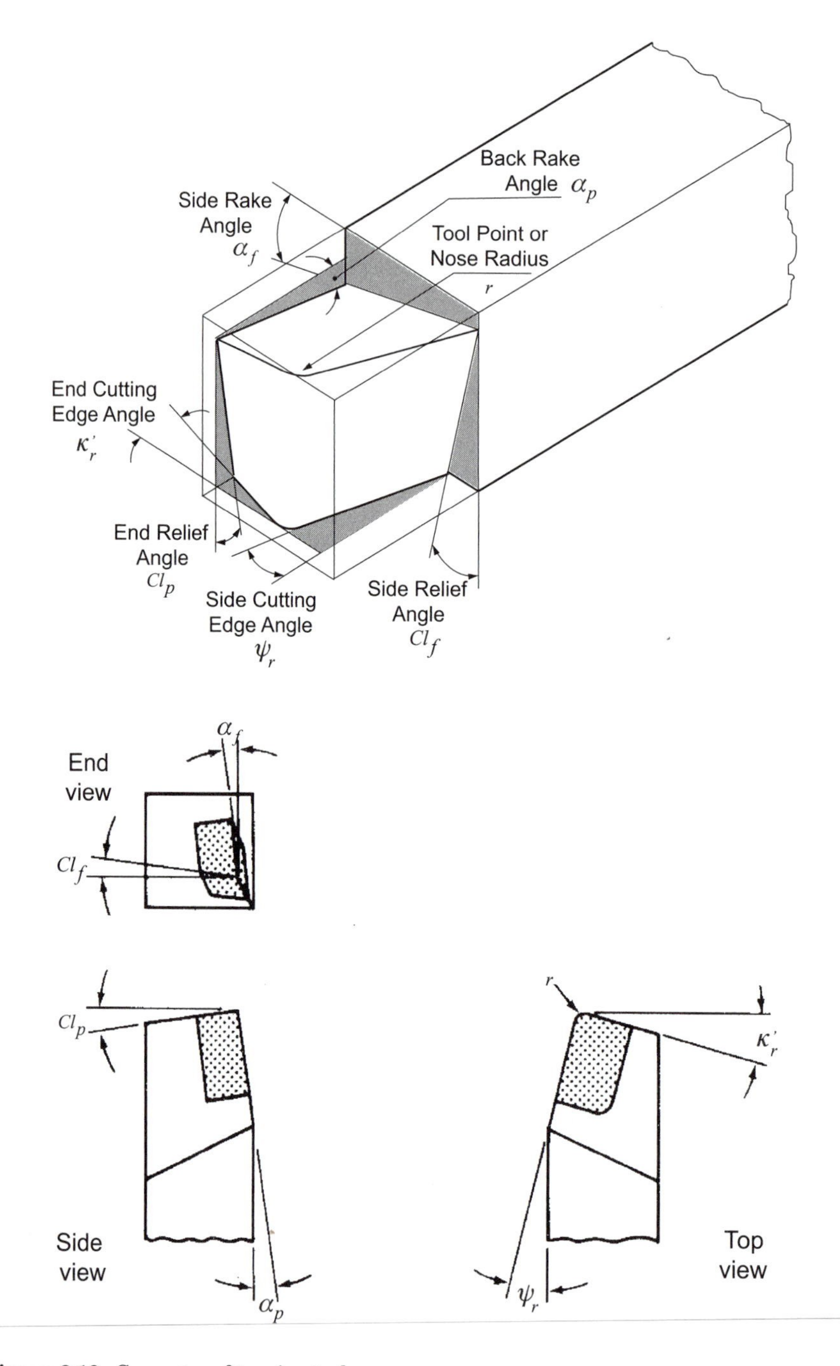

Figure 2.12: Geometry of turning tool.

where i, α_o, and α_n are the equivalent oblique, orthogonal, and normal rake angles, respectively. The equivalent oblique (i) and normal rake (α_n) angles must be evaluated for oblique cutting tools before using the cutting constants given in Eq. (2.63) for predicting the cutting forces. For simplicity, the chip

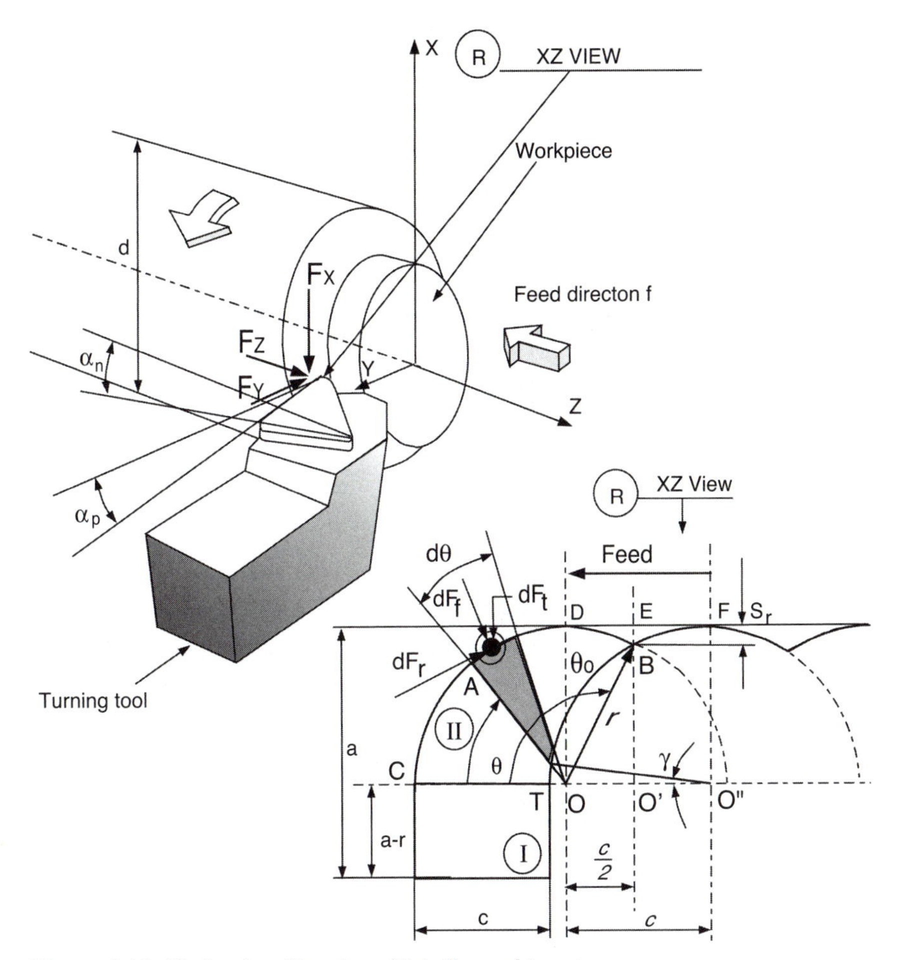

Figure 2.13: Mechanics of turning with bull-nosed inserts.

flow angle can be assumed to be equal to the oblique angle as suggested by Stabler [103].

Prediction of Cutting Forces in Turning

The cutting forces in turning can be predicted by transforming the orthogonal cutting parameters to an oblique turning geometry by using the angle transformations given in Eq. (2.64) and the oblique cutting constants given in Eq. (2.63). The process can be explained by taking a turning tool with a corner radius r, side rake angle of α_f, and back rake angle of α_p, as shown in Figure 2.13. The diameter of the workpiece is d, and the radial depth of cut is a, which is larger than the tool nose radius (r). The feed rate is c, which corresponds to the linear travel of the cutting tool in one revolution of the spindle. The chip has a uniform thickness in region I where the radial depth is less than the radius (r). However, the chip has a varying thickness in region II owing to the corner radius of the tool. The cutting forces can be predicted by applying the classical oblique cutting transformation proposed by Armarego

([24, 27]) as explained in Section 2.5. The transformation is applied separately to each region.

Region I. The radial depth is less than the corner radius of the tool (i.e., $0 < y < r$). The chip thickness is constant and equal to the feed rate (i.e., $h = c$). The cutting forces in x, y, z coordinates are parallel to the oblique cutting forces F_t, F_r, F_f respectively, and they are given as follows:

$$\left.\begin{aligned} F_{xI} &= F_{tI} = K_{tc}c(a-r) + K_{te}(a-r), \\ F_{yI} &= F_{rI} = K_{rc}c(a-r) + K_{re}(a-r), \\ F_{zI} &= F_{fI} = K_{fc}c(a-r) + K_{fe}(a-r). \end{aligned}\right\} \tag{2.65}$$

The cutting constants (K_{tc}, K_{rc}, K_{fc}) are evaluated from Eqs. (2.63) by using the orthogonal cutting parameters (ϕ_n, τ_s, β_a) obtained from orthogonal cutting tests. However, because the tool has both side and back rake angles, the equivalent oblique angle (i) and normal rake angle (α_n) must be evaluated from Eq. (2.64). The normal friction angle is evaluated from

$$\beta_n = \tan^{-1}(\tan\beta_a \cos i). \tag{2.66}$$

Note that the side cutting angle (or approach angle) is zero for this particular tool in region I. The edge force K_{re} can be taken equal to zero because it does not theoretically exist in the orthogonal cutting test measurements.

Region II. In this region the chip thickness reduces continuously, and the oblique cutting forces change their directions around the curved chip segment. The most accurate evaluation can be handled by dividing the chip into small differential elements with an angular increment $d\theta$. The center of the chip's outer surface curvature is O, and the center of its inner curvature is O''. The tool nose radius is r, and the total angular contact is $\angle COB = \theta_0$. The area of differential (dA) chip can be approximated by

$$dA \approx \overline{AT}dS, \tag{2.67}$$

where $dS = rd\theta, \overline{AT} = \overline{AO} - \overline{TO}, \overline{AO} = r$, $\overline{OO''} = c$, $\overline{TO''} = r$, and

$$\overline{TO} = \sqrt{c^2 + r^2 - 2cr\cos\gamma}. \tag{2.68}$$

Using the law of sines,

$$\frac{\overline{OO''}}{\sin[\pi - (\pi - \theta + \gamma)]} = \frac{\overline{TO''}}{\sin(\pi - \theta)},$$

the following relationship is obtained:

$$\gamma = \theta - \sin^{-1}\left[\frac{c}{r}\sin(\pi - \theta)\right]. \tag{2.69}$$

The instantaneous chip thickness at the position defined by angle θ is given by

$$\overline{AT} = h(\theta) = r - \sqrt{c^2 + r^2 - 2cr\cos\gamma}, \tag{2.70}$$

and the corresponding differential chip area is

$$dA_i = h(\theta)rd\theta. \tag{2.71}$$

The tangential ($dF_{t,II}$), radial ($dF_{r,II}$), and feed ($F_{f,II}$) forces acting on the differential chip element are given by the following:

$$\left.\begin{aligned} dF_{t,II} &= K_{tc}(\theta)dA + K_{te}dS = [K_{tc}(\theta)h(\theta) + K_{te}]rd\theta, \\ dF_{r,II} &= K_{rc}(\theta)dA + K_{re}dS = [K_{rc}(\theta)h(\theta) + K_{re}]rd\theta, \\ dF_{f,II} &= K_{fc}(\theta)dA + K_{fe}dS = [K_{fc}(\theta)h(\theta) + K_{fe}]rd\theta. \end{aligned}\right\} \qquad (2.72)$$

Because the oblique geometry varies as a function of approach angle θ, the cutting constants at each differential element must be evaluated for an accurate prediction of the cutting forces. Substituting the back rake, side rake, and instantaneous approach angle ($\psi_r = \theta$) in Eq. (2.64), we can identify the equivalent oblique angle (i) and normal rake angle α_n, which are then used in evaluating the oblique cutting constants given in Eq. (2.63). The edge forces are considered to be the same as those found from orthogonal cutting tests, and the radial component of the edge force is assumed to be zero (i.e., $K_{re} = 0$). The differential oblique cutting forces can be resolved in the x, y, z directions of the lathe, in which the forces can be measured experimentally using a dynamometer as follows:

$$\left.\begin{aligned} dF_{x,II} &= dF_{t,II}, \\ dF_{y,II} &= -dF_{f,II}\sin\theta + dF_{r,II}\cos\theta, \\ dF_{z,II} &= dF_{f,II}\cos\theta + dF_{r,II}\sin\theta. \end{aligned}\right\} \qquad (2.73)$$

Integrating the differential cutting forces along the curved chip segment gives the total cutting forces contributed by the chip in region II as follows:

$$F_{q,II} = \int_0^{\theta_0} dF_{q,II}, \quad q = x, y, z, \qquad (2.74)$$

where the approach angle limit is $\theta_0 = \pi - \cos^{-1}(c/2r)$. Because the cutting constants and chip thickness are functions of the instantaneous approach angle θ, the continuous integration in Eq. (2.74) is not trivial. It is much more practical to use constant average cutting coefficients by assuming a mean approach angle of $\pi/2$ in Eqs. (2.64) and (2.63) or by digitally integrating the forces along the curved chip segment. The curved chip is divided into $K = \theta_0/\Delta\theta$ small chip segments, and the cutting forces contributed by each small segment are evaluated and summed numerically:

$$F_{q,II} = \sum_{k=0}^{K} dF_{q,II}, \quad q = x, y, z, \qquad (2.75)$$

where the instantaneous approach angle for each segment is $\theta_k = k\Delta\theta$.

The total cutting forces acting on the tool are evaluated by summing the forces generated in regions I and II. Thus,

$$F_q = F_{q,I} + F_{q,II}, \quad q = x, y, z. \qquad (2.76)$$

The torque (T) and power (P) predicted by

$$T = F_x\left(\frac{d-a}{2}\right), \quad P = F_x V, \qquad (2.77)$$

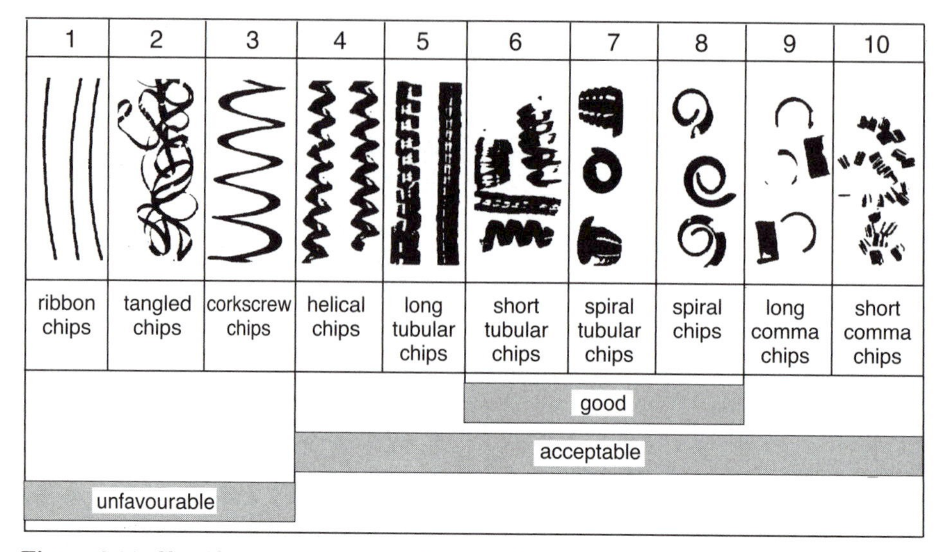

Figure 2.14: Classification of chip forms. Source: Kluft et al. CIRP [60].

where d is the diameter of the shaft in [m] and V is the cutting speed in [m/s]. The units of torque and power are [Nm] and [W], respectively. It is also important to select a suitable feed rate that does not leave larger feed marks on the surface than specified by the tolerance of the part. The amplitude of the feed mark (R_s) can be evaluated from Figure 2.13 as

$$R_s = r\left(1 - \cos\left(\sin^{-1}\left(\frac{c}{2r}\right)\right)\right). \tag{2.78}$$

Similar analysis can be applied to other turning tools by dividing their cutting edges into small oblique elements as evaluated here. The prediction of cutting forces, torque, power, and cutting constants are important in sizing machine tools for a particular operation or for selecting cutting speeds, feeds, and depth of cuts to avoid tool breakage and chatter vibrations. It is equally important to predict the direction of the chip flow and the type of chip produced by the particular tool geometry used at certain cutting conditions [54]. The chips produced in continuous turning processes may be of a ribbon, tangled, or corkscrew type, which is unfavorable (see Fig. 2.14 [60]). They can rub the finished workpiece or entangle around the tool, and it is hard to remove them from the machine tool with mechanized chip conveyors. Such unfavorable chips also cause danger for the operator's safety; productivity of the machining operation suffers from the scratched finished workpiece surfaces; and the chips can cause tool breakage because they can become entangled around the cutting edge. The basic factors that influence the chip form are the workpiece material, the tool geometry, the cutting fluid, the dynamic properties of the machine tool, and the cutting conditions. Chip breakers, which are clamped on the rake face of the tool or formed on inserts, are used to break long chips. Chip breakers disturb the free-flow direction of the chip, and they force the chip to curl toward the

Figure 2.15: A vertical CNC milling machine. Source: Hitachi Seiki Co., Ltd.

workpiece or tool, thus creating tensile stresses in the chip that lead to its breakage.

2.7 MECHANICS OF MILLING PROCESSES

A sample vertical CNC machining center, which is capable of milling, drilling, and tapping, is shown in Figure 2.15. The milling operation is an intermittent cutting process using a cutter with one or more teeth. A milling cutter is held in a rotating spindle, while the workpiece clamped on the table is linearly moved toward the cutter. A sample face milling operation and various inserted milling cutters are shown in Figure 2.16. Each milling tooth therefore traces a trochoidal path [71, 72], producing varying but periodic chip thickness at each tooth passing interval. Various milling operations are shown in Figure 2.17. Depending on the workpiece geometry, different milling cutters and machines are used. In this section, the mechanics of the milling process are presented for simple face milling operations. Mechanics of other milling operations are modeled by geometrically extending the mechanics of face milling.

The standard tool geometry in milling is shown in Figure 2.18. Double-negative tools are shock resistant in heavy-duty face milling operations. Rigid and high-power milling machines are suitable for heavy machining with negative cutters. For accurate and light milling operations, double-positive milling

Figure 2.16: A face milling operation and various inserted milling cutters. Source: Mitsubishi Materials.

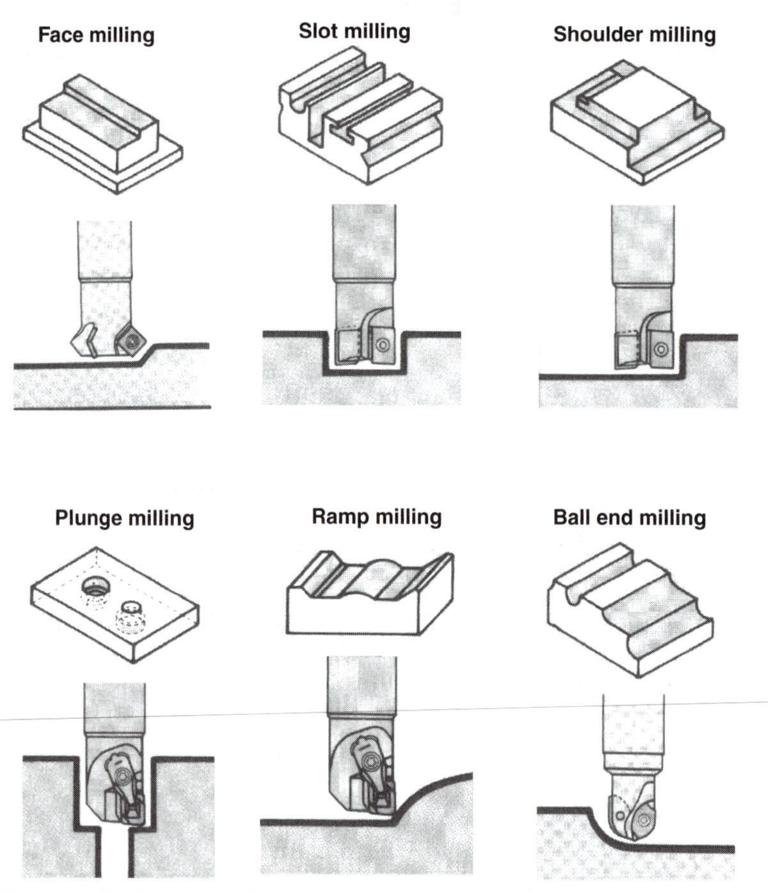

Figure 2.17: Various milling operations. Source: Mitsubishi Materials.

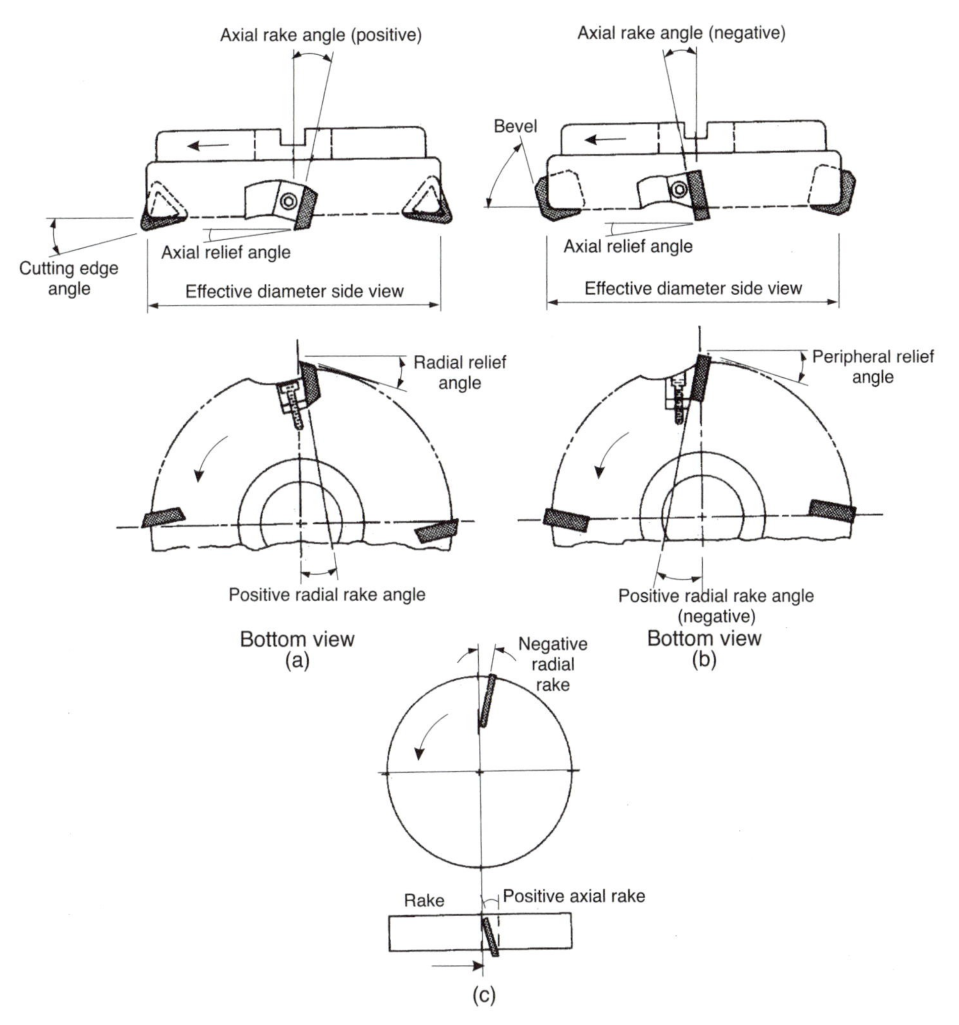

Figure 2.18: Standard face milling cutter geometry.

cutters are ideal. Negative – positive tools produce a good surface finish and are efficient in removing the chips from the insert pockets. There are three types of milling operations used in practice:

- *face milling operations*, in which entry and exit angles of the milling cutter relative to the workpiece are nonzero;
- *up-milling operations*, in which the entry angle is zero and the exit angle is nonzero; and
- *down-milling operations*, in which the entry angle is not zero and the exit angle is zero.

Both up- and down-milling operations are called peripheral or end milling operations. The geometry of chip formation in milling is shown in Figure 2.19. Unlike in turning processes, in milling the instantaneous chip thickness (h)

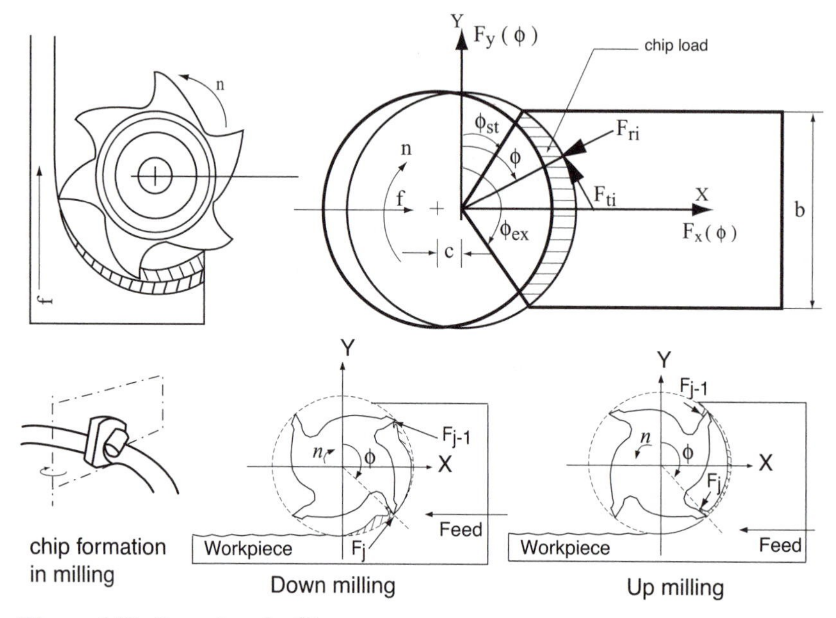

Figure 2.19: Geometry of milling process.

varies periodically as a function of time-varying immersion. The chip thickness variation can be approximated as

$$h(\phi) = c \sin\phi, \tag{2.79}$$

where c is the feed rate (mm/rev-tooth) and ϕ is the instantaneous angle of immersion. First, the helix angle is considered to be zero, which is the case in face milling operations with inserted cutters. Tangential ($F_t(\phi)$), radial ($F_r(\phi)$), and axial ($F_a(\phi)$) cutting forces are expressed as a function of varying uncut chip area ($ah(\phi)$) and edge contact length (a) as follows:

$$\left.\begin{aligned} F_t(\phi) &= K_{tc} a h(\phi) + K_{te} a, \\ F_r(\phi) &= K_{rc} a h(\phi) + K_{re} a, \\ F_a(\phi) &= K_{ac} a h(\phi) + K_{ae} a, \end{aligned}\right\} \tag{2.80}$$

where K_{tc}, K_{rc}, and K_{ac} are the cutting force coefficients contributed by the shearing action in tangential, radial, and axial directions, respectively, and K_{te}, K_{re}, and K_{ae} are the edge constants. If we assume zero nose radius and zero approach angle on the inserts, the axial components of the cutting forces become zero ($F_a = 0$). Otherwise, their influence must be modeled as presented in the section on turning (Section 2.6). The cutting coefficients are assumed to be constant for a tool–work material pair, and they can be evaluated either mechanistically from milling tests or by using the classical oblique cutting transformations given by Eqs. (2.64) and (2.63). They are sometimes expressed as a nonlinear function (Eq. 2.45) of the instantaneous or average chip thickness h_a [46]. The average chip thickness per revolution is calculated from the swept zone as

$$h_{\text{a}} = \frac{\int_{\phi_{\text{st}}}^{\phi_{\text{ex}}} c \sin\phi \, d\phi}{\phi_{\text{ex}} - \phi_{\text{st}}} = -c \frac{\cos\phi_{\text{ex}} - \cos\phi_{\text{st}}}{\phi_{\text{ex}} - \phi_{\text{st}}}. \tag{2.81}$$

The instantaneous cutting torque (T_{c}) on the spindle is where D is the diameter of the milling cutter. Horizontal (i.e., feed), normal, and axial components of the cutting forces acting on the cutter are derived from the equilibrium diagram shown in Figure 2.19 as follows:

$$\left.\begin{aligned} F_x(\phi) &= -F_{\text{t}} \cos\phi - F_{\text{r}} \sin\phi, \\ F_y(\phi) &= +F_{\text{t}} \sin\phi - F_{\text{r}} \cos\phi, \\ F_z(\phi) &= +F_{\text{a}}. \end{aligned}\right\} \tag{2.82}$$

It must be noted that the cutting forces are produced only when the cutting tool is in the cutting zone, that is,

$$F_x(\phi),\ F_y(\phi),\ F_z(\phi) > 0 \ \text{when} \ \phi_{\text{st}} \leq \phi \leq \phi_{\text{ex}},$$

where ϕ_{st} and ϕ_{ex} are the cutter entry and exit angles, respectively. Another important point is that there may be more than one tooth cutting simultaneously depending on the number of teeth on the cutter and the radial width of cut. The tooth spacing angle ϕ_{p} (or cutter pitch angle) is given as

$$\phi_{\text{p}} = \frac{2\pi}{N},$$

where N is the number of teeth on the cutter. There will be more than one tooth cutting simultaneously when the swept angle ($\phi_{\text{s}} = \phi_{\text{ex}} - \phi_{\text{st}}$) is larger than the cutter pitch angle (i.e., $\phi_{\text{s}} > \phi_{\text{p}}$). When more than one tooth cuts simultaneously, the contribution of each tooth to total feed and normal forces must be considered. It must also be noted that, because each tooth will be away from its neighboring tooth by the amount of pitch angle, the uncut chip thickness removed by each cutting edge will be different at an instantaneous position of the cutter. We can formulate the total feed, normal, and axial forces as

$$F_x = \sum_{j=1}^{N} F_{xj}(\phi_j), \quad F_y = \sum_{j=1}^{N} F_{yj}(\phi_j), \quad F_z = \sum_{j=1}^{N} F_{zj}(\phi_j), \tag{2.83}$$

whenever $\phi_{\text{st}} \leq \phi_j \leq \phi_{\text{ex}}$. Each term in the summation block represents the contribution of each tooth to the cutting forces. If the tooth j is out of the immersion zone, it contributes zero to total forces. The instantaneous resultant cutting force on the cutter (or workpiece) is given as

$$F = \sqrt{F_x^2 + F_y^2 + F_z^2}. \tag{2.84}$$

Instantaneous cutting torque on the spindle is

$$T_{\text{c}} = \frac{D}{2} \cdot \sum_{j=1}^{N} F_{\text{t}j}(\phi_j) \longrightarrow \phi_{\text{st}} \leq \phi_j \leq \phi_{\text{ex}}, \tag{2.85}$$

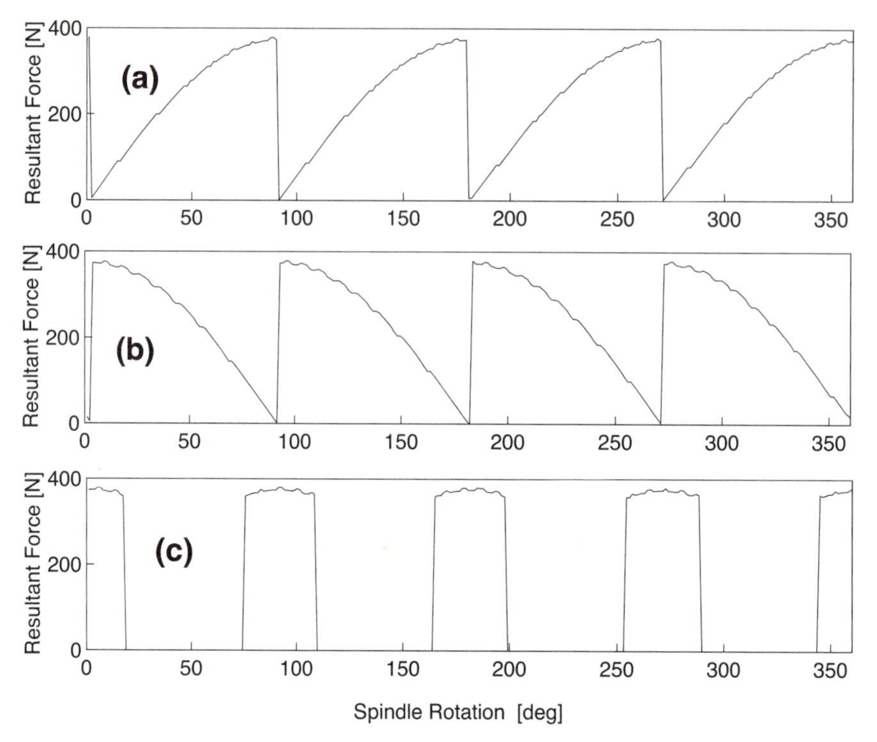

Figure 2.20: Simulated resultant cutting forces. Face milling cutter: $N = 4$ teeth, $a = 2$ mm, $c = 0.1$ mm/rev/ tooth. $K_{tc} = 1800$ N/mm^2, $K_{rc} = 540$ N/mm^2. (a) Half-immersion up-milling, (b) half-immersion down-milling, (c) center face milling with $\phi_{st} = 75$ deg, $\phi_{ex} = 105$ deg.

where D is the diameter of the cutter. The cutting power (P_t) drawn from the spindle motor is

$$P_t = V \cdot \sum_{j=1}^{N} F_{tj}(\phi_j) \longrightarrow \phi_{st} \leq \phi_j \leq \phi_{ex}, \tag{2.86}$$

where $V = \pi D n$ is the cutting speed and n is the spindle speed. For a given set of cutting conditions, the engineer may be required to predict the maximum cutting power, torque, and cutting forces required from the machine tool spindle and feed drives. The cutting forces, torque, and power are uniformly periodic at tooth passing frequency. Periodic cutting forces dynamically load and unload the machine tool structure, workpiece, and the cutter at each tooth period. Typical resultant cutting forces for three types of milling operations are shown in Figure 2.20. Half-immersion (i.e., $b = D/2$) up- and down-milling forces have opposite trends. The chip load starts with zero and gradually increases to maximum at the exit in up-milling; hence, forces have the same trend. However, the tooth experiences maximum chip load during entry followed by a gradual decrease of the chip load and, hence, the cutting forces. Manufacturing engineers are advised to use up-milling operations for heavy metal removal rates where the shock loading is reduced. For light finish cuts, down-milling is preferred to obtain a smooth surface finish. Center face milling has severe interrupted, pulse loading of the machine tool, which is not advised for light machines and positive tools. Pulse loading of the machine may resonate various structural vibration modes and cause transient vibrations during each entry and exit.

2.7.1 Mechanics of Helical End Mills

Periodic loading causes cyclic mechanical and thermal stresses on the tool, which leads to a shorter tool life. Helical end mills are used to dampen the sharp variations in the oscillatory components of the milling forces, and they are used when the depth of cut is large, but the width of cut is small. Their primary function is peripheral milling, where the walls of parts are the target finished surface. A typical end milling cutter with helical flutes is shown in Figure 2.21. The helix on the cutter provides a gradually increasing chip load along the helical flutes of the end mill [59]. If the helix angle on the cutter is β, a point on the axis of the cutting edge will be lagging behind the end point of the tool. The lag angle (ψ) at the axial depth of cut (z) is found as (Fig. 2.21)

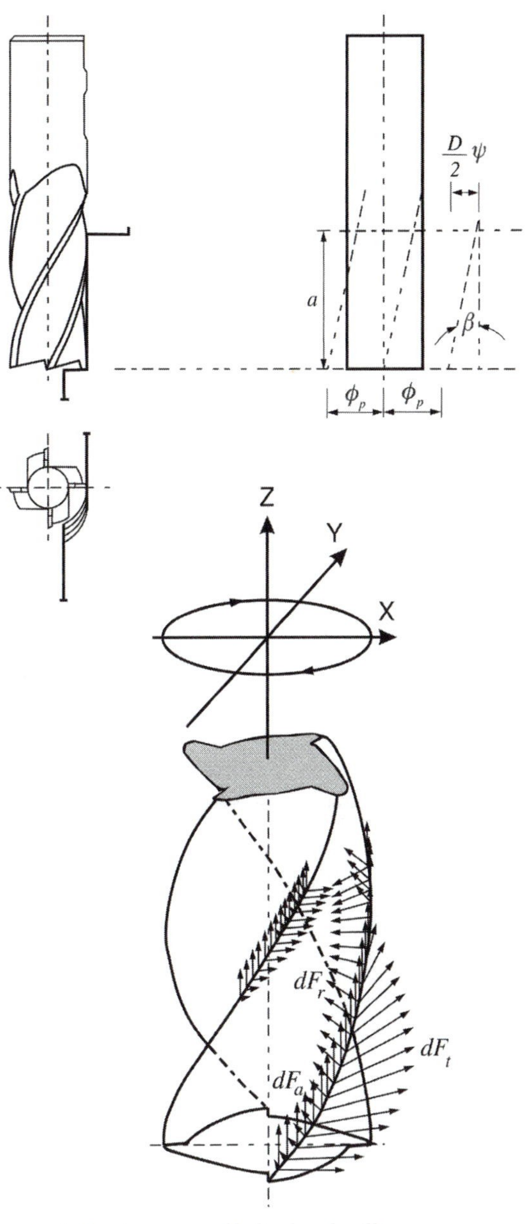

Figure 2.21: Geometry of helical end milling.

$$\tan\beta = \frac{D\psi}{2z} \quad \text{and}$$

$$\psi = \frac{2z\tan\beta}{D}. \tag{2.87}$$

When the bottom point of a reference flute of the end mill is at immersion angle ϕ, a cutting edge point that is axially z [mm] above will have an immersion angle of $(\phi - \psi)$. Obviously, the chip thickness removed along the flute's axis will also be different at each point. A general pseudocode of a milling force

TABLE 2.2. Pseudocode for Milling Force Simulation Algorithm

Inputs	
Cutting conditions	$a, c, n, \phi_{st}, \phi_{ex}$
Tool geometry	D, N, β
Cutting constants	$K_{tc}, K_{rc}, K_{te}, K_{re}$
Integration angle	$\Delta\phi$
Integration height	Δa
Outputs	
Cutting force history	$F_x(\phi), F_y(\phi), F(\phi)$
Cutting torque and power history	$T_c(\phi), P_c(\phi)$
Variables	
$\phi_p = \frac{2\pi}{N}$	Cutter pitch angle
$K = \frac{2\pi}{\Delta\phi}$	Number of angular integration steps
$L = \frac{a}{\Delta a}$	Number of axial integration steps
$i = 1$ to K	Angular integration loop
$\phi(i) = \phi_{st} + i\Delta\phi$	Immersion angle of flute's bottom edge
$F_x(i) = F_y(i) = F_t(i) = 0.0$	Initialize the force integration registers
$k = 1$ to N	Calculate the force contributions of all teeth
$\phi_1 = \phi(i) + (k-1)\phi_p$	Immersion angle for tooth k
$\phi_2 = \phi_1$	Memorize the present immersion
$j = 1$ to L	Integrate along the axial depth of cut
$a(j) = j \cdot \Delta a$	Axial position
$\phi_2 = \phi_1 - \frac{2\tan\beta}{D} a(j)$	Update the immersion angle due helix
if $\phi_{st} \le \phi_2 \le \phi_{ex}$	If the edge is cutting,
	then
$h = c \sin\phi_2$	Chip thickness at this point
$\Delta F_t = \Delta a(K_{tc}h + K_{te})$	Differential tangential force
$\Delta F_r = \Delta a(K_{rc}h + K_{re})$	Differential radial force
$\Delta F_x = -\Delta F_t \cos\phi_2 - \Delta F_r \sin\phi_2$	Differential feed force
$\Delta F_y = \Delta F_t \sin\phi_2 - \Delta F_r \cos\phi_2$	Differential normal force
$F_x(i) = F_x(i) + \Delta F_x$	Sum the cutting forces
$F_y(i) = F_y(i) + \Delta F_y$	contributed by the all
$F_t(i) = F_t(i) + \Delta F_t$	'active edges
else	
next j	
next k	
Resulting cutting force values at immersion angle $\phi(i)$	
$F(i) = \sqrt{F_x^2(i) + F_y^2(i)}$	Resultant cutting force
$T_c(i) = \frac{D}{2} F_t(i)$	Cutting torque
next i	
Plot $F_x(i), F_y(i), F_i, T_c(i)$ with varying immersion $\phi(i)$	
stop	
end	

simulation program is given in Table 2.2. The input variables set by the user are helix, entry, and exit angles, axial depth of cut, the number of teeth, feed rate, spindle speed, cutter diameter, and cutting constants. The cutter is rotated with small incremental angles. At each incremental rotation, the cutting forces are integrated axially along the sliced differential elements from the bottom of the flute toward the final axial depth of cut.

Sample experimental and simulation results for helical cylindrical and helical ball end milling of a titanium alloy are given in Figure 2.22 [15, 35]. The cutting coefficients are obtained by transforming orthogonal cutting constants of titanium given in Table 2.1 to oblique end mill geometry (see Eq. 2.63). The agreement between the experimental and simulation results are very satisfactory owing to the careful evaluation of shear stress, shear angle, and friction coefficient in a series of orthogonal tests, as well as accurate modeling of geometry and oblique transformation of mechanics. More complex milling cutters and operations can also be modeled by using the same mechanics approach, by designing a generalized, parametric milling cutter geometry as presented by Altintas and Lee [15].

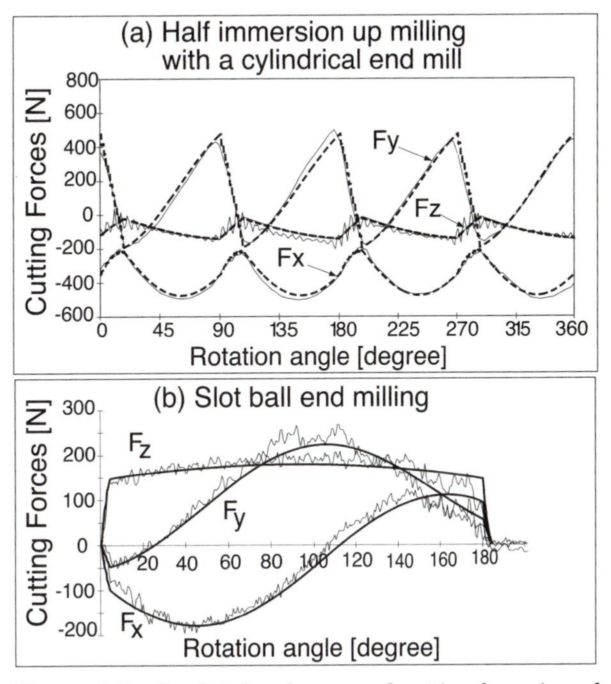

Figure 2.22: Simulated and measured cutting forces in end milling of Ti_6Al_4V alloy. Feed rate = 0.05 mm/rev/tooth. (a) $R_0 = 9.05$ mm, $V = 30m/$ min, $N_t = 4$ flutes, $\alpha_n = 12$ deg, axial depth of cut = 5.08 mm/rev, $i_o = 30$ deg. (b) Single flute ball end mill with radius $R_0 = 9.525$ mm, nominal helix angle $i_0 = 30$ deg, axial depth of cut = 1.27 mm, spindle speed $n = 269$ rev/min.

2.8 ANALYTICAL MODELING OF END MILLING FORCES

The discrete simulation of cutting forces in end milling was explained in the previous section. The accuracy of the cutting force prediction strongly depends on the selected digital integration interval. When the axial depth of cut is large in helical end milling operations, the differential element height in the axial direction must be very small to avoid numerical oscillations on the cutting force waveforms. When the cutting forces are used to predict the vibrations of the end mill or workpiece, the numerical oscillations lead to faulty simulation of vibrations. In addition, an accurate prediction of force distribution along the end mill and flexible thin webs is necessary to predict the dimensional form errors left on the finish surface. When kinematics and certain properties of the milling process are considered, it is possible to derive semianalytical expressions for end milling forces [23]. An end mill with a helix angle of β, diameter of D, and N number of flutes is assumed (see Fig. 2.21). The axial depth of cut (a) is constant, and the immersion is measured clockwise from the normal (y) axis. Assuming that the bottom end of one flute is designated as the reference immersion angle ϕ, the bottom end points of the remaining flutes are at angles $\phi_j(0) = \phi + j\phi_{\mathrm{p}}$; $j = 0, 1, 2 \ldots (N-1)$. At an axial depth of cut z the lag angle is $\psi = k_\beta z$, where $k_\beta = (2 \tan \beta)/D$. The immersion angle for flute j at axial depth of cut z therefore is

$$\phi_j(z) = \phi + j\phi_{\mathrm{p}} - k_\beta z. \tag{2.88}$$

Tangential ($dF_{\mathrm{t},j}$), radial ($dF_{\mathrm{r},j}$), and axial ($dF_{\mathrm{a},j}$) forces acting on a differential flute element with height dz are expressed similar to Eq. (2.80) as follows:

$$\left.\begin{aligned} dF_{\mathrm{t},j}(\phi, z) &= [K_{\mathrm{tc}} h_j(\phi_j(z)) + K_{\mathrm{te}}]dz, \\ dF_{\mathrm{r},j}(\phi, z) &= [K_{\mathrm{rc}} h_j(\phi_j(z)) + K_{\mathrm{re}}]dz, \\ dF_{\mathrm{a},j}(\phi, z) &= [K_{\mathrm{ac}} h_j(\phi_j(z)) + K_{\mathrm{ae}}]dz, \end{aligned}\right\} \tag{2.89}$$

where the chip thickness is

$$h_j(\phi, z) = c \sin \phi_j(z). \tag{2.90}$$

The directions of the cutting forces are aligned with the cutter axis. The cutting constants can be evaluated from Eq. (2.63) by accepting the helix angle as the oblique angle of the end mill (i.e., $i = \beta$). The elemental forces are resolved into feed (x), normal (y), and axial (z) directions using the transformation as follows:

$$\begin{aligned} dF_{x,j}(\phi_j(z)) &= -dF_{\mathrm{t},j} \cos \phi_j(z) - dF_{\mathrm{r},j} \sin \phi_j(z), \\ dF_{y,j}(\phi_j(z)) &= +dF_{\mathrm{t},j} \sin \phi_j(z) - dF_{\mathrm{r},j} \cos \phi_j(z), \\ dF_{z,j}(\phi_j(z)) &= +dF_{\mathrm{a},j}. \end{aligned} \tag{2.91}$$

Substituting the differential forces (Eq. 2.89) and the chip thickness (Eq. 2.90) into Eqs. (2.91) leads to

$$\left.\begin{aligned} dF_{x,j}(\phi_j(z)) &= \Big\{\frac{c}{2}[-K_{\mathrm{tc}} \sin 2\phi_j(z) - K_{\mathrm{rc}}(1 - \cos 2\phi_j(z))] \\ &\quad + [-K_{\mathrm{te}} \cos \phi_j(z) - K_{\mathrm{re}} \sin \phi_j(z)]\Big\}dz, \\ dF_{y,j}(\phi_j(z)) &= \Big\{\frac{c}{2}[K_{\mathrm{tc}}(1 - \cos 2\phi_j(z)) - K_{\mathrm{rc}} \sin 2\phi_j(z)] \\ &\quad + [K_{\mathrm{te}} \sin \phi_j(z) - K_{\mathrm{re}} \cos \phi_j(z)]\Big\}dz, \\ dF_{z,j}(\phi_j(z)) &= [K_{\mathrm{ac}} c \sin \phi_j(z) + K_{\mathrm{ae}}]dz. \end{aligned}\right\} \tag{2.92}$$

The differential cutting forces are integrated analytically along the in-cut portion of the flute j in obtain the total cutting force produced by the flute as follows:

$$F_q(\phi_j(z)) = \int_{z_{j,1}}^{z_{j,2}} dF_q(\phi_j(z))dz, \quad q = x, y, z, \tag{2.93}$$

where $z_{j,1}(\phi_j(z))$ and $z_{j,2}(\phi_j(z))$ are the lower and upper axial engagement limits of the in-cut portion of the flute j. The integrations are carried out by noting $\phi_j(z) = \phi + j\phi_\mathrm{p} - k_\beta z$, $d\phi_j(z) = -k_\beta dz$. Thus,

$$\left.\begin{aligned}
F_{x,j}(\phi_j(z)) &= \left\{\frac{c}{4k_\beta}\left[-K_\mathrm{tc}\cos 2\phi_j(z) + K_\mathrm{rc}[2\phi_j(z) - \sin 2\phi_j(z)]\right]\right.\\
&\qquad \left. + \frac{1}{k_\beta}[K_\mathrm{te}\sin\phi_j(z) - K_\mathrm{re}\cos\phi_j(z)]\right\}_{z_{j,1}(\phi_j(z))}^{z_{j,2}(\phi_j(z))},\\
F_{y,j}(\phi_j(z)) &= \left\{\frac{-c}{4k_\beta}[K_\mathrm{tc}(2\phi_j(z) - \sin 2\phi_j(z)) + K_\mathrm{rc}\cos 2\phi_j(z)]\right.\\
&\qquad \left. + \frac{1}{k_\beta}[K_\mathrm{te}\cos\phi_j(z) + K_\mathrm{re}\sin\phi_j(z)]\right\}_{z_{j,1}(\phi_j(z))}^{z_{j,2}(\phi_j(z))},\\
F_{z,j}(\phi_j(z)) &= \frac{1}{k_\beta}[K_\mathrm{ac}c\cos\phi_j(z) - K_\mathrm{ae}\phi_j(z)]_{z_{j,1}(\phi_j(z))}^{z_{j,2}(\phi_j(z))}.
\end{aligned}\right\} \quad (2.94)$$

The axial integration limits $z_{j,1}$ and $z_{j,2}$ are required for each flute to implement the cutting force model. When extruded axially, each $[\phi_\mathrm{st}, \phi_\mathrm{ex}]$ cutter arc segment defines an immersion section that can interact with a helical tooth j in five distinct ways (Fig. 2.23). The lag angle at full axial depth of cut $z = a$ is $\psi_\mathrm{a} = k_\beta a$.

The following computer algorithm is used in determining the axial integration boundaries:

- If $\phi_\mathrm{st} < \phi_j(z = 0) < \phi_\mathrm{ex}$, then $z_{j,1} = 0$;
  If $\phi_\mathrm{st} < \phi_j(z = a) < \phi_\mathrm{ex}$, then $z_{j,2} = a$.
  If $\phi_j(z = a) < \phi_\mathrm{st}$, then $z_{j,2} = (1/k_\beta)(\phi + j\phi_\mathrm{p} - \phi_\mathrm{st})$.
- If $\phi_j(z = 0) > \phi_\mathrm{ex}$ and $\phi_j(z = a) < \phi_\mathrm{ex}$, then $z_{j,1} = (1/k_\beta)(\phi + j\phi_\mathrm{p} - \phi_\mathrm{ex})$;
  If $\phi_j(z = a) > \phi_\mathrm{st}$, then $z_{j,2} = a$.
  If $\phi_j(z = a) < \phi_\mathrm{st}$, then $z_{j,2} = (1/k_\beta)(\phi + j\phi_\mathrm{p} - \phi_\mathrm{st})$.
  If $\phi_j(z = 0) > \phi_\mathrm{ex}$ and $\phi_j(z = a) > \phi_\mathrm{ex}$, the flute is out of cut.

The values for the integration limits $z_{j,1}$ and $z_{j,2}$ are taken from the cases

Figure 2.23: Helical flute-part face integration zones and differential forces on the cutting tool at a particular rotation angle ϕ and depth z.

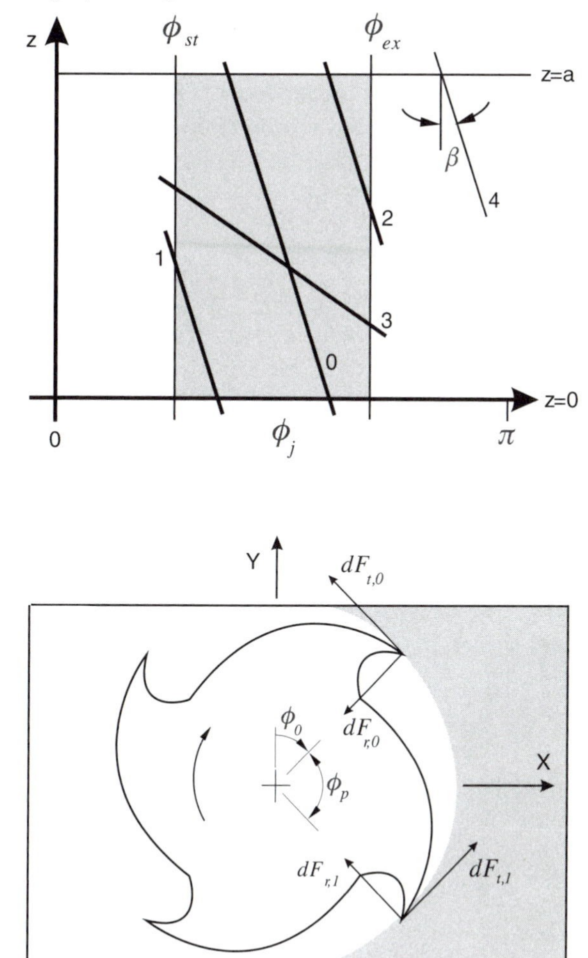

listed above substituted into Eq. (2.94), and the resulting expressions can be further simplified for an efficient computation in computer programs. Note that, to use the expressions, flute $j = 0$ must be aligned at $\phi = 0$ in the beginning of the algorithm. The remaining flutes must be indexed $(j = 1, 2, \ldots, N-1)$ from the reference tooth at pitch angle (ϕ_p) intervals. The cutting forces contributed by all flutes are calculated and summed to obtain the total instantaneous forces on the cutter at immersion ϕ as follows:

$$F_x(\phi) = \sum_{j=0}^{N-1} F_{x_j}; \quad F_y(\phi) = \sum_{j=0}^{N-1} F_{y_j}; \quad F_z(\phi) = \sum_{j=0}^{N-1} F_{z_j}. \tag{2.95}$$

The resultant cutting force acting on the milling cutter is

$$F(\phi) = \sqrt{F_x(\phi)^2 + F_y(\phi)^2 + F_z(\phi)^2}. \tag{2.96}$$

Thus, computationally inefficient and approximate digital integration methods are eliminated by the closed-form expressions of the instantaneous cutting forces. The closed-form expressions can be used for process planning, investigating the interaction between tool–workpiece structure and the milling process, and predicting finish surface. The algorithm can be efficiently implemented to the CAD/CAM systems for milling process simulation [99].

2.8.1 Mechanistic Identification of Cutting Constants in Milling

Using orthogonal cutting parameters such as shear angle, shear stress, and friction coefficient to determine oblique milling constants is desired for modeling a variety of milling cutter geometries (see Eqs. 2.64 and 2.63). However, some cutting tools may have complex cutting edges, and the evaluation of cutting constants by creating a time-consuming orthogonal cutting database may not be possible. In such cases, a quick method of calibrating the milling tools, the mechanistic approach, is used [35]. A set of milling experiments are conducted at different feed rates, but at constant immersion and axial depth of cut. The average forces per tooth period are measured. To avoid the influence of runout on the measurements, the total force per spindle revolution is collected and divided by the number of teeth on the cutter. The experimentally evaluated average cutting forces are equated to analytically derived average milling force expressions, which leads to the identification of cutting constants. Because the total material removed per tooth period is constant with or without helix angle, the average cutting forces are independent of helix angle. Replacing $dz = a, \phi_j(z) = \phi$, and $k_\beta = 0$ in Eqs. (2.92) and integrating them over one revolution and dividing by the pitch angle $(\phi_p = 2\pi/N)$ yields to average milling forces per tooth period,

$$\overline{F}_q = \frac{1}{\phi_p} \int_{\phi_{st}}^{\phi_{ex}} F_q(\phi) d\phi, \tag{2.97}$$

because the flute cuts only within the immersion zone (i.e., $\phi_{st} \leq \phi \leq \phi_{ex}$). Integrating the instantaneous cutting forces leads to

$$\left.\begin{aligned}
\overline{F}_x &= \Big\{\tfrac{Nac}{8\pi}\left[K_{tc}\cos 2\phi - K_{rc}[2\phi - \sin 2\phi]\right] \\
&\quad + \tfrac{Na}{2\pi}[-K_{te}\sin\phi + K_{re}\cos\phi]\Big\}_{\phi_{st}}^{\phi_{ex}}, \\
\overline{F}_y &= \Big\{\tfrac{Nac}{8\pi}\left[K_{tc}(2\phi - \sin 2\phi) + K_{rc}\cos 2\phi\right] \\
&\quad - \tfrac{Na}{2\pi}[K_{te}\cos\phi + K_{re}\sin\phi]\Big\}_{\phi_{st}}^{\phi_{ex}}, \\
\overline{F}_z &= \tfrac{Na}{2\pi}\left[-K_{ac}c\cos\phi + K_{ae}\phi\right]_{\phi_{st}}^{\phi_{ex}}.
\end{aligned}\right\} \tag{2.98}$$

Full-immersion (i.e., slotting) milling experiments are most convenient; here, the entry and exit angles are $\phi_{st} = 0$ and $\phi_{ex} = \pi$, respectively. When full-immersion conditions are applied to Eqs. (2.98), the average forces per tooth period are simplified as follows:

$$\left.\begin{aligned}
\overline{F}_x &= -\tfrac{Na}{4}K_{rc}c - \tfrac{Na}{\pi}K_{re}, \\
\overline{F}_y &= +\tfrac{Na}{4}K_{tc}c + \tfrac{Na}{\pi}K_{te}, \\
\overline{F}_z &= +\tfrac{Na}{\pi}K_{ac}c + \tfrac{Na}{2}K_{ae}.
\end{aligned}\right\} \tag{2.99}$$

The average cutting forces can be expressed by a linear function of feed rate (c) and an offset contributed by the edge forces as follows:

$$\overline{F}_q = \overline{F}_{qc}c + \overline{F}_{qe} \quad (q = x, y, z). \tag{2.100}$$

The average forces at each feed rate are measured, and the cutting edge components ($\overline{F}_{qc}, \overline{F}_{qe}$) are estimated by a linear regression of the data. Finally, the cutting force coefficients are evaluated from Eqs. (2.99) and (2.100) as follows:

$$\begin{aligned}
K_{tc} &= \tfrac{4\overline{F}_{yc}}{Na}, \quad K_{te} = \tfrac{\pi\overline{F}_{ye}}{Na}, \\
K_{rc} &= \tfrac{-4\overline{F}_{xc}}{Na}, \quad K_{re} = \tfrac{-\pi\overline{F}_{xe}}{Na}, \\
K_{ac} &= \tfrac{\pi\overline{F}_{zc}}{Na}, \quad K_{ae} = \tfrac{2\overline{F}_{ze}}{Na}.
\end{aligned} \tag{2.101}$$

The procedure is repeated for each cutter geometry; hence, the milling force coefficients can not be predicted before the testing of newly designed cutters using mechanistic models. However, oblique cutting transformation using basic orthogonal cutting parameters can predict the cutting constants before the cutter is manufactured.

2.9 MECHANICS OF DRILLING

A sample twist drill, which is used in drilling holes, is shown in Figure 2.24. A twist drill has a chisel edge at the bottom and two helical cutting lips with a

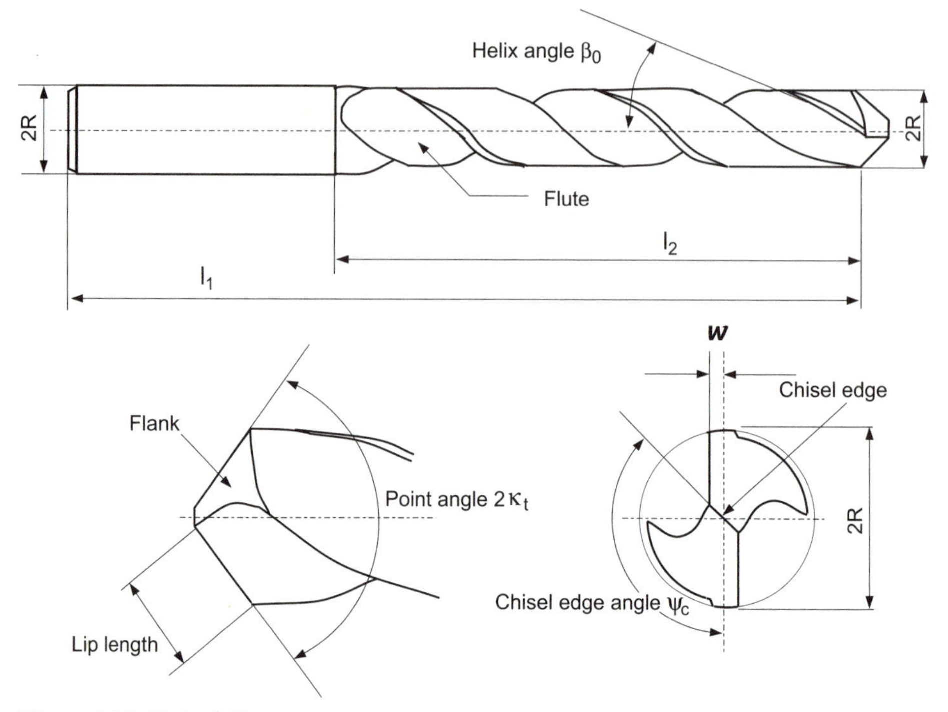

Figure 2.24: Twist drill geometry.

taper angle (κ_t) that meet with the flutes with a helix angle of β_0. The helical flutes do not cut; they are used to evacuate the chips from the drilled hole. The chisel has a width of $2w$ and an edge angle of ψ_c. The cutting lips have an offset from the drill center because of the chisel edge. The lips expand the hole by removing the material with a constant chip thickness (h) as the drill is fed into the material at a feed rate of c [mm/rev]. The thrust force that is used to push the drill into the work material and the torque applied to the drill and spindle drive are required to evaluate the mechanics of the drilling operations. The mechanics of drilling must be analyzed separately for the chisel and cutting lip regions.

Chisel Edge

The chisel edge does not cut but only spreads the material sideways by an indentation mechanism. Instead of using the laws of cutting, the mechanics of indentation must be used. If the process is simplified as in a hardness test, then the thrust force acting on the chisel edge can be simplified as

$$\text{THRUST}_i = F_{z,i} = A_{ch} H_B, \tag{2.102}$$

where H_B is the Brinell hardness of the work material and A_{ch} is the instantaneous indentation area of the chisel edge. A_{ch} is evaluated as the product of chisel length $(2w/\sin(\pi - \psi_c))$ and the contact length of the spread material with the lip $(c/(2\cos\gamma_t))$. Considering both sides of the chisel contact, we have an area of contact for indentation of

$$A_{ch} = \frac{2wc}{\sin(\pi - \psi_c)\cos\kappa_t}. \tag{2.103}$$

Approximating the chisel force by the simple indentation method is not reliable for an accurate analysis. The chisel edge geometry and indentation mechanism is rather complex and requires detailed geometric modeling, and experimental calibration of various empirical factors, as well [26]. Recently, the chisel geometry has been improved significantly by tool manufacturers to minimize the skidding of the drill on the work surface during penetration. For practical considerations, the chisel forces can be assumed to be about 10 to 15 percent of the lip cutting forces, and the torque can be neglected because the chisel width ($2w$) is rather small.

Cutting Lip

The geometry of the cutting lip of a drill is quite complex. To use the orthogonal to oblique cutting transformation, it is necessary to identify the helix, normal rake, and oblique angles at cutting points along the lip. Because of the offset caused by the chisel and the varying diameter, the helix, normal rake, and oblique angles vary from the chisel–lip to lip–flute intersection. The following treatment of the drill geometry is based on the work presented by Galloway [47] and Armarego and Brown [25].

The geometric model of the drill shown in Figure 2.25 is used in explaining the oblique cutting model. The drill axis is aligned with z, the lip that has an offset from the drill axis because chisel edges is aligned parallel to x, and the axis normal to the lip is parallel to y in a Cartesian coordinate system with a center at the drill tip O.

First consider the bottom of the flute where the lips and chisel edge meet (i.e., the plane of $z = 0$). The chisel web offset from the drill axis is w, and the chisel edge angle is ψ_c (see the detailed view in Fig. 2.26). The radial distance between the drill center and the point where the lip intersects with the chisel edge is

$$r(0) = \frac{w}{\sin(\pi - \psi_c)}, \tag{2.104}$$

with coordinates $x(0) = r(0)\cos(\pi - \psi_c), y(0) = w, z = 0$.

The lip and helical flute intersect at elevation $z = a$ and the drill diameter is R (see detail (b) in Fig. 2.26). The radial distance between the lip's outermost point and drill center is $r(a) = R$, with coordinates

$$\left.\begin{aligned} &x(a) = R\cos\theta(a), \quad \theta(a) = \sin^{-1}(w/R), \\ &y(a) = w, \\ &z = a, \end{aligned}\right\} \tag{2.105}$$

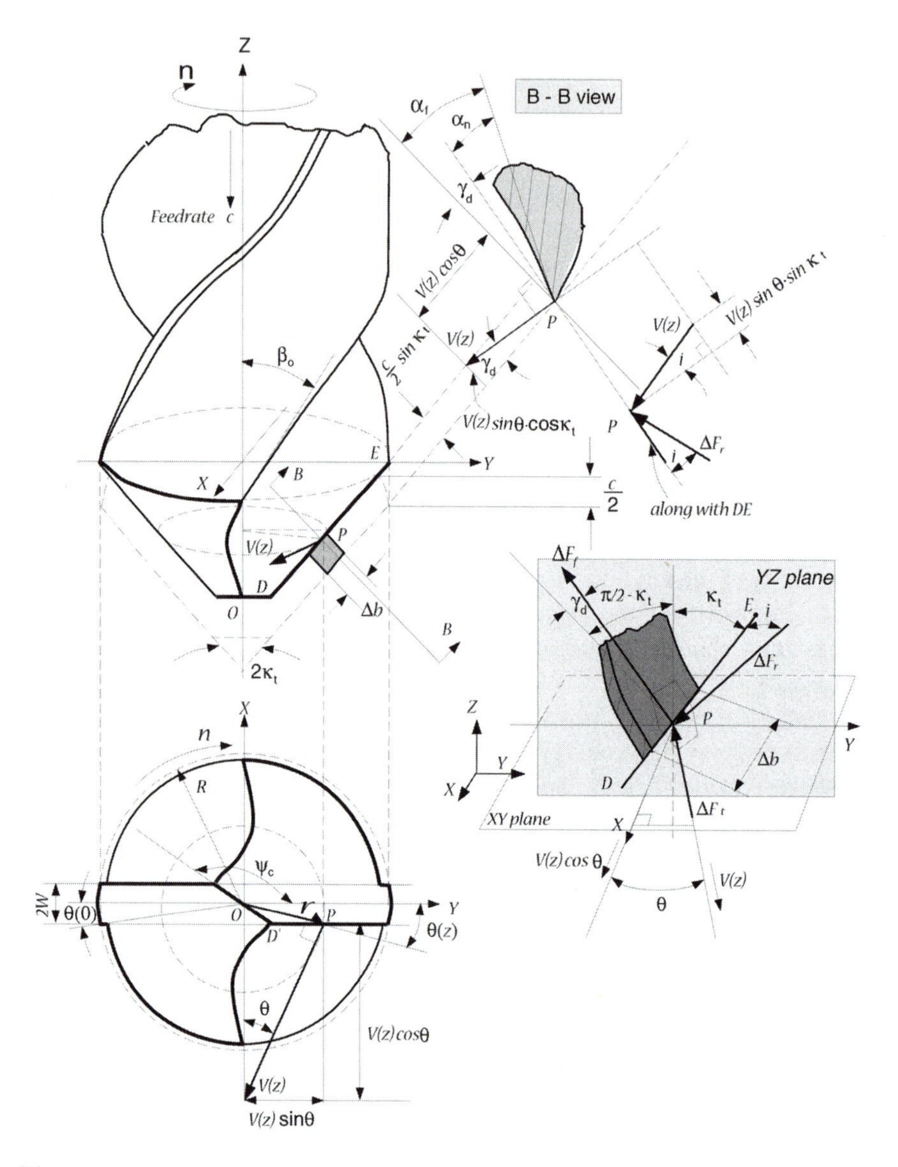

Figure 2.25: Mechanics model of twist drill.

where R is the drill radius. The projected length of the whole cutting lip on the xy plane at elevation $z = a$ becomes

$$\overline{D'E} = \overline{DE} \sin \kappa_t = b \sin \kappa_t, \tag{2.106}$$

where $b = \overline{DE}$ is the length of lip that cuts the work material. Although the helix angle changes along the lip, a nominal helix angle (β_0) of the drill is defined for the flutes at the cylindrical region ($r(a) = R$) as follows:

$$\tan \beta_0 = \frac{2\pi R}{L_p}, \tag{2.107}$$

where L_p is the constant pitch length of the helix.

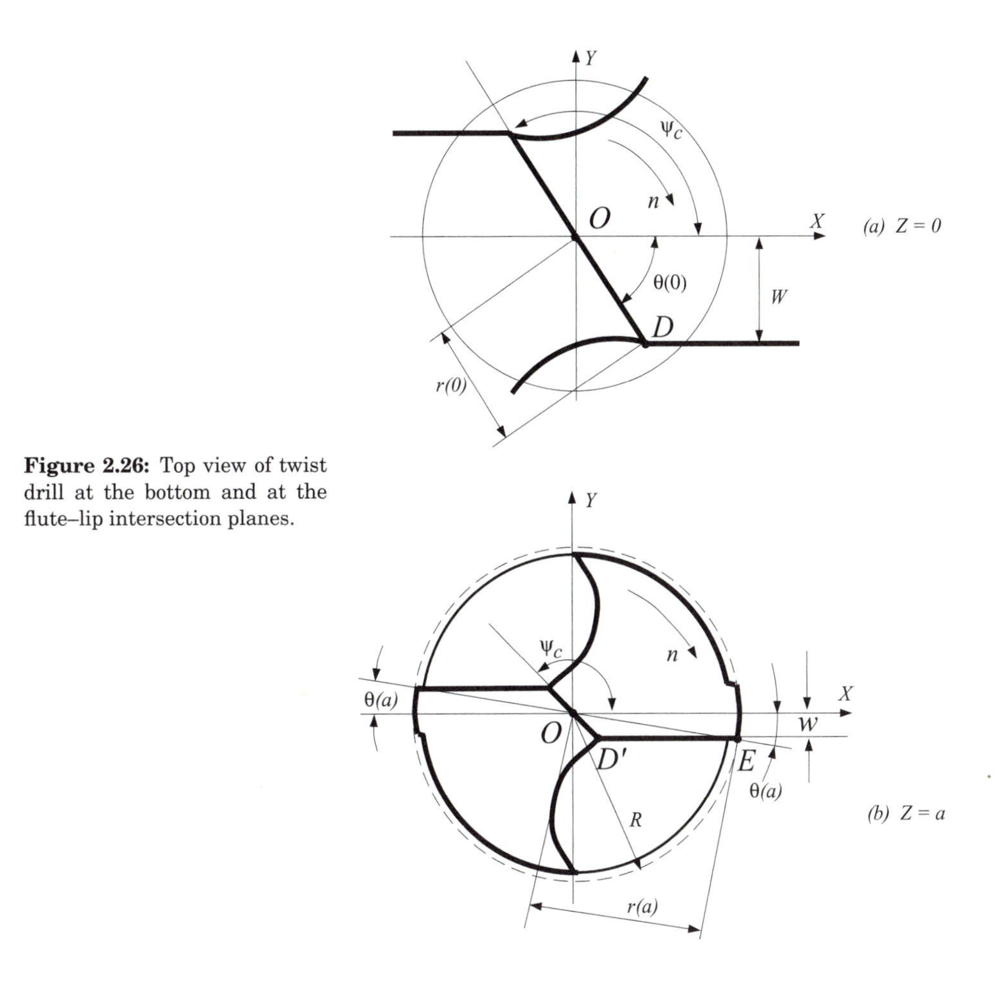

Figure 2.26: Top view of twist drill at the bottom and at the flute–lip intersection planes.

A point $P(x, y, z)$ on the cutting lip is considered in Figure 2.25. The projection of the cutting lip on the xy plane at elevation z is $\overline{D'P} = z \tan\kappa_t$. The radial distance between the axis of the drill and point P is

$$r(z) = \sqrt{y(0)^2 + [x(0) + \overline{D'P}]^2} = \sqrt{w^2 + [w\cot(\pi - \psi_c) + z\tan\kappa_t]^2} \quad (2.108)$$

with coordinates

$$x(z) = r(z)\cos(\theta(z)), \quad y(z) = r(z)\sin(\theta(z)). \quad (2.109)$$

The local helix angle at point P can be expressed as

$$\beta(z) = \frac{2\pi r(z)}{L_p}, \quad (2.110)$$

which has a different value at each elevation z. The cutting velocity (V) is perpendicular to the radius $r(z)$ and has the following components in the xy plane:

$$V_x(z) = V \cdot \cos\theta(z), \quad V_y(z) = V \cdot \sin\theta(z). \quad (2.111)$$

The projection of the cutting velocity on the cutting lip, which has a taper angle of κ_t, is

$$V_t = V_y \sin \kappa_t = V \sin \theta(z) \sin \kappa_t. \tag{2.112}$$

The oblique angle (i) is defined between the cutting velocity and the normal to the cutting edge as follows:

$$\sin i = \frac{V_t}{V} = \sin \theta(z) \sin \kappa_t. \tag{2.113}$$

The cutting mechanics are defined on the normal plane, which is perpendicular to the cutting edge. If a normal plane to cutting edge is considered at point P, the velocity (V) has a component perpendicular to the tapered cutting edge ($V_y(z) \cos \kappa_t$) and a component parallel to V_y. The angle between the two velocity components is

$$\tan \gamma_d = \frac{V_y \cos \kappa_t}{V_x} = \tan \theta(z) \cos \kappa_t. \tag{2.114}$$

The effective rake angle (α_f) is evaluated by Armarego and Brown by considering a point inside the lip and along the cutting velocity and projecting it on the differential. This yields [25]

$$\tan \alpha_f = \frac{\tan \beta(z) \cos \theta(z)}{\sin \kappa_t - \tan \beta(z) \sin \theta(z) \cos \kappa_t}. \tag{2.115}$$

The normal rake angle is found from the geometry via

$$\alpha_n = \alpha_f - \gamma_d. \tag{2.116}$$

Although the drill geometry is complex, one can take the geometric relationships (2.114–2.116) for granted and predict the cutting forces along the lip. If the drill lip is broken into small differential elements with height dz and width Δb, each differential lip removes a chip with an area of

$$dA(z) = \Delta b \cdot h, \tag{2.117}$$

where the chip thickness (h) removed by one of the two flutes and the chip width (Δb) are

$$h = \frac{c}{2} \sin \kappa_t, \quad \Delta b = \frac{dz}{\cos \kappa_t}. \tag{2.118}$$

The cutting forces in tangential (parallel to the cutting velocity) chip flow and radial directions are expressed as

$$\left.\begin{aligned} dF_t(z) &= K_{tc}(z) dA + K_{te} \Delta b, \\ dF_f(z) &= K_{fc}(z) dA + K_{fe} \Delta b, \\ dF_r(z) &= K_{rc}(z) dA + K_{re} \Delta b, \end{aligned}\right\} \tag{2.119}$$

where the cutting coefficients for each element at elevation z are different because of varying helix, normal rake, and oblique angles. Edge constants K_{te}, K_{fe}, and K_{re} are evaluated experimentally, and cutting constants K_{tc}, K_{fc},

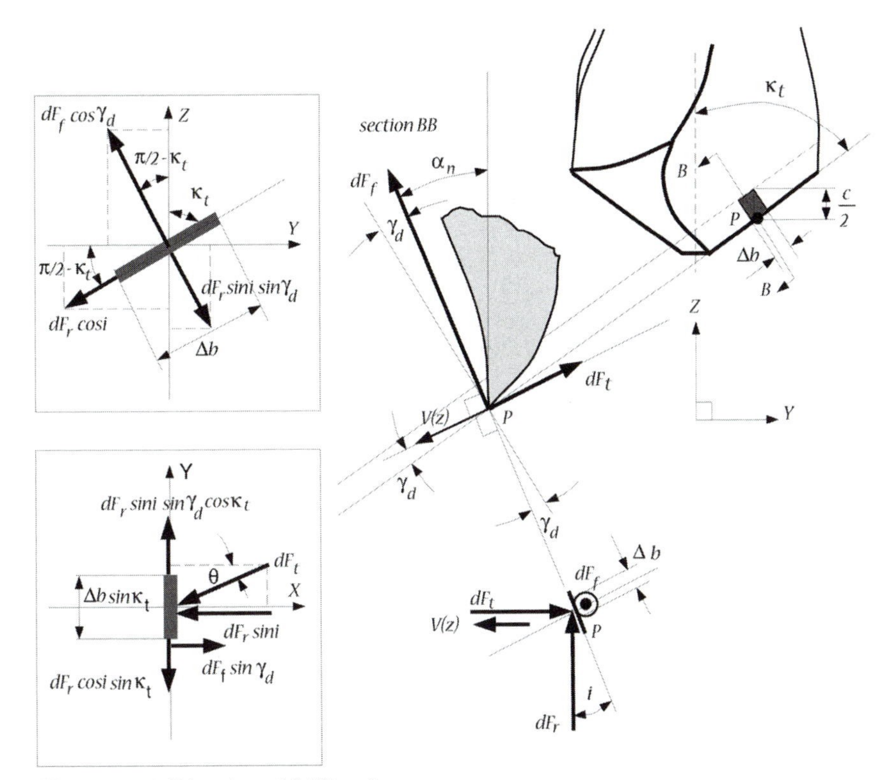

Figure 2.27: Direction of drilling forces.

and K_{rc} are evaluated from an orthogonal cutting database using the oblique transformations given by Eqs. (2.63).

The components of the elemental cutting force components $(dF_{\mathrm{t}}, dF_{\mathrm{f}}, dF_{\mathrm{r}})$ can be evaluated in the x, y, z directions as shown in Figure 2.27 as follows:

$$\left.\begin{aligned} dF_x(z) &= [dF_{\mathrm{f}} \sin\gamma_{\mathrm{d}} - dF_{\mathrm{t}} \cos\theta - dF_{\mathrm{r}} \sin i], \\ dF_y(z) &= [dF_{\mathrm{r}}(\sin i \cdot \sin\gamma_{\mathrm{d}} \cdot \cos\kappa_{\mathrm{t}}) - dF_{\mathrm{f}} \cos\gamma_{\mathrm{d}} \cdot \cos\kappa_{\mathrm{t}} - dF_{\mathrm{t}} \sin\theta], \\ dF_z(z) &= [dF_{\mathrm{f}} \cos\gamma_{\mathrm{d}} \cdot \sin\kappa_{\mathrm{t}} - dF_{\mathrm{r}}(\cos i \cdot \cos\kappa_{\mathrm{t}} + \sin i \cdot \sin\gamma_{\mathrm{d}} \cdot \sin\kappa_{\mathrm{t}})]. \end{aligned}\right\} \tag{2.120}$$

The total thrust and torque exerted on the drill can be evaluated by summing the contribution of all lip elements that number $M = b/\Delta b = b\cos\kappa_{\mathrm{t}}/dz)$. The total thrust force and torque produced by the two lips (i.e., region II) can be expressed as

$$\left.\begin{aligned} \mathrm{THRUST}_{ii} &= 2\sum_{m=1}^{M} dF_z(z), \\ \mathrm{TORQUE}_{ii} &= 2\sum_{m=1}^{M} dF_{\mathrm{t}}(z) \cdot r(z). \end{aligned}\right\} \tag{2.121}$$

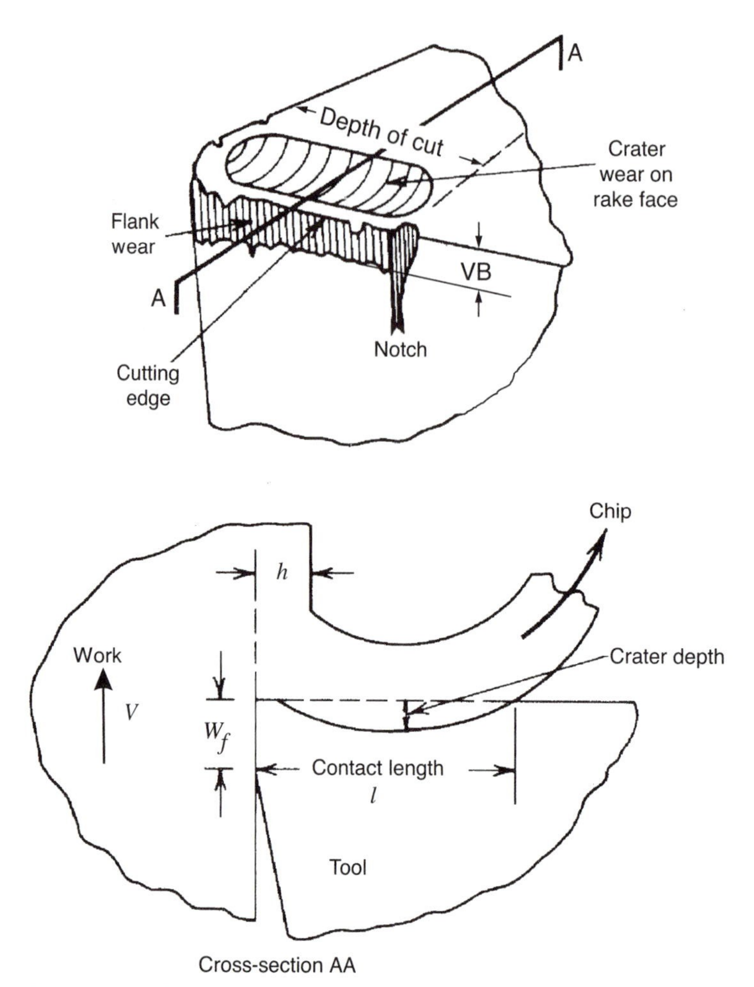

Figure 2.28: The types of tool wear and breakage.

The total thrust force exerted on the drill is found by summing the contributions of the chisel and lip forces as follows:

$$\text{THRUST} = \text{THRUST}_i + \text{THRUST}_{ii}. \tag{2.122}$$

The torque produced by the chisel edge can be neglected.

The mechanics of chisel and difficulty in modeling the oblique geometry complicate the mechanics of drilling. Readers are referred to a detailed study presented by Armarego et al. [26]. The grinding of twist drills requires accurate modeling of chisel and lip geometry, as well as rake, helix, and clearance angles along the lip and chisel because these strongly influence the cutting mechanics [37], vibrations [91], and tool wear [55].

2.10 TOOL WEAR AND TOOL BREAKAGE

Cutting tools can be used only when their edges produce parts with specified surface finish and dimensional tolerance. When the quality of the cutting edge

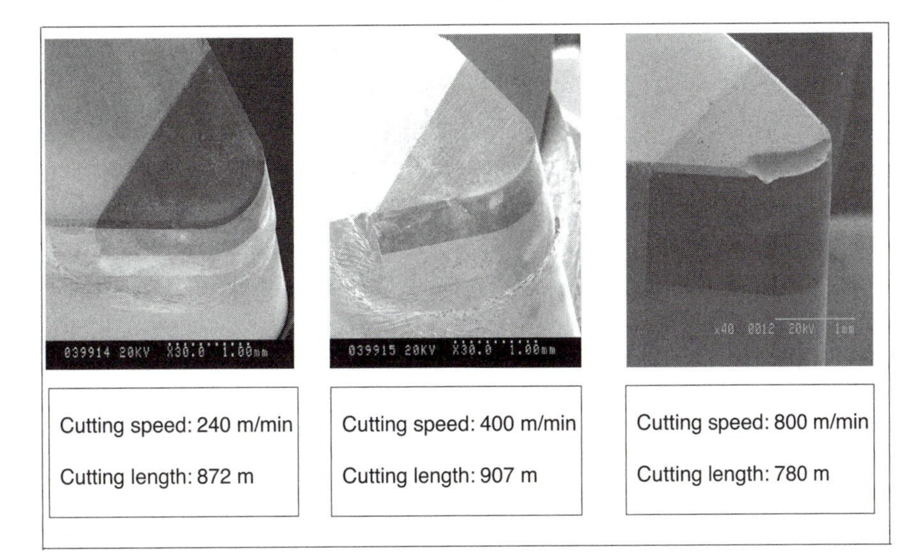

Figure 2.29: Wear of CBN tools in turning P20 mold steel at cutting speeds 240 m/min, 400 m/min, and 800 m/min.

is lost because of wear or breakage (Fig. 2.28), the tool reaches its life limit and must be replaced by a new one. Cutting tools experience various forms of wear and breakage during machining. Tool wear is defined as a gradual loss of tool material at workpiece material and tool contact zones [62]. The moving chip is in contact with the part of the rake face where crater wear occurs (see Fig. 2.29). The cutting edge generates the finish surface, and as it wears, the flank face of the tool starts rubbing against the finished surface leading to flank wear (see Fig. 2.30). A piece of the cutting edge may suddenly fracture during machining because of thermal or mechanical overloading of the cutting edge, increased cutting forces caused by significant tool wear, or weakening of the edge because of thermally or mechanically induced microcracks. The objective of the machining engineer is to select a cutting tool material and geometry, as well as optimal cutting conditions (i.e., feed, speed, depth of cut, and lubricant), that lead to the most economical machining time without violating the surface finish and tolerance of the workpiece and wear and breakage limits of the cutting tool simultaneously. The identification of conditions leading to optimal machining is called a *machinability study*, and its simple fundamentals are briefly introduced in the following section.

Figure 2.30: Flank wear and built-up edge.

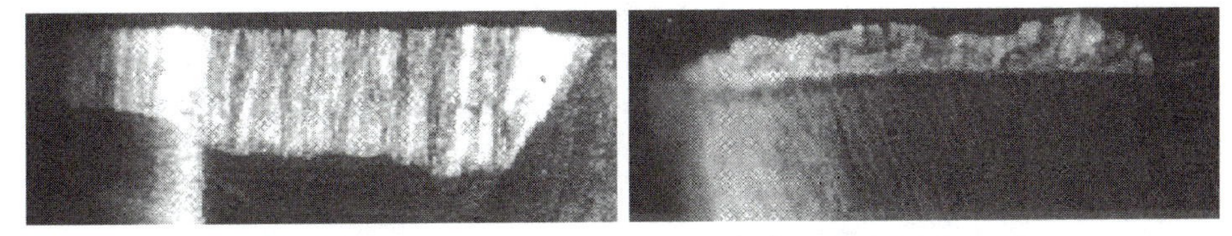

2.10.1 Tool Wear

Several wear mechanisms may occur simultaneously, or one of them may dominate the process. They can be listed as *abrasion, adhesion, diffusion, fatigue*, and *chemical* wear.

Abrasion Wear

Abrasion occurs when a harder material (i.e., the tool) shears away small particles from the softer work material. However, softer work material also removes small particles from the tool material, although at a smaller rate. The hard tool particles are caught between the hard tool and soft work material, and this causes additional abrasion wear. Tool and work materials contain carbides, oxides, and nitrides with hard microstructures; these cause abrasion wear during machining.

Adhesion Wear

When there is a relative motion between the two bodies that are under the normal load, fragments of softer work material adhere to the harder tool. The adhered material is unstable, and it separates from the cutting tool and tears small fragments of the tool material. The typical example in metal cutting is a built-up edge, which usually occurs at low cutting speeds when part of the chip material welds to the cutting edge (see Fig. 2.30). Depending on the size and stability of the built-up edge, either the forces decrease because the effective rake angle becomes positive or the lumped built-up edge dulls the tool and increases the forces. An unstable, large built-up edge occurs close to the cutting edge at low speeds where the tool–chip interface temperature is low (i.e., less than austenitic temperature). The material is still strong at this point and difficult to move over the rake face. As the chip moves over the rake face, the chip–tool interface temperature increases, leading to a softer chip, which is easier to move. As the cutting speed is increased, the magnitude and length of the built-up edge becomes smaller and localizes close to the cutting edge. Predicting the tool–chip interface temperature is therefore important in identifying cutting speeds where the built-up edge is minimum [83].

Diffusion Wear

When the temperatures of the tool and work materials increase at the contact zones, the atoms in the two materials become restive and migrate to the opposite material where the concentration of the same atom is lower. Typically, in a tool material such as tungsten carbide (WC), where carbide (C) provides the hardness, whereas cobalt (Co) binds the WC grains, carbon diffuses to the moving steel chips, which have a lower concentration of the same atoms. Progressive diffusion of tool materials into the chip gradually leads to a weakened cutting edge and the eventual chipping or breakage of the tool.

Oxidation Wear

The atoms in the cutting tool and/or work material form new molecules at the contact boundary where the area is exposed to the air (i.e., oxygen). Tungsten and cobalt in the cutting tool are oxidized close to the work surface–cutting tool flank, which leads to a notch wear on the cutting tool. Depending on the tool–work materials, tool geometry, and cutting conditions, one wear mechanism may be dominant, but all of them may occur simultaneously but at different rates. The wear of the tool localizes at two regions where the tool is in contact with the work material. The chip moves over the rake face until it leaves the contact area, where crater wear occurs. The freshly cut surface contacts the flank face of the tool, where the flank wear is observed (see Fig. 2.28).

Crater Wear

Crater wear occurs at the tool–chip contact area where the tool is subject to a friction force of the moving chip under heavy loads and high temperatures (Fig. 2.29). At higher speeds (i.e., turning P20 mold steel at $v = 250$ m/min cutting speed), the temperature on the rake face of a carbide tool may reach over 1,000°C. At these high temperatures, the atoms in the tool continuously diffuse to the moving chip. The temperature is greatest near the midpoint of the tool–chip contact length, where the greatest amount of crater wear occurs because of intensive diffusion. As the crater wear approaches the cutting edge, it weakens the wedge and causes chipping of the tool. Crater wear can be minimized by selecting a tool material that has the least affinity to the workpiece material in terms of diffusion. The use of lubricants also reduces the wear. The lubricant penetrates between the chip and the tool, reducing the friction force and thus the temperature (see Eq. 2.27). As a result, the activity of diffusion is reduced leading to less crater wear. The chemical affinity between the workpiece material (mostly iron Fe) and the tool material (mostly tungsten-carbide WC) can be reduced using Al_2O_3, TiN, and TiC coatings on the tools. The thickness of the coating layer is typically 3 to 5 μm. The coating materials have low friction coefficients and have strong chemical stability at high temperatures. They act as a temperature barrier between the tool's core carbide material and the moving chip. However, they can not be easily deposited on the sharp edges of the tool, and they may crack under heavy interrupted cutting conditions. The use of coated tools may significantly increase productivity by reducing tool wear at high speeds. Care must be taken when selecting coating materials that must not have chemical affinity with the contents of the workpiece material at elevated temperatures.

Flank Wear

Flank wear is caused by friction between the flank face (primary clearance face) of the tool and the machined workpiece surface (Fig. 2.30). At the tool flank–workpiece surface contact area, tool particles adhere to the workpiece surface and are periodically sheared off. Adhesion of the tool and workpiece materials increases at higher temperatures. Abrasive wear occurs when hard inclusions of work material or escaped tool particles scratch the flank and workpiece surface as they move across the contact area. Although adhesive and abrasive wear mechanisms are predominant in flank wear, some diffusion wear also exists.

After selecting the lubricant, cutting tool material, and geometry to be used in machining a particular work material, the best engineering approach is to identify a machinability speed that corresponds to the critical temperature limits of both cutting tool and work materials. The temperature can be estimated approximately by using the orthogonal cutting mechanics expressions given in this chapter (see Eq. 2.32). Alternatively, more accurate analytical and experimental techniques can be used to predict the tool–work material interface temperature as presented by Oxley [83] and Trent [115]. The tool–work material interface temperature must be below the diffusion and melting limits of the materials used in the cutting tool; otherwise, binding materials such as cobalt in the WC or TiN and Al_2O_3 in CBN tools may diffuse toward the chip leading to a chain of diffusion, adhesion, and abrasive wear mechanisms. It is also advantageous to have a temperature that corresponds to at least that of the recrystallization phase of the work material where the strain hardening diminishes or even to that where the material becomes almost like a liquid with a significantly reduced yield shear strength. The built-up edge is avoided, and cutting forces and, hence, the stresses in the cutting tool are reduced when the material softens at elevated temperature, provided that the diffusion limits of the material within the cutting tool are not exceeded.

A sample investigation conducted by Ren and Altintas is summarized here as an example. A P20 mold steel with 34 Rc hardness is turned with a medium grain CBN tool without any lubricant. The CBN cutting tool had a 0.1 mm wide 15° negative chamfer angle with binders of TiN and Al_2O_3. P20 compositions are: 0.33%C, 0.3%Si, 1.4%Mn, 1.8%Cr, 0.8%Ni, 0.2%Mo, and 0.008%S. P20 is widely used for injection molding dies. CBN cutting tools are the hardest (87Rc) after diamond, and owing to their high thermal and abrasion resistance and low friction coefficient, they are widely used in high-speed machining of hardened tool and die steels. With the use of the analytical cutting mechanics models, the average temperature at the shear plane and tool–chip interface are estimated for a given speed and feed [90]. It is found that the most optimal cutting speed in turning P20 with the CBN tool used is about 500 m/min, which corresponds to an average temperature of 1,400°C on the tool–chip interface. The diffusion limit for the CBN binding material (Al_2O_3) is about 1,600°C, and the melting temperature of P20 steel is about 1,300°C. For ISO S10 carbide tools, the diffusion limit of cobalt binding is about 1,300°C, which requires a cutting speed of less than 300 m/min without lubricant. A series of orthogonal cutting tests are conducted to verify the predictions, and three sample CBN tool pictures are shown in Figure 2.29. The photographs are taken with a SEM. The CBN tool experienced the least amount of wear at a cutting speed of 240 m/min, where the tool–chip interface temperature is predicted to be about 1,150°C.

The tool still does not exhibit any wear at 400 m/min after cutting a length of 907 m of work material. However, the temperature is very close to the melting limit of P20 steel, and chemical wear such as oxidation of Fe in steel with CBN tool materials may occur. When the speed is increased to 800 m/min, which corresponds to about 1,600°C at the tool–chip interface, severe crater wear can be observed from the third tool after a shorter machining length of 780 m. At this speed, the bindings of the CBN tool diffuse to the chip. The temperature prediction study indicates that the most optimal cutting speed for this particular medium-grained CBN tool is about 500 m/min.

The classical wear mechanisms explained above are applicable to continuous machining operations such as turning. The milling process is a more complex operation because of its intermittent nature, and, thus, additional factors affect tool wear in these operations. In milling, the cutting forces periodically change as the chip thickness varies. The tooth periodically enters and exits from the workpiece; hence, it experiences stress and temperature cycling during cutting. As the tool enters the workpiece, it is heated. The tool starts cooling when it exits from the workpiece to the cutting environment (i.e., air or lubricant shower). The cooling period continues until it reenters the workpiece. This periodic thermal cycle produces alternating compressive (heating cycle) and tensile (cooling) stresses on the tool that may exceed its strength. Even if the thermal stress amplitudes are not large enough to break the tool suddenly, the thermal stress cycling causes gradual fatigue failure and wear of the tool. All of the wear mechanisms are therefore related to the tool–workpiece materials, the cutting forces acting on the tool, and the temperature in the contact regions. The temperature is a direct function of the relative speed and friction force between the materials in contact. Higher speed results in more friction energy, which then increases the temperature at the rake face–chip and flank face–workpiece surface contact zones. Thus, the cutting speed has the strongest influence on the wear mechanism. As flank wear increases, the tool–workpiece contact area becomes larger and, hence, rubbing on the workpiece surface becomes stronger. This results in a poor surface finish as well as high friction forces and temperatures, which may eventually lead to tool breakage. Flank wear occurs at the expense of losing a portion of the sharp cutting edge; therefore, the accuracy of the finished workpiece dimension suffers equally to this amount. In practice, it is more important to control flank wear than crater wear. However, the volume that can be worn away before the total destruction of the tool occurs is much greater for crater wear than for flank wear, if there is not a strong chemical affinity between the tool and workpiece materials.

The flank wear land is measured as the width of wear land (VB) on the primary clearance face (Fig. 2.28). A typical tool life curve is shown in Figure 2.31. The history of flank wear with machining time may be split into three regions. The very sharp edge of the tool is worn soon after the cutting starts. This is followed by a gradual, approximately linear tool wear development with increasing cutting time. After the wear land (VB) reaches a critical limit, the

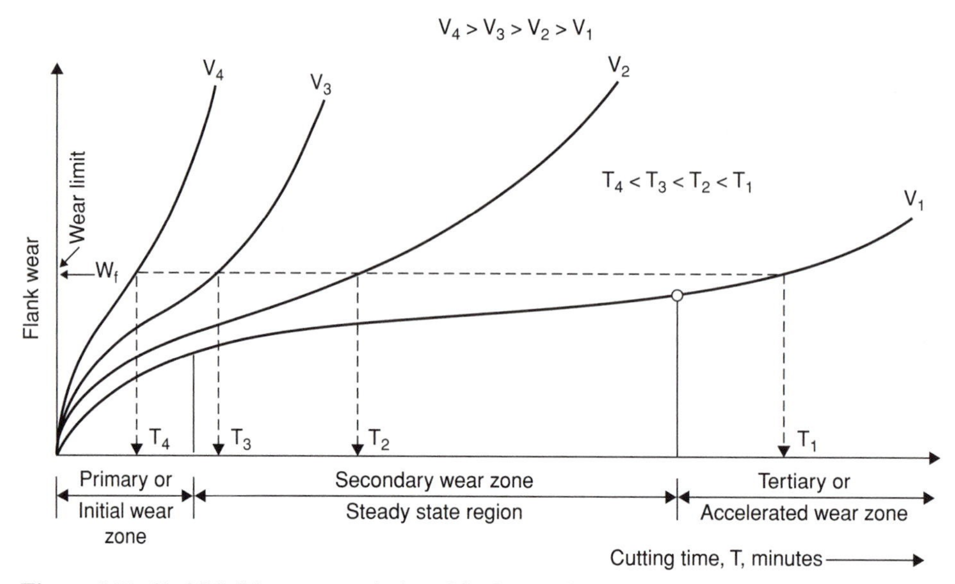

Figure 2.31: Tool life (T) curves: variation of flank wear land (VB) with time at different cutting speeds (V).

flank wear exponentially increases. The tool must be replaced before reaching the critical limit (VB_{lim}) to avoid catastrophic tool failure. The corresponding cutting time to VB_{lim} is called the *tool life* and was expressed first by Taylor [108] as a function of cutting conditions as follows:

$$T_{\text{t}} = C_{\text{t}} V^{-p'} c^{-q'}, \tag{2.123}$$

where T_{t} [min] is the tool life, V [m/min] is the cutting speed, and c [mm/rev] is the feed rate. C_{t}, p', and q' are constants for a given tool–workpiece material pair and are identified from machinability tests.

Example. The following measurements are obtained from the machinability tests, which are conducted by turning normalized AISI-1045 steel using grade K21 carbide inserts.

Test No.	Cutting speed V (m/min)	Feed rate (mm/rev)	Measured tool Life T_t (min)
1	100	0.2	80
2	200	0.2	10
3	200	0.1	40

From tests 1 and 2,

$$p' = \frac{\ln(T_{t1}/T_{t2})}{\ln(V_2/V_1)} = 3.$$

From tests 2 and 3,

$$q' = \frac{\ln(T_{t3}/T_{t2})}{\ln(c_2/c_3)} = 2.$$

Substituting p' and q' into Eq. (2.123) gives the third parameter as

$$C_t = 3.2 \cdot 10^6.$$

The resulting empirical Taylor tool life equation for this particular workpiece material–tool couple becomes

$$3.2 \cdot 10^6 = T_t V^3 c^2.$$

For constant feed rate (c) or constant speed (V), logarithmic tool life charts can be prepared (Eq. 2.123) that can be used for appropriate selection of feeds and speeds for a desired tool life performance. It can be observed from the example that the feed rate has far less influence than the cutting speed (i.e., $V \gg c \rightarrow q' > p'$). The ideal situation is to monitor the progress of tool wear online during machining, so that the tool is replaced only when the wear limit is reached. However, there is no reliable and practical in-process tool wear measurement system available at the present. The topic is the subject of ongoing intensive research efforts. It is a well-known fact that the cutting forces increase, especially in the feed direction, as the flank wear grows. Friction between the flank and finish surface produces additional forces to the shearing. These forces have greater influence in the direction normal to the flank surface (i.e., the feed direction). However, correlating the increase in cutting force with tool wear is still difficult, since the cutting forces may also increase because of changes in the workpiece geometry and material properties of the workpiece.

2.10.2 Tool Breakage

Tool fracture is defined as the loss of a major portion of the tool wedge, which terminates the total cutting ability of the tool. Chipping of the tool (i.e., the loss of small particles from the cutting edge of the tool) is undesirable but does not prevent cutting totally. Chipping does, however, increase the friction on both the rake and flank faces of the chipped tool (Fig. 2.28). If it remains undetected, chipping eventually leads to total breakage of the tool.

Metal cutting tools are made of brittle materials that can fail under excessive chip loading (i.e., large feed rate) and fatigue owing to cyclic mechanical and thermal stresses. Experiments carried out by Zorev [120] showed the profile of tangential and normal load distribution on the tool–chip contact zone (see Fig. 2.2). Zorev found that, when the chip moves away from the cutting edge, it sticks to the rake face first; this is followed by a sliding motion until it separates from the tool. Thus, the tangential load in the sticking zone is equal to the yield shear stress of the material. In the sliding zone, the friction coefficient is constant and equal to the coefficient of friction between the tool and workpiece material. In the analysis of simplified orthogonal cutting mechanics, loading is assumed to be linear using an average friction coefficient (μ_a) between the chip and the tool's rake face. Loladze [69] used a photoelasticity method to measure the stress distribution within the tool wedge during cutting. He showed that after the tool–chip contact ends, the compressive stress zone in the tool is replaced by tensile stresses (see Fig. 2.2). Changes in the cutting conditions, especially in the uncut chip thickness, vary the stress distributions. As the uncut chip thickness is increased, the tensile stress zone expands and its magnitude increases.

Therefore, if the feed rate is increased until the principal tensile stresses reach the fracture limit (i.e., ultimate tensile strength of the brittle cutting tool materials), the tool starts to crack. Both carbide and high-speed steel (HSS) tools may plastically fail, as well, because of the high temperatures at high cutting speeds. The tool loses its strength, deforms plastically, and is sheared off by cutting forces at the elevated temperatures. The plastic failure of the tool is most common in machining low-machinability, heat-resistant metals such as the titanium and nickel alloys used in the aerospace industry. The chips of these alloys do not easily dissipate the heat away from the tool contact zone, thus causing large temperature loading of the tools. The friction energy, or heat created at the tool–chip contact zone, is directly proportional to the cutting speed (see Eq. 2.27). Therefore, whereas a free machining steel can be cut with a carbide tool at a typical cutting speed of 200 m/min, titanium or nickel alloy can be machined only at about 60 m/min. HSS tools can not be used in machining these alloys. Most of the coating materials used on carbide tools use TiC or TiN, which have chemical affinity to titanium-alloyed workpieces. With TiC/Al_2O_3-coated carbide tools in turning nickel-based alloys, the flank wear rate is reduced substantially because (1) TiC acts as a heat barrier between the chip and the substrate carbide tool and (2) Al_2O_3 reduces the friction coefficient. The subjects of the fundamental mechanics of cutting, machinability, and prediction of cutting forces for different operations are only briefly introduced here. Each subject requires a dedicated text for in-depth understanding of the physics as well as the methods used in modeling. Readers are referred to the well-established texts authored by Oxley [83], Armarego and Brown [25], Shaw [96], Trent [115], and Boothroyd [31].

2.11 PROBLEMS

1. A set of orthogonal tests are conducted to identify the shear angle, average friction coefficient, and shear stress of P20 mold steel that has a hardness of 34Rc. The cutting conditions and measured forces and chip thicknesses are given in Table 2.3. The cutting tool was an S10 grade plunge turning tool with a zero rake angle. The width of cut (i.e., the width of disk) was $b = 5$ mm, and the cutting speed was $V = 240$ m/min. The properties of P20 steel are given as: specific coefficient $c_s = 460$ Nm/kg°C, specific density $\rho = 7{,}800$ kg/m^3, thermal conductivity $c_t = 28.74$ [W/mC].

   a. Evaluate the cutting coefficients K_{tc} and K_{fc} [N/mm^2] and edge force constants K_{te} and K_{fe} [N/mm] by a linear regression of the measured forces.
   b. Evaluate the shear angle (ϕ_c), shear stress (τ_s), and average friction coefficient (β_a) for each test, and express them as an empirical function of uncut chip thickness (h) to form an orthogonal cutting database.

TABLE 2.3. Orthogonal Cutting Test Conditions and Measurements in Plunge Turning of P20 Mold Steel

Feed Rate c [mm/rev]	Tangential Force F_t [N]	Feed Force F_r [N]	Measured Chip Thickness h_c [mm]
0.02	350	290	0.050
0.03	480	350	0.058
0.04	590	400	0.074
0.05	690	440	0.083
0.06	790	480	0.102
0.07	890	505	0.116
0.08	980	540	0.131

c. Predict the cutting force coefficients (K_{tc}, K_{fc}) using empirically expressed τ_s, ϕ_c, and β_a, and compare them against the values identified from mechanistic linear regression of the forces.
d. Evaluate the shear strain and strain rate for each test at the primary shear zone.
e. Identify the average temperature at the primary shear zone and tool–chip interface.
f. Using the orthogonal to oblique transformation, express the cutting constants for a helical end mill that has 30° and 5° of helix and normal rake angles, respectively.

2. A P20 mold steel shaft with a diameter of $D = 57$ mm is turned with an S10 grade carbide tool with a side rake angle of $\alpha_f = -5^\circ$, a back rake angle of $\alpha_p = -5^\circ$, and zero approach angle. The nose radius of the insert is $r = 0.8$ mm. The radial depth of cut, the feed rate, and the cutting speed are $a = 1$ mm, $c = 0.06$ mm/rev, and $V = 240$ m/min, respectively. The orthogonal dry cutting parameters of P20 mold steel cut by an S10 carbide tool with a rake angle $\alpha_r = -5^\circ$ are given as follows:
Shear stress $\tau_s = 1400h + 0.327V + 507$ [N/mm^2],
Friction coefficient $\beta_a = 33.69 - 12.16h - 0.0022V$ [degree],
Chip ratio $r_c = 2.71h + 0.00045V + 0.227$,
where the units are in h [mm] and V [m/min].

a. Evaluate the distribution of chip thickness along the curved chip length.
b. Evaluate the cutting coefficients along the curved chip length using the oblique cutting transformation.
c. Evaluate the distribution of tangential, radial, and feed cutting forces (F_t, F_r, F_f) along the curved chip.
d. Find the total cutting forces (F_x, F_y, F_z), torque, and power required to turn the shaft.

3. Consider an end mill with $N = 8$ flutes, rake angle $\alpha_r = 5$ deg, +0 deg helix angle, $D = 20$ mm diameter with $L = 40$ mm overhang from the cantilevered tool holder. The end mill is used in half-immersion down-milling of Al-7050

TABLE 2.4. Measured Average Milling Forces for Al7075

Feed Rate [mm/th]	$\bar{F}_x$ [N]	$\bar{F}_y$ [N]	$\bar{F}_z$ [N]
0.025	−69.4665	64.3759	−10.2896
0.050	−88.1012	99.1383	−20.5814
0.100	−105.6237	155.7100	−33.9852
0.150	−118.1107	210.4644	53.4833
0.200	−130.6807	263.4002	73.1515

alloy with a feed rate of $c = 0.1$ mm/rev/tooth and the axial depth of cut is $a = 30$ mm. The orthogonal cutting parameters of Al7050 are given as follows: Shear stress $\tau_s = 250$ MPa, shear angle $\phi_c = 20 + \alpha_r$, average friction coefficient is $\mu_a = 0.35$ The edge cutting forces are assumed to be negligible. The Young Modules of carbide end mill is $E = 200$ GPa with the mass density of $\rho = 7{,}860$ kg/m^3.

a. Identify the cutting force coefficients of the end mill.
b. Assuming that the total cutting forces act at the free end of the tool ($z = 0$), calculate the maximum deflection mark left on the finish surface of the part at the bottom of the surface ($z = 0$).
c. Consider that the damping ratio is $\zeta = 2\%$, what will be the maximum amplitude of forced vibration in the y direction when the tooth passing frequency matches with the natural frequency of the end mill?

4. An end mill with 100 mm diameter, 4 flutes, 30° helix angle, and 10° rake angles is used in the peripheral milling of aerospace wing support frames. The material is Al-7075. A series of full-immersion (i.e., slot) milling experiments are conducted at constant speed (2,500 rev/min) and axial depth of cut ($a = 1.5$ mm), but at a series of feed rates. The cutting forces are measured and the average forces per tooth period are given in Table 2.4. Calculate the cutting constants (K_{tc}, K_{rc}, K_{ac}) and edge constants (K_{te}, K_{re}, K_{ae}) by assuming the force model given in Eq. (2.80).
5. Write a simulation program for general milling operations. The program must be flexible to test various face and end milling operations. Use the cutting constants identified from the previous question given above. Neglect the influence of helix angle on cutting constants. Your simulation program must be able to cover at least one full revolution of the cutter. Show the graphical results for the milling cases given in Table 2.5.

Feed rate (c) = 0.1 mm/tooth,	Spindle speed (n) = 6,000 rev/min,
Gauge length (l) = 54.5 mm,	Cutter diameter (d) = 19.05 mm,
No. of flutes (N) = 4,	Young's modulus (E) = $2 \cdot 10^5$ MPa.

TABLE 2.5. Simulation Cases

Case No.	*a* (mm)	β (deg)	ϕ_{st} (deg)	ϕ_{ex} (deg)
i	5.00	0	0	120
ii	40.0	30	0	90
iii	40.0	30	0	180

Plot the following variables on the same graph and provide short comments for the simulation results:

a. F_x and F_y cutting forces, all cases.
b. Resultant cutting force and torque, all cases.

6. A twist drill with a diameter $2R = 19.05$ mm, chisel edge angle $\psi_c = 125^\circ$, chisel width $2w = 0.78$ mm, and nominal helix angle of $\beta_0 = 30^\circ$ is used in drilling a titanium alloy Ti_6Al_4V. Predict the thrust, torque, and power of the drilling operation when the drill feed and spindle speed are $c = 0.1$ mm/rev and $n = 200$ rev/min, respectively. Use the orthogonal cutting parameters of Ti_6Al_4V given in Table 2.1.
7. A CBN tool is used in dry turning of P20 mold steel with 34Rc hardness. The cutting edge has a 15° chamfer edge, which acts like a negative rake angle. A number of turning tests are conducted at different feeds and speeds. Each test is continued until a cutting length of l_m is reached, and the average flank wear (VB) is recorded. By the use of a least-squares method, the following relationship was identified from an actual cutting test data:

$$VB = 0.0023568 V^{1.33} c^{0.7423} l_m^{0.796},$$

where the units are flank wear VB [mm], cutting speed V [m/min], feed rate c [mm/rev], and machining length l_m [m]. Predict the tool change time when the cutting conditions are selected as $V = 240$ m/min and $c = 0.06$ mm/rev and the maximum allowable wear is $VB = 0.2$ mm. The diameter of the shaft is 100 mm.

CHAPTER THREE

STRUCTURAL DYNAMICS OF MACHINES

3.1 INTRODUCTION

Machine tools are called *machine making machines*. Various machining and forming operations are executed by a variety of machine tools to produce mechanical parts. To maintain specified tolerances, the machine tools must have greater accuracy than the tolerances of the manufactured parts. The precision of a machine tool is affected by the positioning accuracy of the cutting tool with respect to the workpiece and the relative structural deformations between them. The engineering analysis and modeling of relative static and dynamic deformations between the cutting tool and workpiece are covered in this chapter.

3.2 MACHINE TOOL STRUCTURES

A machine tool system has three main groups of parts: mechanical structures, drives, and controls. The components can be observed from the horizontal computer numerically controlled (CNC) machining center shown in Figure 3.1.

Mechanical Structure

The structure consists of stationary and moving bodies. The stationary bodies include beds, columns, bridges, and gear box housings. They usually carry moving bodies, such as tables, slides, spindles, gears, bearings, and carriages. The structural design of machine tool parts requires high rigidity, thermal stability, and damping. In general, the dimensions of machine tools are overestimated to minimize static and dynamic deformations during machining. The general design of machine tool structures will not be covered in this text. Instead, it is assumed that the relative static and dynamic compliance between the tool and the workpiece is measured experimentally or predicted with analytical methods. The effect of structural compliance on the accuracy of the machined workpiece and machining performance is presented.

Drives

Moving mechanisms are grouped into spindle and feed drives in machine tools. The spindle drive provides sufficient angular speed, torque, and power to a rotating spindle shaft, which is held in the spindle housing with roller or magnetic bearings. Low- to medium-speed spindle shafts are connected to

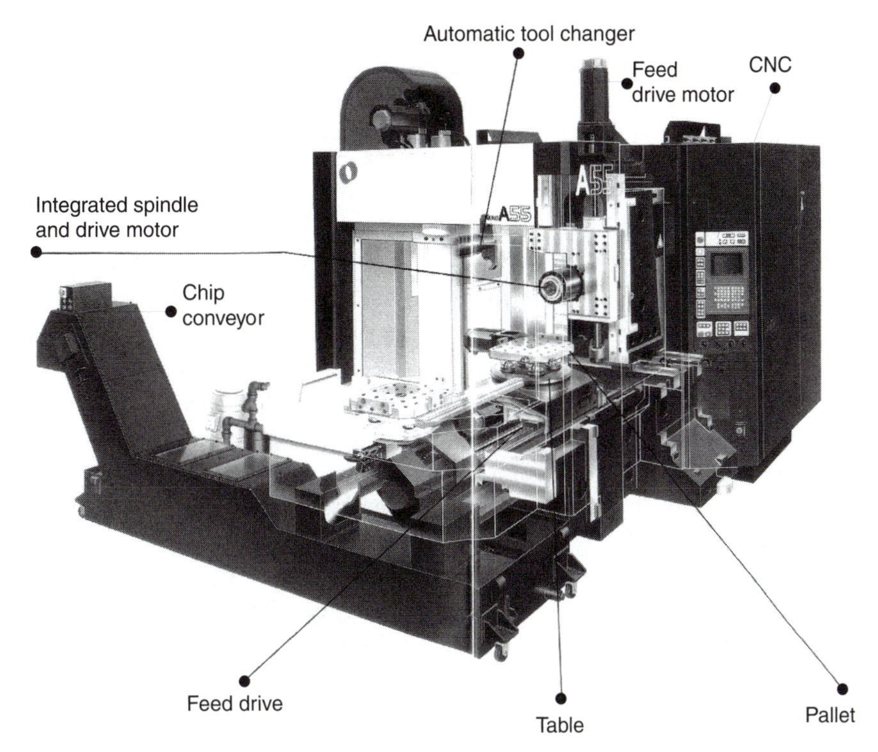

Figure 3.1: A horizontal machining center. Source: Makino Milling Machine Co., Ltd.

the electric motor via V belts. There may be a single-step gear reducer and a clutch between the electric motor and spindle shaft. High-speed spindles ($n > 15{,}000$ rev/min) may have electric motors that are built into the spindle to reduce the inertia and friction produced by the motor–spindle shaft coupling. In a typical and versatile machining center, the high- and low-speed spindles can be swapped in a short time. The feed drives carry the table or the carriage. In general, the table is connected to the nut, and the nut houses a leadscrew. The screw is connected to the drive motor either directly or via a gear system depending on the feed speed, inertia, and torque reduction requirements. Conventional machine tools have numerous gear reduction steps to obtain a desired feed speed. Each feed screw has a dedicated drive motor in CNC machine tools. Very high speed machine tools may use linear direct motors and drives without the feed screw and nut, thus avoiding excessive inertia and friction contact elements.

Controls

The control parts include motors, amplifiers, switches, and computers that are used to energize the electrical parts in a controlled sequence and time. Conventional machine tools mostly contain relays, limit switches, and operator-controlled potentiometers and directional control switches. CNC machine tools have power servoamplifiers, opto-isolated ON/OFF and limit switches, and a computer unit equipped with emergency, control, and operator interface units. The feed velocity and positioning accuracy of feed drives depend on the torque and power delivery of the servomotors and on the feed drive servocontrol algorithms executed in the CNC unit of the machine tool, which is covered in Chapters Five and Six.

3.3 DIMENSIONAL FORM ERRORS IN MACHINING

The source of relative deformations between the cutting tool and the workpiece at the contact point may be due to thermal, weight, and cutting loads. The machine tool has rotating and moving elements, which are heated by friction energy. The temperature increase is never uniform among the machine elements, owing to both their varying thermal coefficients and the location of heat sources. The machine tool table spindle head, on an overhang portal frame or on a vertical column, changes location as the coordinate of the cutting point varies on the workpiece. The moving weights change the stiffness or the relative displacement between the tool and workpiece at the cutting point. Thermal and weight deformations can be measured at various positions of the machine tool table or spindle head and stored in the compensation registers of the CNC unit. When the machine tool travels, the errors are read from the registers and additional motions are ordered by the CNC to compensate for the weight and thermal deformation errors during machining.

The deformation errors caused only by the cutting forces will be presented in this text. As the cutting tool travels along a tool path, there may be variations in the magnitude and the direction of cutting forces, as well as the relative stiffness between the cutting tool and the workpiece [61]. The relative displacements cause deviations from the desired dimensions of the workpiece, causing *dimensional form errors*.

3.3.1 Form Errors in Cylindrical Turning

A typical cylindrical turning operation is shown in Figure 3.2. If the cutting tool has a nonzero approach angle or a significant nose radius, the turning operation has an oblique geometry where the radial force perpendicular to the finished surface is not zero. Any flexibility in the direction of radial force produces a relative displacement between the workpiece and the cutting tool at a point where the finished surface is generated. The structure can be modeled as an elastic circular shaft supported by a pin at the tail stock center and a pin support and resistive bending moment in the chuck at the spindle side. For simplicity, an average diameter (d) of the shaft can be used to account for the radial depth of the cut removed. If the radial cutting force F_{r} is applied l distance away from the chuck (see Fig. 3.2), the radial deflection of the beam at an axial location x is

$$y(x) = \frac{1}{EI}\left[-\frac{R_{\mathrm{c}}}{6}x^3 + \frac{F_{\mathrm{r}}}{6}\langle x-l\rangle^3 + \frac{M_{\mathrm{c}}}{2}x^2\right], \tag{3.1}$$

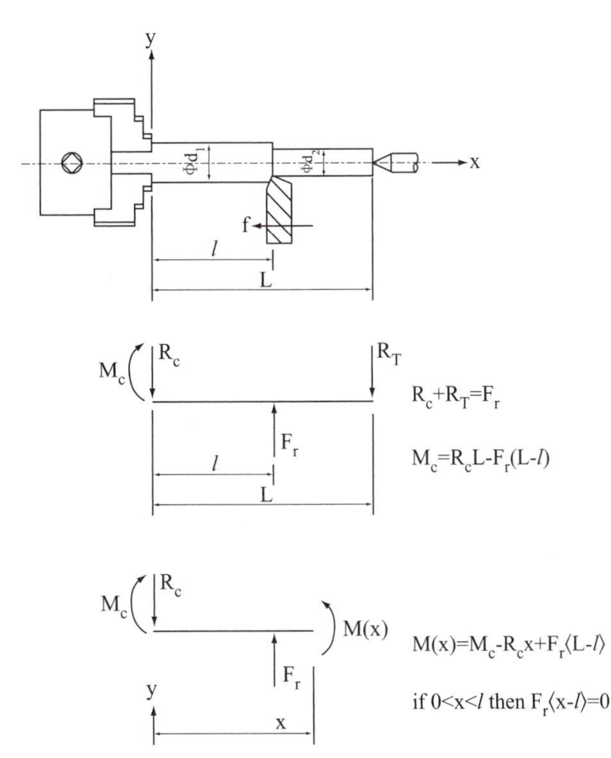

Figure 3.2: Form errors in cylindrical turning of a shaft.

where E is the modulus of elasticity and $I = (\pi d^4)/64$ is the moment of inertia of the workpiece. The support force R_c and bending moment M_c at the chuck are given by

$$R_c = \frac{F_r(L-l)}{2L^3}[3L^2 - (L-l)^2], \qquad M_c = \frac{F_r l(L-l)(2L-l)}{2L^2}.$$

The deflection, which is passed on the turned shaft as a dimensional error, occurs at the point where the tool is in contact with the workpiece (i.e., $x = l$). The radial deflection $y(l)$ is defined from the axis drawn between the center of the tail stock and that of the spindle (i.e., machine tool axis). Because the bar stock is pushed away by amount $y(l)$ in the radial direction from the stationary tool, the actual radial depth of cut will be $a - y(l)$. Hence, the radius of the finished workpiece will have $y(l)$[mm] amount of extra stock or *form error* at location $x = l$. Note that as the tool travels toward the chuck, the stiffness of the beam varies and, hence, the form error left on the finished surface will be different along the workpiece axis. The maximum form error occurs at the middle of the workpiece ($x = L/2$). Note that the finished workpiece will not be cylindrical any more, but will have a barrel shape. When the barrel-type workpiece is turned again in the subsequent pass, the effective axial depth of cut will no longer be uniform but will vary along the bar axis, that is, $a_{\text{effective}} = a + y(l)$. Because the amplitude of the cutting force is proportional to the axial depth of cut, the radial cutting force and the stiffness of the workpiece

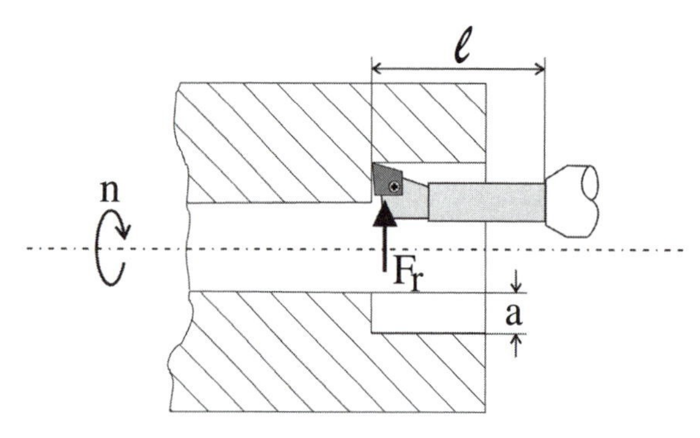

Figure 3.3: Bore form errors produced by a boring bar.

will vary along the bar axis, thus still leaving a form error. It is therefore recommended to have a small depth of cut and feed rate (i.e., small force) in the finishing operations. The process-planning engineer should calculate the feed and axial depth of cut according to the tolerance and stiffness in the center of the workpiece.

3.3.2 Boring Bar

A general diagram of a boring operation is shown in Figure 3.3. The boring bar resembles a cantilevered elastic beam, and it is held in a variety of configurations depending on the machine tool and workpiece setup. The boring bar is held in the stationary tail stock of a lathe, and the workpiece is attached to the rotating spindle. Enlarging the internal diameter of a cylindrical shaft is a typical example for a boring operation on a lathe. The boring bar can be held on the spindle of a horizontal boring or machining center, but it is offset by an amount equal to the radius of the bore to be opened. The offset is adjusted with an *eccentric* attached to the spindle. Large bores of prismatic parts are opened on the boring centers by placing the workpiece on the machine tool table [56].

If the length of the boring bar is l, the deflection at the workpiece–cutting tool contact point will be

$$\delta = \frac{F_r l^3}{3EI} = \frac{F_r}{k_r}, \tag{3.2}$$

where F_r is the radial force and $k_r = (3EI)/l^3$ is the radial stiffness of the boring bar. The radial deflection is passed as a form error to the bore, and its magnitude depends on the amplitude of the cutting force and the stiffness in the radial direction. If a cylindrical boring bar is used, $I = \pi d^4/64$ and $k_r = (3\pi E d^4)/(64 l^3)$. If a boring bar has a rectangular section with a width of b and a height of h, the moment of inertia is $I = (bh^3)/12$. The stiffness of the rectangular boring bar becomes

$$k_r = \frac{Ebh^3}{4l^3}.$$

To increase the radial stiffness, the cutting tool must be mounted on the face of boring bar section that has a width of h. If the ratio is $h/b > 1$, the magnitude of the form error left on the surface is reduced by h^3 times by simply orienting the rectangular bar in the stiffer direction.

3.3.3 Form Errors in End Milling

The walls of parts are produced by the periphery of end mills, and, thus, end milling is also called *peripheral milling*. The finished workpiece surface is perpendicular to the direction of feed in end milling. If the feed and normal directions are aligned with the Cartesian x and y axes, respectively, any deflection in the y axis may produce a static form error. End mills can be considered as an elastic cylindrical beam, cantilevered to the spindle through collet and chuck. They are generally the most flexible part in the machine tool system, because their aspect ratio, the ratio of diameter over the gauge length from the collet, is rather small.

The form errors produced by helical end mills are rather complex [98]. The cutting forces are not constant but vary with the rotation of the end mill. Furthermore, the helix angle of the flutes produces additional variation on the distribution of cutting forces along the cutter z axis [58].

First, consider an end mill with straight flutes (i.e., zero helix angle) for simplicity in explaining the surface generation mechanism. Here, the deflections perpendicular to the finished surface (in the normal direction y) are important. The static deflection caused by a normal force (F_y) at the free end of the end mill is given by [32]

$$\delta_y = \frac{F_y}{k}, \tag{3.3}$$

where $k = (3EI)/l^3$ and $I = (\pi d^4)/64$. The effective diameter of the cutter is d, and the gauge distance is l from the collet. The effective diameter of the cutter can be found by scaling its outer diameter by 0.8–0.85 to take flutes into account. The cutting force is proportional to the chip thickness h, that is,

$$F_y(\phi) = K_{tc} a h(\phi)[\sin(\phi) - K_r \cos(\phi)],$$

where ϕ is the immersion angle measured from the y axis, the chip thickness is $h(\phi) = c \sin\phi$, and c is the feed per tooth. The edge forces are neglected for simplicity and $K_r = K_{rc}/K_{tc}$. The tooth generates the surface when it is in contact with it, or when it is on the normal y axis. When the cutting tool lies on the y axis, the chip thickness is always zero. This situation occurs when the tool is at the entry in up-milling ($\phi = 0$) or at the exit in down-milling ($\phi = \pi$) (see Fig. 3.4). Thus, when only one tooth is present in the cutting zone, regardless of the flexibility of the end mill or magnitude of the cut, the surface form errors will be zero in end milling with straight flutes. This is a common situation in finishing cuts where the radial immersion angle is usually much less than the cutter pitch angle. However, if there are two or more teeth cutting simultaneously, the cutting force will not be zero when one of the teeth is aligned with the y axis, because the other flutes cut a chip within the immersion zone. In this case, the tooth either deflects toward the surface in up-milling, thus causing an *overcut form error*, or deflects away from the surface in down-milling, causing an *undercut form error*.

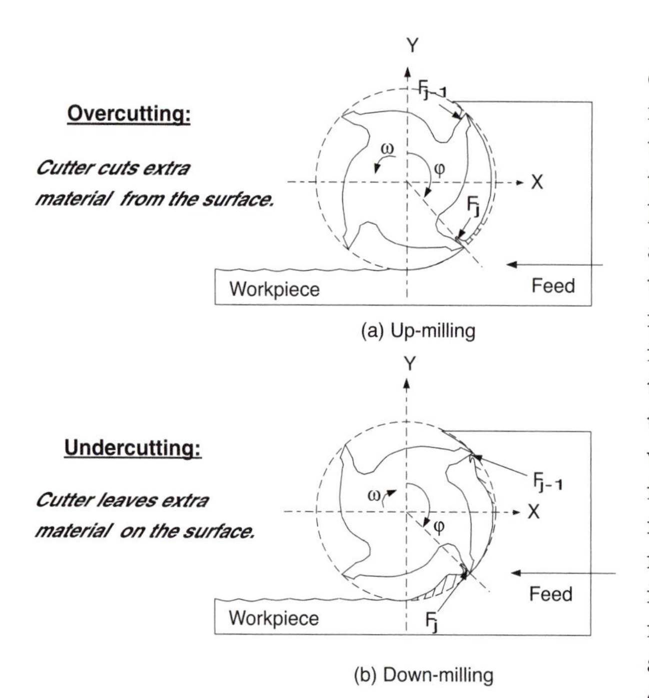

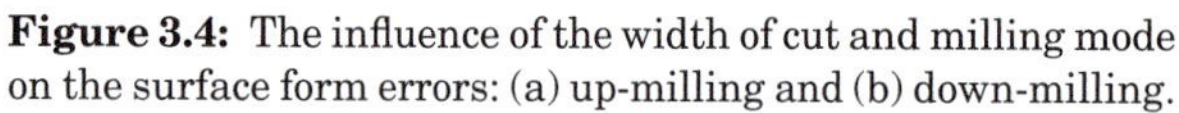

Figure 3.4: The influence of the width of cut and milling mode on the surface form errors: (a) up-milling and (b) down-milling.

Generating the surface becomes complex when the end mill has helical flutes. Even when there is only one flute in cut, there will always be form errors left on the surface [106]. Consider a flute's edge at the bottom of the cutter aligned with the normal y axis, thus having a zero immersion (i.e., $\phi(z=0)=0$). As the end mill rotates, the tip of the flute moves to immersion ϕ, whereas an upper point on the flute, which has a z coordinate measured from the tip, will be right on the y axis, hence, generating the surface. Because the normal cutting force will not be zero at this instant, the elastic end mill displacement will produce a form error on the surface. As the cutter rotates, the cutting edge point that generates the surface will move upward along the flute because of the helix angle. Depending on the number of flutes and width of cut, there may be more than one cutting edge point aligned at the y axis or in contact with the finished surface. The contact points can be calculated by equating the instantaneous immersion angle ($\phi_j(z)=\phi+(j-1)\phi_{\mathrm{p}}-k_\beta z$, with $k_\beta=(2\tan\beta)/d$) to zero in up-milling and to π in down-milling [99]:

$$z=\frac{\phi+(j-1)\phi_{\mathrm{p}}}{k_\beta} \quad \text{(up-milling)},$$

$$z=\frac{-\pi+(\phi+(j-1)\phi_{\mathrm{p}})}{k_\beta} \quad \text{(down-milling)},$$

where β is the helix angle, $j=1,2,\ldots N-1$ is the flute index, and $\phi_{\mathrm{p}}=(2\pi)/N$ is the cutter pitch angle. An algorithm that can predict the surface form errors can be integrated to the cutting force prediction program presented in the previous chapter. The cutter can be divided into M number of small disk elements within the axial depth of cut a (see Fig. 3.5), and it can be rotated at increments $\Delta\phi$ (i.e., $\phi=0,\Delta\phi,2\Delta\phi,\ldots,\phi_{\mathrm{p}}$). Each differential element has an axial depth of cut $\Delta z=a/M$, and the influence of the helix angle may be neglected by selecting small elements. The differential cutting force produced by element m is given by (Fig. 3.5)

$$\Delta F_{y,m}(\phi)=K_{\mathrm{tc}}c\Delta z\sum_{j=0}^{N-1}[\sin\phi_j(z)-K_{\mathrm{r}}\cos\phi_j(z)]\sin\phi_j, \tag{3.4}$$

where K_{t} and K_{r} are cutting constants and c is the feed rate per tooth.

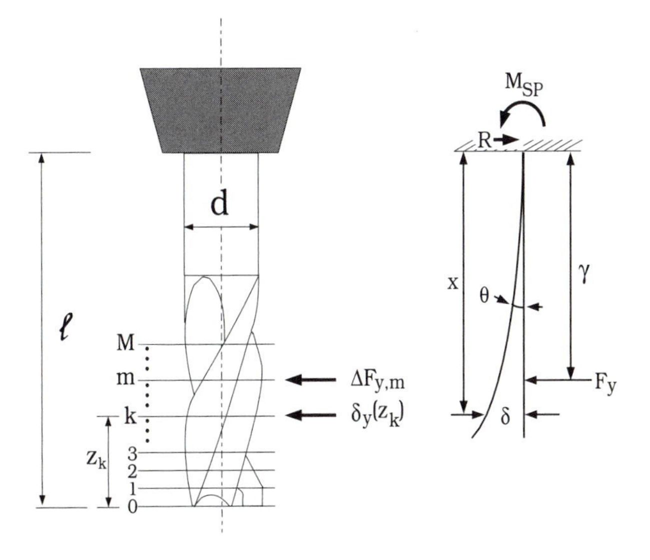

Figure 3.5: Static deformation model of an end mill.

The immersion angle for the element m is $\phi_j(m) = \phi + (j-1)\phi_\mathrm{p} - k_\beta \cdot m\Delta z$. The elemental cutting forces may be lumped at the upper boundaries of disks. The deflection in the y direction at the contact point z_k caused by the force applied at element m is given by the cantilevered beam formulation as [33]

$$\delta_y(z_k, m) = \begin{cases} \frac{\Delta F_{y,m} \nu_k^2}{6EI}(3\nu_m - \nu_k), & 0 < \nu_k < \nu_m, \\ \frac{\Delta F_{y,m} \nu_m^2}{6EI}(3\nu_k - \nu_m), & \nu_m < \nu_k, \end{cases} \tag{3.5}$$

where E is Young's Modulus, I is the area moment of inertia of the tool, and $\nu_k = l - z_k$, with l being the gauge length of the cutter measured from the collet face. The area moment of the tool is calculated by using an equivalent tool radius of $R_\mathrm{e} = 0.80R$, where 0.8 is the approximate scale factor due to flutes. The total static deflection at axial contact point z_k is calculated by the superposition of the deflections produced by all M elemental forces on the end mill as follows:

$$\delta_y(z_k) = \sum_{m=1}^{M} \delta_y(z_k, m). \tag{3.6}$$

At the points where the cutting edge is in contact with the finished surface, the deflection $\delta_y(z_k)$ is imprinted as a dimensional error on the workpiece. By rotating the cutter, the dimensional errors along the finished surface wall can be generated and displayed on the plane of the cutter axis and surface normal (y, z). The three-dimensional topography of the workpiece is generated by simply repositioning the cutter along the feed direction (x). Figure 3.6 shows a predicted surface error map of a workpiece that has a varying width of cut along the feed direction. A detailed analysis of dimensional errors in end milling can be found in Reference [33].

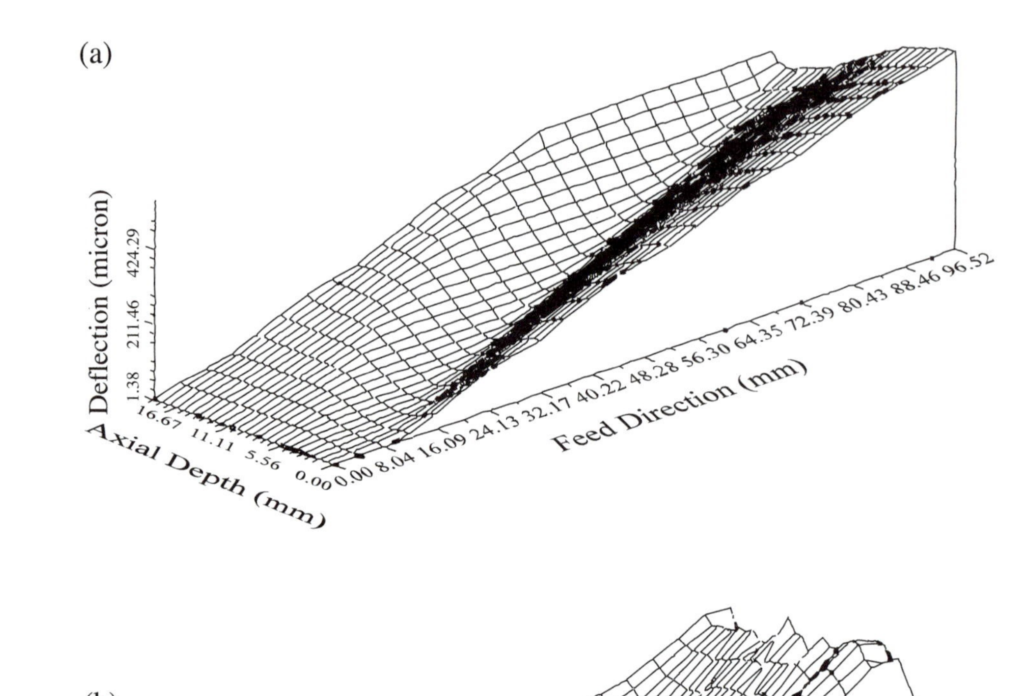

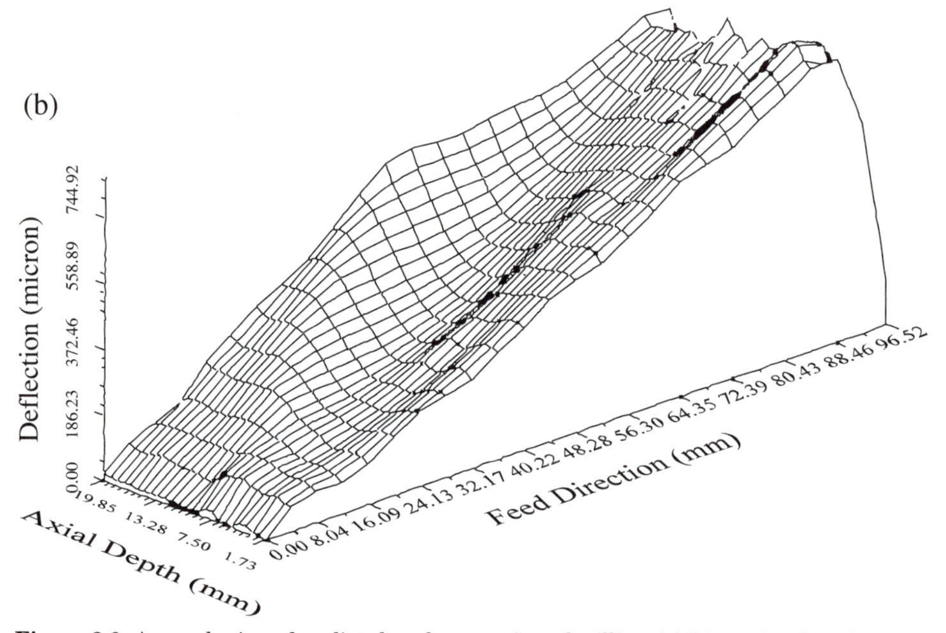

Figure 3.6: A sample view of predicted surface area in end milling. (a) Dimensional surface error simulation results; (b) experimental results.

3.4 STRUCTURAL VIBRATIONS IN MACHINING

Machine tool vibrations play an important role in hindering productivity during machining. Excessive vibrations accelerate tool wear and chipping, cause poor surface finish, and may damage the spindle bearings [57]. A brief review of basic vibration theory is provided first. As experimental modal analysis techniques are most readily used in modern machining facilities, a practical review of modal analysis theory and its practice in machining is then presented. The

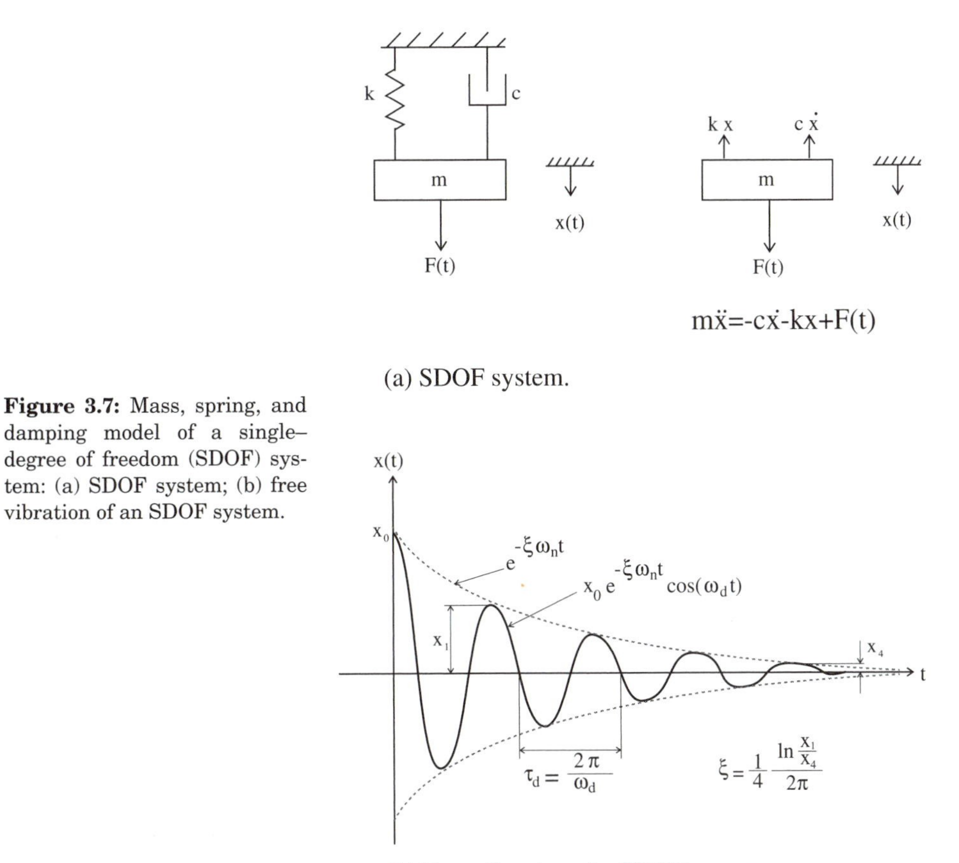

Figure 3.7: Mass, spring, and damping model of a single–degree of freedom (SDOF) system: (a) SDOF system; (b) free vibration of an SDOF system.

basics of vibration engineering should enable the reader to understand machine tool vibrations and their avoidance in practice.

3.4.1 Fundamentals of Free and Forced Vibrations

A simple structure with a single–degree of freedom (SDOF) system can be modeled by a combination of mass (m), spring (k), and damping (c) elements as shown in Figure 3.7. When an external force $F(t)$ is exerted on the structure, its motion is described by the following differential equation:

$$m\ddot{x} + c\dot{x} + kx = F(t) \quad \text{or} \quad \ddot{x} + 2\zeta\omega_n\dot{x} + \omega_n^2 x = \frac{\omega_n^2}{k}F(t). \tag{3.7}$$

If the system receives a hammer blow for a very short duration, or when it is at rest and statically deviates from its equilibrium and lets go, the system experiences free vibrations. The amplitude of vibrations decays with time as a function of the system's damping constant. The frequency of the vibrations is mainly dominated by the stiffness and the mass and is lightly influenced by the viscous damping constant, which is very small in mechanical structures.

When the damping constant is zero ($c = 0$), the system oscillates at its natural frequency as follows:

$$\omega_\mathrm{n} = \sqrt{\frac{k}{m}}.$$

A damping ratio is defined as $\zeta = c/2\sqrt{km}$, which is always less than one in mechanical structures. In most metal structures $\zeta < 0.05$ or even less. The damped natural frequency of the structure is defined by

$$\omega_\mathrm{d} = \omega_\mathrm{n}\sqrt{1-\zeta^2}.$$

Assuming that the mass is *free* from any external force and displaced statically by an amount of x_0 and that the system is released, the free vibration motion can be described by

$$x(t) = x_0 e^{-\zeta\omega_\mathrm{n} t}\cos\ \omega_\mathrm{d} t.$$

The period between each vibration wave is $\tau_\mathrm{d} = 2\pi/\omega_\mathrm{d}$, which is a simple way of estimating the system's damped natural frequency from the *free* or *transient* vibration measurements. The damping constant can be approximated from the ratio of decaying maximum amplitudes of first and nth successive waves with the following equation:

$$\zeta = \frac{1}{n}\left(\ln\frac{x_1}{x_\mathrm{n}}\right)\Big/ 2\pi.$$

When an external force $F(t)$ is present, the system experiences *forced vibrations*. When a constant force $F(t) = F_0$ is applied to the structure, the system experiences a short-lived *free* or *transient* vibration and then stabilizes at a static deflection $x_\mathrm{st} = F_0/k$.

The general response of the structure can be evaluated by solving the differential equation of the motion. The Laplace transform of the equation of motion with initial displacement $x(0)$ and vibration velocity $x'(0)$ under externally applied force $F(t)$ is expressed as

$$\mathcal{L}\left(\ddot{x} + 2\zeta\omega_\mathrm{n}\dot{x} + \omega_\mathrm{n}^2 x\right) = \mathcal{L}\left(\frac{\omega_\mathrm{n}^2}{k}F(t)\right)$$

$$s^2x(s) - sx(0) - x'(0) + 2\zeta\omega_\mathrm{n}sx(s) - 2\zeta\omega_\mathrm{n}x(0) + \omega_\mathrm{n}^2x(s) = \frac{\omega_\mathrm{n}^2}{k}F(s). \tag{3.8}$$

The system's general response, the vibrations of the structure with a SDOF dynamics, can be expressed as the following:

$$x(s) = \frac{\omega_\mathrm{n}^2}{k}\frac{1}{s^2 + 2\zeta\omega_\mathrm{n}s + \omega_\mathrm{n}^2}F(s) + \frac{(s + 2\zeta\omega_\mathrm{n})x(0) - x'(0)}{s^2 + 2\zeta\omega_\mathrm{n}s + \omega_\mathrm{n}^2}. \tag{3.9}$$

The transfer function of the system is represented by neglecting the effect of initial conditions that will eventually disappear as transient vibrations.

$$\Phi(s) = \frac{x(s)}{F(s)} = \frac{\omega_\mathrm{n}^2}{k}\frac{1}{s^2 + 2\zeta\omega_\mathrm{n}s + \omega_\mathrm{n}^2}. \tag{3.10}$$

The characteristic equation of the system has complex conjugate roots (p, p^*) because the mechanical systems are underdamped ($\zeta \ll 1$) and vibrate as follows:

$$s^2 + 2\zeta\omega_{\mathrm{n}}s + \omega_{\mathrm{n}}^2 = (s-p)(s-p^*) = 0$$
$$p = -\zeta\omega_{\mathrm{n}} + j\omega_{\mathrm{d}}, \quad p^* = -\zeta\omega_{\mathrm{n}} - j\omega_{\mathrm{d}}.$$

Example 1. Response of the system to step load $(F(t) = F_0 \to t \geq 0, \mathcal{L}(F_0) = F_0/s)$ with zero initial conditions $(x(0) = x'(0) = 0)$.

$$x(s) = \frac{1}{k}\frac{\omega_{\mathrm{n}}^2}{s^2 + 2\zeta\omega_{\mathrm{n}}s + \omega_{\mathrm{n}}^2}\frac{F_0}{s} = \frac{F_0}{k}\left(\frac{A}{s} + \frac{Bs + C}{s^2 + 2\zeta\omega_{\mathrm{n}}s + \omega_{\mathrm{n}}^2}\right)$$

$$\omega_{\mathrm{n}}^2 = As^2 + 2\zeta\omega_{\mathrm{n}}s + \omega_{\mathrm{n}}^2 + Bs^2 + Cs$$

$$A = \lim_{s\to 0} s\frac{\omega_{\mathrm{n}}^2}{s\left[s^2 + 2\zeta\omega_{\mathrm{n}}s + \omega_{\mathrm{n}}^2\right]} = 1, B = -A = 1, C = -2\zeta\omega_{\mathrm{n}}$$

$$x(s) = \frac{1}{k}\frac{\omega_{\mathrm{n}}^2}{s^2 + 2\zeta\omega_{\mathrm{n}}s + \omega_{\mathrm{n}}^2}\frac{F_0}{s} = \frac{F_0}{k}\left(\frac{1}{s} - \frac{s + 2\zeta\omega_n}{(s + \zeta\omega_{\mathrm{n}})^2 + \omega_{\mathrm{d}}^2}\right)$$

$$x(s) = \frac{F_0}{k}\left(\frac{1}{s} - \frac{\zeta\omega_n}{(s + \zeta\omega_{\mathrm{n}})^2 + \omega_{\mathrm{d}}^2} - \frac{s + \zeta\omega_n}{(s + \zeta\omega_{\mathrm{n}})^2 + \omega_{\mathrm{d}}^2}\right). \tag{3.11}$$

Noting that $\mathcal{L}^{-1}(\frac{s+a}{(s+a)^2+b^2}) = e^{-at}\cos bt$ and $\mathcal{L}^{-1}(\frac{b}{(s+a)^2+b^2}) = e^{-at}\sin bt$,

$$x(t) = \mathcal{L}^{-1}x(s) = \frac{F_0}{k}\left(1 - \frac{\zeta\omega_n}{\omega_{\mathrm{d}}}e^{-\zeta\omega_n t}\sin\omega_{\mathrm{d}}t - e^{-\zeta\omega_n t}\cos\omega_{\mathrm{d}}t\right).$$

Considering that $\sin(a + \phi) = \sin a\cos\phi + \cos a\sin\phi = \sqrt{1-\zeta^2}\cos\omega_{\mathrm{d}}t + \zeta\sin\omega_{\mathrm{d}}t \to \tan\phi = \sqrt{1-\zeta^2}/\zeta$, the response of the system to a step load can be evaluated as follows:

$$x(t) = \frac{F_0}{k}\left[1 - e^{-\zeta\omega_n t}\frac{1}{\sqrt{1-\zeta^2}}\sin(\omega_{\mathrm{d}}t + \phi)\right] \to \phi = \tan^{-1}\frac{\sqrt{1-\zeta^2}}{\zeta}. \tag{3.12}$$

Example 2. Free vibrations with initial displacement $x(0) = x_0$. (External force $F(t) = 0$, initial velocity $x'(0) = 0$).

$$x(s) = \frac{s + 2\zeta\omega_{\mathrm{n}}}{(s + \zeta\omega_{\mathrm{n}})^2 + \omega_{\mathrm{d}}^2}x_0 = \left[\frac{\zeta\omega_n}{(s + \zeta\omega_{\mathrm{n}})^2 + \omega_{\mathrm{d}}^2} + \frac{s + \zeta\omega_n}{(s + \zeta\omega_{\mathrm{n}})^2 + \omega_{\mathrm{d}}^2}\right]x_0$$

$$x(t) = x_0 e^{-\zeta\omega_n t}\frac{1}{\sqrt{1-\zeta^2}}\sin(\omega_{\mathrm{d}}t + \phi) \to \phi = \tan^{-1}\frac{\sqrt{1-\zeta^2}}{\zeta}. \tag{3.13}$$

Example 3. The continuous system is sampled at $T[sec]$ discrete time intervals. By substituting Euler's approximation $(s \approx (1 - z^{-1})/T)$ to the Laplace domain-based transfer function, the discrete transfer function of the system can be evaluated in z domain.

$$\Phi(s=\frac{1-z^{-1}}{T})=\frac{\omega_{\mathrm{n}}^2}{k}\frac{1}{s^2+2\zeta\omega_{\mathrm{n}}s+\omega_{\mathrm{n}}^2}\bigg|_{s=\frac{1-z^{-1}}{T}}$$

$$\Phi(z^{-1})=\frac{\omega_{\mathrm{n}}^2}{k}\frac{1}{\left(\frac{1-z^{-1}}{T}\right)^2+2\zeta\omega_{\mathrm{n}}\left(\frac{1-z^{-1}}{T}\right)+\omega_{\mathrm{n}}^2}$$

$$\Phi(z^{-1})=\frac{b_0}{z^{-2}+a_1z^{-1}+a_0}. \tag{3.14}$$

where $b_0=\dfrac{T^2\omega_{\mathrm{n}}^2}{k}$, $a_1=-2\left(\zeta\omega_{\mathrm{n}}T+1\right)$, and $a_0=1+2\zeta\omega_{\mathrm{n}}T+T^2\omega_{\mathrm{n}}^2$.

The vibration of the machine with a SDOF dynamics can be evaluated at discrete time intervals as $(z^{-1}x(k)=x(k-1))$,

$$\Phi(z^{-1})=\frac{b_0}{z^{-2}+a_1z^{-1}+a_0}=\frac{x(k)}{F(k)}$$

$$\left(z^{-2}+a_1z^{-1}+a_0\right)x(k)=b_0F(k)$$

$$x(k-2)+a_1x(k-1)+a_0x(k)=b_0F(k)$$

$$x(k)=\frac{1}{a_0}\left[-x(k-2)-a_1x(k-1)\right]+\frac{b_0}{a_0}F(k) \tag{3.15}$$

where $x(k)$ is the vibration of the structure when the machine is loaded with force $F(k)$ at time intervals $t=kT\to k=0,1,2,...,k$. The difference equation can be solved recursively at each time interval k. The approximate transformation methods from continuous to discrete time domain are listed below, and the errors associated from each approximation are given in Figure 3.8.

Method	s	z	Integral approximations
Euler (Backward)	$\frac{z-1}{zT}=\frac{1-z^{-1}}{T}$	$\frac{1}{1-Ts}$	$\omega(k)\cdot T$
Forward	$\frac{z-1}{T}=\frac{1-z^{-1}}{z^{-1}T}$	$1+Ts$	$\omega(k-1)\cdot T$
Tustin (Trapezoidal)	$\frac{2(z-1)}{T(z+1)}=\frac{2(1-z^{-1})}{T(1+z^{-1})}$	$\frac{2+sT}{2-sT}$	$\frac{T}{2}\left[\omega(k-1)\cdot T+\omega(k)\right]$

Assuming that the external force is harmonic (i.e., can be represented by sine or cosine functions or their combinations), we can write

$$\ddot{x}+2\zeta\omega_{\mathrm{n}}\dot{x}+\omega_{\mathrm{n}}^2x=\frac{\omega_{\mathrm{n}}^2}{k}F_0\sin(\omega t). \tag{3.16}$$

Thus, the system experiences *forced* vibrations at the same frequency ω of the external force, but with a time or *phase delay*. Let us assume that transient vibrations caused by initial loading have diminished and the system is at *steady-state* operation. Then,

$$x(t) = X\sin(\omega t + \phi),$$

which is called frequency response of the structure. It is mathematically more convenient to use complex harmonic functions in forced vibrations. The harmonic force can be expressed by $F(t) = F_0 e^{j\omega t}$. The corresponding harmonic response is $x(t) = Xe^{j(\omega t + \phi)}$, and when this is substituted into the equation of motion (3.16) we get the frequency response function (FRF) of the system as follows:

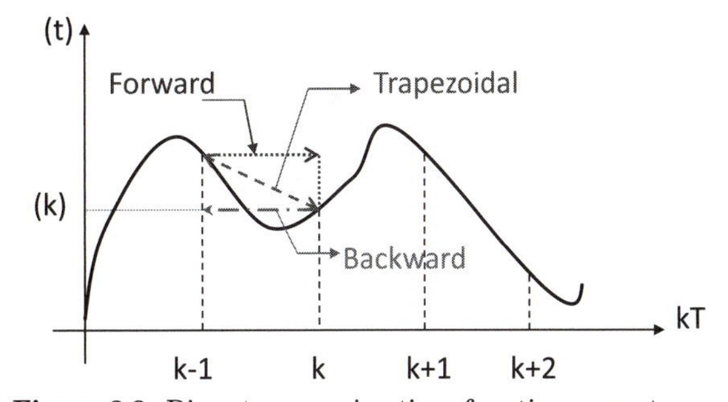

Figure 3.8: Discrete approximation of continuous systems with Euler (backward), forward, and trapezoidal (Tustin) approximations.

$$(\omega_n^2 - \omega^2 + j2\zeta\omega_n)Xe^{j\phi}e^{j\omega t} = \frac{\omega_n^2}{k}F(t) = \frac{\omega_n^2}{k}F_0 e^{j\omega t}$$

$$\Phi(\omega) = \frac{X(\omega)}{F_0(\omega)} = \frac{\omega_n^2}{k}\frac{1}{\omega_n^2 - \omega^2 + j2\zeta\omega_n}. \quad (3.17)$$

Alternatively, the FRF of the system can be obtained by simply replacing the Laplace operator ($s = j\omega$) in the transfer function (Eq. 3.10) which is equivalent to obtaining the steady-state response of the system to harmonic excitation at excitation frequency ω.

The resulting amplitude and phase of the harmonic vibrations are

$$|\Phi(\omega)| = \left|\frac{X}{F_0}\right| = \frac{\omega_n^2}{k}\frac{1}{\sqrt{(\omega_n^2-\omega^2)^2+(2\zeta\omega\omega_n)^2}} = \frac{1}{k}\frac{1}{\sqrt{(1-r^2)^2+(2\zeta r)^2}},$$
$$\phi = \tan^{-1}\frac{-2\zeta\omega\omega_n}{\omega_n^2-\omega^2} = \tan^{-1}\frac{-2\zeta r}{1-r^2}, \quad (3.18)$$

respectively, where the excitation to natural frequency ratio is $r = \omega/\omega_n$. Equation (3.18) is called the *frequency response function* (FRF), or *receptance* of the SDOF structure, and its graphical illustration is shown in Figure 3.9. The FRF ($\Phi(\omega)$) can be separated into real ($G(\omega)$) and imaginary ($H(\omega)$) components of $\frac{X}{F_0}e^{j(\phi-\alpha)}$ as follows:

$$G(\omega) = \frac{1-r^2}{k[(1-r^2)^2+(2\zeta r)^2]},$$
$$H(\omega) = \frac{-2\zeta r}{k[(1-r^2)^2+(2\zeta r)^2]}, \quad (3.19)$$

and

$$\Phi(\omega) = G(\omega) + jH(\omega).$$

Note that at resonance ($\omega = \omega_n$, $r = 1$), $G(\omega_n) = 0$, $H(\omega_n) = -1/(2k\zeta)$.

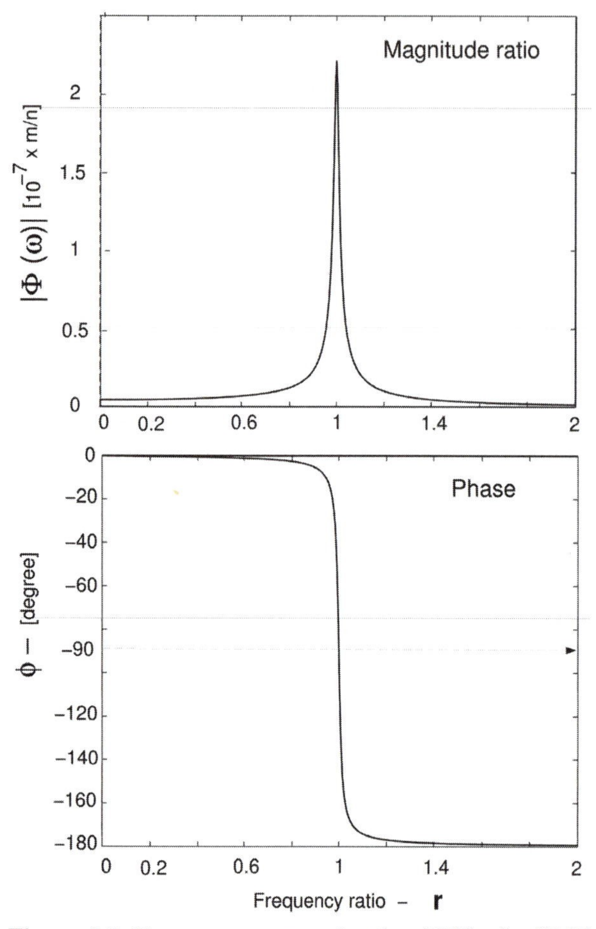

Figure 3.9: Frequency response function (FRF) of a SDOF system.

The real and imaginary parts of the transfer function are illustrated separately in Figure 3.10, and by a polar plot in Figure 3.11. At zero frequency crossing, the real part is equal to the static flexibility ($1/k$). As the excitation frequency approaches the natural frequency (i.e., $r = 1$), the system resonates, the amplitude of the vibrations become maximum, and the phase angle approaches -90 degrees. The time delay between the excitation and response can be evaluated by $t_{\mathrm{d}} = \phi/\omega$ at the harmonic excitation frequency ω. If the excitation frequency continues to increase, the phase angle approaches -180 degrees, or the delay becomes half a period of excitation. The amplitude of the vibrations decreases, because the physical structure can not respond to high-frequency disturbances. The damping ratio, stiffness, and natural frequency can be estimated from the transfer function measured with a Fourier analyzer. At zero excitation frequency ($\omega = 0$), the magnitude of $\Phi(\omega)$, and the real receptance $G(\omega)$ are equal to the static flexibility ($1/k$). Care must be taken in reading this value at low frequency because of the poor sensitivity of measurements taken with velocity and acceleration sensors. An extrapolation of the transfer function from higher frequencies where the resonance levels off can be considered as an alternative method to estimate the stiffness. Displacement sensors provide more accurate measurement of static flexibility. The maximum magnitude of $\Phi(\omega)$ occurs at $\omega = \omega_{\mathrm{n}}\sqrt{1 - 2\zeta^2}$. The real receptance $G(\omega)$ has two extrema at

$$\begin{aligned} \omega_1 &= \omega_{\mathrm{n}}\sqrt{1 - 2\zeta} \rightarrow G_{\max} = \tfrac{1}{4k\zeta(1-\zeta)}, \\ \omega_2 &= \omega_{\mathrm{n}}\sqrt{1 + 2\zeta} \rightarrow G_{\min} = -\tfrac{1}{4k\zeta(1+\zeta)}. \end{aligned} \tag{3.20}$$

In practical machinery, the external excitations are usually periodic but not harmonic. Any periodic force can be represented by its harmonic components. When the external force $F(t)$ (i.e., milling force) is periodic with a

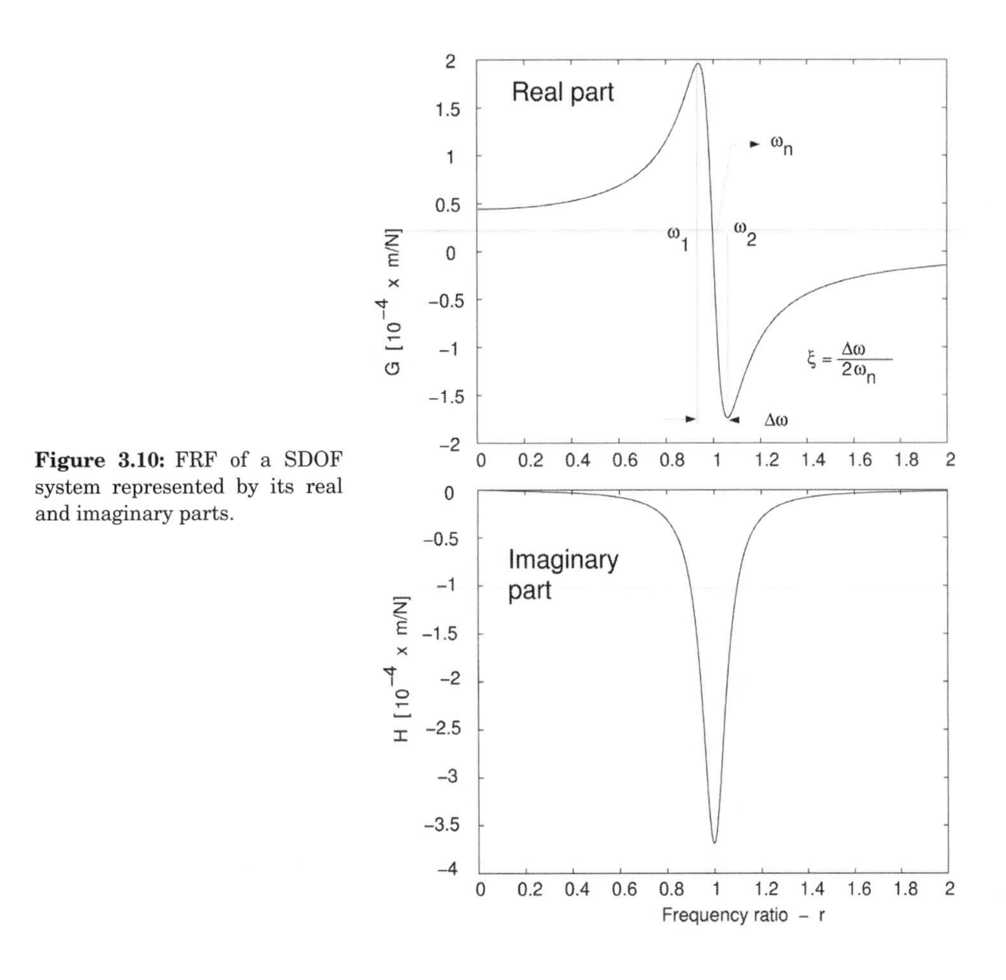

Figure 3.10: FRF of a SDOF system represented by its real and imaginary parts.

period of $\tau = 2\pi/\omega$ (i.e., tooth period), it can be expanded in a Fourier series as follows:

$$F(t) = \frac{a_0}{2} + \sum_{n=1}^{\infty} a_{\rm n} \cos n\omega t + \sum_{n=1}^{\infty} b_{\rm n} \sin n\omega t, \qquad (3.21)$$

where n is the harmonic of the fundamental frequency ω. The exact solution of Fourier coefficients can be found by taking continuous integrals, which require a mathematical representation of the periodic forcing function $F(t)$. Because practical external excitations, such as milling forces, have periodic but irregular wave forms, a discrete numerical technique is used to calculate the Fourier coefficients. Assuming that the periodic excitation is uniformly digitized at T [s] intervals for N times per period τ (i.e., $\tau = NT$) we have

$$a_0 = \tfrac{2}{N}\textstyle\sum_{i=1}^{N} F_i,$$
$$a_{\rm n} = \tfrac{2}{N}\textstyle\sum_{i=1}^{N} F_i \cos \frac{n2\pi t_i}{\tau}, n = 1, 2, 3, \ldots,$$
$$b_{\rm n} = \tfrac{2}{N}\textstyle\sum_{i=1}^{N} F_i \sin \frac{n2\pi t_i}{\tau}, n = 1, 2, 3, \ldots.$$

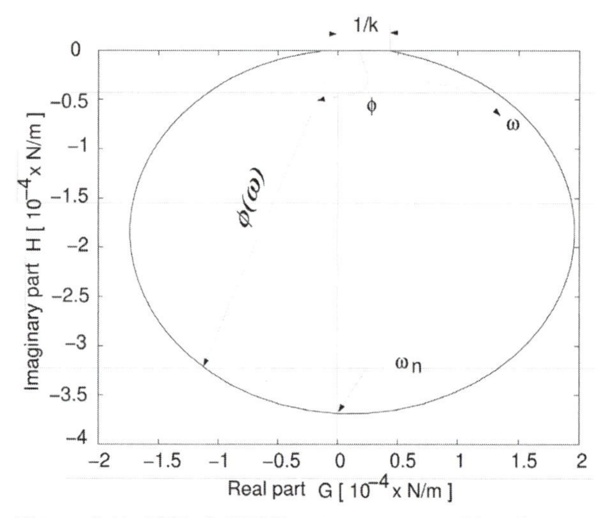

Figure 3.11: FRF of a SDOF system represented in polar coordinates.

In experimental analysis, F_i corresponds to the ith sample of the force by the Fourier analyzer. An alternative discrete Fourier series representation of the periodic function is

$$F(t) = \sum_{n=0}^{N} c_n e^{-j\alpha_n} e^{jn\omega t}, \quad (3.22)$$

where

$$c_n = \sqrt{a_n^2 + b_n^2}, \quad \alpha_n = \tan^{-1}\frac{b_n}{a_n}.$$

The plot of c_n, α_n versus frequency is called a *Fourier spectrum*. For a periodic but nonharmonic excitation, the steady-state response of an SDOF system can be calculated by superposing the vibration produced by each harmonic component of the periodic excitation as follows:

$$x(t) = \sum_{n=0}^{N} \frac{c_n}{k\sqrt{(1-n^2r^2)^2 + (2\zeta nr)^2}} e^{j(n\omega t - \alpha_n - \phi_n)}. \quad (3.23)$$

In most practical cases, the external periodic excitation can be approximated by only its first four to five harmonic (n) components, and the higher harmonics usually may not have sufficient energy to cause considerable influence on the vibrations. Milling forces, which are periodic at tooth passing frequency, can be represented by Fourier series components. If a harmonic of the milling force is close to one of the natural modes of the structure, an alternative spindle speed may be selected to avoid forced vibrations.

3.4.2 Oriented Frequency Response Function

The static and dynamic deformations between the tool and cut surface determine the accuracy and reliability of the manufactured component. A typical machine tool can be described by a series of masses interconnected by springs in different directions [61]. The resultant cutting force is transmitted to the machine via the springs and masses. The superposition of the displacements produced by all springs in the direction perpendicular to the cut surface determine the resulting dimensional accuracy and chip volume removed from the machined part.

Consider a common mass (m) connected to the rigid ground by a series of springs as shown in Figure 3.12. Each spring (i) and the mass define an SDOF system in each independent direction indicated by its angular orientation (θ_i) from the cut surface direction y. The resultant cutting force F has angle β with the cut surface direction y. The displacement produced in direction y by all springs that feel the force F is required.

The force transmitted to spring i is

$$F_i = F\cos(\theta_i - \beta).$$

If the transfer function in each spring direction is $\Phi_i(\omega) = x_i/F_i$, the corresponding displacement of spring i is

$$x_i = F_i\Phi(\omega) = F\cos(\theta_i - \beta)\Phi_i(\omega).$$

The displacement in direction y produced by spring i is

$$\begin{aligned} y_i &= x_i \cos\theta_i \\ &= F\cos\theta_i\cos(\theta_i - \beta)\Phi(\omega). \end{aligned}$$

Superposing the components of vibrations x_i in direction y gives the resultant vibrations. The FRF between the cutting force F and the resultant vibration y, called the *cross* or *oriented* FRF by Koenigsberger and Tlusty [61], is

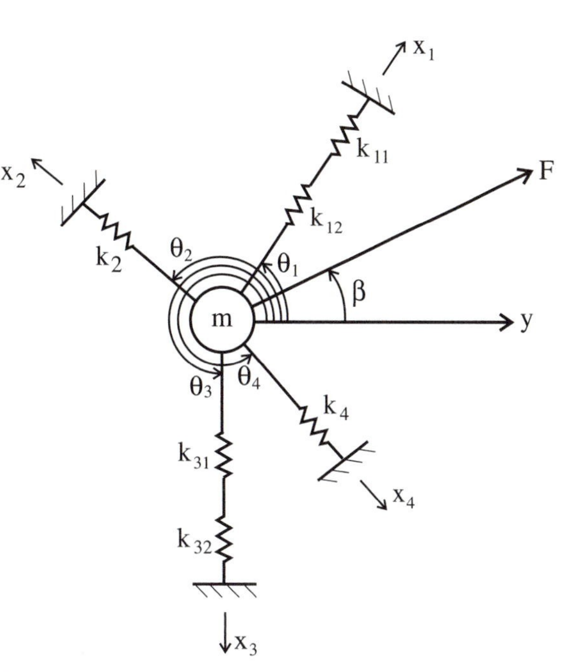

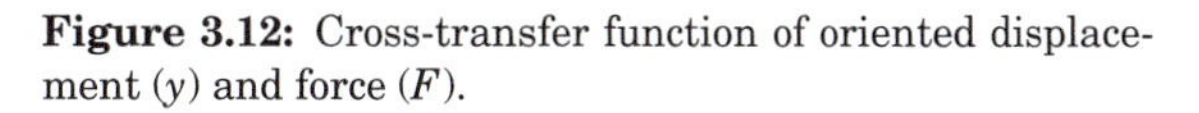

Figure 3.12: Cross-transfer function of oriented displacement (y) and force (F).

$$\Phi_{yF}(\omega) = \frac{y}{F} = \sum_{i=1}^{4} u_{di}\Phi_i(\omega), \tag{3.24}$$

where $u_{di} = \cos\theta_i\cos(\theta_i - \beta)$ is the *directional factor* in direction i. When the mass is neglected, the preceeding formulation can be used to calculate the resultant static deformation in direction y.

3.4.3 Design and Measurement Coordinate Systems

There are three coordinate systems in the analysis of machine tool structures that are modeled by a multiple number of spring, damping, and mass elements. These are *design, local,* and *modal* coordinate systems. The masses, spring, and damping constants may be defined in any of the three coordinate systems, depending on the convenience of either computation or physical interpretation. The modal coordinates do not have any physical meaning. They are used to analyze the strength and behavior of the whole structure at a particular natural frequency, as explained in this section.

The design and local coordinates are explained by the use of a simple structure represented by two springs connected in a series (see Fig. 3.13). Assume that forces F_1 and F_2 are applied on the two connected springs, and the displacements are measured at fixed locations (i.e., at the initial points 1 and 2

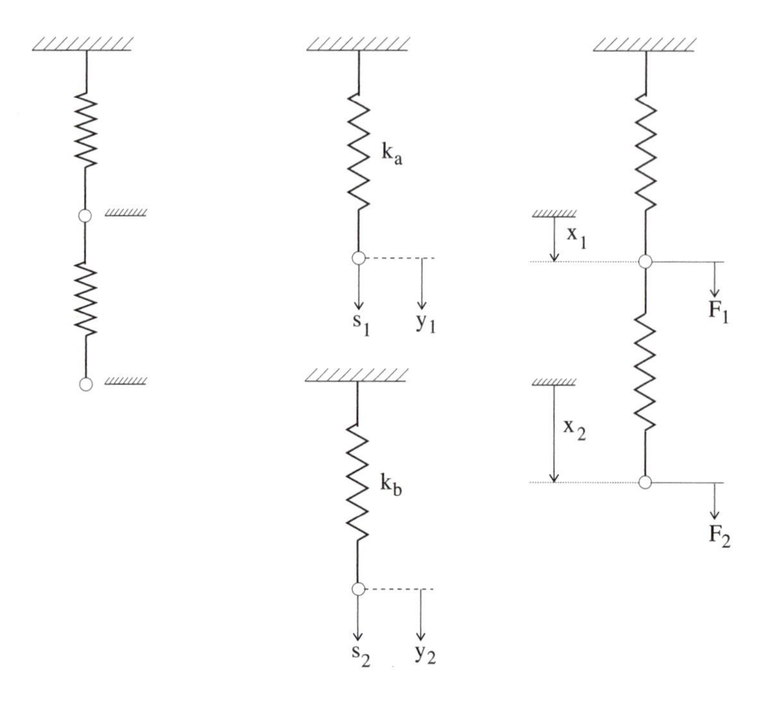

(a) Connected springs (b) Design coordinates (c) Local coordinates

Figure 3.13: Displacements and forces defined in local and design coordinates: (a) connected springs; (b) design coordinates; (c) local coordinates.

where the springs were undeformed). The displacements are x_1 and x_2, and the reference points for the measurements are fixed. The displacements x_1 and x_2 are defined in *measurement* or *local coordinates*. However, when the relative elongation of each spring between its two ends is measured, the corresponding displacement is defined in *design coordinates*. In other words, whereas the local displacement represents the absolute change in the coordinate of a point on the structure, the design displacement represents elongation or compression of an individual spring [111].

Let the forces be S_1 and S_2 and the displacements be y_1 and y_2 in design coordinates. Thus,

$$S_1 = k_a y_1, \qquad S_2 = k_b y_2,$$

or, in general matrix form,

$$\begin{Bmatrix} S_1 \\ S_2 \end{Bmatrix} = \begin{bmatrix} k_a & 0 \\ 0 & k_b \end{bmatrix} \begin{Bmatrix} y_1 \\ y_2 \end{Bmatrix},$$

where k_a and k_b are the design stiffness values for individual springs. For simplicity, we can use the following matrix notation:

$$\{S\} = [K_y]\{y\}, \tag{3.25}$$

where $\{S\}$ and $\{y\}$ are force and displacement vectors and $[K_y]$ is the stiffness matrix in the design coordinates. The relation between the design and local displacements can be expressed as

$$x_1 = y_1 \rightarrow y_1 = x_1,$$

$$x_2 = y_1 + y_2 \rightarrow y_2 = -x_1 + x_2,$$

or

$$\begin{Bmatrix} y_1 \\ y_2 \end{Bmatrix} = \begin{bmatrix} 1 & 0 \\ -1 & 1 \end{bmatrix} \begin{Bmatrix} x_1 \\ x_2 \end{Bmatrix} \rightarrow \{y\} = [T]\{x\}, \tag{3.26}$$

where matrix $[T]$ is the transformation matrix between the *local* and *design* displacements. A similar expression can be written for the forces, that is,

$$S_2 = F_2 \rightarrow F_2 = S_2,$$

$$S_1 = F_1 + F_2 \rightarrow F_1 = S_1 - S_2,$$

or

$$\{F\} = [T]^T\{S\}. \tag{3.27}$$

The stiffness, mass, and damping elements can be transformed from one coordinate system to another by using the transformation matrix $[T]$. Substituting Eqs. (3.25) and (3.26) into (3.27) yields

$$\{F\} = [T]^T\{S\} = [T]^T[K_y]\{y\} = [T]^T[K_y][T]\{x\}.$$

Noting that $\{F\} = [K_x]\{x\}$, we can obtain the local stiffness matrix from the design coordinates as

$$[K_x] = [T]^T[K_y][T]. \tag{3.28}$$

Similar transformations can be applied to the damping and mass matrices, that is, $[C_x] = [T]^T[C_y][T]$, $[M_x] = [T]^T[M_y][T]$.

Note that, depending on the displacements and forces, the units may be different in the design and local coordinates [111]. For example, the displacements and forces may be defined in [mm] and [N] in local coordinates at specific locations of a structure, whereas the displacements may be angular if produced by torsional loads and springs in the design coordinates. Although all the experimental measurements are done in local coordinates, the design engineer is interested in determining individual weak elements defined in the design coordinates.

3.4.4 Analytical Modal Analysis for Multi–Degree-of-Freedom Systems

Machine tools have multiple degrees of freedom (DOF) in various directions. The vibrations between the cutting tool and generated workpiece surface are the main interest, because they affect the accuracy of the surface finish, the chip thickness removal, and cutting forces exciting the machine tool. In the

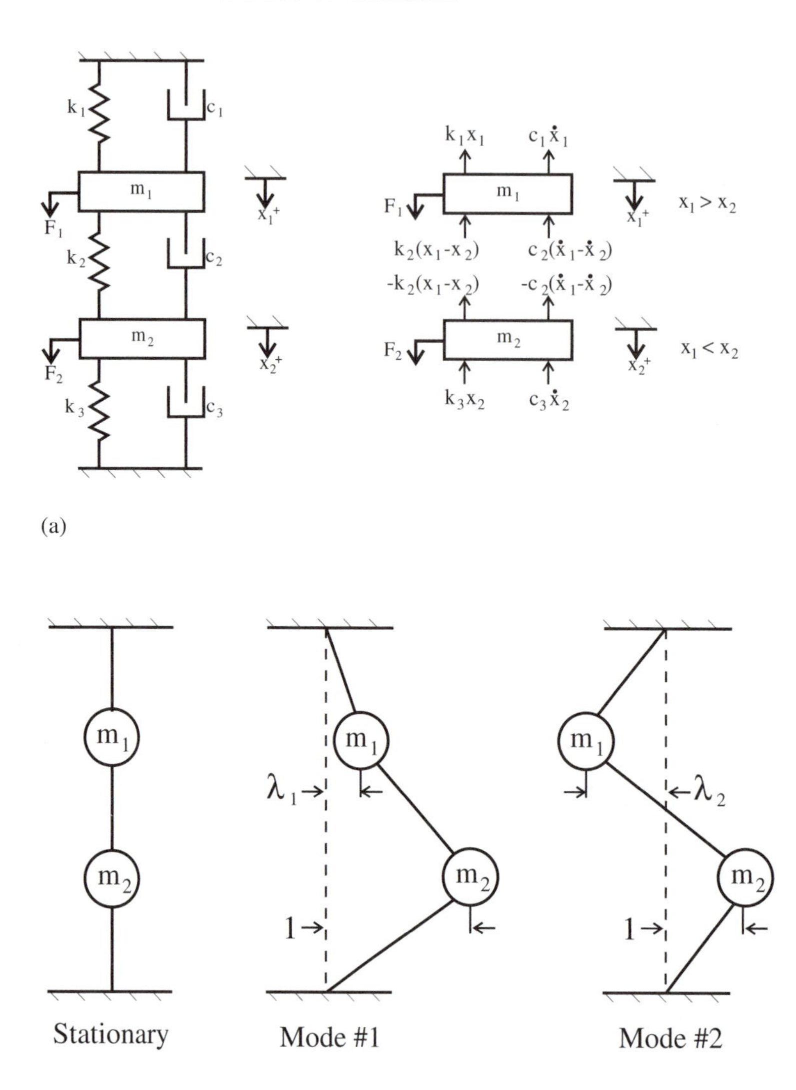

Figure 3.14: (a) Mathematical model of a sample 2-DOF system; (b) mode shapes of a 2-DOF system.

following, the basic principles of modal analysis are reviewed by using the 2-DOF system shown in Figure 3.14 as an example [89].

From Newton's second law, the equations of motion for masses m_1 and m_2 in local or measurement coordinates (x_1, x_2) can be written as

$$m_1\ddot{x}_1 = F_1 - c_1\dot{x}_1 - c_2(\dot{x}_1 - \dot{x}_2) - k_1x_1 - k_2(x_1 - x_2),$$

$$m_2\ddot{x}_2 = F_2 - c_2(\dot{x}_2 - \dot{x}_1) - k_2(x_2 - x_1) - k_3x_2 - c_3\dot{x}_2.$$

Rearranging the equations in matrix form yields

$$[M_x]\{\ddot{x}\} + [C_x]\{\dot{x}\} + [K_x]\{x\} = \{F\}, \tag{3.29}$$

where the mass, stiffness, and damping matrices are

$$[M_x] = \begin{bmatrix} m_1 & 0 \\ 0 & m_2 \end{bmatrix}, \quad [K_x] = \begin{bmatrix} k_1 + k_2 & -k_2 \\ -k_2 & k_2 + k_3 \end{bmatrix}, \quad [C_x] = \begin{bmatrix} c_1 + c_2 & -c_2 \\ -c_2 & c_2 + c_3 \end{bmatrix},$$

respectively. The displacement vector $\{x\}$ and force vector $\{F\}$ are defined as

$$\{x\} = \begin{Bmatrix} x_1(t) \\ x_2(t) \end{Bmatrix}, \qquad \{F\} = \begin{Bmatrix} F_1(t) \\ F_2(t) \end{Bmatrix}.$$

The solution to the preceeding set of differential equations is first obtained for an undamped free vibration case (i.e., $c_1 = c_2 = c_3 = 0$ and $\{F\} = \{0\}$) as follows:

$$[M_x]\{\ddot{x}\} + [K_x]\{x\} = \{0\}. \tag{3.30}$$

The undamped system has a general solution of

$$\{x(t)\} = \{X\} \sin(\omega t + \psi),$$

where $\{X\}$ and ψ are constants and ω is the natural frequency of the system. Substituting the displacement vector $(x\{t\})$ and its second derivative as an acceleration vector $\{\ddot{x}\} = -\omega^2\{X\} \sin(\omega t + \psi)$, Eq. (3.30) becomes

$$([K_x] - \omega^2[M_x])\{X\} = \{0\}, \tag{3.31}$$

or

$$\begin{bmatrix} k_1 + k_2 - \omega^2 m_1 & -k_2 \\ -k_2 & k_2 + k_3 - \omega^2 m_2 \end{bmatrix} \begin{Bmatrix} X_1 \\ X_2 \end{Bmatrix} = \begin{Bmatrix} 0 \\ 0 \end{Bmatrix}.$$

The determinant of these simultaneous algebraic equations must be zero for a nontrivial solution. Letting $s = \omega^2$, we have

$$\begin{vmatrix} k_1 + k_2 - sm_1 & -k_2 \\ -k_2 & k_2 + k_3 - sm_2 \end{vmatrix} = 0,$$

or

$$s^n + a_1 s^{n-1} + \cdots + a_n = 0, \tag{3.32}$$

where n is the system's number of degrees of freedom. For the 2-DOF system example,

$$s^2 - \left(\frac{k_1 + k_2}{m_1} + \frac{k_2 + k_3}{m_2} \right) s + \frac{k_1 k_2 + k_2 k_3 + k_1 k_3}{m_1 m_2} = 0.$$

This polynomial has two real values of s: $s_1 = \omega_{\mathrm{n}1}^2$ and $s_2 = \omega_{\mathrm{n}2}^2$, where $\omega_{\mathrm{n}1}$ and $\omega_{\mathrm{n}2}$ are the natural frequencies of the system. Superposing the contribution of each solution or mode, we get

$$\begin{Bmatrix} x_1(t) \\ x_2(t) \end{Bmatrix}_1 = \begin{Bmatrix} X_1 \\ X_2 \end{Bmatrix}_1 \sin(\omega_{\mathrm{n}1} t + \Psi_1) + \begin{Bmatrix} X_1 \\ X_2 \end{Bmatrix}_2 \sin(\omega_{\mathrm{n}2} t + \Psi_2), \tag{3.33}$$

where $\{P\}_{1,2} = \{X_1\ X_2\}^T_{1,2}$ are the *eigenvectors* or *mode shapes* associated with the fundamental (ω_{n1}) and second natural frequencies (ω_{n2}), respectively. X_{ik} is the displacement of node i, and Ψ_k is the phase contributed by the natural mode k. The solution of Eq. (3.31) gives only the ratio of amplitudes at each node. It is customary to normalize them with respect to a reference coordinate (i.e., X_2). Substituting ω_{n1} and ω_{n2} in Eq. (3.31) and rearranging yields

$$\left(\frac{X_1}{X_2}\right)_1 = \lambda_1 = \frac{k_2}{k_1+k_2-\omega_{n1}^2 m_1} = \frac{k_2+k_3-\omega_{n1}^2 m_2}{k_2},$$

$$\left(\frac{X_1}{X_2}\right)_2 = \lambda_2 = \frac{k_2}{k_1+k_2-\omega_{n2}^2 m_1} = \frac{k_2+k_3-\omega_{n2}^2 m_2}{k_2}.$$

Substituting $\lambda_{1,2} = (X_1/X_2)_{1,2}$ and setting $X_{21} = Q_1$ and $X_{22} = Q_2$ into the transient vibrations (Eq. 3.33) gives

$$\begin{Bmatrix} x_1(t) \\ x_2(t) \end{Bmatrix} = \begin{bmatrix} \lambda_1 & \lambda_2 \\ 1 & 1 \end{bmatrix} \begin{Bmatrix} Q_1 \sin(\omega_{n1}t + \Psi_1) \\ Q_2 \sin(\omega_{n2}t + \Psi_2) \end{Bmatrix},$$

or, in vector form,

$$\{x(t)\} = [\{P\}_1\ \{P\}_2] \begin{Bmatrix} q_1(t) \\ q_2(t) \end{Bmatrix} = [P]\{q(t)\}, \tag{3.34}$$

where $\{P\}_1 = \{\lambda_1\ 1\}^T$ and q_1 are the first mode shape and modal displacement contributed by the first mode. The physical interpretation of a mode shape can be explained from Figure 3.14b. The natural mode number causes one unit of displacement at mass m_2 and λ_1 units at mass m_1. $[P]$ is the complete modal matrix, which has dimension $[n \times n]$ for an n degrees of freedom system. However, the modal matrix does not have to be square. The number of rows is equal to the number of coordinate points on the machine, and each column represents a mode.

Because the modes are orthogonal to each other, they have the following properties:

$$\{P\}_1^T [M_x]\{P\}_2 = 0,$$

whereas

$$\{P\}_1^T [M_x]\{P\}_1 = m_{q1},$$

where m_{q1} is the modal mass associated with the first mode. When the orthogonality principle is applied similarly to the remaining mode shapes, the local mass and stiffness matrices are transformed into modal coordinates as follows:

$$\left.\begin{aligned} [M_q] &= [P]^T[M_x][P], \\ [K_q] &= [P]^T[K_x][P]. \end{aligned}\right\} \tag{3.35}$$

The resulting modal mass ($[M_q]$) and modal stiffness ($[K_q]$) matrices are diagonal, and each diagonal element represents the modal mass or modal stiffness associated with a mode. Note that when the system has a proportional damping (i.e., $[C_x] = \alpha_1[M_x] + \alpha_2[K_x]$, where α_1 and α_2 are empirical constants), the

determined [111]. Consider the general machine tool structure shown in Figure 3.15. The relative displacement between the tool and workpiece is $(x_t - x_w)$, where x_t and x_w are displacements of the tool and workpiece, respectively. The cutting forces acting on the tool (F_t) and the workpiece (F_w) have the same magnitude of F_0, but they have opposing directions (i.e., $F_t = -F_w$). The displacement and force vectors for an n-DOF system can be expressed in local coordinates [111] as follows:

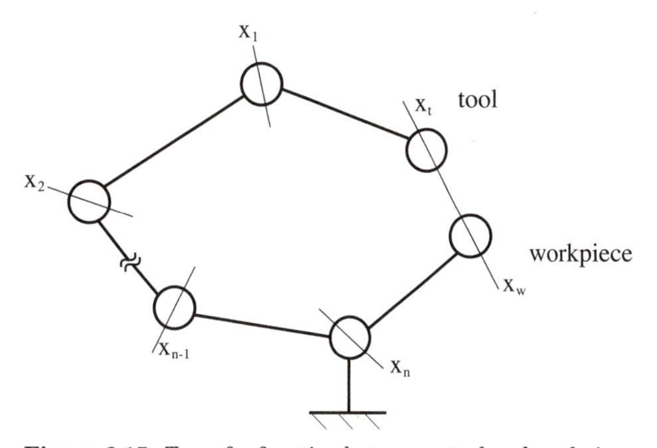

Figure 3.15: Transfer function between a tool and workpiece on the machine tools.

$$
\begin{aligned}
\{x\} &= \{x_1, x_2, \ldots, x_t; \quad x_w, \ldots, x_n\}, \\
\{F\} &= \{0, 0, \ldots, 1; \quad -1, \ldots, 0\}F_0.
\end{aligned}
\tag{3.44}
$$

The equation of motion for the system is

$$
[M_x]\{\ddot{x}\} + [C_x]\{\dot{x}\} + [K_x]\{x\} = \{F\},
$$

where the local mass $[M_x]$, damping $[C_x]$, and stiffness $[K_x]$ matrices are square with dimensions $[n \times n]$, and the force vector $\{F\}$ has a dimension of $[n \times 1]$. The solution of the eigenvalue problem leads to a modal matrix $[P]$ with a dimension of $[n \times n]$ as follows:

$$
[P] = \begin{bmatrix}
P_{11} & P_{12} \ldots P_{1t} & P_{1w} \ldots P_{1n} \\
P_{21} & P_{22} \ldots P_{2t} & P_{2w} \ldots P_{2n} \\
\vdots & \vdots\vdots & \vdots\vdots \\
P_{t1} & P_{t2} \ldots P_{tt} & P_{tw} \ldots P_{tn} \\
P_{w1} & P_{w2} \ldots P_{wt} & P_{ww} \ldots P_{wn} \\
\vdots & \vdots\vdots & \vdots\vdots \\
P_{n1} & P_{n2} \ldots P_{nt} & P_{nw} \ldots P_{nn}
\end{bmatrix},
$$

or

$$
[P] = [\{P\}_1, \{P\}_2, \ldots, \{P\}_t, \{P\}_w, \ldots \{P\}_n],
$$

where each column represents a mode shape $\{P\}$ of the n-DOF system structure. From the modal coordinate transformation equation $(x = [P]\{q\})$, the tool and workpiece displacements can be calculated by using the associated rows t and w of the above modal matrix:

$$
\begin{aligned}
\{x_t\} &= \{P_{t1}\, P_{t2} \ldots P_{tt}\, P_{tw} \ldots P_{tn}\}\{q_1\, q_2 \ldots q_t\, q_w \ldots q_n\}^T, \\
\{x_w\} &= \{P_{w1}\, P_{w2} \ldots P_{wt}\, P_{ww} \ldots P_{wn}\}\{q_1\, q_2 \ldots q_t\, q_w \ldots q_n\}^T.
\end{aligned}
\tag{3.45}
$$

Substituting the modal force ($\{R\} = [P]^T\{F\}$) into the modal displacement vector ($\{q\} = [\Phi_q]\{R\}$), and remembering that all elements of the force vector $\{F\}$ are zero, with the exception of the two corresponding to coordinates $x_{\rm t}$ and $x_{\rm w}$, yields

$$\{q\} = [\Phi_q]\begin{Bmatrix} P_{\rm t1} - P_{\rm w1} \\ P_{\rm t2} - P_{\rm w2} \\ \vdots \\ P_{\rm tn} - P_{\rm wn} \end{Bmatrix} F_0 = \begin{Bmatrix} \Phi_{q1}(P_{\rm t1} - P_{\rm w1}) \\ \Phi_{q2}(P_{\rm t2} - P_{\rm w2}) \\ \vdots \\ \Phi_{qn}(P_{\rm tn} - P_{\rm wn}) \end{Bmatrix} F_0.$$

Substitution of these modal displacements into Eq. (3.45) gives the local displacements of the tool and workpiece, respectively:

$$\begin{aligned} x_{\rm t} &= F_0 \textstyle\sum_{i=1}^{n} \Phi_{qi} P_{{\rm t}i}(P_{{\rm t}i} - P_{{\rm w}i}), \\ x_{\rm w} &= F_0 \textstyle\sum_{i=1}^{n} \Phi_{qi} P_{{\rm w}i}(P_{{\rm t}i} - P_{{\rm w}i}). \end{aligned} \tag{3.46}$$

When the machine tool structure is excited by a harmonically varying force at the cutting point, the relative FRF between the tool and workpiece becomes

$$\frac{x_{\rm t}(\omega) - x_{\rm w}(\omega)}{F_0(\omega)} = \sum_{i=1}^{n} \Phi_{qi}(P_{{\rm t}i} - P_{{\rm w}i})^2. \tag{3.47}$$

If the 2-DOF system is assumed to represent the tool and workpiece structures (i.e., $x_1 \equiv x_{\rm t}$ and $x_2 \equiv x_{\rm w}$), the force vector becomes $\{F\} = \{1 \; -1\}F_0$. The relative FRF between the tool and the workpiece in this particular example is found to be

$$\frac{x_1 - x_2}{F_0} = \Phi_{q1}(\lambda_1 - 1)^2 + \Phi_{q2}(\lambda_2 - 1)^2,$$

according to the outlined procedure.

3.5 MODAL TESTING OF MACHINE STRUCTURES

The FRF of mechanical structures can be measured by using modal testing techniques. The structure is excited by exerting force, and the response is measured with vibration sensors as shown in Figure 3.16. Sample impact test measurements and a flow chart of the signal-processing algorithm used in evaluating the FRF of structures are shown in the Figure 3.17. The theory of signal processing, excitation techniques and instruments, and practical difficulties in measuring the FRF of mechanical structures are explained in the following sections.

3.5.1 Theory of Frequency Response Testing

Consider that the excitation force applied to the structure is $F(t)$, and the resulting vibration is $x(t)$. The Fourier transforms of both force and vibration

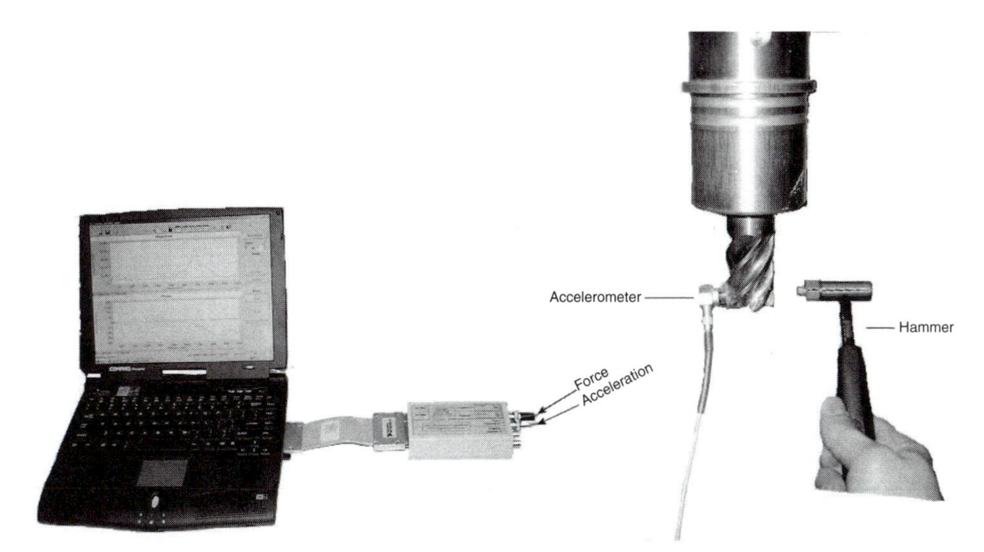

Figure 3.16: Measurement of FRFs with the use of a hammer instrumented with a force sensor and an accelorometer attached to an end mill.

can be evaluated in a continuous time domain as follows:

$$\mathcal{F}\{x(t)\} = \int_{-\infty}^{+\infty} x(t)e^{-j\omega t}dt, \ \mathcal{F}\{F(t)\} = \int_{-\infty}^{+\infty} F(t)e^{-j\omega t}dt. \quad (3.48)$$

However, applied force and the resulting vibrations are measured during the limited measurement time t_l,

$$X(j\omega) = \frac{1}{t_l}\int_{0}^{t_l} x(t)e^{-j\omega t}dt, \ F(j\omega) = \frac{1}{t_l}\int_{0}^{t_l} F(t)e^{-j\omega t}dt, \quad (3.49)$$

where $X(j\omega)$ and $F(j\omega)$ are the power spectra of vibration and force, respectively. The vibration and force are measured at discrete time intervals (T_s) by using data acquisition boards attached to a computer or a dedicated Fourier analyzer. If we wish to have a frequency resolution of ω_r in the measurement, it is necessary to collect measurements for a total time of $t_l = 2\pi/\omega_r$ with N number of data samples,

$$N = \frac{t_l}{T_s}, \longrightarrow t_l = \frac{2\pi}{\omega_r} = NT_s. \quad (3.50)$$

The highest-frequency content of the measured signal is equal to Nyquist frequency as follows:

$$\omega_m = \frac{2\pi/T_s}{2} = \frac{1}{2}\frac{2\pi N}{t_l} = \frac{N}{2}\omega_r. \quad (3.51)$$

The continuous integrals given in Eq. (3.49) must be replaced by their discrete time equivalents, as follows:

$$t \to nT_s, \;\; \omega \to k\omega_r, \;\; dt \to T_s$$

$$X(k\omega_r) = \frac{1}{t_l}\sum_{n=0}^{N-1} x(nT_s)e^{-jk\omega_r t}T_s = \frac{T_s}{NT_s}\sum_{n=0}^{N-1} x(nT_s)e^{-jk\frac{2\pi}{NT_s}nT_s}$$
$$= \frac{1}{N}\sum_{n=0}^{N-1} x(nT_s)e^{-jk\frac{2\pi}{N}n}, \quad k = 0, 1, \ldots, \frac{N}{2}$$

Alternatively, the vibration and force can be decomposed into their real and imaginary parts as follows:

$$\left.\begin{aligned} X(k\omega_r) &= \tfrac{1}{N}\sum_{n=0}^{N-1} x(nT_s)\left[\cos\tfrac{2\pi k}{N}n - j\sin\tfrac{2\pi k}{N}n\right] \\ F(k\omega_r) &= \tfrac{1}{N}\sum_{n=0}^{N-1} F(nT_s)\left[\cos\tfrac{2\pi k}{N}n - j\sin\tfrac{2\pi k}{N}n\right] \end{aligned}\right\}, \quad k = 0, 1, \ldots, \frac{N}{2}, \tag{3.52}$$

where $X(k\omega)$ and $F(k\omega)$ are the discrete Fourier transforms or *spectra* of measured vibration ($x(nT_s)$) and force ($F(nT_s)$), respectively. Here, the fundamental frequency is $\omega_r = 2\pi f_r = 2\pi/t_l$, n corresponds to sampling counter, and k is the frequency counter. $X(k=0)$ and $F(k=0)$ correspond to the average values of vibration and force measurements, respectively. The vibration and force measurements can be mathematically reconstructed by substituting spectra into Fourier series representation of the signals.

$$x(t) = \sum_{k=0}^{N/2} X(k\omega_r)e^{-jk\omega_r t}, \;\; F(t) = \sum_{k=0}^{N/2} X(k\omega_r)e^{-jk\omega_r t}, \;\; t = (0, 1, .., N)T_s. \tag{3.53}$$

As can be seen from Eq. (3.52), spectra $X(k\omega)$ and $F(k\omega)$ have real and imaginary parts, and their complex conjugates are $X^*(k\omega)$ and $F^*(k\omega)$, respectively.

The FRF of the machine is $\Phi(j\omega)$, and can be expressed by dividing the Fourier transform of vibration $x(t)$ by the Fourier transform of force $F(t)$ as follows:

$$\Phi(j\omega) = \frac{X(jk\omega_r)}{F(jk\omega_r)} = \frac{X(j\omega)}{F(j\omega)}, \tag{3.54}$$

where $(j\omega)$ indicates that the spectra are complex numbers and $\omega = k\omega_r$. Eq. (3.54) is not used in evaluating the FRF from measurements because of the presence of noise in the sensor signals. Assuming that the spectrum of the vibration measurement noise is $N(j\omega)$ and the force measurement noise is $M(j\omega)$, the measured FRF ($\Phi_m(j\omega)$) will be

$$\Phi_m(j\omega) = \frac{X(j\omega) + N(j\omega)}{F(j\omega) + M(j\omega)}, \tag{3.55}$$

which is not as accurate as the ideal, noise-free system given in Eq. (3.54). The influence of the noise can be attenuated by introducing the *cross-power spectrum* [50]. The cross-power spectrum ($S_{xF}(j\omega)$) of vibration and force is obtained by taking Fourier transforms of the two measurements separately and multiplying them as follows:

$$S_{xF}(j\omega) = X(j\omega) \cdot F^*(j\omega), \tag{3.56}$$

where $F^*(j\omega)$ is the complex conjugate of force spectrum $F(j\omega)$. Multiplying both sides of Eq. (3.55) by the complex conjugate of the force measurement spectrums shown in the denominator gives

$$\begin{aligned} \Phi_m(j\omega) &= \frac{X(j\omega) + N(j\omega)}{F(j\omega) + M(j\omega)} \times \frac{F^*(j\omega) + M^*(j\omega)}{F^*(j\omega) + M^*(j\omega)} \\ &= \frac{X(j\omega)F^*(j\omega) + X(j\omega)M^*(j\omega) + N(j\omega)F^*(j\omega) + N(j\omega)M^*(j\omega)}{F(j\omega)F^*(j\omega) + F(j\omega)M^*(j\omega) + M(j\omega)F^*(j\omega) + M(j\omega)M^*(j\omega)}. \end{aligned} \tag{3.57}$$

The actual force exerted on the structure and the vibration are not correlated to noise terms in the measurements; hence, the cross-power spectrum terms containing noise must be zero.

$$X(j\omega)M^*(j\omega) = N(j\omega)F^*(j\omega) = N(j\omega)M^*(j\omega) \simeq 0.$$

$$F(j\omega)M^*(j\omega) = M(j\omega)F^*(j\omega) \simeq 0$$

The measured FRF is now reduced to

$$\begin{aligned} \Phi_m(j\omega) &= \frac{X(j\omega)F^*(j\omega)}{F(j\omega)F^*(j\omega) + M(j\omega)M^*(j\omega)} \\ &= \frac{S_{xF}(j\omega)}{S_{FF}(j\omega) + S_{mm}(j\omega)}, \end{aligned} \tag{3.58}$$

where the cross-power spectrum of vibration and force is $S_{xF}(j\omega) = X(j\omega)F^*(j\omega)$, the autospectrum of force is $S_{FF}(j\omega) = F(j\omega)F^*(j\omega)$, and the autospectrum of force sensor signal noise is $S_{mm}(j\omega) = M(j\omega)M^*(j\omega)$. By dividing both sides of Eq. (3.58) by the autospectrum of force, ($S_{FF}(j\omega)$) yields

$$\Phi_m(j\omega) = \frac{\Phi(j\omega)}{1 + S_{mm}(j\omega)/S_{FF}(j\omega)}, \tag{3.59}$$

where $\Phi(j\omega)$ is the desired FRF of the structure. Eq. (3.59) indicates that if the noise-to-force signal ratio is much smaller than one ($\frac{S_{mm}(j\omega)}{S_{FF}(j\omega)} \ll 1$), the measured FRF will be close to the actual FRF, that is, $\Phi_m(j\omega) \simeq \Phi(j\omega)$. Thus, instead of simply dividing the Fourier spectra of vibration by force, as in Eq. (3.54), the cross-power spectrum of the measurements is divided by the autospectrum of force in measuring FRFs in practice.

$$\Phi(j\omega) \simeq \frac{S_{xF}(j\omega)}{S_{FF}(j\omega)}. \tag{3.60}$$

To further attenuate the noise and smooth out the measurements, averages of several power spectra are used in practice. Typically, an average of ten measurements may be sufficient to obtain a reliable FRF of a system.

$$\Phi_1(j\omega) = \frac{\frac{1}{N_m}\sum_{n=1}^{N_m}\left[X(j\omega)\cdot F^*(j\omega)\right]}{\frac{1}{N_m}\sum_{n=1}^{N_m}\left[F(j\omega)\cdot F^*(j\omega)\right] + \frac{1}{N_m}\sum_{n=1}^{N_m}\left[M(j\omega)\cdot M^*(j\omega)\right]}$$
$$= \frac{\overline{S_{xF}(j\omega)}}{\overline{S_{FF}(j\omega)} + \overline{S_{mm}(j\omega)}}, \quad (3.61)$$

where $\overline{S_{xF}(j\omega)}$ is the average cross-power spectrum of displacement and force, and $\overline{S_{FF}(j\omega)}$ is the average power spectrum of force, respectively.

Alternatively, we could have used the power spectrum of the vibration by multiplying Eq. (3.55) by $X^*(j\omega)$ and obtain FRF as follows:

$$\Phi_2(j\omega) = \frac{\overline{S_{xx}(j\omega)} + \overline{S_{nn}(j\omega)}}{\overline{S_{xF}(j\omega)}}. \quad (3.62)$$

If the measurements are made under ideal conditions and the structure is linear, both Eqs. (3.61) and (3.62) should give the same FRF value, $\Phi(j\omega) = \Phi_1(j\omega) = \Phi_2(j\omega)$. The accuracy of the measurement carried out can be checked by observing *coherence function* (γ_{xF}^2), which is defined by the ratio of the two FRF estimates as follows:

$$\gamma_{xF}^2 = \frac{\Phi_1(j\omega)}{\Phi_2(j\omega)} = \frac{\overline{S_{xF}(j\omega)S_{xF}(j\omega)}}{[\overline{S_{FF}(j\omega)} + \overline{S_{mm}(j\omega)}][\overline{S_{xx}(j\omega)} + \overline{S_{nn}(j\omega)}]}.$$

If the products of noise and signal terms are zero, the coherence function must be unity, and $\Phi_1(j\omega) = \Phi_2(j\omega)$. If the coherence function is unity at frequency ω, then the vibrations recorded at this frequency are due to applied input force. If the coherence is zero, then the vibrations recorded are not due to the applied force but to other sources; hence, the measurement is not acceptable. The coherence function of the measurement is evaluated by the following:

$$\gamma_{xF}^2 = \frac{\left|\overline{S_{xF}(j\omega)}\right|^2}{\overline{S_{FF}(j\omega)S_{xx}(j\omega)}}, \quad (3.63)$$

where all spectra contain the measured signals that have noise buried in them. The coherence function (Eq.3.63) is evaluated in modal tests, and the quality of the measurements is assessed by observing the deviation of coherence function from the unity. Alternatively, it is important to compare the power spectra of the applied force and resulting vibrations as well. If the power spectrum of the applied force is close to being zero where the vibration spectra show large peaks due to the presence of natural frequency, again the measurement is not acceptable because their ratio (i.e., FRF) will have infinite flexibility at this point, which is not possible in reality.

3.5.2 Experimental Procedures in Modal Testing

The structure is excited either by using shakers or impact hammers. Shakers can deliver force with a controlled amplitude and frequency by using feedback-controlled amplifiers. The controller can deliver sinusoidal or random forces through its reciprocating shaft attached to the structure via a flexible bar. A force sensor is inserted between the reciprocating shaft of the shaker and flexible bar. Typically, piano wire is used as a flexible bar, so that the flexibility of the connection system does not influence the structural parameters of the measured machine. If the machine has a play, such as in the spindle or feed drives, the structure is preloaded so that it remains in a linear elastic zone during measurements. Otherwise, the structure does not move when the force is applied at certain frequencies. The excitation force can have sinusoidal or random waveform. If sinusoidal force is used, the system must be excited at one frequency, and the ratio of displacement and force magnitudes, and the phase of the vibration relative to the force must be measured at each frequency. The procedure must continue until the frequency range of interest, where the natural modes lie, is covered. Some shakers come with a controller that can sweep all the desired frequencies.

Alternatively, an impact hammer instrumented with a force transducer can be used in exciting the machine (see Fig. 3.16). Although electromagnetic or electrodynamic shakers can deliver force at the desired frequency and amplitude, it is more time consuming to set them up on the machine. Impact hammers are much easier and quicker to set up, but at the expense of less accurate excitation of the machine structure at the desired frequency. Whereas machine builders can use shakers and impact hammers together for a comprehensive analysis of the machines, shop engineers usually have to use an impact hammer for quick identification of machine tool dynamics, so that chatter vibrations can be avoided by selecting proper speeds. The vibration of the machine can be measured by using an accelerometer or a noncontact displacement sensor. An accelerometer can be attached to the machine by using wax, glue, magnet, or screw. The lighter the structure is, the smaller the mass of the accelerometer should be to avoid adding extra mass to the system. When the accelerometer is used, the measured FRF will be $\Phi_a(j\omega) = \ddot{x}/F$. To convert the FRF from acceleration to displacement, the measurement made with the accelerometer must be divided by $(j\omega)^2$, that is, $\Phi(j\omega) = [\ddot{x}/(j\omega)^2]/F$. Note that the acceleration and displacement has the following relationship if the excitation is harmonic: $x(t) = Xe^{j\omega t}$, $\frac{d^2}{dt^2}x(t) = \ddot{x}(t) = (j\omega)^2 Xe^{j\omega t}$. Noncontact displacement sensors can be capacitance, inductive, or laser type. Although laser displacement sensors can be quite accurate with a linear response over a wide frequency range within one-centimeter distance, they are costly and time consuming to set up on the machine. Capacitive or inductive sensors have less frequency bandwidth and a linear response only within a millimeter displacement range. However, vibration amplitudes are typically less than a millimeter; hence, capacitive or inductive displacement sensors are widely used in practice, especially at the low-frequency modes where the accelerometer is not effective because of the scaling of the measurements by $(j\omega)^2$.

If an impact hammer is used, a force waveform similar to a half sine wave is exerted on the structure. The magnitude and duration of the force waveform depends on the mass and tip of the hammer. Larger hammers provide force with a long time duration, which is useful to excite large structures with low natural frequencies. Small hammers provide a more narrow force waveform with less energy; hence, they are suitable to excite lighter structures having higher natural frequencies. Tips can be screwed to the force sensor attached to the impact hammer. Tip materials can range from hard steel, aluminum, bronze, polyvinyl chloride, and rubber. The harder the tip is, the narrower the force waveform will be; hence, providing a larger excitation frequency range. Softer tips provide longer contact with the structure and provide a smaller frequency excitation range.

The impact-testing technique requires preprocessing of the hammer force and vibration signal measurements before their spectra are evaluated. The contact time between the hammer and the structure is typically very short, and significantly shorter than the duration of machine vibrations picked up by the vibration sensors. The force signal can be box windowed by setting the box amplitude to unity during the duration of the contact force, and zero at the remaining time. The box window ensures the elimination of noise after the contact force is zero. However, this technique creates distortion of the Fourier transform; hence, an alternative method that has smoother decay to zero is preferred. The force signal is multiplied by unity during its duration and tapers off to zero with an exponential function that has a duration of one-sixteenth of the total sample time. The vibration measurement must also be multiplied by an exponentially decaying window so that the vibrations diminish within the measured time. In addition, if the vibrations die out long before the measurement ends, the remaining noise may have a poor effect on the signal processing. The exponential window typically decays from unity to 0.05 in the total sample time. Typical force and vibration measurements, along with the signal-processing flow diagram, are given in Figure 3.17.

3.6 EXPERIMENTAL MODAL ANALYSIS FOR MULTI–DEGREE-OF-FREEDOM SYSTEMS

Transfer functions of existing multi–degree-of-freedom (MDOF) systems are identified by structural dynamic tests. The physical machine is modeled by discrete lumped masses connected with linear and/or torsional springs. Good engineering judgment and experience must be applied to the observation of the physical structure and the evaluation of its technical drawing. Further reading and measurement experience are necessary to acquire advanced knowledge in the area. The reference books from Ewins [44] and training courses provided by the manufacturers of Fourier analyzers and commercial modal analysis software systems are useful.

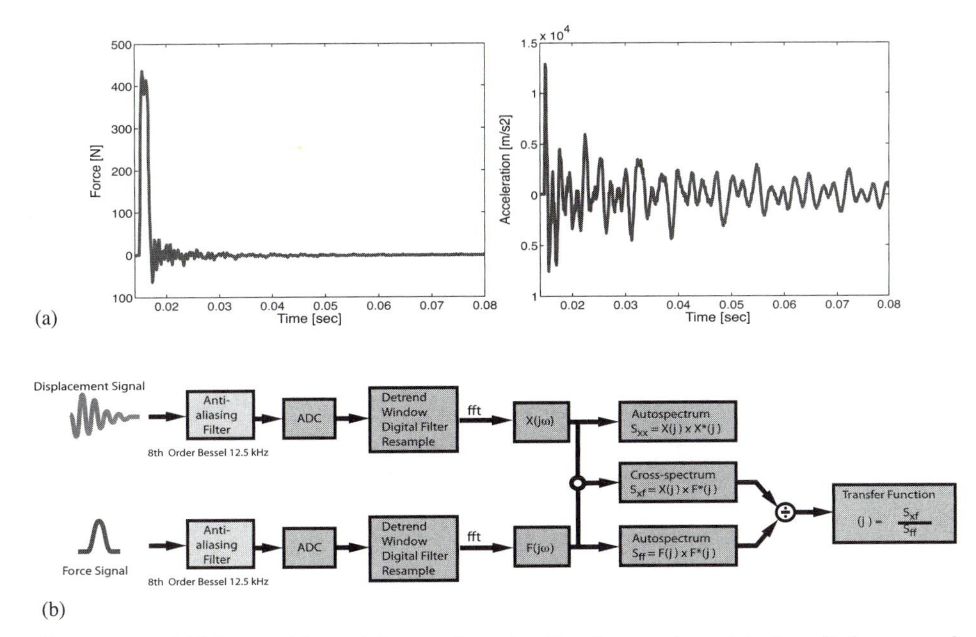

Figure 3.17: (a) Measured impact force and acceleration from an impact test applied on an end mill attached to a milling spindle. (b) The flow chart of measurement and signal processing for FRF measurement with impact testing (courtesy of MAL Inc. CUTPRO System).

The fundamentals of experimental modal analysis are briefly introduced here.

Concept of Residues

The transfer function of a SDOF system Eq. (3.7) can be expressed in the Laplace domain as follows:

$$h(s) = \frac{X(s)}{F(s)} = \frac{1/m}{s^2 + 2\zeta\omega_n s + \omega_n^2}, \tag{3.64}$$

where $s^2 + 2\zeta\omega_n s + \omega_n^2$ is the characteristic equation of the system that has two complex conjugate roots:

$$s_1 = -\zeta\omega_n + j\omega_d, \qquad s_1^* = -\zeta\omega_n - j\omega_d.$$

The transfer function (3.64) can be expressed by its partial fraction expansion as follows:

$$h(s) = \frac{r}{s - s_1} + \frac{r^*}{s - s_1^*} = \frac{\alpha + \beta s}{s^2 + 2\zeta\omega_n s + \omega_n^2}, \tag{3.65}$$

where the *residue* is

$$r = \sigma + j\nu, \qquad r^* = \sigma - j\nu, \tag{3.66}$$

and the corresponding parameters are

$$\alpha = 2(\zeta\omega_n\sigma - \omega_d\nu), \qquad \beta = 2\sigma. \tag{3.67}$$

The residues may have real and imaginary parts depending on the damping and the number of modes in the system. However, the residues have distinct values for a SDOF system as follows:

$$r = \lim_{s=s_1}(s - s_1)\frac{1/m}{(s-s_1)(s-s_1^*)} = \frac{1/m}{s_1 - s_1^*} = \frac{1/m}{2j\omega_d},$$

$$r^* = \lim_{s=s_1^*}(s - s_1^*)\frac{1/m}{(s-s_1)(s-s_1^*)} = \frac{1/m}{s_1^* - s_1} = -\frac{1/m}{2j\omega_d}.$$

Hence, the real part of the residues must be zero ($\sigma = 0$ and $\beta = 0$) for SDOF systems. Note that for a unit mass the residue must have a value of $r = 1/(2j\omega_d)$. Note that the transfer functions are alternatively represented with the following residue notation in the literature as well:

$$h(s) = \frac{r'}{2j(s-s_1)} + \frac{r'^*}{2j(s-s_1^*)} = \frac{\alpha' + \beta' s}{s^2 + 2\zeta\omega_n s + \omega_n^2}, \tag{3.68}$$

where the *residue* is

$$r' = \sigma' + j\nu', \qquad r'^* = \sigma' - j\nu', \tag{3.69}$$

and the corresponding parameters are

$$\alpha' = \zeta\omega_n\nu' - \omega_d\sigma', \qquad \beta = \nu'.$$

Hence, the imaginary part of the residues must be zero ($\nu' = 0$ and $\beta' = 0$) for SDOF systems. Note that, for a unit mass, the residue must have a value of $r' = 1/\omega_d$.

One can switch from one notation to the next with the following residue transformation:

$$r = r'/(2j) \rightarrow \sigma = \nu'/2, \qquad \nu = \sigma'/2.$$

Some of the commercial modal analysis packages use this notation.

Transfer Function of a MDOF System

The equation of motion for a MDOF system is represented in s domain by taking the Laplace transform of (3.29) as follows:

$$([M]s^2 + [C]s + [K])\{X(s)\} = \{F(s)\}, \tag{3.70}$$

or

$$[B(s)]\{X(s)\} = \{F(s)\}.$$

The transfer function matrix of the MDOF system is then

$$[H(s)] = \frac{\{X(s)\}}{\{F(s)\}} = \frac{\text{adj}[B(s)]}{|[B(s)]|}, \tag{3.71}$$

where $|[B(s)]|$ is the characteristic equation and the solution of $|[B(s)]| = 0$ gives the eigenvalues of the MDOF system. Note that the transfer function matrix $[H(s)]$ has $[n \times n]$ dimension for an n-DOF system, and all its elements have a common denominator of $|[B(s)]|$.

The transfer function of the example 2-DOF system is

$$[H(s)] = \begin{bmatrix} h_{11}(s) & h_{12}(s) \\ h_{21}(s) & h_{22}(s) \end{bmatrix},$$

where

$$h_{11}(s) = \left[\frac{r_{11,1}}{s - s_1} + \frac{r^*_{11,1}}{s - s^*_1}\right]_{\text{mode 1}} + \left[\frac{r_{11,2}}{s - s_2} + \frac{r^*_{11,2}}{s - s^*_2}\right]_{\text{mode 2}},$$

or

$$h_{11}(s) = \left[\frac{\alpha_{11,1} + \beta_{11,1}s}{s^2 + 2\zeta_1\omega_{\text{n},1}s + \omega^2_{\text{n},1}}\right]_{\text{mode 1}} + \left[\frac{\alpha_{11,2} + \beta_{11,2}s}{s^2 + 2\zeta_2\omega_{\text{n},2}s + \omega^2_{\text{n},2}}\right]_{\text{mode 2}}. \quad (3.72)$$

The elements h_{il} in the transfer function matrix are obtained from experimental measurements. The denominator consists of modal parameters $(\zeta, \omega_n)_k$ for each mode k. When the 2-DOF structure is excited by force $\{F\} = \{F_1\ F_2\}$ and the vibration is measured at point 1, we have

$$h_{11} = \frac{x_1}{F_1} \to F_2 = 0; \qquad h_{12} = \frac{x_1}{F_2} \to F_1 = 0.$$

For example, the accelerometer is attached to point 1, and the structure is hit with an instrumented hammer at point 1 only to measure h_{11} and is hit at point 2 to measure h_{12}. By attaching the accelerometer at point 2 and exciting the structure at points 1 and 2, the respective transfer function elements h_{21} and h_{22} are measured with Fourier analyzers. Note that all the elements have common denominators, but different residues or numerators. The transfer function matrix is symmetric for linear systems (i.e., $h_{12} \equiv h_{21}$). It will be proven later that the measurement of only one row or column of the transfer function is sufficient to model the complete transfer function matrix by using the symmetry and modal matrix properties. The measured transfer functions are stored in the frequency domain by the analyzers, usually in the form of real and imaginary components at each frequency, although the analyzers have built-in transformation routines to display the measurement data in time or other frequency domain representations. The measured frequency domain transfer function data (h_{il}) is transferred to a digital computer loaded with modal analysis software. For a given number of natural modes, the modal analysis system scans the transfer function data for the dominant maximum resonance amplitudes and the corresponding frequencies where the real part of the transfer function is zero. These frequencies are the natural frequencies of the system. The system then fits a curve to the data with a denominator having a $(2 \times n)$-order polynomial. With further numerical processing, the transfer function curve is fitted to n independent second-order differential equations as shown in Eq. (3.72). Thus, the numerical values of natural frequency, damping, and residue for each mode are estimated from the curve fitting [2]. In general form, the element in row i and column l of the transfer function matrix $[H(s)]$ is measured as

$$h_{il} = \sum_{k=1}^{n} \frac{\alpha_{il,k} + \beta_{il,k} s}{s^2 + 2\zeta_k \omega_{n,k} s + \omega_{n,k}^2}, \tag{3.73}$$

where $\omega_{d,k}$, $\omega_{n,k}$, and ζ_k are the damped and undamped natural frequencies and the modal damping ratio for mode k of the system, respectively. The frequency response of the structure can be obtained by replacing $s = j\omega$, where the excitation frequency ω can be scanned in a range covering all natural frequencies. The same equation can also be used in a time domain simulation of forced and chatter vibrations by simply using a bilinear approximation, $s = [2(1 - z^{-1}]/[\delta t(1 + z^{-1})]$, where δt is the digital integration time interval and z^{-1} is the backward time-shift operator (i.e., $z^{-1}x(t) = x(t - \delta t)$). The complete transfer function can be represented in the following matrix form:

$$[H(s)] = \sum_{k=1}^{n} \frac{[R]_k}{s^2 + 2\zeta_k \omega_{n,k} s + \omega_{n,k}^2}, \tag{3.74}$$

where each element in the $[n \times n]$-dimensional matrix $[R]_k = [\alpha + \beta s]_k$ reflects the residues of mode k at row i and column l.

The mode shapes of the system are found from the estimated residues. Remembering from (3.43) that the displacement vector can be expressed by its mode shapes and modal transfer functions,

$$\{x\} = \left(\sum_{k=1}^{n} \{P\}_k \{P\}_k^T \Phi_{q,k} \right) \{F\}.$$

Thus,

$$[H(s)] = \sum_{k=1}^{n} \{P\}_k \{P\}_k^T \Phi_{q,k}. \tag{3.75}$$

The eigenvectors $\{P\}_k$ can be scaled arbitrarily as stated before. Combining Eqs. (3.74) and (3.75), we have

$$[H(s)] = \sum_{k=1}^{n} \frac{\{P\}_k \{P\}_k^T}{m_{q,k}} \frac{1}{s^2 + 2\zeta_k \omega_{n,k} s + \omega_{n,k}^2} = \sum_{k=1}^{n} \frac{[R]_k}{s^2 + 2\zeta_k \omega_{n,k} s + \omega_{n,k}^2}.$$

Note that the modal mass for mode k using the unscaled modal matrix is

$$m_{q,k} = \{P\}_k^T [M_x] \{P\}_k.$$

Thus, $(\{P\}_k^T \{P\}_k)/m_{q,k}$ represents the normalization of each eigenvector with the square root of the modal mass (i.e., $\{u\}_k = \{P\}_k / \sqrt{m_{q,k}}$). The identified residues, therefore, have the following relationship with the mode shapes:

$$\left(\{P\}_k \{P\}_k^T \right) / m_{q,k} \equiv \{u\}_k \{u\}_k^T = [R]_k, \tag{3.76}$$

where $\{u_k\}$ corresponds to the normalized mode shape giving a unity modal mass. In other words, the mass is unity when the following transformation is used:

$$\{u\}_k^T [M_x]\{u\}_k = 1.$$

This procedure is a mathematically convenient way to simplify the identification of mode shapes, modal stiffness, and modal damping constants of the structure.

The residue matrix for a particular mode k can be expressed in the following general form:

$$[R]_k = \begin{bmatrix} u_1u_1 & u_1u_2 & \dots & u_1u_l & \dots & u_1u_\text{n} \\ u_2u_1 & u_2u_2 & \dots & u_2u_l & \dots & u_2u_\text{n} \\ \vdots & \vdots & \vdots & \vdots & \vdots & \\ u_lu_1 & u_lu_2 & \dots & u_lu_l & \dots & u_lu_\text{n} \\ \vdots & \vdots & \vdots & \vdots & \vdots & \\ u_\text{n}u_1 & u_\text{n}u_2 & \dots & u_\text{n}u_l & \dots & u_\text{n}u_\text{n} \end{bmatrix}_k .$$

If we take column or row l of the residue matrix for mode k, we have

$$\begin{Bmatrix} R_{1l} \\ R_{2l} \\ \vdots \\ R_{ll} \\ \vdots \\ R_{\text{n}l} \end{Bmatrix}_k = \begin{Bmatrix} u_1u_l \\ u_2u_l \\ \vdots \\ u_lu_l \\ \vdots \\ u_nu_l \end{Bmatrix}_k ,$$

where $k = 1, 2, \dots, n$ for n number of modes. Starting first with the solution of u_l where the excitation and measurement points match, the mode shapes can be calculated from only one row or column of transfer function measurements. The mode shape vector for mode k is found as follows:

$$\left.\begin{aligned} u_{l,k} &= \sqrt{R_{ll,k}} \\ u_{1,k} &= \frac{R_{1l,k}}{u_{l,k}} \\ u_{2,k} &= \frac{R_{2l,k}}{u_{l,k}} \\ \vdots &= \vdots \\ u_{\text{n},k} &= \frac{R_{\text{n}l,k}}{u_{l,k}} \end{aligned}\right\} . \tag{3.77}$$

This procedure is repeated for all modes to construct the full modal matrix of the structure. The modal matrix of the system consists of mode shapes in the columns of an $[n \times m]$ matrix,

$$[U] = [\{u\}_1 \; \{u\}_2 \dots \{u\}_\text{m}], \tag{3.78}$$

where m is the total number of modes in the system and n is the number of measurement points or coordinates on the structure. Note that the modal matrix does not have to be a square matrix. For example, there may be two or three measurement points with a different number of modes. This depends on the structure and the number of measurement points used for vibration analysis. The identified transfer function can be used to analyze the general behavior of the machine tool under different loading conditions and to study machine tool stability for chatter vibrations during cutting as presented in the following section.

Note that, because the elements of matrix $[R] = [\alpha + \beta s]$ are complex numbers, the resulting mode shapes will be complex and depend on the modal frequency (ω_{d}). It is possible to obtain simplified real mode shapes from the residues:

$$R_{il,k} = \alpha_{il,k} + \beta_{il,k} s \qquad \leftarrow \quad s = j\omega_{\mathrm{d}}. \tag{3.79}$$

For the mode shapes to be real, the imaginary part of $R_{il,k}$ must be zero (i.e., $\beta_{il,k} = 2\sigma_{il,k} = 0$) or the real part of the residues must be zero ($\sigma_{il,k} = 0$). This corresponds to structures having proportional damping, where the damping $[C]$ is a linear combination of mass $[M]$ and stiffness $[K]$, that is, $c = \eta_m m + \eta_k k$, where η_m, η_k are constant numbers. The residues become $r = j\nu, r^* = -j\nu$, and $\alpha_{il,k} = 2\omega_{\mathrm{d},k}\nu_{il,k}$. For example, the modal parameter in Eq. (3.77) becomes $u_{l,k} = \sqrt{R_{ll}} = \sqrt{2\omega_{\mathrm{d}k}\nu_{lk}}$.

Example 1. A 2-DOF system shown in Figure 3.18 is analyzed experimentally. The masses lumped at the middle and the tip are $m_0 = 0.76$ kg each. The beam dimensions are $l = 450$ mm, $b = 25.4$ mm, $h = 5$ mm, and the specific mass density of the steel is $\rho = 7{,}860$ kg/m^3. The mass of the beam becomes $m_{\mathrm{b}} = bhl\rho = 0.45$ kg. An accelerometer was used to measure the vibrations, and an impact hammer instrumented with a piezoelectric force transducer was used to hit the structure (see Fig. 3.16). Transducers are connected to amplifiers and power units to convert the measured charge signals to amplified voltage. The amplified force signal line is connected to the input channel, and the amplified accelerometer signal is connected to the output channel of a Fourier analyzer. The calibration factors of both sensors are entered into the analyzer to obtain correct measurement units. The transfer function h_{11} is measured by hitting the structure at point 1, where the accelerometer is mounted. The experiment is repeated ten times, and a spectral average of the measured FRFs is accepted from the analyzer. The cross-FRF element h_{12} is measured by hitting the structure at point 2, whereas the accelerometer is still mounted at point 1. The measured FRFs (h_{11}, h_{12}) are transferred to a computer and processed by modal analysis software.

The modal analysis software provided the following information for the two modes:

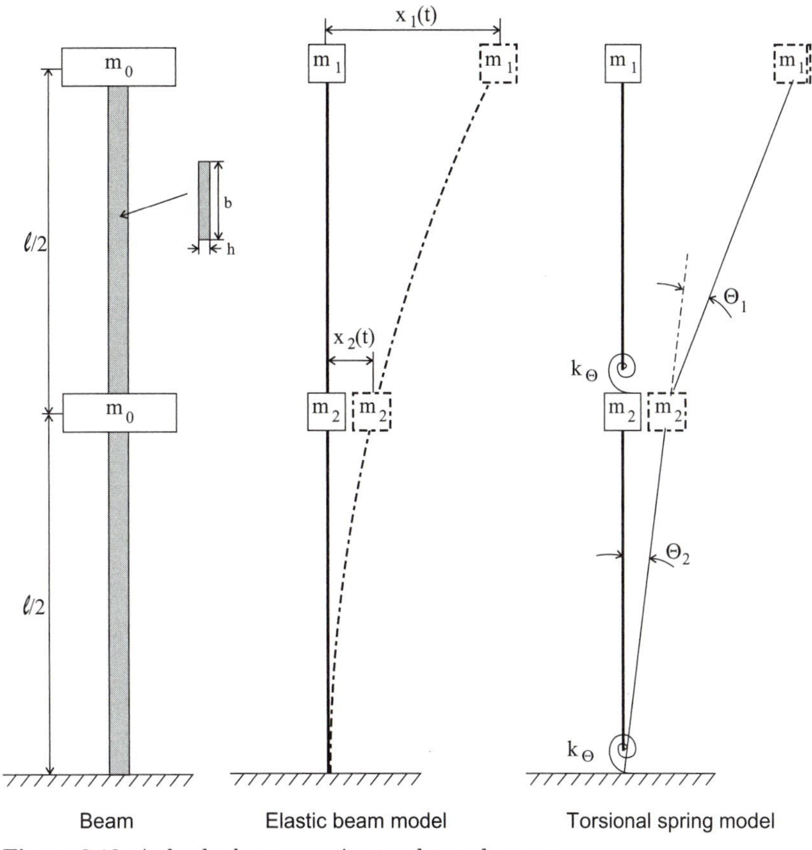

Figure 3.18: A slender beam carrying two lumped masses.

Estimated	**Mode 1**	**Mode 2**
ω_d (rad/s)	53.0041	349.9659
ζ	0.0113	0.0047
ω_n (Hz)	8.433	55.677
Residues for h_{11} (m/N)	$1.8303 \times 10^{-3} - j6.9526 \times 10^{-3}$	$1.1546 \times 10^{-5} - j1.4518 \times 10^{-4}$
Residues for h_{12} (m/N)	$2.0721 \times 10^{-4} - j3.0926 \times 10^{-3}$	$4.6012 \times 10^{-5} + j3.7015 \times 10^{-4}$

Note that, when the measurements are made with an acceleromoter, residues must be scaled to displacement units before processing them (i.e., by dividing the residues by $\omega_d^2 e^{j\pi}$ for each mode). By substituting $\alpha = 2(\zeta\omega_n\sigma - \omega_d\nu)$, $\beta = 2\sigma$, the following direct (h_{11}) and cross (h_{12}) transfer functions are obtained:

$$h_{11}(s) = \frac{0.7392 + 0.00366s}{s^2 + 2\zeta_1\omega_{n1}s + \omega_{n1}^2} + \frac{0.1017 + 2.3092e - 05s}{s^2 + 2\zeta_2\omega_{n2}s + \omega_{n2}^2},$$

$$h_{12}(s) = \frac{0.3281 + 4.1442e - 04s}{s^2 + 2\zeta_1\omega_{n1}s + \omega_{n1}^2} + \frac{-0.2589 + 9.2024e - 05s}{s^2 + 2\zeta_2\omega_{n2}s + \omega_{n2}^2}.$$

By replacing $s = j\omega$ and sweeping all frequencies of interest, the curve-fitted FRFs of the structure can be reconstructed. The complex mode shapes for the first mode can be extracted from the residues as follows:

$$\begin{aligned} u_{11} &= \sqrt{\alpha_{11,1} + \beta_{11,1}s}|_{s=j\omega_{d1}} &&= \sqrt{0.7392 + j0.00366 * 53.0041} \\ & &&= 0.8670 + j0.1119, \\ u_{21} &= \tfrac{\alpha_{12,1}+\beta_{12,1}s}{u_{11}}|_{s=j\omega_{d1}} &&= \tfrac{0.3281+j4.1442e-04*53.0041}{0.8670+j0.1119} \\ & &&= 0.3754 - j0.0231. \end{aligned} \tag{3.80}$$

For the second mode, we obtain

$$\begin{aligned} u_{12} &= \sqrt{\alpha_{11,2} + \beta_{11,2}s}|_{s=j\omega_{d1}} &&= \sqrt{-0.2589 + j2.3092e - 05 * 349.9659} \\ & &&= 0.3191 + j0.0127 \\ u_{22} &= \tfrac{\alpha_{12,2}+\beta_{12,2}s}{u_{12}}|_{s=j\omega_{d2}} &&= \tfrac{-0.2589+j9.2024e-05*349.9659}{0.3191+j0.0127} \\ & &&= -0.8062 + j0.1329. \end{aligned} \tag{3.81}$$

The resulting complex modal matrix for unity modal mass is

$$[U] = \left[\begin{Bmatrix} 0.8670 + j0.1119 \\ 0.3754 - j0.0231 \end{Bmatrix}_1 \quad \begin{Bmatrix} 0.3191 + j0.0127 \\ -0.8062 + j0.1329 \end{Bmatrix}_2 \right].$$

Note that the complex part of the numerator can be neglected (i.e., $\beta = 0$) for simplicity, which leads to real mode shapes.

Because both modal masses are unity (i.e., $m_{q,1} = m_{q,2} = 1$) the modal stiffness values are

$$\begin{aligned} k_{q,1} &= \omega_{\mathrm{n},1}^2 = 53.0041^2 = 2{,}809\,\mathrm{N/m}, \\ k_{q,2} &= \omega_{\mathrm{n},2}^2 = 349.9659^2 = 122{,}480\,\mathrm{N/m}, \end{aligned}$$

and the modal damping constants are

$$\begin{aligned} c_{q,1} &= 2\zeta_1\omega_{\mathrm{n},1} = 2 \times 0.0113 \times 53.0041 = 1.1928\,\mathrm{N/ms}^{-1}, \\ c_{q,2} &= 2\zeta_2\omega_{\mathrm{n},2} = 2 \times 0.0047 \times 349.9659 = 3.2741\,\mathrm{N/ms}^{-1}. \end{aligned}$$

The transfer functions of two modes in modal coordinates are

$$\begin{aligned} \Phi_{q,1} &= \tfrac{1}{m_{q,1}s^2+c_{q,1}s+k_{q,1}} = \tfrac{1}{s^2+1.1928s+2809}, \\ \Phi_{q,2} &= \tfrac{1}{m_{q,2}s^2+c_{q,2}s+k_{q,2}} = \tfrac{1}{s^2+3.2741s+122480}, \end{aligned}$$

and when substituted in Eq. (3.74), the transfer function matrix is found as

$$\begin{Bmatrix} X_1(s) \\ X_2(s) \end{Bmatrix} = \begin{bmatrix} u_{11}^2\Phi_{q,1} + u_{12}^2\Phi_{q,2} & u_{11}u_{21}\Phi_{q,1} + u_{12}u_{22}\Phi_{q,2} \\ u_{11}u_{21}\Phi_{q,1} + u_{12}u_{22}\Phi_{q,2} & u_{21}^2\Phi_{q,1} + u_{22}^2\Phi_{q,2} \end{bmatrix} \begin{Bmatrix} F_1(s) \\ F_2(s) \end{Bmatrix},$$

where $m_{q,1} = m_{q,2} = 1$, and real mode shapes (real(u)) are considered. The mass, damping, and stiffness matrices in local coordinates can be obtained from the inverse transformation by the use of the modal matrix as follows:

$$[M_x] = [U]^{-T}[I]U^{-1} = \begin{bmatrix} 1.1797 & -0.1018 \\ -0.1018 & 1.2732 \end{bmatrix} \text{(kg)}.$$

The actual mass of the beam (m_b) can be distributed to points 1 and 2 as follows:

$$m_1 = m_0 + m_b/4 = 0.76 + 0.45/4 = 0.873\,\text{kg},$$

$$m_2 = m_0 + m_b/2 = 0.76 + 0.45/2 = 0.985\,\text{kg}.$$

The analytical actual masses are approximately 0.873 kg and 0.985 kg; the predicted masses in the diagonal locations of the mass matrix (1.1797 kg, 1.2732 kg) are slightly different, but the ratios are similar. The difference is due to errors in the measurement, noise, frequency resolution, accelerometer, and force sensor. The nonzero off-diagonal masses represent the presence of some dynamic coupling in the system as follows:

$$[C_x] = [U]^{-T}[C_q]U^{-1} = \begin{bmatrix} 1.8447 & -1.1320 \\ -1.1320 & 3.8524 \end{bmatrix} \text{(N/ms}^{-1}\text{)},$$

$$[K_x] = [U]^{-T}[K_q]U^{-1} = \begin{bmatrix} 0.2847e+05 & -0.5839e+05 \\ -0.5839e+05 & 1.3776e+05 \end{bmatrix} \text{(N/m)}.$$

The stiffness values are quite reasonable; they reflect the losses in the clamping of the beam on the ground and measurement errors with the accelerometer.

Example 2. Consider a slender beam with the measured FRFs shown in Figure 3.19, and the values of the measurement points are given in the table.

$\Phi(\mu$ m/N)	*G*(668 Hz)	*G*(812 Hz)	*H*(741 Hz)	*G*(3,812 Hz)	*G*(3,862 Hz)	*H*(3,838 Hz)
Φ_{11}	+0.4386	−0.3292	−0.7613	+0.6935	−0.6959	−1.3915
Φ_{12}	+0.3751	−0.2995	−0.6695	+0.1996	−0.2076	−0.4060

The mode shapes and FRFs at the two measurement points can be evaluated as follows. The modal parameters can be identified from the real and imaginary parts around resonance frequencies.

$\omega_{n1} = 741$ [Hz] $= 2\pi 741$ [rad/s] $= 4655.8$ [rad/s], $\zeta_1 = \frac{812-688}{2\times 741} = 0.084$,
$k_1 = \frac{-1}{2\zeta_1 H_{11,1}} = \frac{1}{2\times 0.084\times 0.7613} = 7.8187[\text{N}/\mu\text{m}]$, $m_1 = \frac{k_1}{\omega_{n1}^2} = \frac{7.8187\times 10^6}{4197.2^2} =$ $0.3607[kg]$

$\omega_{n2} = 3{,}838$ [Hz]$=2\pi 3838$ [rad/s] $= 24{,}115$ [rad/s], $\zeta_2 = \frac{3{,}862-3{,}812}{2\times 3{,}838} = 0.0065$
$k_2 = \frac{-1}{2\zeta_2 H_{11,2}} = \frac{1}{2\times 0.0065\times 1.3915} = 55.28$ [N/μm], $m_2 = \frac{k_2}{\omega_{n2}^2} = \frac{55.28\times 10^6}{24{,}115^2} = 0.0951$ [kg]

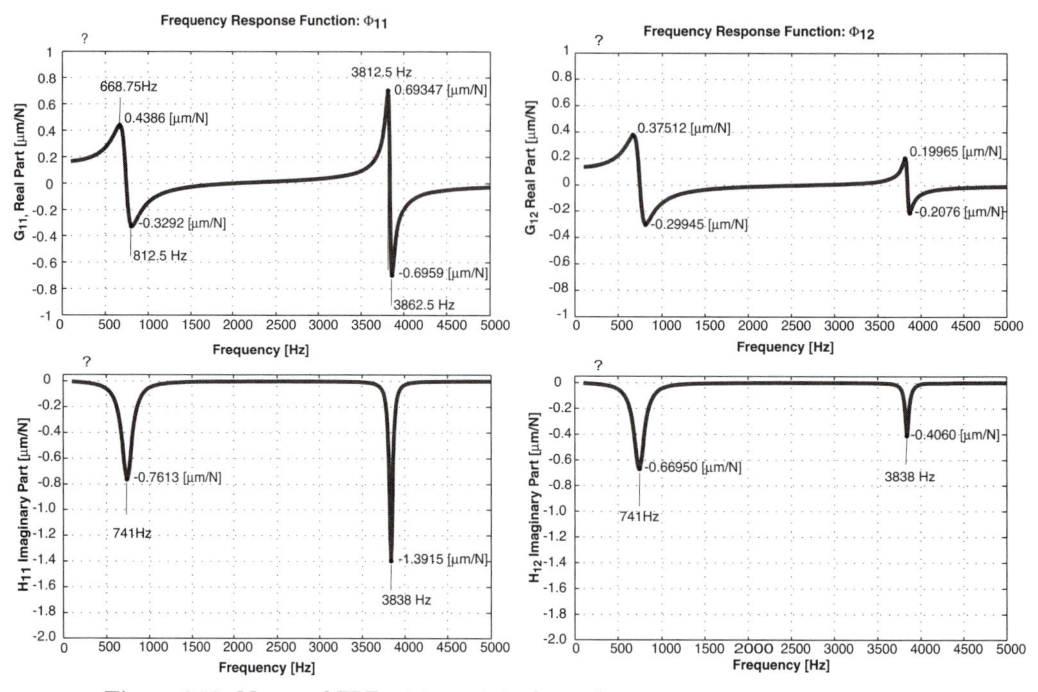

Figure 3.19: Measured FRFs at two points along a beam.

Note that the stiffness and mass are associated with the mode number one when the force and vibrations are observed at point 1.

$$\Phi_{11}(s) = \frac{u_{11}u_{11}}{s^2 + 2\zeta_1\omega_{n1}s + \omega_{n1}^2} + \frac{u_{12}u_{12}}{s^2 + 2\zeta_2\omega_{n2}s + \omega_{n2}^2}$$

$$\Phi_{11}(j\omega) = \frac{u_{11}u_{11}}{\omega_{n1}^2 - \omega^2 + j2\zeta_1\omega_{n1}\omega} + \frac{u_{12}u_{12}}{\omega_{n2}^2 - \omega^2 + j2\zeta_2\omega_{n2}\omega}$$

Substituing $s = j\omega_{n1}$ leads to a negligible contribution from ω_{n2}, and the first part becomes equal to

$$\Phi_{11}(\omega = \omega_{n1}) \approx H_{11,1} = \frac{u_{11}u_{11}}{j2\zeta_1\omega_{n1}^2} = \frac{-ju_{11}u_{11}}{2\zeta_1\omega_{n1}^2},$$

$$u_{11} = \sqrt{-2\zeta_1\omega_{n1}^2 H_{11,1}} = \sqrt{2 \times 0.084 \times 4655.8^2 \times 0.7613 \times 10^{-6}} = 1.6651$$

$$\Phi_{11}(\omega = \omega_{n2}) \approx H_{11,2} = \frac{-ju_{12}u_{12}}{2\zeta_2\omega_{n2}^2}$$

$$u_{12} = \sqrt{-2\zeta_2\omega_{n2}^2 H_{11,2}} = \sqrt{2 \times 0.006\,5 \times 24{,}115^2 \times 1.3915 \times 10^{-6}} = 3.2434$$

The cross-transfer function between points 1 and 2 are as follows:

$$\Phi_{12} = \frac{u_{11}u_{21}}{s^2 + 2\zeta_1\omega_{n1}s + \omega_{n1}^2} + \frac{u_{12}u_{22}}{s^2 + 2\zeta_2\omega_{n2}s + \omega_{n2}^2}$$

$$\Phi_{12}(\omega = \omega_{n1}) \approx H_{12,1} = \frac{-ju_{11}u_{21}}{2\zeta_1\omega_{n1}^2} \quad u_{21} = \frac{2\zeta_1\omega_{n1}^2 H_{12,1}}{-ju_{11}}$$

$$u_{21} = \frac{2 \times 0.084 \times 4655.8^2 \times (-j0.6695 \times 10^{-6})}{-j1.6651} = 1.4643$$

$$\Phi_{12}(\omega = \omega_{n2}) \approx H_{12,2} = \frac{-ju_{12}u_{22}}{2\zeta_2\omega_{n2}^2} \quad u_{22} = \frac{2\zeta_2\omega_{n2}^2 H_{12,2}}{-ju_{12}}$$

$$u_{22} = \frac{2 \times 0.0065 \times 24{,}115^2 \times (-j0.4060 \times 10^{-6})}{-j3.2434} = 0.9463$$

$$\text{Modal matrix } U = \begin{bmatrix} \begin{Bmatrix} 1.6651 \\ 1.4643 \end{Bmatrix}_{\text{mode 1}} & \begin{Bmatrix} 3.2434 \\ 0.9463 \end{Bmatrix}_{\text{mode 2}} \end{bmatrix}$$

The direct transfer functions of the beam at the two measurement points can be identified from modal matrix and modal parameters.

The transfer function at point 1 is as follows:

$$\frac{x_1}{F_{y1}} = \frac{u_{11}u_{11}}{s^2 + 2\zeta_1\omega_{n1}s + \omega_{n1}^2} + \frac{u_{12}u_{12}}{s^2 + 2\zeta_2\omega_{n2}s + \omega_{n2}^2}$$

$$= \frac{(1.6551)^2}{s^2 + 2 \times 0.084 \times 4655.8s + 4655.8^2} + \frac{(3.2434)^2}{s^2 + 2 \times 0.0065 \times 24{,}115s + 24{,}115^2}$$

$$\frac{x_1}{F_{y1}} = \frac{2.7724}{s^2 + 782.18s + 21.677 \times 10^6} + \frac{10.5192}{s^2 + 313.5s + 581.533225 \times 10^6}.$$

The expression can be checked by inserting $s = j\omega_{n1}, H_{11,1} = \frac{2.7724}{782.18 \times 4655.8} = -j0.7613 \times 10^{-6}$, which is the same as the measured value given in the table.

The transfer function at point 2 is as follows:

$$\frac{x_2}{F_{y2}} = \frac{u_{11}u_{21}}{s^2 + 2\zeta_1\omega_{n1}s + \omega_{n1}^2} + \frac{u_{12}u_{22}}{s^2 + 2\zeta_2\omega_{n2}s + \omega_{n2}^2}$$

$$= \frac{1.6651 \times 1.4643}{s^2 + 2 \times 0.084 \times 4655.8s + 4655.8^2} + \frac{3.2434 \times 0.9463}{s^2 + 2 \times 0.0065 \times 24{,}115s + 24{,}115^2}$$

$$\frac{x_2}{F_{y2}} = \frac{2.4381}{^2 + 782.18s + 21.677 \times 10^6} + \frac{3.0692}{s^2 + 313.5s + 581.533225 \times 10^6}.$$

3.7 IDENTIFICATION OF MODAL PARAMETERS

Consider the dynamics of a system with n modes that is represented by the following FRF:

$$\Phi_{pq}(\omega) = \sum_{k=1}^{n} \left(\frac{\alpha_k + j\omega\beta_k}{-\omega^2 + \omega_{n,k}^2 + j2\zeta_k\omega_{nk}\omega} \right) = \sum_{k=1}^{n} \left(\frac{A_{pq,k}}{j\omega - \lambda_k} + \frac{A_{pq,k}^*}{j\omega - \lambda_k^*} \right) \quad (3.82)$$

where $(A_{pq,k}, A^*_{pq,k})$ and (λ_k, λ^*_k) are the complex conjugate residues and eigenvalues of the system, respectively. The estimation of modal parameters is conducted by considering a limited frequency range where the FRF for one mode is expressed as

$$\Phi_R(\omega) = \Phi_{\text{low}}(\omega) + \frac{A_{pq,k}}{j\omega - \lambda_k} + \frac{A^*_{pq,k}}{j\omega - \lambda^*_k} + \Phi_{\text{hi}}(\omega) \tag{3.83}$$

where $\Phi_{\text{low}}(\omega) = \frac{\alpha_{\text{low}} + j\omega\beta_{\text{low}}}{-\omega^2 + \omega^2_{n,\text{low}}}$ is the residual effects of lower modes and called the residual inertia and $\Phi_{li}(\omega) = \frac{\alpha_{\text{hi}} + j\omega\beta_{\text{hi}}}{-\omega^2 + \omega^2_{n,\text{hi}}}$ is the residual effects of higher modes and called the residual flexibility.

The residual modes are modeled as undamped second-order linear systems as follows:

$$\begin{aligned}\Phi_R(\omega) &= G_r(\omega) + jH_r(\omega) \\ &= \frac{\alpha_{\text{low}} + j\omega\beta_{\text{low}}}{-\omega^2 + \omega^2_{n,\text{low}}} + \frac{\alpha_k + j\omega\beta_k}{-\omega^2 + \omega^2_{n,k} + j2\zeta_k\omega_{nk}\omega} + \frac{\alpha_{\text{hi}} + j\omega\beta_{\text{hi}}}{-\omega^2 + \omega^2_{n,\text{hi}}},\end{aligned} \tag{3.84}$$

where $\omega_{n,\text{low}}$ and $\omega_{n,\text{hi}}$ are used to account for the lower- and higher-frequency modes, respectively. The estimation requires the identification of mode shapes, damping ratio, and natural frequency for each mode. The identification starts with the prediction of system eigenvalues $(\lambda_k = -\zeta_k\omega_{nk} + j\omega_{dk})$. Multiplying both sides of Eq. (3.84) by $(-\omega^2 + \omega^2_{n,k} + j2\zeta_k\omega_{nk}\omega)$,

$$\begin{aligned}&\Phi_R(\omega)\left[-\omega^2 + \omega^2_{n,k} + j2\zeta_k\omega_{nk}\omega\right] \\ &= \frac{(\alpha_{\text{low}} + j\omega\beta_{\text{low}})\left(-\omega^2 + \omega^2_{n,k} + j2\zeta_k\omega_{nk}\omega\right)}{-\omega^2 + \omega^2_{n,\text{low}}} \\ &\quad + \alpha_k + j\omega\beta_k + \frac{(\alpha_{\text{hi}} + j\omega\beta_{\text{hi}})\left(-\omega^2 + \omega^2_{n,k} + j2\zeta_k\omega_{nk}\omega\right)}{-\omega^2 + \omega^2_{n,\text{hi}}} \\ &= \frac{-\omega^2\alpha_{\text{low}} + \alpha_{\text{low}}\omega^2_{n,k} + j2\alpha_{\text{low}}\zeta_k\omega_{nk}\omega - j\omega^3\beta_{\text{low}} + j\omega\omega^2_{n,k}\beta_{\text{low}} - 2\zeta_k\omega_{nk}\omega^2\beta_{\text{low}}}{-\omega^2 + \omega^2_{n,\text{low}}} \\ &\quad + (\alpha_k + j\omega\beta_k) \\ &\quad + \frac{-\alpha_{\text{hi}}\omega^2 + \alpha_{\text{hi}}\omega^2_{n,k} + j2\alpha_{\text{hi}}\zeta_k\omega_{nk}\omega - j\omega^3\beta_{\text{hi}} + j\omega\omega^2_{n,k}\beta_{\text{hi}} - 2\zeta_k\omega_{nk}\omega^2\beta_{\text{hi}}}{-\omega^2 + \omega^2_{n,\text{hi}}} \\ &\cong \frac{\sigma_1 + j\nu_1}{-\omega^2 + \omega^2_{n,\text{low}}} + (\alpha_k + j\omega\beta_k) + \frac{\sigma_2 + j\nu_2}{-\omega^2 + \omega^2_{n,\text{hi}}} = \frac{C_1}{-\omega^2 + \omega^2_{n,\text{low}}} + (\alpha_k + j\omega\beta_k) \\ &\quad + \frac{C_2}{-\omega^2 + \omega^2_{n,\text{hi}}},\end{aligned} \tag{3.85}$$

where $C_1 = \sigma_1 + j\nu_1$ and $C_2 = \sigma_2 + j\nu_2$ are complex residual constants. As long as residual terms account for the effects of out of band frequencies, they can be expressed by any convenient mathematical term. The estimated residuals do not have any physical meaning as long as they can be used to estimate the natural frequency and damping ratio (i.e., pole) of the interested mode. Eq. (3.85) can be separated into its real and imaginary parts as follows:

$$\Phi_R(\omega)\left[-\omega^2+\omega_{n,k}^2+j2\zeta_k\omega_{nk}\omega\right]=\left[G_r(\omega)+jH_r(\omega)\right]\left[-\omega^2+\omega_{n,k}^2+j2\zeta_k\omega_{nk}\omega\right]$$

$$-\omega^2G_r(\omega)+\omega_{n,k}^2G_r(\omega)-2\zeta_k\omega_{nk}\omega H_r(\omega)=\alpha_k+\frac{\sigma_1}{-\omega^2+\omega_{n,\mathrm{low}}^2}+\frac{\sigma_2}{-\omega^2+\omega_{n,\mathrm{hi}}^2}$$

$$-\omega^2H_r(\omega)+\omega_{n,k}^2H_r(\omega)+2\zeta_k\omega_{nk}\omega G_r(\omega)=\omega\beta_k+\frac{\nu_1}{-\omega^2+\omega_{n,\mathrm{low}}^2}+\frac{\nu_2}{-\omega^2+\omega_{n,\mathrm{hi}}^2}. \tag{3.86}$$

Considering that when FRF is measured, the user sets the lower ($\omega_{n,\mathrm{low}}$) and upper ($\omega_{n,\mathrm{hi}}$) frequency bounds of the mode, and the FRF contains the real $G_r(\omega)$ and imaginary ($H_r(\omega)$) parts of the measurement at each frequency (ω). The unknowns are eigenvalues (ζ_k, ω_{nk}) and residues (α_k, β_k; σ_1, ν_1; σ_2, ν_2) that need to be identified. Eq. (3.86) can be organized in matrix form as follows:

$$\begin{bmatrix} -\omega H_r(\omega) & G_r(\omega) & -1 & 0 & \frac{-1}{-\omega^2+\omega_{n,\mathrm{low}}^2} & \frac{-1}{-\omega^2+\omega_{n,\mathrm{hi}}^2} & 0 & 0 \\ \omega G_r(\omega) & H_r(\omega) & 0 & -\omega & 0 & 0 & \frac{-1}{-\omega^2+\omega_{n,\mathrm{low}}^2} & \frac{-1}{-\omega^2+\omega_{n,\mathrm{hi}}^2} \end{bmatrix} \begin{Bmatrix} 2\zeta_k\omega_{nk} \\ \omega_{n,k}^2 \\ \alpha_k \\ \beta_k \\ \sigma_1 \\ \sigma_2 \\ \nu_1 \\ \nu_2 \end{Bmatrix} \tag{3.87}$$

$$=\begin{Bmatrix} \omega^2G_r(\omega) \\ \omega^2H_r(\omega) \end{Bmatrix} \tag{3.88}$$

The modal parameters estimation matrix (Eq. 3.87) can be expressed as follows:

$$\left[A(\omega)\right]\{P_m\}=\{B(\omega)\}, \tag{3.89}$$

where $\{P_m\}$ contains unknown modal parameters. Because we have a number of frequency points around the interested modal frequency range,

$$\begin{bmatrix} A(\omega_1) \\ A(\omega_2) \\ .. \\ .. \\ A(\omega_n) \end{bmatrix}\{P_m\}=\begin{Bmatrix} B(\omega_1) \\ B(\omega_2) \\ .. \\ .. \\ B(\omega_n) \end{Bmatrix} \tag{3.90}$$

the unknown parameter vector $\{P\}$ can be easily evaluated by using the linear least-squares solution as follows:

$$\{P\} = \left([A]^T [A]\right)^{-1} [A]^T \{B\} \tag{3.91}$$

Only ζ_k and $\omega_{n,k}$ are retained here, and the rest of the residue variables are discarded in the least-squares identification at this stage. The above expression, where only the eigenvalues are estimated, is applied to only one set of measurements or FRF, because the poles are global and the same for all other FRF measurements.

The residues or mode shapes can now be estimated from the original FRF Eq. (3.84) by separating the terms into real and imaginary parts as follows:

$$G_r + jH_r = \frac{\alpha_{\text{low}} + j\omega\beta_{\text{low}}}{D_l} + \frac{\alpha_k + j\omega\beta_k}{E_1 + jE_2} + \frac{\alpha_{\text{hi}} + j\omega\beta_{\text{hi}}}{D_h} \tag{3.92}$$

$$(G_r + jH_r) = \frac{(\alpha_{\text{low}} + j\omega\beta_{\text{low}})(E_1 + jE_2) D_h + D_l (\alpha_k + j\omega\beta_k) D_h + D_l (E_1 + jE_2)(\alpha_{\text{hi}} + j\omega\beta_{\text{hi}})}{D_l D_h (E_1 + jE_2)}, \tag{3.93}$$

where

$$D_l = -\omega^2 + \omega_{n,\text{low}}^2$$
$$E_1 = -\omega^2 + \omega_{nk}^2$$
$$E_2 = 2\zeta_k \omega_{nk} \omega$$
$$D_h = -\omega^2 + \omega_{n,\text{hi}}^2$$

$$\begin{aligned}&(G_r + jH_r)\left[D_l D_h (E_1 + jE_2)\right] \\ &= (\alpha_{\text{low}} + j\omega\beta_{\text{low}})(E_1 + jE_2) D_h + D_l (\alpha_k + j\omega\beta_k) D_h + D_l (E_1 + jE_2)(\alpha_{\text{hi}} + j\omega\beta_{\text{hi}}).\end{aligned} \tag{3.94}$$

Simplifying the right-hand side of the equation,

$$\begin{aligned}&(\alpha_{\text{low}} + j\omega\beta_{\text{low}})(E_1 + jE_2) D_h + D_l (\alpha_k + j\omega\beta_k) D_h + D_l (E_1 + jE_2)(\alpha_{\text{hi}} + j\omega\beta_{\text{hi}}) \\ &= (\alpha_{\text{low}} + j\omega\beta_{\text{low}})(D_h E_1 + jD_h E_2) + D_l D_h \alpha_k + j\omega D_l D_h \beta_k + (D_l E_1 + jD_l E_2)(\alpha_{\text{hi}} + j\omega\beta_{\text{hi}}) \\ &= D_h E_1 \alpha_{\text{low}} + jD_h E_2 \alpha_{\text{low}} + j\omega D_h E_1 \beta_{\text{low}} - \omega D_h E_2 \beta_{\text{low}} + D_l D_h \alpha_k \\ &\quad + j\omega D_l D_h \beta_k + D_l E_1 \alpha_{\text{hi}} + j\omega D_l E_1 \beta_{\text{hi}} + jD_l E_2 \alpha_{\text{hi}} - D_l E_2 \omega\beta_{\text{hi}} \\ &= D_l D_h \alpha_k + D_h E_1 \alpha_{\text{low}} - \omega D_h E_2 \beta_{\text{low}} + D_l E_1 \alpha_{\text{hi}} - D_l E_2 \omega\beta_{\text{hi}} \\ &\quad + j(\omega D_l D_h \beta_k + D_h E_2 \alpha_{\text{low}} + \omega D_h E_1 \beta_{\text{low}} + D_l E_2 \alpha_{\text{hi}} + \omega D_l E_1 \beta_{\text{hi}})\end{aligned} \tag{3.95}$$

Similarly, collecting the real and imaginary terms on the left-hand side,

$$\begin{aligned}(G_r + jH_r) D_l D_h (E_1 + jE_2) &= D_l D_h E_1 G_r + jD_l D_h G_r E_2 + jD_l D_h E_1 H_r - H_r D_l D_h E_2 \\ &= D_l D_h E_1 G_r - H_r D_l D_h E_2 + j(D_l D_h G_r E_2 + D_l D_h E_1 H_r).\end{aligned} \tag{3.96}$$

The real and imaginary parts of both the right and left sides must be equal as follows:

$$D_l D_h \alpha_k + E_1 D_h \alpha_{\text{low}} - \omega E_2 D_h \beta_{\text{low}} + D_l E_1 \alpha_{\text{hi}} - \omega D_l E_2 \beta_{\text{hi}} = D_l E_1 D_h G_r - D_l E_2 D_h H_r$$
$$\omega D_l D_h \beta_k + E_2 D_h \alpha_{\text{low}} + \omega D_h E_1 \beta_{\text{low}} + D_l E_2 \alpha_{\text{hi}} + \omega D_l E_1 \beta_{\text{hi}} = D_l E_2 D_h G_r + D_l E_1 D_h H_r. \tag{3.97}$$

The unknowns are the residue terms $\{P_m\}^T = \{\alpha_k, \beta_k, \alpha_{\text{low}}, \beta_{\text{low}}, \alpha_{\text{hi}}, \beta_{\text{hi}}\}^T$. The unknown parameter vector $\{P_m\}$ can be isolated in both real and imaginary parts as follows:

$$\begin{bmatrix} D_l D_h & 0 & E_1 D_h & -\omega E_2 D_h & D_l E_1 & -\omega D_l E_2 \\ 0 & \omega D_l D_h & E_2 D_h & \omega E_1 D_h & D_l E_2 & \omega D_l E_1 \end{bmatrix} \begin{Bmatrix} \alpha_k \\ \beta_k \\ \alpha_{\text{low}} \\ \beta_{\text{low}} \\ \alpha_{\text{hi}} \\ \beta_{\text{hi}} \end{Bmatrix} \tag{3.98}$$

$$= \begin{Bmatrix} D_l E_1 D_h G_r - D_l E_2 D_h H_r \\ D_l E_2 D_h G_r + D_l E_1 D_h H_r. \end{Bmatrix}. \tag{3.99}$$

For each measurement around the mode, residues α_k and β_k are estimated by the use of linear least squares similar to the estimation of eigenvalues. The remaining terms ($\alpha_{\text{low}}, \beta_{\text{low}}, \alpha_{\text{hi}}, \beta_{\text{hi}}$) are discarded. The complex residue for mode k, $A_{pq,k}$ is estimated as follows:

$$A_{pq,k} = \frac{\beta_k}{2} + j\left(\frac{\alpha_k - \beta_k \omega_{n,k} \zeta_k}{-2\omega_{dk}}\right). \tag{3.100}$$

Although the pole of the system is global and valid for all measurements, the residues must be estimated for each measurement, because each constitutes an element in mode shape. When the system is considered to have a proportional damping, the real part of the residue is zero ($\beta_k = 0$); hence, the residue becomes

$$A_{pq,k} = -j\frac{\alpha_k}{2\omega_{dk}}. \tag{3.101}$$

3.7.1 Global Nonlinear Optimization of Modal Parameter Identification

The estimation of each mode individually often provides sufficient accuracy when the residual mode effects are considered. However, when there is a strong coupling between the individual modes, a global parameter optimization that considers all the modes and all frequency response measurements simultaneously would yield more accurate results. A nonlinear least-squares optimization method with a steepest descent algorithm can be used where all the parameters are varied until a specified error criterion is satisfied.

By construing the transfer function ($H_{pq}(s)$) in the Laplace domain,

$$H_{pq}(s) = \sum_{k=1}^{n}\left(\frac{\sigma_k + j\nu_k}{s - s_1} + \frac{\sigma_k - j\nu_k}{s - s_1^*}\right) = \sum_{k=1}^{n}\left[\frac{(\sigma_k + j\nu_k)\left(s - s_1^*\right) + (\sigma_k - j\nu_k)(s - s_1)}{(s - s_1)\left(s - s_1^*\right)}\right]$$

$$= \sum_{k=1}^{n}\left[\frac{(\sigma_k + j\nu_k)(s - (-\zeta_k\omega_{nk} - j\omega_d)) + (\sigma_k - j\nu_k)(s - (-\zeta_k\omega_{nk} + j\omega_d))}{(s - (-\zeta_k\omega_{nk} + j\omega_d))\cdot(s - (-\zeta_k\omega_{nk} - j\omega_d))}\right]_1$$

$$= \sum_k = 1^n\left[\frac{2\sigma_k s + 2\sigma_k\zeta_k\omega_n k + j\omega_d\sigma_k + j\nu_k\zeta_k\omega_n k - \nu_k\omega_d - j\sigma_k\omega_d - j\nu_k\zeta_k\omega_n k - \nu_k\omega_d}{s^2 - \left(-\zeta_k\omega_n k - j\omega_d - \zeta_k\omega_n k + j\omega_d\right)s + \left(\zeta_k^2\omega_n k^2 + \omega_d^2\right)}\right]$$

$$= \sum_k = 1^n\left[\frac{2\sigma_k s + 2\sigma_k\zeta_k\omega_n k - 2\nu_k\omega_d}{s^2 - \left(-2\zeta_k\omega_n k\right)s + \left(\zeta_k^2\omega_n k^2 + \omega_d^2\right)}\right]$$

$$= \sum_k = 1^n\left[\frac{2\sigma_k s + 2\sigma_k\zeta_k\omega_n k - 2\nu_k\omega_d}{s^2 - \left(-2\zeta_k\omega_n k\right)s + \left(\zeta_k^2\omega_n k^2 + \omega_n k^2 - \zeta_k^2\omega_n k^2\right)}\right]$$

$$= \sum_k = 1^n\left[\frac{2\sigma_k s + 2\sigma_k\zeta_k\omega_n k - 2\nu_k\omega_d}{s^2 + 2\zeta_k\omega_n ks + \omega_n k^2}\right]. \tag{3.102}$$

Substituting $s = j\omega$, the FRF is evaluated as follows:

$$H_{pq}(j\omega) = \sum_{k=1}^{n}\left[\frac{2\left(\sigma_k\zeta_k\omega_{nk} - \nu_k\omega_d\right) + j2\sigma_k\omega}{\omega_{nk}^2 - \omega^2 + j2\zeta_k\omega\omega_{nk}}\right].$$

The FRF is expressed in terms of the modal damping ratio (ζ_k), natural frequency of each mode (ω_{nk}), and complex parts of the mode residues σ_k, ν_k. For each mode, there are four modal parameters (ω_{nk}, ζ_k, σ_k, and ν_k) that need to be identified. If the number of points where the measurements are taken is N, the total number of modal parameters that need to be identified for M number of modes becomes $M(2N+2)$. The $M(2N+2)$ estimation parameters (ω_{nk}, ζ_k, σ_k, and ν_k) are represented as p_j, where $j = 1, 2, \ldots, M(2N+2)$. If the measured response is denoted as $H_{pq}(\omega)$, and the estimated function is $\widetilde{H}_{pq}(\omega, p)$, the cost function used to evaluate the curve fitting, J, is the summed magnitudes of the difference $\Delta H_{pq}(\omega, p) = H_{pq}(\omega) - \widetilde{H}_{pq}(\omega, p)$ between measured and estimated responses. For all n data points and N number of measurement points, the cost function becomes

$$J(\omega, p) = \sum_{q=1}^{N}\sum_{i=1}^{n}\left\{\left[\mathrm{Re}\,\Delta H_{pq}(\omega_i, p)\right]^2 + \left[\mathrm{Im}\,\Delta H_{pq}(\omega_i, p)\right]^2\right\}.$$

The parameters are updated in the direction of negative gradient ($-\nabla J$) in the steepest descent approach. The partial fractions are individually evaluated by numerically approximating the slope over small interval of parameters (p). The parameters are updated for each iteration as

$$p_j = p_j - S\frac{\partial J}{\partial P_j} \quad j = 1, 2, \ldots, M(2N+2),$$

where S is the step size, which is selected for the optimum decrease of the cost function J. Typically, the cost function is first evaluated with a step size of $S = 1$. If the cost function is decreased, S is incremented by a factor γ ($\gamma > 1$), as new $S_{\text{new}} = \gamma S_{\text{prev}}$ until a minimum J is found in the direction $(-\nabla J)$. Alternatively, if the cost function increased, S is decremented as $S_{\text{new}} = S_{\text{prev}}/\gamma$ until a step size is found causing J to decrease. Note that, while moving in the direction $(-\nabla J)$, the cost function will always decrease given a small enough interval. The algorithm continues until a relative change in J between iterations reaches an acceptably small value (ε) as follows:

$$\frac{|J_{\text{new}} - J_{\text{old}}|}{J_{\text{new}}} < \varepsilon.$$

When the proportional damping is assumed ($\sigma_k = 0$), each mode shape element in the eigen vector becomes

$$u_{k,p} = \sqrt{-2\nu_k \omega_{d,k}} \quad \rightarrow u_{k,q} = \frac{-2\nu_k \omega_{d,k}}{u_{k,p}} \quad q = 1, 2, \ldots, N,$$

where the mode shape coefficient for measurement p, the residue must be negative for $u_{k,p}$ to be a real value. This enforces that the structure will always move in the direction of applied force at the point of impact.

3.8 RECEPTANCE COUPLING OF END MILLS TO SPINDLE-TOOL HOLDER ASSEMBLY

The measurement of each tool and tool holder on the spindles may be too costly and time consuming in a production environment. The receptance coupling method allows the analytical assembly of a linear, free–free end mill or end mill–holder units to the spindle that always remain the same. Although the spindle is measured using experimental modal tests, the free–free end mill or holder assembly can be modeled by using beam theory or finite element methods. The two substructures then can be assembled mathematically, thus avoiding time-consuming impulse test measurements on the production machines [94]. However, the stiffness and damping at the assembly joints of two substructures need to be identified first.

The machine tool assembly (structure AB) is divided into two substructures as shown in Figure 3.20. Substructure A represents the tool holder, and substructure B represents the remaining machine tool assembly up to the tool holder flange. The two structures are rigidly connected at point 2.

Consider the FRF of end mill (A) at two free ends (1, 2) as follows:

$$\begin{Bmatrix} X_1 \\ X_{A,2} \end{Bmatrix} = \begin{bmatrix} H_{A,11} & H_{A,12} \\ H_{A,21} & H_{A,22} \end{bmatrix} \begin{Bmatrix} F_1 \\ F_{A,2} \end{Bmatrix}, \tag{3.103}$$

where $X_1, X_{A,2}$ are the displacement vectors with both translational and angular displacement components. F_1 and $F_{A,2}$ are applied on the structure at points 1 and 2, respectively. $H_{A,ij}$ terms are the FRFs between points i and j.

Similarly, the FRF of the substructure (B) at its free end (2) is

$$\{X_{B,2}\} = [H_{B,22}] \cdot \{F_{B,2}\}. \tag{3.104}$$

In view of the rigid coupling of the two structures A and B at point 2, the equilibrium and compatibility conditions at point 2 are

$$\begin{aligned} F_2 &= F_{A,2} + F_{B,2} \\ X_2 &= X_{A,2} = X_{B,2}, \end{aligned} \tag{3.105}$$

which is used in coupling the spindle (A) with the free–free model of the substructure (B). By letting

$$H_2 = H_{A,22} + H_{B,22} \tag{3.106}$$

and substituting Eq. (3.106) into (3.104) gives

$$X_2 = H_{B,22}F_{B,2} = H_{A,21}F_1 + H_{A,22}(F_2 - F_{B,2}). \tag{3.107}$$

By rearranging Eq. (3.107), the forces on structure B are as follows:

$$F_{B,2} = (H_{B,22} + A_{A,22})^{-1}(H_{A,21}F_1 + H_{A,22}F_2) = (H_2)^{-1}(H_{A,21}F_1 + H_{A,22}F_2). \tag{3.108}$$

Finally, the displacements at points 1 and 2 can be expressed as functions of FRFs and applied forces F_1, F_2 as follows:

$$\begin{aligned} X_1 &= H_{A,11}F_1 + H_{A,12}(F_2 - F_{B,2}) \\ &= H_{A,11}F_1 + H_{A,12}F_2 - H_{A,12}(H_2)^{-1} \cdot (H_{A,21}F_1 + H_{A,22}F_2) \\ &= (H_{A,11} - H_{A,12}(H_2)^{-1}H_{A,21})F_1 + (H_{A,12} - H_{A,12}(H_2)^{-1}H_{A,22})F_2 \\ X_2 &= H_{A,21}F_1 + H_{A,22}(F_2 - F_{B,2}) \\ &= H_{A,21}F_1 + H_{A,22}F_2 - H_{A,22}(H_2)^{-1}(H_{A,21}F_1 + H_{A,22}F_2) \\ &= (H_{A,21} - H_{A,22}(H_2)^{-1}H_{A,21})F_1 + (H_{A,22} - H_{A,22}(H_2)^{-1}H_{A,22})F_2 \end{aligned} \tag{3.109}$$

Equation (3.109) can be rearranged in a matrix form as follows:

$$\begin{Bmatrix} X_1 \\ X_2 \end{Bmatrix} = \begin{bmatrix} (H_{A,11} - H_{A,12}H_2^{-1}H_{A,21}) & (H_{A,12} - H_{A,12}H_2^{-1}H_{A,22}) \\ (H_{A,21} - H_{A,22}H_2^{-1}H_{A,21}) & (H_{A,22} - H_{A,22}H_2^{-1}H_{A,22}) \end{bmatrix} \times \begin{Bmatrix} F_1 \\ F_2 \end{Bmatrix}, \tag{3.110}$$

where $H_2 = H_{A,22} + H_{B,22}$.

Equation (3.110) represents the receptance coupling of the spindle with holder up to the flange and holder-tool stickout structures. The receptances of the free–free-tool holder–tool assembly, $H_{A,11}, H_{A,12}$, and $H_{A,22}$ are modeled by using the finite element model and the receptance of the spindle at point 2, $H_{B,22}$ is obtained through the inverse receptance coupling method as explained next.

The following cross and direct receptances are obtained from Eqs. (3.109) and (3.110):

$$\left.\begin{aligned}\frac{X_1}{F_1} &= H_{11} = H_{A,11} - H_{A,12}(H_2)^{-1}H_{A,21}\\ \frac{X_2}{F_1} &= H_{12} = H_{A,21} - H_{A,22}(H_2)^{-1}H_{A,21}\\ \frac{X_1}{F_2} &= H_{21} = H_{A,21} - H_{A,12}(H_2)^{-1}H_{A,22}\\ \frac{X_2}{F_2} &= H_{22} = H_{A,22} - H_{A,22}(H_2)^{-1}H_{A,22}\end{aligned}\right\}. \tag{3.111}$$

Each FRF contains both translational and rotational displacements, thus Eqs. (3.111) can be expanded as follows:

$$\begin{Bmatrix} x_1 \\ \theta_1 \end{Bmatrix} = \begin{bmatrix} h_{11,ff} & h_{11,fM} \\ h_{11,Mf} & h_{11,MM} \end{bmatrix} \begin{Bmatrix} f_1 \\ M_1 \end{Bmatrix} = [H_{11}] \begin{Bmatrix} f_1 \\ M_1 \end{Bmatrix} \tag{3.112}$$

$$\begin{Bmatrix} x_2 \\ \theta_2 \end{Bmatrix} = \begin{bmatrix} h_{12,ff} & h_{12,fM} \\ h_{12,Mf} & h_{12,MM} \end{bmatrix} \begin{Bmatrix} f_1 \\ M_1 \end{Bmatrix} = [H_{12}] \begin{Bmatrix} f_1 \\ M_1 \end{Bmatrix}$$

$$\begin{Bmatrix} x_1 \\ \theta_1 \end{Bmatrix} = \begin{bmatrix} h_{21,ff} & h_{21,fM} \\ h_{21,Mf} & h_{21,MM} \end{bmatrix} \begin{Bmatrix} f_2 \\ M_2 \end{Bmatrix} = [H_{21}] \begin{Bmatrix} f_2 \\ M_2 \end{Bmatrix}$$

$$\begin{Bmatrix} x_2 \\ \theta_2 \end{Bmatrix} = \begin{bmatrix} h_{22,ff} & h_{22,fM} \\ h_{22,Mf} & h_{22,MM} \end{bmatrix} \begin{Bmatrix} f_2 \\ M_2 \end{Bmatrix} = [H_{22}] \begin{Bmatrix} f_2 \\ M_2 \end{Bmatrix}.$$

By substituting Eqs. (3.112) into Eqs. (3.111), direct and cross-transfer functions at points 1 and 2 with both rotational and translational degrees of freedom are found as follows:

$$\begin{aligned}
[H_{11}] &= \begin{bmatrix} h_{11,ff} & h_{11,fM} \\ h_{11,Mf} & h_{11,MM} \end{bmatrix} = \begin{bmatrix} h_{A11,ff} & h_{A11,fM} \\ h_{A11,Mf} & h_{A11,MM} \end{bmatrix} - \begin{bmatrix} h_{A12,ff} & h_{A12,fM} \\ h_{A12,Mf} & h_{A12,MM} \end{bmatrix} [H_2]^{-1} \begin{bmatrix} h_{A21,ff} & h_{A21,fM} \\ h_{A21,Mf} & h_{A21,MM} \end{bmatrix} \\
[H_{21}] &= \begin{bmatrix} h_{21,ff} & h_{21,fM} \\ h_{21,Mf} & h_{21,MM} \end{bmatrix} = \begin{bmatrix} h_{A21,ff} & h_{A21,fM} \\ h_{A21,Mf} & h_{A21,MM} \end{bmatrix} - \begin{bmatrix} h_{A12,ff} & h_{A12,fM} \\ h_{A12,Mf} & h_{A12,MM} \end{bmatrix} [H_2]^{-1} \begin{bmatrix} h_{A22,ff} & h_{A22,fM} \\ h_{A22,Mf} & h_{A22,MM} \end{bmatrix} \\
[H_{22}] &= \begin{bmatrix} h_{22,ff} & h_{22,fM} \\ h_{22,Mf} & h_{22,MM} \end{bmatrix} = \begin{bmatrix} h_{A22,ff} & h_{A22,fM} \\ h_{A22,Mf} & h_{A22,MM} \end{bmatrix} - \begin{bmatrix} h_{A22,ff} & h_{A22,fM} \\ h_{A22,Mf} & h_{A22,MM} \end{bmatrix} [H_2]^{-1} \begin{bmatrix} h_{A22,ff} & h_{A22,fM} \\ h_{A22,Mf} & h_{A22,MM} \end{bmatrix},
\end{aligned} \tag{3.113}$$

where

$$\begin{aligned}
[H_2]^{-1} &= \left(\begin{bmatrix} h_{A22,ff} & h_{A22,fM} \\ h_{A22,Mf} & h_{A22,MM} \end{bmatrix} + \begin{bmatrix} h_{B22,ff} & h_{B22,fM} \\ h_{B22,Mf} & h_{B22,MM} \end{bmatrix} \right)^{-1} \\
&= \begin{bmatrix} h_{A22,ff} + h_{B22,ff} & h_{A22,Mf} + h_{B22,Mf} \\ h_{A22,Mf} + h_{B22,Mf} & h_{A22,MM} + h_{B22,MM} \end{bmatrix}^{-1} \\
&= \begin{bmatrix} h_{22,ff} & h_{22,Mf} \\ h_{22,Mf} & h_{22,MM} \end{bmatrix}^{-1} \\
&= \frac{1}{h_{22,Mf}^2 - h_{22,MM}h_{22,ff}} \begin{bmatrix} -h_{22,MM} & h_{22,Mf} \\ h_{22,Mf} & -h_{22,ff} \end{bmatrix}.
\end{aligned}$$

The first elements in the three matrices $[H11]$, $[H12]$, and $[H22]$ in Eqs. (3.113) along with the equation for the reciprocity condition yields the following four sets of nonlinear equations:

$$
\begin{aligned}
h_{11,ff} &= h_{A11,ff} - \frac{1}{(h_{2,ff}\cdot h_{2,MM} - h_{2,fM}\cdot h_{2,Mf})}[(h_{A12,ff}\cdot h_{2,MM} - h_{A12,fM}\cdot h_{2,MF})\,h_{A21,ff} + \cdots \\
&\quad \cdots + (h_{A12,fM}\cdot h_{2,ff} - h_{A12,ff}\cdot h_{2,fM})\,h_{A21,Mf}] \qquad (3.114)\\
h_{12,ff} &= h_{A12,ff} - \frac{1}{(h_{2,ff}\cdot h_{2,MM} - h_{2,fM}\cdot h_{2,Mf})}[(h_{A12,ff}\cdot h_{2,MM} - h_{A12,fM}\cdot h_{2,MF})\,h_{A22,ff} + \cdots \\
&\quad \cdots + (h_{A12,fM}\cdot h_{2,ff} - h_{A12,ff}\cdot h_{2,fM})\,h_{A22,Mf}] \\
h_{22,ff} &= h_{A22,ff} - \frac{1}{(h_{2,ff}\cdot h_{2,MM} - h_{2,fM}\cdot h_{2,Mf})}[(h_{A22,ff}\cdot h_{2,MM} - h_{A22,fM}\cdot h_{2,MF})\,h_{A22,ff} + \cdots \\
&\quad \cdots + (h_{A22,fM}\cdot h_{2,ff} - h_{A22,ff}\cdot h_{2,fM})\,h_{A22,Mf}] \\
h_{2,fM} &= h_{2,Mf}.
\end{aligned}
$$

The four unknowns are, $h_{2,ff}, h_{2,fM}, h_{2,Mf}$, and $h_{2,MM}$, which are the receptances of the assembly at point 2 and need to be solved. The terms $h_{11,ff}$, $h_{12,ff}$, and $h_{22,ff}$ are obtained by three impact hammer tests at points 1 and 2. The FRFs of the free–free substructure A are obtained through the finite element method. This system of nonlinear equations is symbolically solved by using MAPLE®. The translational and rotational degrees of freedom of FRFs at point B can be obtained as follows:

$$
\begin{aligned}
h_{B22,ff} &= h_{2,ff} - h_{A22,ff} \\
h_{B22,fM} &= h_{B22,Mf} = h_{2,fM} - h_{A22,fM} \qquad (3.115)\\
h_{B22,MM} &= h_{2,MM} - h_{A22,MM}.
\end{aligned}
$$

The spindle dynamics that include the rotational degrees of freedom are stored in a matrix as shown below:

$$
H_{B,22} = \begin{bmatrix} h_{B22,ff} & h_{B22,fM} \\ h_{B22,Mf} & h_{B22,MM} \end{bmatrix}. \qquad (3.116)
$$

3.8.1 Experimental Procedure

A shrink fit HSK 63 tool holder with a tool is attached to the spindle as shown in Figure 3.21. Three impact modal tests are performed at points 1 and 2: the direct FRF at point 1; $h_{11,ff}$, cross FRF at points 1 and 2; $h_{12,ff}$ and direct FRF measurement at point 2; and $h_{22,ff}$ as shown in Figure 3.20. Structure A, the stickout of the tool-holder, is modeled by using the finite element method based on Timoshenko beams.

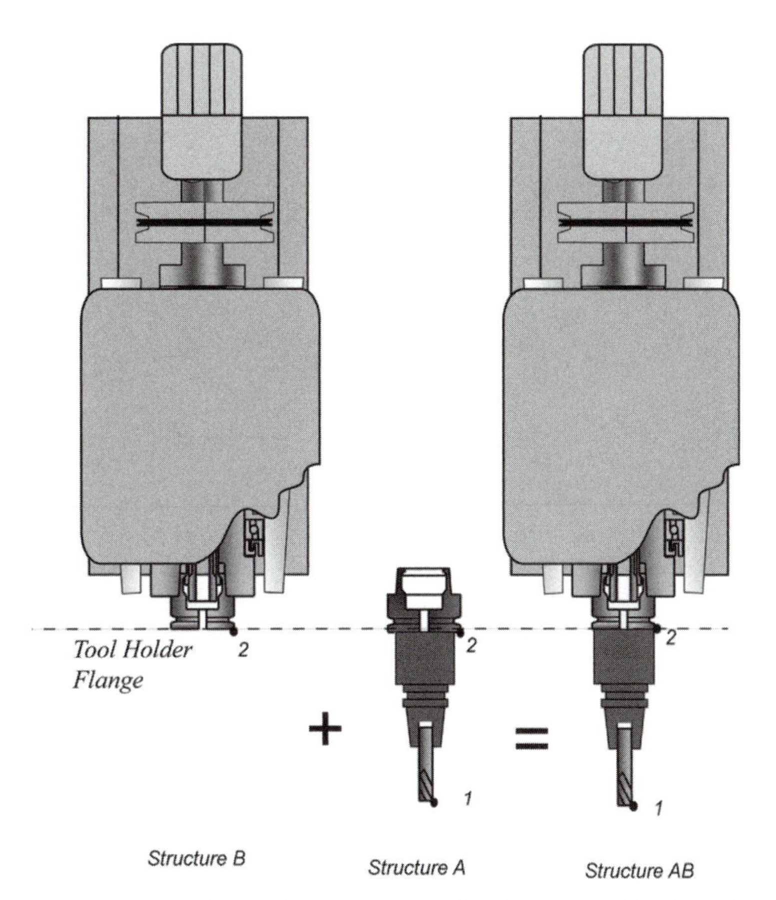

Figure 3.20: Modeling of receptance coupling between tool holder and spindle.

The tool-holder and spindle with HSK 63A taper assembly is coupled by using Eqs. (3.106) and (3.111) as follows:

$$H_{11} = \left[H_{A,11} - H_{A,12} \cdot \left(H_{B,22} + H_{A,22}\right)^{-1}\right] \cdot H_{A,21}. \tag{3.117}$$

The fluted section of the end mill is considered to be 80% of the total diameter in the finite element model and the tool–tool holder connection in the shrink-fit is assumed to be rigid. The tool–tool holder model is in the free–free condition as the rigid body modes play an important role in the coupling between the structures. The damping ratio used for the finite element model is 1 to 3 percent, which was verified by several impact tests.

The proposed receptance coupling method is experimentally evaluated on a horizontal machining center. The spindle with HSK 63 interface was identified by using a short shrink-fit holder with a gauge length of 60 mm first. The FRF of another tool holder with 140 mm gauge length (Fig. 3.21) was estimated by using the proposed receptance coupling technique. The tool–tool holder connection in the shrink-fit is modeled as a rigid connection and the fluted tool is considered to be 80% of the total shank diameter. The predicted FRFs at

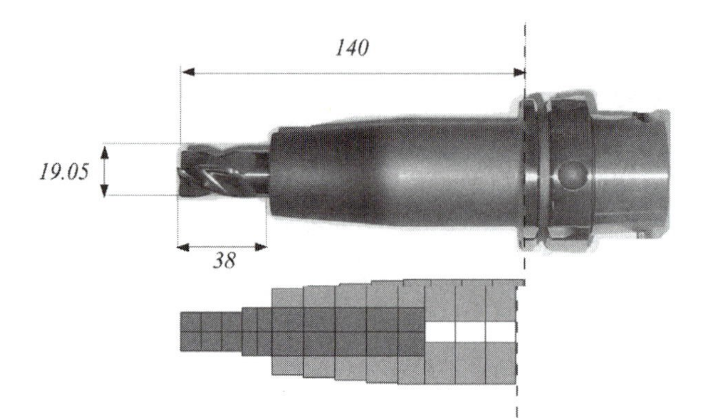

Figure 3.21: Finite element model of HSK 63 tool holder and end mill assembly.

the tool tip are compared with the experiments in both x and y directions, as shown in Figure 3.22, which has acceptable accuracy for use in chatter stability prediction methods. The predicted FRFs are also able to model the higher frequencies more accurately, which are difficult to measure accurately via impulse modal tests because of the loss of spectral strength of the hammers at the high-frequency range dominated by the flexible tool modes.

3.9 PROBLEMS

1. A cylindrical AISI 4340 shaft with an elastic modulus of $E = 200$ GPa is placed between the chuck and the tail stock center of an engine lathe. The diameter of the shaft is $d = 30$ mm, the length of the bar is $L = 20$ mm, and the radial depth of cut is $a = 0.5$ mm. The radial cutting force (F_r) coefficient of the oblique turning tool is $K_r = 500$ MPa, where $F_r = K_r ah$, and h is the chip load.
   a. Plot the dimensional errors left on the whole shaft as the tool is fed from the tail stock center toward the chuck. Assume that the tail stock and chuck supports are rigid.
   b. The stiffnesses of the chuck and tail stock are measured as $k_{sp} = 50{,}000$ N/mm and $k_{ts} = 30{,}000$ N/mm, respectively. Plot the dimensional errors again by including the stiffnesses of both supports.
2. A slender four-fluted high-speed steel (HSS) end mill with a diameter of $d = 19.05$ mm and gauge length of $L = 100$ mm is used in half-immersion up- and down-milling of 7075 aluminum alloy. The tangential and radial force cutting constants are given as $K_t = 1{,}200$ MPa and $K_r = 0.3$, where $F_t = K_t ah$ and $K_r = F_r/F_t$. The elastic modulus of HSS is $E = 204$ GPa. Plot and compare the dimensional errors left on both up- and down-milled surfaces for the axial depth of cuts $a = 10$, 20, and 30 mm.
3. Two small rectangular steel parts, with identical masses of $m_0 = 0.76$ kg, are attached to the middle and the end of a slender steel cantilever beam as shown in Figure 3.18. The dimensions of the steel bar are given as

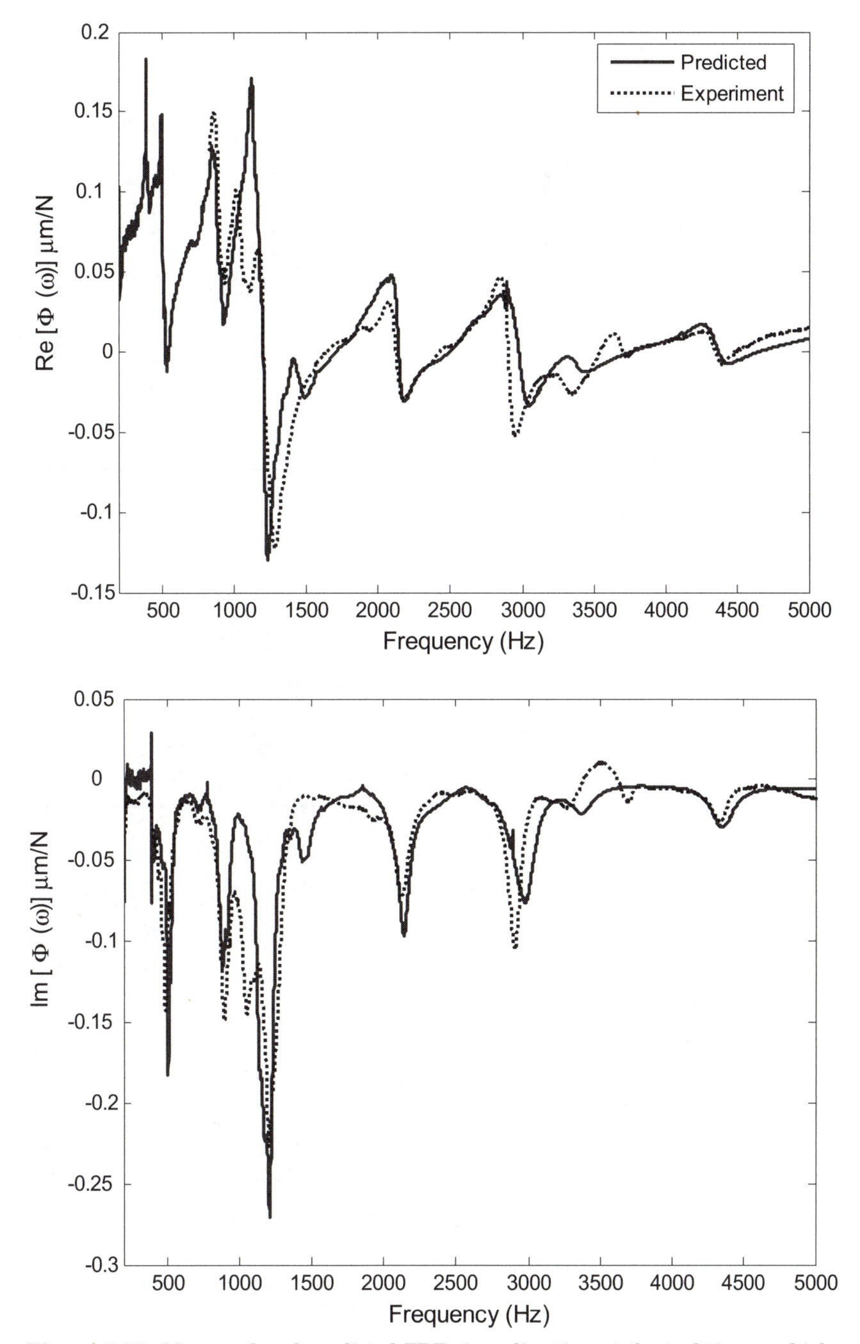

Figure 3.22: Measured and predicted FRFs in x direction at the tool tip on a high-speed machining center.

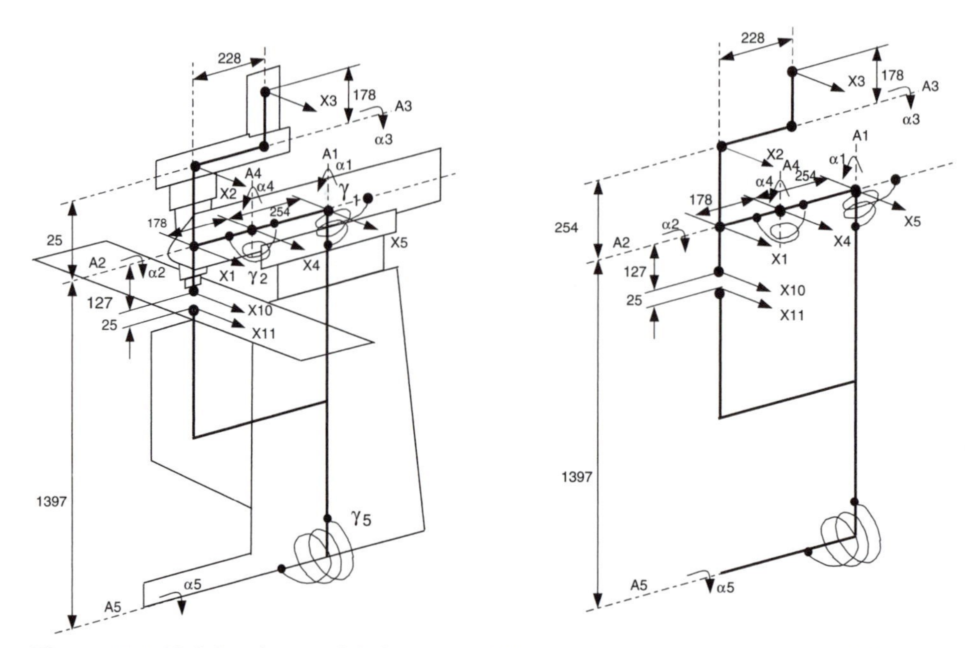

Figure 3.23: Modal analysis model of a vertical milling machine presented by Tlusty and Moriwaki [111]. The measurements are given in millimeters.

$l = 450$ mm, $b = 25.4$ mm, and $h = 5$ mm. Young's modulus (E) and the mass density (ρ) of the steel are $E = 204$ GPa and $\rho = 7{,}860$ kg/m^3. Assume that half of the bar mass is added to the middle and a quarter is added to the end point of the beam, and assume that the assembly is approximated as a 2-DOF system when it vibrates in its most flexible direction x. Develop a comprehensive computer program to solve the following:

a. Obtain the stiffness matrix in local (x) coordinates using the flexibility method.

b. Derive the equation of motion in local coordinates using Newton's law. Obtain the natural frequencies, modal matrix, and mode shapes. Plot the real mode shapes of the beam.

c. Assume that the beam is modeled by two torsional springs at the base and at the middle. Using the torsional design stiffness of the beams at these points, express the equation of motion using the Lagrangian formulation. Use torsional displacements (θ_1, θ_2) as design coordinates. Solve for the natural frequencies and mode shapes. Comment on the differences between the solutions in local and design coordinates.

d. Express the transformation matrix between the local (x) and design (θ) coordinates. Obtain local mass and stiffness matrices using design stiffness, design mass, and coordinate transformation matrices.

e. Express the modal mass and modal stiffness matrices. Predict the free vibrations of masses 1 and 2 when the tip of the bar is displaced 1 mm and released.

4. Assume that the same beam given in the previous question is analyzed by an experimental modal analysis. The measurements are analyzed by an experimental modal analysis software, and the identified parameters of the structure are given in Example 1.
   a. Estimate the local mass, damping, and stiffness matrices.
   b. Estimate the torsional design spring constants.
   c. Plot the direct transfer function at points 1 and 2 and the cross-transfer function.
   d. Compare the transfer functions obtained from analytical and experimental modal analysis methods.
5. A vertical milling machine was modeled by Tlusty and Moriwaki [111] as shown in Figure 3.23. $x_1, \ldots, x_5$ are local coordinates. Design coordinates are the rotational angles $\alpha_1, \ldots, \alpha_5$ with corresponding design torsional springs $\gamma_1, \ldots, \gamma_5$. The design coordinates (α_i) are relative type, which exist as relative rotations around axes $A_1, \ldots, A_5$. From the measurements and analysis made, the local stiffness, mass, and damping matrices are identified and given below.

$[K_x]$: Local Stiffness Matrix [m/N]

+3.0383e+08	−1.4309e+08	+4.7365e+07	−2.6006e+08	+5.9897e+07
−1.4309e+08	+1.2182e+08	−3.7234e+07	+5.8508e+07	−7.9269e+06
+4.7365e+07	−3.7234e+07	+1.7730e+07	−2.7861e+07	−1.8665e+03
−2.6006e+08	+5.8508e+07	−2.7861e+07	+4.2966e+08	−2.0022e+08
+5.9897e+07	−7.9269e+06	−1.8665e+03	−2.0022e+08	+1.6220e+08

$[M_x]$: Local Mass Matrix [kg]

+7.3779e+01	−1.1939e+01	+4.3726e−05	+5.5505e−04	−7.3919e−04
−1.1939e+01	+3.9957e+01	+3.4192e+00	−2.9887e−04	+1.7549e−04
+4.3726e−05	+3.4192e+00	+2.7603e+01	−3.6039e−05	−2.2129e−04
+5.5505e−04	−2.9887e−04	−3.6039e−05	+4.8745e+01	−2.7705e+01
−7.3919e−04	+1.7549e−04	−2.2129e−04	−2.7705e+01	+4.7240e+02

$[C_x]$: Local Damping Matrix [N/m/s]

+8.6796e+03	−4.5561e+03	+1.1930e+03	−3.4018e+03	−1.3886e+03
−4.5561e+03	+5.5794e+03	−1.3945e+03	−5.1046e+02	+4.4820e+02
+1.1930e+03	−1.3945e+03	+1.1798e+03	−8.6024e+01	−1.0541e+03
−3.4018e+03	−5.1046e+02	−8.6024e+01	+9.8393e+03	−5.6767e+03
−1.3886e+03	+4.4820e+02	−1.0541e+03	−5.6767e+03	+1.0013e+04

   a. Compute eigenvalues, eigenvectors, and damping ratios of individual modes for the system. Draw diagrams showing mode shapes imposed on the sketch of the model of the machine. All the significant flexibilities of the machine occur in the directions $x_1, \ldots, x_5$, whereas the flexibilities in the other two directions perpendicular to x are neglected.

b. Now consider the relative motion between tool x_{10} and workpiece x_{11}. Assume that the table and knee are rigidly connected with the column and participate with x_5 in the freedom α_5. The tool is a rigid extension of the rigid connection between coordinates x_1 and x_2. With this in mind, replace the coordinate system $x_1, \ldots, x_5$ by the coordinate system $x_{10}, x_2, x_3, x_4, x_{11}$. Transform the local k_x, m_x, and c_x to this new coordinate system. Compute the relative transfer function between points x_{10} and x_{11}. Plot the real part and Nyquist (i.e., polar) plot of the relative tool–workpiece transfer function.
c. As an extension to the identification problem, write the transformation matrix $[C]$ between the local and design coordinates, and transform the local stiffness $[K_x]$ to $[K_\alpha]$ to find the design stiffness of the five torsional springs.

CHAPTER FOUR

MACHINE TOOL VIBRATIONS

4.1 INTRODUCTION

Machine tools experience both forced and self-excited vibrations during machining operations. The cutting forces can be periodic, as in the case of milling. The nonsymmetric teeth in drilling, unbalance, or shaft runout in turning and boring can also produce periodically varying cutting forces. In all cases, the cutting forces can be periodic at tooth- or spindle-passing frequencies, which may have strong harmonics up to four to five times the tooth- or spindle-passing frequencies. If any of the harmonics coincide with one of the natural frequencies of the machine and/or workpiece structure, the system exhibits forced vibrations. The forced vibrations can simply be solved by applying the predicted cutting or disturbance forces on the transfer function of the structure by the use of the solution of ordinary differential equations in the time domain. However, self-excited, chatter vibrations are the most detrimental for the safety and quality of the machining operations, which are covered in this chapter.

Machine tool chatter vibrations result from a self-excitation mechanism in the generation of chip thickness during machining operations. One of the structural modes of the machine tool–workpiece system is initially excited by cutting forces. A wavy surface finish left during the previous revolution in turning, or by a previous tooth in milling, is removed during the succeeding revolution or tooth period, which also leaves a wavy surface owing to structural vibrations [112]. Depending on the phase shift between the two successive waves, the maximum chip thickness may grow exponentially while oscillating at a chatter frequency that is close to, but not equal to, a dominant structural mode in the system. The growing vibrations increase the cutting forces and may chip the tool and produce a poor, wavy surface finish. The self-excited chatter vibrations may be caused by mode-coupling or regeneration of the chip thickness [114]. The mode-coupling chatter occurs when there are vibrations in two directions in the plane of cut. The regenerative chatter results from phase differences between the vibration waves left on both sides of the chip and occurs earlier than the mode-coupling chatter in most machining cases. Hence, the fundamentals of regenerative chatter vibrations are explained in the following section by the use of a simple orthogonal cutting process as an example. However, when the cross-coupling of vibration modes is considered, the mode coupling is inherently covered by the stability models presented in the chapter.

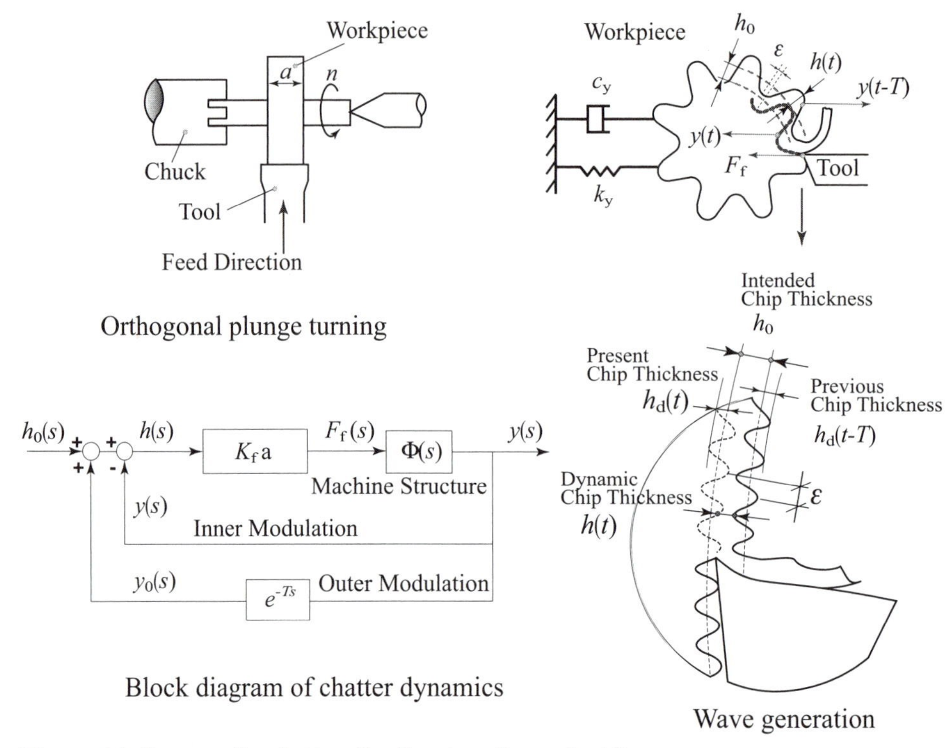

Figure 4.1: Regenerative chatter vibrations in orthogonal cutting.

4.2 STABILITY OF REGENERATIVE CHATTER VIBRATIONS IN ORTHOGONAL CUTTING

The fundamentals of regenerative chatter in classical orthogonal cutting are first presented by neglecting the process damping. The dimensionless analysis of the stability is derived next to illustrate the relationship between the lobes, spindle speed, and damping ratio. The effect of process damping at lower cutting speeds is also explained analytically and experimentally.

4.2.1 Stability of Orthogonal Cutting

Assume that a flat-faced orthogonal grooving tool is fed perpendicular to the axis of a cylindrical shaft held between the chuck and the tail stock center of a lathe (see Fig. 4.1). The shaft is flexible in the direction of feed, and the feed cutting force (F_f) causes it to vibrate. The initial surface of the shaft is smooth without waves during the first revolution, but the tool starts leaving a wavy surface behind because of the bending vibrations of the shaft in the feed direction y, which is in the direction of the radial cutting force (F_f). When the second revolution starts, the surface has waves both inside the cut where the tool is cutting (i.e., inner modulation, $y(t)$) and on the outside surface of the cut owing to vibrations during the previous revolution of cut (i.e., outer modulation, $y(t-T)$). Hence, the resulting dynamic chip thickness is no longer constant, but it varies as a function of vibration frequency and the speed of the workpiece. The general dynamic chip thickness can be expressed as follows:

$$h(t) = h_0 - [y(t) - y(t - T)], \tag{4.1}$$

where h_0 is the intended chip thickness, which is equal to the feed rate of the machine, and $[y(t) - y(t - T)]$ is the dynamic chip thickness produced owing to vibrations at the present time t and one spindle revolution period (T) before. Assuming that the workpiece is approximated as a single–degree-of-freedom system in the radial direction, the equation of motion of the system can be expressed as the following:

$$\begin{aligned} m_y\ddot{y}(t) + c_y\dot{y}(t) + k_y y(t) &= F_f(t) = K_f a h(t) \\ &= K_f a[h_0 + y(t - T) - y(t)], \end{aligned} \tag{4.2}$$

where the feed cutting force is proportional to the cutting constant in the feed direction (K_f), with width of cut a and the dynamic chip load $h(t)$. Because the forcing function on the right-hand side depends on the present and past solutions of vibrations $(y(t), y(t - T))$ on the left side of the equation, the chatter vibration expression is a delay differential equation. Furthermore, if the vibration is too large (i.e., $y(t) - y(t - T) > h_0$), then the tool jumps out of cut, thus producing a zero chip thickness and zero cutting force. In addition, the influence of vibration marks left on the surface during the previous revolutions may further complicate the computation of exact chip thickness because of the tool jumping out of cut at various instances and revolutions (i.e., a multiple regenerative effect). The cutting constant K_f may also change depending on the magnitude of instantaneous chip thickness and the orientation of the vibrating tool or workpiece, which adds an additional difficulty in the dynamic cutting process. When the flank face of the tool rubs against the wavy surface left behind, additional process damping is added to the dynamic cutting process, and this attenuates the chatter vibrations. Because the whole process is too complex and nonlinear to model correctly analytically, time domain, numerical methods are widely used to simulate the chatter vibrations in machining. However, a clear understanding of chatter stability is still important and is best explained by using a linear stability theory. The stability of chatter vibrations is analyzed by using the linear theory of Tobias [113], Tlusty and Polacek [112], and Merrit [77]. Nonlinearities such as the tool jumping out of cut [110], multiple regeneration, process damping, and nonlinear cutting constant are neglected in linear stability analysis.

The chatter vibration system can be represented by the block diagram shown in Figure 4.1, where the parameters of the dynamic cutting process are shown in the Laplace domain. Input to the system is the desired chip thickness h_0, and the output of the feedback system is the current vibration $y(t)$ left on the inner surface. In the Laplace domain, $y(s) = \mathcal{L}y(t)$, and the vibration imprinted on the outer surface during the previous revolution is $e^{-sT}y(s) = \mathcal{L}y(t - T)$, where

T is the spindle period. The dynamic chip thickness in the Laplace domain is

$$h(s) = h_0 - y(s) + e^{-sT}y(s) = h_0 + (e^{-sT} - 1)y(s), \tag{4.3}$$

which produces a dynamic cutting force of

$$F_\mathrm{f}(s) = K_\mathrm{f} a h(s). \tag{4.4}$$

The cutting force excites the structure and produces the current vibrations as follows:

$$y(s) = F_\mathrm{f}(s)\Phi(s) = K_\mathrm{f} a h(s)\Phi(s), \tag{4.5}$$

where the transfer function of the single degree of workpiece structure is

$$\Phi(s) = \frac{y(s)}{F_\mathrm{f}(s)} = \frac{\omega_\mathrm{n}^2}{k_y\left(s^2 + 2\zeta\omega_\mathrm{n} s + \omega_\mathrm{n}^2\right)}.$$

Substituting $y(s)$ into $h(s)$ yields

$$h(s) = h_0 + (e^{-sT} - 1)K_\mathrm{f} a h(s)\Phi(s),$$

and the resulting transfer function between the dynamic and reference chip loads becomes

$$\frac{h(s)}{h_0(s)} = \frac{1}{1 + (1 - e^{-sT})K_\mathrm{f} a\Phi(s)}. \tag{4.6}$$

The stability of this closed-loop transfer function is determined by the roots (s) of its characteristic equation, that is,

$$1 + (1 - e^{-sT})K_\mathrm{f} a\Phi(s) = 0.$$

Let the root of the characteristic equation be $s = \sigma + j\omega_\mathrm{c}$. If the real part of the root is positive ($\sigma > 0$), the time domain solution will have an exponential term with positive power (i.e., $e^{+|\sigma|t}$). The chatter vibrations will grow indefinitely, and the system will be unstable. A negative real root ($\sigma < 0$) will suppress the vibrations with time (i.e., $e^{-|\sigma|t}$), and the system is stable with chatter vibration–free cutting. When the real part is zero ($s = j\omega_\mathrm{c}$), the system is critically stable, and the workpiece oscillates with a constant vibration amplitude at chatter frequency ω_c. Note that the chatter vibration frequency does not equal the natural frequency of the structure, because the characteristic equation of the dynamic cutting process has additional terms beyond the structure's transfer function. However, the chatter vibration frequency is still close to the natural mode of the structure. For critical borderline stability analysis ($s = j\omega_\mathrm{c}$), the characteristic function becomes

$$1 + (1 - e^{-j\omega_\mathrm{c} T})K_\mathrm{f} a_\mathrm{lim}\Phi(j\omega_\mathrm{c}) = 0, \tag{4.7}$$

where a_{lim} is the maximum axial depth of cut for chatter vibration–free machining. The transfer function can be partitioned into real and imaginary parts (i.e., $\Phi(j\omega_c) = G + jH$). Rearranging the characteristic equation with real and complex parts yields

$$\{1 + K_f a_{\mathrm{lim}}[G(1 - \cos\omega_c T) - H \sin\omega_c T]\} \\ + j\{K_f a_{\mathrm{lim}}[G \sin\omega_c T + H(1 - \cos\omega_c T)]\} = 0.$$

Both real and imaginary parts of the characteristic equation must be zero. If the imaginary part is considered first, then

$$G \sin\omega_c T + H(1 - \cos\omega_c T) = 0$$

and

$$\tan\psi = \frac{H(\omega_c)}{G(\omega_c)} = \frac{\sin\omega_c T}{\cos\omega_c T - 1}, \tag{4.8}$$

where ψ is the phase shift of the structure's transfer function. Using the trigonometric identity $\cos\omega_c T = \cos^2(\omega_c T/2) - \sin^2(\omega_c T/2)$ and $\sin\omega_c T = 2\sin(\omega_c T/2)\cos(\omega_c T/2)$, we have

$$\tan\psi = \frac{\cos(\omega_c T/2)}{-\sin(\omega_c T/2)} = \tan[(\omega_c T)/2 - (3\pi)/2]$$

and

$$\omega_c T = 3\pi + 2\psi \rightarrow \psi = \tan^{-1}\frac{H}{G}. \tag{4.9}$$

Note that calculation of the phase angle ψ from the transfer function must be correctly done with digital computers, as explained in Figure 4.2.

The spindle speed (n [rev/s]) and the chatter vibration frequency (ω_c) have a relationship that affects the dynamic chip thickness. Let us assume that the chatter vibration frequency is ω_c [rad/s] or f_c [Hz]. The number of vibration waves left on the surface of the workpiece is

$$f_c\,[\mathrm{Hz}] \cdot T\,[\mathrm{s}] = \frac{f_c}{n} = k + \frac{\epsilon}{2\pi}, \tag{4.10}$$

where k is the integer number of waves and $\epsilon/2\pi$ is the fractional wave generated. The angle ϵ represents the phase difference between the inner and outer modulations. Note that, if the spindle and vibration frequencies have an integer ratio, the phase difference between the inner and outer waves on the chip surface will be zero or 2π; hence, the chip thickness will be constant despite the presence of vibrations. In this case, the inner ($y(t)$) and outer ($y(t-T)$) waves are parallel to each other. If the phase angle is not zero, the chip thickness changes continuously. Consider (k') the integer number of full vibration cycles and the phase shift

$$2\pi f_c T = 2k\pi + \epsilon, \tag{4.11}$$

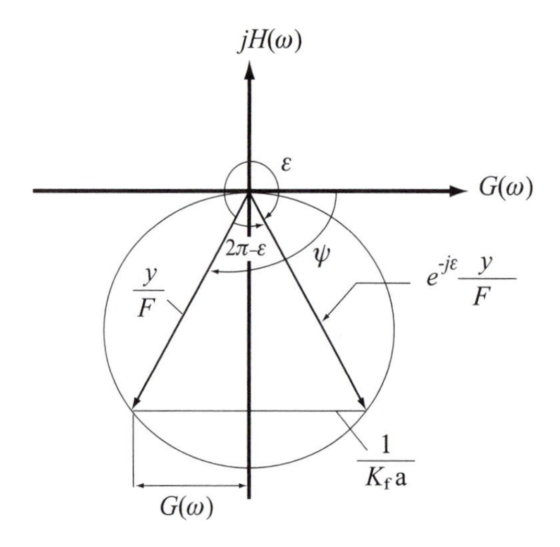

$G > 0, H < 0 \rightarrow \psi = -\tan^{-1}|(H/G)|.$
$G < 0, H < 0 \rightarrow \psi = -\pi + \tan^{-1}|(H/G)|.$
$G < 0, H > 0 \rightarrow \psi = -\pi - \tan^{-1}|(H/G)|.$
$G > 0, H > 0 \rightarrow \psi = -2\pi + \tan^{-1}|(H/G)|.$

Figure 4.2: Calculation of phase angle ψ from the polar plots.

where the phase shift between the inner and outer waves is $\epsilon = 3\pi + 2\psi$. The corresponding spindle period (T [s]) and speed (n [rev/min]) is found to be

$$T = \frac{2k\pi + \epsilon}{2\pi f_c} \rightarrow n = \frac{60}{T}. \tag{4.12}$$

The critical axial depth of cut can be found by equating the real part of the characteristic equation to zero as follows:

$$1 + K_f a_{lim}[G(1 - \cos\omega_c T) - H \sin\omega_c T] = 0,$$

or

$$a_{lim} = \frac{-1}{K_f G[(1 - \cos\omega_c T) - (H/G)\sin\omega_c T]}.$$

Substituting $H/G = (\sin\omega_c T)/(\cos\omega_c T - 1)$ and rearranging this equation yields

$$a_{lim} = \frac{-1}{2K_f G(\omega_c)}. \tag{4.13}$$

Note that, because the depth of cut is a physical quantity, the solution is valid only for negative values of the real part of the transfer function ($G(\omega_c)$). The chatter vibrations may occur at any frequency where $G(\omega_c)$ is negative. If a_{lim} is selected by using the minimum value of $G(\omega_c)$, the avoidance of chatter is guaranteed at any spindle speed. The expression indicates that the axial depth of cut is inversely proportional to the flexibility of the structure and to the cutting constant of the workpiece material. The harder the work material is, the larger the cutting constant K_f will be, thus reducing the axial depth of cut. Similarly, flexible machine tool or workpiece structures will also reduce the axial depth of cut or the productivity.

The above stability expression was first obtained by Tlusty and Polacek [112]. Tobias [113] and Merrit [77] presented similar solutions. Tobias presented stability charts indicating chatter vibration–free spindle speeds and axial depth of cuts. Assuming that the transfer function of the structure at the cutting point (Φ) and cutting constant K_f are known or measured, the procedure for plotting the stability lobes can be summarized in the following:

- Select a chatter frequency (ω_c) at the negative real part of the transfer function.
- Calculate the phase angle of the structure at ω_c (Eq. 4.8).
- Calculate the critical depth of cut from Eq. (4.13).
- Calculate the spindle speed from Eq. (4.12) for each stability lobe $k = 0, 1, 2, \ldots$.

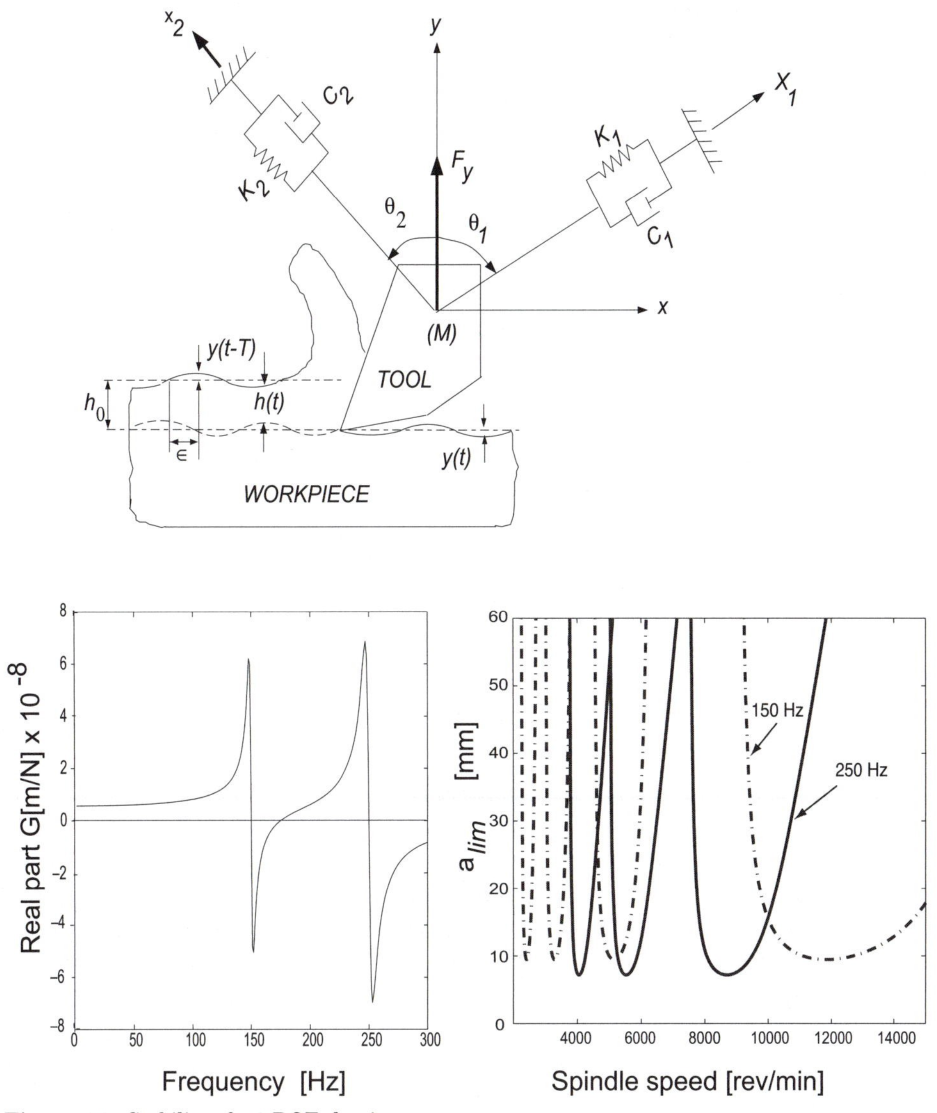

Figure 4.3: Stability of a 2-DOF shaping process.

- Repeat the procedure by scanning the chatter frequencies around the natural frequency of the structure.

If the structure has multiple degrees of freedom, a transfer function of the system oriented in the direction of chip thickness must be considered for Φ. The computation of oriented transfer function was given in Section 3.4.2. In that case, the negative real part of the complete transfer function around all dominant modes must be scanned by using the same procedure outlined for the orthogonal cutting process.

Example. Stability of a Shaping System with Two Degrees of Freedom. A shaping process with a two–degree-of-freedom (2-DOF) system is shown in Figure 4.3.

The system flexibilities in directions x_1 and x_2 are given as

$$x_1 \leftarrow \omega_{n1} = 250\,\text{Hz}, \quad \zeta_1 = 1.2\%, \quad k_1 = 2.26 \times 10^8\,\text{N/m},$$
$$x_2 \leftarrow \omega_{n2} = 150\,\text{Hz}, \quad \zeta_2 = 1.0\%, \quad k_2 = 2.13 \times 10^8\,\text{N/m}.$$

The cutting force is $F_y = K_f a h(t)$, where the cutting constant $K_f = 1{,}000$ MPa. The flexibilities are oriented with $\theta_1 = 30^\circ$ and $\theta_2 = -45^\circ$ from the y axis. Plot the stability lobe of the system.

Answer. The real part of the oriented transfer function between the displacement in the y direction and the cutting force F_y is

$$\begin{aligned}\Re\Phi(j\omega) = G(j\omega) &= \frac{y(j\omega)}{F_y(j\omega)} \\ &= \cos^2\theta_1 \frac{1-r_1^2}{k_1\left[\left(1-r_1^2\right)^2 + (2\zeta_1 r_1)^2\right]} + \cos^2\theta_2 \frac{1-r_2^2}{k_2\left[\left(1-r_2^2\right)^2 + (2\zeta_2 r_2)^2\right]},\end{aligned}$$

where $r_{1,2} = \omega/\omega_{n1,2}$. The oriented transfer function and the corresponding stability lobes are given in Figure 4.3.

4.2.2 Dimensionless Analysis of Stability Lobes in Orthogonal Cutting

The stability lobes are used to identify the chatter free, productive cutting conditions. The understanding of the governing physics behind the lobes, and their mathematical relationship with the dynamics of the machine are important to apply the theory to various machining operations and machine tool design [7]. The same chatter stability theory given in the previous section is presented in a different mathematical format by Insperger and Stepan [53] in the following derivations where the speed is normalized with the natural frequency of a single–degree-of-freedom system.

The equation of motion for a dynamic cutting process represented by Eq. (4.2) is expressed as follows:

$$\frac{d^2y(t)}{dt^2} + 2\zeta\omega_n \frac{dy(t)}{dt} + \omega_n^2 y(t) = \frac{\omega_n^2}{k_y} K_f a \left[h_0 - y(t) + y(t-T)\right]. \tag{4.14}$$

Let the static chip load be dropped ($h_0 = 0$) to consider only the dynamic component that affects the stability. To transform the system dynamics into a dimensionless form, let time t be replaced by $t = \tau/\omega_n$ as follows:

$$\tau \to \omega_n t, \quad \delta = \omega_n T, \qquad d\tau = \omega_n dt \to \frac{d\tau}{dt} = \omega_n$$

$$\frac{dy}{dt} = \frac{dy}{d\tau}\frac{d\tau}{dt} = \frac{dy}{d\tau}\omega_n, \quad \frac{d^2y}{dt^2} = \frac{d}{dt}\left(\frac{dy}{d\tau}\omega_n\right) = \frac{d^2y}{d\tau^2}\omega_n^2. \tag{4.15}$$

By substituting angular distances $\delta = \omega_n T$, $\tau \to \omega_n t$ into the dynamic cutting equation,

$$\omega_n^2 \frac{d^2 y(\tau)}{d\tau^2} + 2\zeta\omega_n^2 \frac{dy(\tau)}{d\tau} + \omega_n^2 y(\tau) = \frac{\omega_n^2}{k_y} K_f a\left[-y(\tau) + y(\tau - \delta)\right]$$

$$\frac{d^2 y(\tau)}{d\tau^2} + 2\zeta \frac{dy(\tau)}{d\tau} + \left[1 + \frac{K_f a}{k_y}\right] y(\tau) = \frac{K_f a}{k_y}\left[y(\tau - \delta)\right]$$

$$\frac{d^2 y(\tau)}{d\tau^2} + 2\zeta \frac{dy(\tau)}{d\tau} + \left[1 + w\right] y(\tau) = w\left[y(\tau - \delta)\right], \tag{4.16}$$

where the dimensionless gain of the system is $w = K_f a / k_y$. The Laplace transform of the delayed system is

$$s^2 y(s) + 2\zeta s y(s) + \left[1 + w\right] y(s) = w e^{-\delta s} y(s), \tag{4.17}$$

which leads to the characteristic equation of dynamic cutting as follows:

$$s^2 + 2\zeta s + \left[1 + w\right] - w e^{-\delta s} = 0 \tag{4.18}$$

To find the critically stable cutting conditions, the roots of the characteristic equation (Eq. 4.18) must have zero real part, and the imaginary part must indicate the corresponding vibration frequency (ω_c) during machining. By substituting $s = ir$, where $r = \omega_c / \omega_n$ is the dimensionless, ratio of chatter frequency over natural frequency at the critical stability condition, and $e^{-i\delta r} = \cos \delta r - i \sin \delta r$, the characteristic equation becomes

$$-r^2 + i2\zeta r + \left[1 + w\right] - w \cos \delta r + iw \sin \delta r = 0 \tag{4.19}$$

$$\text{Real part} \qquad 1 - r^2 + w = w \cos \delta r$$

$$\text{Imaginary part} \qquad 2\zeta r = -w \sin \delta r$$

By taking the squares of both real and imaginary parts and summing them ($(1 - r^2 + w)^2 + (2\zeta r)^2 = w^2$) leads to the dimensionless gain of

$$w = -\frac{\left(1 - r^2\right)^2 + 4\zeta^2 r^2}{2(1 - r^2)}.$$

The physical depth of cut at the critically stable conditions, i.e., on the stability lobes, are given by

$$w = \frac{K_f a}{k_y} \to a = -\frac{k_y}{K_f} \frac{\left(1 - r^2\right)^2 + 4\zeta^2 r^2}{2(1 - r^2)}. \tag{4.20}$$

Substituting $\sin\delta r = 2\tan\frac{\delta r}{2}/\left[1+\tan^2(\frac{\delta r}{2})\right]$ into the imaginary part ($2\zeta r = -w\sin\delta r$) of the characteristic Eq. (4.19),

$$w = -\frac{2\zeta r}{\sin\delta r} = -\frac{1+\tan^2(\frac{\delta r}{2})}{\tan\frac{\delta r}{2}}\zeta r,$$

which is substituted again into the real part ($1 - r^2 + w = w\cos\delta r$) of the characteristic Eq. (4.19) with $\cos\delta r = (1-\tan^2\frac{\delta r}{2})/(1+\tan^2\frac{\delta r}{2})$ as follows:

$$1 - r^2 - \frac{1+\tan^2(\frac{\delta r}{2})}{\tan\frac{\delta r}{2}}\zeta r = -\frac{1+\tan^2(\frac{\delta r}{2})}{\tan\frac{\delta r}{2}}\zeta r\frac{1-\tan^2\frac{\delta r}{2}}{1+\tan^2(\frac{\delta r}{2})}$$

$$(1-r^2)\tan\frac{\delta r}{2} = \left[1+\tan^2(\frac{\delta r}{2}) - 1 + \tan^2\frac{\delta r}{2}\right]\zeta r$$

$$(1-r^2)\tan\frac{\delta r}{2} = \left[2\tan^2(\frac{\delta r}{2})\right]\zeta r \rightarrow \tan(\frac{\delta r}{2}) = \frac{1-r^2}{2\zeta r}$$

$$\delta = \frac{2}{r}\left[\tan^{-1}\left(\frac{1-r^2}{2\zeta r}\right) + k\pi\right]$$

$$\tan\psi = \frac{-2\zeta r}{1-r^2} \rightarrow \delta = \frac{2}{r}\left[\tan^{-1}(-\cot(\psi)) + k\pi\right]$$

$$\delta = \frac{2}{r}\left[\tan^{-1}\left(-\tan(\frac{\pi}{2}-\psi)\right) + k\pi\right]$$

$$\delta = \frac{2}{r}\left[\psi + \frac{2k-1}{2}\pi\right] \rightarrow \psi = \tan^{-1}\frac{-2\zeta r}{1-r^2}.$$

The corresponding spindle speed on the stability lobe can be found by switching from the dimensionless time delay δ to the real time delay in seconds as follows:

$$\text{Time delay } T = \frac{\delta}{\omega_n} = \frac{2}{\omega_n r}\left[\tan^{-1}\left(\frac{1-r^2}{2\zeta r}\right) + k\pi\right] = \frac{2}{\omega_n r}\left[\psi + \frac{2k-1}{2}\pi\right]$$

$$2\pi n\ [\text{rev/s}]/\omega_n = \frac{2\pi}{\delta} = \frac{2\pi r}{2\left[\tan^{-1}\left(\frac{1-r^2}{2\zeta r}\right) + k\pi\right]} = \frac{2\pi r}{2\left[\psi + \frac{2k-1}{2}\pi\right]}$$

$$\text{Spindle speed } n\ [\text{rev/s}] = \frac{r\omega_n}{2\left[\tan^{-1}\left(\frac{1-r^2}{2\zeta r}\right) + k\pi\right]} = \frac{r\omega_n}{2\left[\psi + \frac{2k-1}{2}\pi\right]}$$

$$\text{Regenerative phase } \varepsilon = \omega T = 2\left[\psi + \frac{2k-1}{2}\pi\right] = 2\psi + (2k-1)\pi.$$

The stability lobes can be derived as follows:

$$\text{The excitation ratio } r = \frac{\omega_c}{\omega_n}$$

$$\text{Critical depth of cut } a = -\frac{k_y}{K_t}\frac{\left(1-r^2\right)^2 + 4\zeta^2 r^2}{2(1-r^2)}$$

$$\text{Spindle speed } n \text{ [rev/s]} = \frac{r\omega_n}{2\left[\tan^{-1}\left(\frac{1-r^2}{2\zeta r}\right) + k\pi\right]} \rightarrow k = 1, 2, \ldots$$

At $r = \omega_c/\omega_n = \sqrt{1+2\zeta}$, the minimum critical depth of cut becomes equal to Tlusty's simple formula $a_{cr} = 2k_y\zeta(1+\zeta)/K_t \approx 2k_y\zeta/K_t$, where $\zeta^2 \approx 0$. Note that the lobes become asymptotically tangent to the absolute, dimensionless depth of cut of $w = 2\zeta(1+\zeta)$ at spindle speeds $n = (\frac{4}{3}, \frac{4}{7}, \frac{4}{11}, \ldots)\omega_n$, where the critical chatter frequency is $\omega_c = \omega_n\sqrt{1+2\zeta}$. By letting $r = \sqrt{1+2\zeta}$, the spindle speeds where the axial depth of cut is minimum can be found.

$$\text{Spindle speed } n \text{ [rev/s]} = \frac{r\omega_n}{2\left[\tan^{-1}\left(\frac{1-1-2\zeta}{2\zeta\sqrt{1+2\zeta}}\right) + k\pi\right]}$$

$$n \text{ [rev/s]} \approx \frac{\omega_n}{2\left[-\frac{\pi}{4} + k\pi\right]} = \frac{\omega_n}{2\pi}\frac{4}{\left[4k-1\right]} \rightarrow k = 1, 2, 3, \ldots$$

$$n \text{ [rev/s]} \approx \frac{\omega_n}{2\pi}\left\{\frac{4}{3}, \frac{4}{7}, \frac{4}{11}, \ldots\right\} = \frac{\omega_n}{2\pi}\{1.333, 0.5714, 0.3636, \ldots\}$$

Example 4. Stability lobes $\zeta = 0.01$.

$r = \omega_c/\omega_n$	$w = -\frac{(1-r^2)^2+4\zeta^2r^2}{2(1-r^2)}$	k	ε	n/ω_c
∞	∞	1	π	200
1.333	0.39	1	π	2.6
$\sqrt{1+2\zeta}$	0.02	1	$\frac{3\pi}{2}$	4/3
1	∞	1	2π	1
100	5000	2	π	66
1.33	0.39	2	π	6.66
$\sqrt{1+2\zeta}$	0.02	2	$\frac{3\pi}{2}$	4/7
1	∞	2	4π	1/2
$\sqrt{1+2\zeta}$	0.02	3	$\frac{3\pi}{2}$	4/11

4.2.3 Chatter Stability of Orthogonal Cutting with Process Damping

The schematic of a simple model of orthogonal cutting is given in Figure 4.4. The traditional regenerative cutting force $(F_x(t))$ at time t is expressed with the

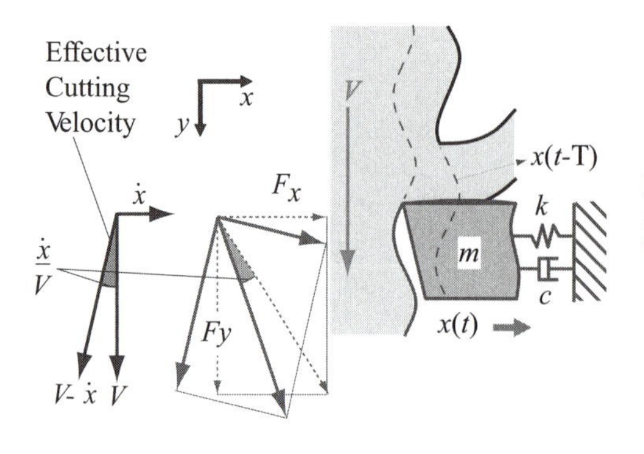

Figure 4.4: The velocity effect on dynamic cutting forces, according to Das and Tobias [42].

velocity effect by Das and Tobias [42] as follows:

$$m\ddot{x}(t) + c\dot{x}(t) + kx(t) = F_x(t)$$

$$F_x(t) = K_f a[h_0 + x(t-T) - x(t)] - K_t a h_0 \frac{\dot{x}}{V}, \tag{4.21}$$

where $x(t)$ and $x(t-T)$ are the inner and outer vibrations, and K_f and K_t are the static cutting force coefficients in feed and cutting speed directions, respectively. V is the cutting velocity, a is the width of cut, h_0 is the feed per revolution, and T is the time delay between the inner and outer vibration waves. The velocity term $((dx/dt)/V)$ introduced by Das and Tobias increases the damping in the system $(c + K_t a h_0 (dx/dt)\ /V$ instead of $c)$ at low cutting speeds. Although Das and Tobias's modified dynamic cutting process model leads to increased stability at the low speeds, it considers only the direction of change in the velocity term and does not consider flank contact. Instead of using the static force coefficient (K_t), the influence of the flank–wave contact is modeled by considering both slope of the waves [12],

$$F_x(t) = a\left\{K_f\left[h_0 - x(t) + x(t-T)\right] - C_i\frac{dx/dt}{V}\right\}, \tag{4.22}$$

or in the Laplace domain,

$$F_x(t) = a\left\{K_f\left[h_0 - \left(1 - e^{-sT}\right)x(s)\right] - \frac{C_i}{V}sx(s)\right\}, \tag{4.23}$$

where the slope (dx/du) is related to the velocity term as follows:

$$\frac{dx}{du} = \frac{dx}{dt}\frac{1}{du/dt} = \frac{dx/dt}{V} \tag{4.24}$$

Although the static cutting force coefficients (K_t) can be easily identified from vibration-free orthogonal cutting tests, the velocity-dependent cutting force coefficient (C_i) can only be identified from a set of dynamic cutting tests, because they represent the contact between the wavy surface material and the flank face of the tool. Alternatively, the contact forces can be predicted by using Hertzian models, but not as accurately as experimental identification from cutting tests. When the cutting tool experiences a harmonic motion $(x(t) = Xe^{j\omega t})$ with an amplitude X and frequency ω during vibrations, the dynamic cutting force equation becomes

$$F_x(t) = K_f a h_0 + aXe^{j\omega t}\left\{-K_f\left[1-e^{-j\omega T}\right] - jC_i\frac{\omega}{V}\right\}. \tag{4.25}$$

The length of the vibration wave imprinted on the surface is $\lambda = V(2\pi/\omega)$, which leads to $\omega/V = 2\pi/\lambda$. The modified dynamic cutting force expression becomes

$$F_x(t) = K_f a h_0 + aXe^{j\omega t}\left\{-K_f\left[1-e^{-j\omega T}\right] - j\frac{2\pi}{\lambda}C_i\right\}, \tag{4.26}$$

which correlates the process damping forces to the vibration wavelength (λ) or the ratio of vibration frequency over cutting speed (ω/V). The effect of cutting speed on the dynamic cutting force can be analyzed from Eq. (4.26). Considering that the critical vibration frequency remains close to the natural frequency of the structure as the cutting speed increases, the wavelength of the vibration marks becomes larger and the flank contact reduces. In other words, the effect of the velocity and acceleration terms in the dynamic cutting force model reduces at high cutting speeds. The wavelength (4.26) decreases, and the process damping increases as the cutting speed decreases. The dynamic cutting force coefficients of materials are identified from a series of cutting tests conducted with a specially instrumented, piezo-actuated dynamic test rig, as explained by Altintas et al. [12].

By neglecting the static cutting force ($K_f a h_0$), the dynamic cutting process equation is derived in the Laplace domain as follows:

$$\left[ms^2 + cs + k\right]x(s) = F_x(s) \rightarrow \Phi_x(s) = \frac{x(s)}{F_x(s)} = \frac{1}{ms^2+cs+k} \tag{4.27}$$

$$F_x(s) = a\left[-K_f\left(1-e^{-sT}\right) - \frac{C_i}{V}s\right]x(s) \rightarrow x(s) = \Phi_x(s)F_x(s), \tag{4.28}$$

which leads to the following characteristic equation:

$$1 + a\left[K_f\left(1-e^{-sT}\right) + \frac{C_i}{V}s\right]\Phi_x(s) = 0. \tag{4.29}$$

By replacing $s = j\omega_c$ and separating the characteristic equation of the system into real and imaginary parts, the stability can be investigated in the frequency domain as follows:

$$\begin{aligned}&\{1 + K_f a[G(1-\cos\omega_c T) - H\left(\sin\omega_c T - \tfrac{C_i}{V}\omega_c\right)]\}\\ &\quad + j\{K_f a[G\left(\sin\omega_c T + \tfrac{C_i}{V}\omega_c\right) + H(1-\cos\omega_c T)]\} = 0.\end{aligned} \tag{4.30}$$

Because the velocity-dependent term ($\frac{C_i}{V}\omega_c$) prevents the application of Tlusty's or Tobias's stability theory, the Nyquist stability criterion is used to construct the stability charts with process damping. The characteristic equation is evaluated by scanning the chatter frequency from zero to the maximum

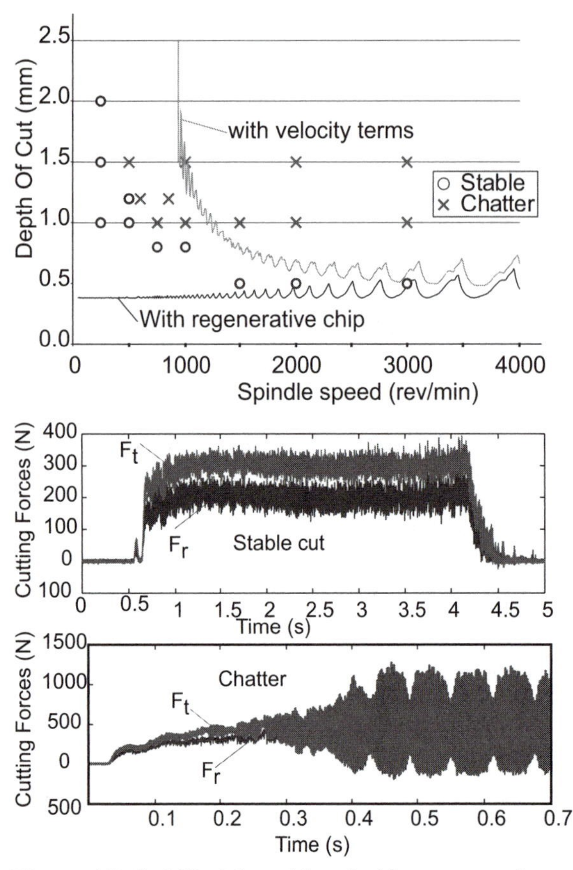

Figure 4.5: Stability lobes with and without process damping terms. Measured cutting forces during stable ($n = 500$ rev/min, $a = 1$ mm) and unstable ($n = 1{,}500$ rev/min, $a = 1$ mm) cutting tests. Material: AISI1045 with a diameter of 35 mm. Process parameters: $m_x = 0.561$ [kg], $k_x = 6.48 \cdot 10^6$ [N/m], $c_x = 145$ [N/m/s], $K_f = 1{,}384$ [N/mm^2], $C_i = 6.9 \cdot 10^6/(2\pi) = 1.1 \cdot 10^6$ [N/m].

possible vibration frequency $\omega_c = 0 \rightarrow \omega_{\max}$. If the polar plot in complex plane encircles origin (4.30), the system is assumed to be unstable and chatter would occur. It must be noted that the characteristic equation represents the closed-loop system; hence, the encirclement condition of the Nyquist criterion is valid around the origin (zero) but not around -1. (If $+1$ is dropped from the real part, then the encirclement around -1 can be used for the stability inspection.) The spindle speed (i.e., spindle period T), depth of cut (a), and chatter frequency (ω_c) are assumed in Eq. (4.30). The Nyquist stability criterion is applied to check whether the process is stable or not, and the chart is constructed by scanning all spindle speeds and depths of cut that lie in the cutting region of interest. The chatter stability diagrams generated with only regenerative and added velocity terms are shown in Figure 4.5. The classical chatter stability law with only regenerative chip thickness is not velocity dependent at the low speeds, where it gives a constant critical depth of cut of 0.4 mm. If the velocity, i.e., process damping term ($C_i \frac{2\pi}{\lambda}$), is included, the stability starts increasing at speeds less than 2,000 (rev/min).

It is well known that the tool wear changes the cutting edge geometry and flank contact with the wavy surface finish. A series of dynamic cutting tests were conducted on stainless steel. Because stainless steel produces high heat, the tools were worn quickly. The dynamic cutting coefficients were identified by using both sharp and worn tools. The corresponding stability charts are shown in Figure 4.6, where the curvature terms are ignored from the dynamic cutting force coefficients. The flank wear was about 0.080 mm. The stability with process damping moved from 1,000 rev/min to 3,000 rev/min, and all the unstable cutting tests with sharp tools were observed to become stable when tested with the worn tool. The chatter occurred only at 3500 rev/min and 1.5-mm depth of cut with the worn tool. It is also shown on Figure 4.6 that the process damping coefficient increases with the tool wear.

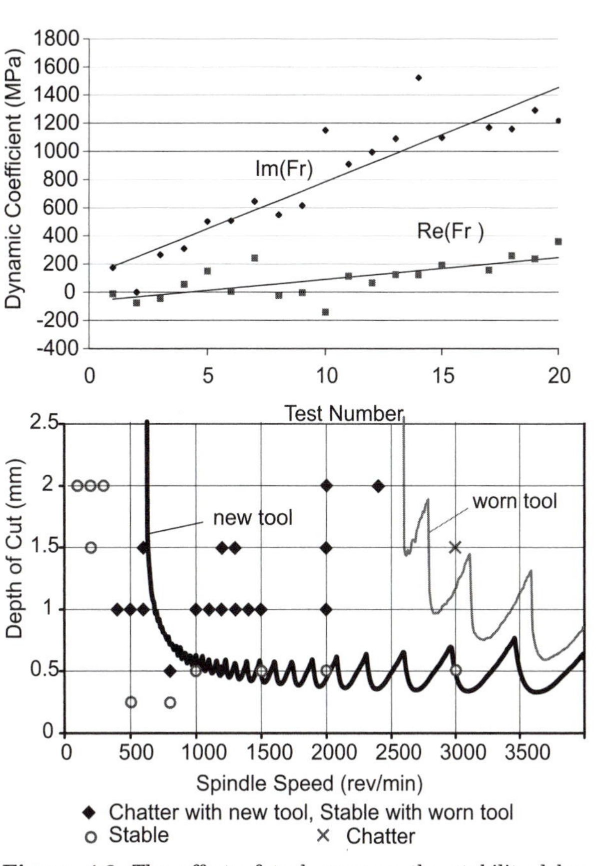

Figure 4.6: The effect of tool wear on the stability lobes with experimental results. Material: Stainless steel SS304 shaft with 35 mm diameter. Feed rate: 0.050 mm/rev. Process parameters: $m_x = 1.742$ [kg], $k_x = 7.92 \cdot 10^6$ [N/m], $c_x = 176.8$ [N/m/s], $K_f = 2{,}585$ [N/mm^2], $C_i = 1.181 \cdot 10^6$ [N/m]. Worn tool with 0.080 mm flank wear $C_i = 4.5856 \cdot 10^6$ [N/m].

4.3 CHATTER STABILITY OF TURNING OPERATIONS

Although the boring bar is usually the most flexible part in hole enlargement operations, the shaft, chuck, tail stock, and tool holder may contribute to the flexibility that leads to chatter in turning operations. However, both operations have similar mechanics and dynamics because of having geometrically defined cutting tool edges. A general diagram of a single-point cutting operation is shown in Figure 4.7. It is customary to model the cutting forces in oblique cutting coordinates, i.e., tangential or cutting speed direction (F_t), chip thickness direction or perpendicular to the cutting edge (F_f), and along the cutting edge (F_r) as shown in Figure 4.7. By neglecting the edge forces, the cutting forces in oblique coordinates are expressed as

$$\left.\begin{aligned} F_{tc} &= K_{tc}bh \\ F_{fc} &= K_{fc}bh \\ F_{rc} &= K_{rc}bh \end{aligned}\right\}. \tag{4.31}$$

where the b and h are the width of cut and uncut chip thickness, respectively. Cutting force coefficients (K_{tc}, K_{fc}, K_{rc}) can be evaluated either from shear stress, shear angle, friction coefficient, and tool geometry, or by mechanistic curve fitting to experimental force data as explained earlier. Alternatively, the resultant cutting force can be constructed from the oblique cutting forces as follows:

$$F_c = \sqrt{F_{tc}^2 + F_{fc}^2 + F_{rc}^2} = bhK_{tc}\sqrt{1 + \left(\frac{K_{fc}}{K_{tc}}\right)^2 + \left(\frac{K_{rc}}{K_{tc}}\right)^2}$$

$$F_c = K_c bh \rightarrow \quad K_c = K_{tc}\sqrt{1 + \left(\frac{K_{fc}}{K_{tc}}\right)^2 + \left(\frac{K_{rc}}{K_{tc}}\right)^2}. \tag{4.32}$$

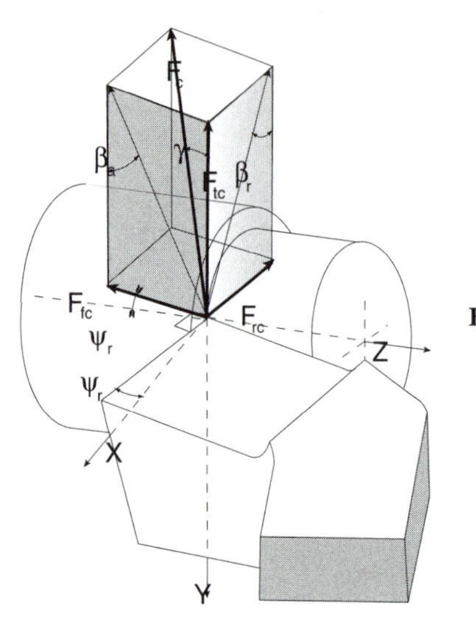

Figure 4.7: Stability of a turning process.

The oblique forces acting on the cutting edge can be expressed as a function of resultant force as follows:

$$
\begin{aligned}
F_{tc} &= F_c \cos\gamma && \rightarrow \cos\gamma = 1 \Big/ \sqrt{1 + \left(\frac{K_{fc}}{K_{tc}}\right)^2 + \left(\frac{K_{rc}}{K_{tc}}\right)^2} \\
F_{fc} &= F_c \cos\gamma \tan\beta_a && \rightarrow \tan\beta_a = \frac{K_{fc}}{K_{tc}} \\
F_{rc} &= F_c \cos\gamma \tan\beta_r && \rightarrow \tan\beta_r = \frac{K_{rc}}{K_{tc}}.
\end{aligned} \tag{4.33}
$$

The cutting forces are projected to the machine tool coordinate system as follows:

$$
\begin{Bmatrix} F_x \\ F_y \\ F_z \end{Bmatrix} = \begin{bmatrix} 0 & -\sin\psi_r & -\cos\psi_r \\ -1 & 0 & 0 \\ 0 & -\cos\psi_r & \sin\psi_r \end{bmatrix} \begin{Bmatrix} \cos\gamma \\ \cos\gamma\tan\beta_a \\ \cos\gamma\tan\beta_r \end{Bmatrix} F_c. \tag{4.34}
$$

It is assumed that the system is flexible in all three machine coordinates, and the vibrations (x, y, z) can be estimated from the cutting forces (F_x, F_y, F_z) and the measured or predicted Frequency Response Function (FRF) matrix at the cutting point $[\Phi]$ as follows:

$$
\begin{Bmatrix} x(j\omega_c) \\ y(j\omega_c) \\ z(j\omega_c) \end{Bmatrix} = \begin{bmatrix} \Phi_{xx}(j\omega_c) & \Phi_{xy}(j\omega_c) & \Phi_{xz}(j\omega_c) \\ \Phi_{yx}(j\omega_c) & \Phi_{yy}(j\omega_c) & \Phi_{yz}(j\omega_c) \\ \Phi_{zx}(j\omega_c) & \Phi_{zy}(j\omega_c) & \Phi_{zz}(j\omega_c) \end{bmatrix} \begin{Bmatrix} F_x(j\omega_c) \\ F_y(j\omega_c) \\ F_z(j\omega_c) \end{Bmatrix} \tag{4.35}
$$

$$
\{X(j\omega_c)\} = \left[\Phi(j\omega_c)\right]\{F(j\omega_c)\},
$$

where the cross-frequency response functions $(\Phi_{xy}, \Phi_{xz}, \Phi_{yz})$ may be negligible if the machine has uncoupled structural dynamics in three orthogonal directions. The resulting vibrations must be projected in the direction of chip thickness that is perpendicular to the cutting edge, so that the deformed chip dynamic thickness can be evaluated as follows:

$$h_d(t) = \left\{ \sin \psi_r \ 0 \ \cos \psi_r \right\} \left\{ \begin{array}{c} -\left[x(t) - x(t-T)\right] \\ -\left[y(t) - y(t-T)\right] \\ -\left[z(t) - z(t-T)\right] \end{array} \right\}, \tag{4.36}$$

where T is the spindle period or time delay term. By assuming that the system is critically stable, the dynamic chip thickness that vibrates at vibration frequency ω_c can be expressed as

$$h_d(j\omega_c) = -(1 - e^{-j\omega_c T}) \left\{ \sin \psi_r \ 0 \ \cos \psi_r \right\} \left\{ \begin{array}{c} x(j\omega_c) \\ y(j\omega_c) \\ z(j\omega_c) \end{array} \right\}. \tag{4.37}$$

By combining Eqs. (4.33)–(4.35) and (4.37), the dynamic chip thickness can be expressed as a function of machine dynamics and tool geometry as follows:

$$\begin{aligned} h_d(j\omega_c) &= -(1 - e^{-j\omega_c T})\Phi_0(j\omega_c)F_c\ (j\omega_c) \rightarrow \ \Phi_0(j\omega_c) \\ &= \sum d_{pq}\ \Phi_{pq}(j\omega_c)\ \ \forall\ \ (p, q) \in (x, y, z), \end{aligned} \tag{4.38}$$

where $\Phi_0(j\omega_c)$ is called the oriented frequency transfer function, and the parameters d_{pq} are called directional factors, which are given below for a stationary cutting edge with an approach angle of ψ_r:

$$\begin{aligned} d_{xx} &= -\cos\gamma\,(\sin^2\psi_r \tan\beta_a + 0.5\sin 2\psi_r \tan\beta_r) \\ d_{xy} &= -\cos\gamma\,\sin\psi_r \\ d_{xz} &= \cos\gamma\,\left(-0.5\sin 2\psi_r \tan\beta_a + \sin^2\psi_r \tan\beta_r\right) \\ d_{yx} &= d_{yy} = d_{yz} = 0 \\ d_{zx} &= -\cos\gamma\,\left(0.5\sin 2\psi_r \tan\beta_a + \cos^2\psi_r \tan\beta_r\right) \\ d_{zy} &= -\cos\gamma\,\cos\psi_r \\ d_{zz} &= \cos\gamma\,\left(-\cos^2\psi_r \tan\beta_a + 0.5\sin 2\psi_r \tan\beta_r\right). \end{aligned}$$

By substituting $F_c = K_c b h_d$, the stability equation is found from the characteristic equation of the system as follows:

$$1 + (1 - e^{-j\omega_c T})K_c b_{\text{lim}}\Phi_0(j\omega_c) = 0. \tag{4.39}$$

The directional factors are used to account the variations in the chip thickness due to vibrations in machine coordinates that are excited by the cutting forces. Koenigsberger and Tlusty [61], Peters and Vanherck [85], Opitz and Bernadi [82], and others applied the real part of the oriented transfer function (Φ_0) to the orthogonal chatter stability given in Eqs. (4.12) and (4.13), and obtained chatter stability lobes for single-point machining operations as follows:

$$b_{\text{lim}} = -\frac{1}{2K_c\,\text{Re}(\Phi_0)}. \tag{4.40}$$

The spindle speed can be evaluated from Eq. (4.12). However, the accuracy of the chatter prediction was not always satisfactory in turning and boring because of several factors. The chatter vibration frequency is typically above $f_c \geq 200$ Hz in single-point machining operations depending on the bar length in boring or the dimensions of the shaft to be turned, whereas the spindle speed is less than 1,500 rev/min or $n \leq 25$ Hz. The integer ratio of chatter frequency over the spindle frequency gives the location of the lobe (k) where the cutting takes place as follows:

$$k = \text{int}\left(\frac{f_c}{n}\right), \quad \varepsilon = 2\pi \cdot \text{frac}\left(\frac{f_c}{n}\right),$$

where ε is the phase shift between the waves left on the surface during two subsequent revolutions. Because the cutting takes place at higher lobes where the spindle speed is low in single-point cutting operations, the operation is always in the process damping region where the flank interferes with the wavy surface leading to friction. In addition, the flank friction changes at every point on the wave, causing the damping to vary harmonically as a function of tool angle, vibration frequency, and cutting speed. The satisfactory prediction of chatter stability of boring and turning in the frequency domain needs modeling of process damping in terms of dynamic or complex cutting coefficients. In addition to process damping, the boring and turning processes have nonlinear dynamics. The chip thickness distribution along the cutting edge depends on the radial depth of cut, the feed rate in the axial direction, the nose radius, and the approach angle. When the tool has a nose radius, there is no constant approach angle. The average direction of the cutting force depends on the feed rate, nose radius, approach angle, and radial depth of cut; hence, Eq. (4.40) can not be used directly. The frequency domain solution of such a nonlinear system can be solved only by linearizing the system around a narrow band of depth of cut and feed, which is not an ideal solution. An alternative method is explained in the next section, which considers both the nose radius and process damping.

4.4 CHATTER STABILITY OF TURNING SYSTEMS WITH PROCESS DAMPING

A typical turning tool having an approach angle (κ_r), and a nose radius (r_ε) is used with a depth of cut (a) and feed rate (c) as shown in Figure 4.8. The distribution of the chip, hence, the force along the cutting edge, depends on the tool geometry and cutting conditions. A convenient mechanistic force model based on the direction of chip flow angle is adopted here [45]. Colwell [41] proposed that the chip flow can be assumed to be normal to the chord, which connects two ends of the cutting edge engaged with the cut, and makes an angle of θ with the feed direction as shown in Figure 4.8. The normal force (F_n) and the side force (F_r) act parallel and normal to the chip flow, respectively. The tangential force (F_t) acts in the direction of the cutting speed, perpendicular to the plane defined by the side and normal forces. It is assumed that the

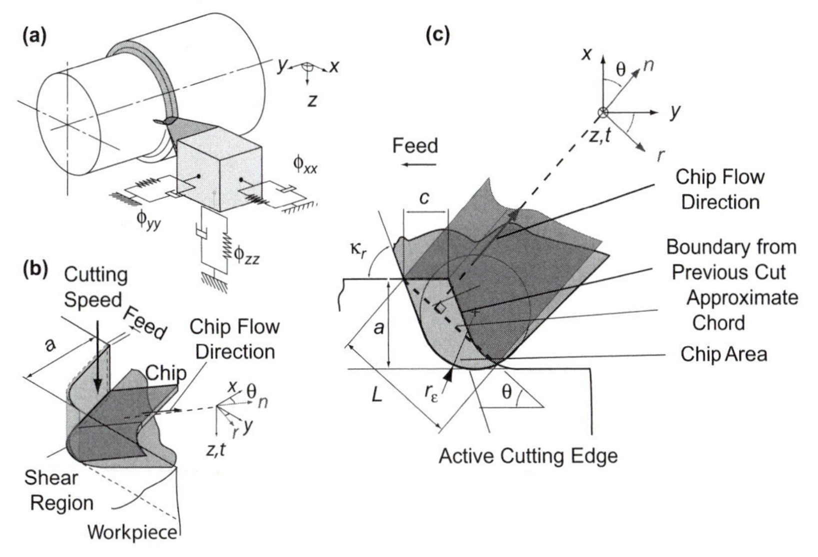

Figure 4.8: Stability of turning with tools having a nose radius.

equivalent chord length acts as a cutting edge, and the nose radius of the tool is neglected. With a tool having an equivalent chord angle of θ, the Cartesian machine coordinate system xyz (the coordinate system for the feed and depth of cut) is shown in Figure 4.8. The vibrations in the measurement $(\vec{i}, \vec{j}, \vec{k})$ and tool $(\vec{n}, \vec{r}, \vec{t})$ coordinate systems are defined as follows:

$$\vec{Q} = x\vec{i} + y\vec{j} + z\vec{k}\ ;\ \vec{S} = N\vec{n} + R\vec{r} + T\vec{t}. \tag{4.41}$$

With the following transformations,

$$\{N\,R\,T\}^T = [C_{nm}]\{x\,y\,z\}^T \leftrightarrows \{x\,y\,z\}^T = [C_{nm}]^T\{N\,R\,T\}^T, \tag{4.42}$$

where

$$[C_{nm}] = \begin{bmatrix} \cos\theta & \sin\theta & 0 \\ -\sin\theta & \cos\theta & 0 \\ 0 & 0 & 1 \end{bmatrix}$$

The structural dynamics of both the machine and part reflected at the tool tip are represented in the Laplace domain as follows:

$$\{Q(s)\} = \begin{Bmatrix} x(s) \\ y(s) \\ z(s) \end{Bmatrix} = \begin{bmatrix} \Phi_{xx}(s) & \Phi_{xy}(s) & \Phi_{xz}(s) \\ \Phi_{yx}(s) & \Phi_{yy}(s) & \Phi_{yz}(s) \\ \Phi_{zx}(s) & \Phi_{zy}(s) & \Phi_{zz}(s) \end{bmatrix} \begin{Bmatrix} F_x(s) \\ F_y(s) \\ F_z(s) \end{Bmatrix} = [\Phi(s)]\{F_m(s)\}, \tag{4.43}$$

where (F_x, F_y, F_z) represents the combined metal cutting and process damping forces acting on the structure in the measurement coordinate (x, y, z) system. Let us separate the forces into metal cutting $(\mathbf{F}_{mc})$ and process damping $(\mathbf{F}_{md})$ as follows:

$$\{F_m(s)\} = \{F_{mc}(s)\} + \{F_{md}(s)\}, \tag{4.44}$$

which are modeled from the cutting and contact mechanics models as follows.

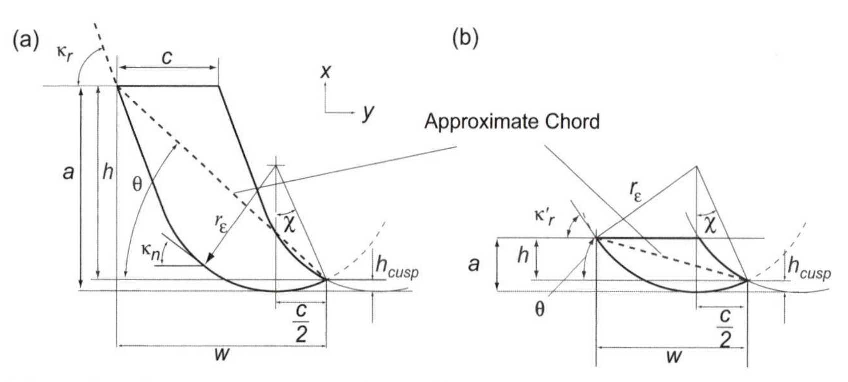

Figure 4.9: Chip geometry produced by a tool having a nose radius and approach angle, (a) $a > r_\varepsilon(1 - \cos\kappa_r)$ and (b) $a \leq r_\varepsilon(1 - \cos\kappa_r)$.

4.4.1 Metal Cutting Forces

The turning tool is assumed to cut along the equivalent chord. The cutting forces are assumed to have the following linear relationship:

$$\left\{F_n\ F_r\ F_t\right\}_c^T = \left\{K_{nA}\ K_{rA}\ K_{tA}\right\}^T Lh_c, \tag{4.45}$$

where L is the equivalent chip width or the chord length, and h_c is the equivalent chip thickness measured perpendicular to the chord as suggested by Colwell [41]. The chord angle (θ) and length (L) depend on the depth of cut (a), feed (c), nose radius (r_ε), and approach angle (κ_r), all of which are shown in Figure 4.9 and defined as:

$$\left.\begin{array}{c} h_{\text{cusp}} \approx \frac{c^2}{8r_\varepsilon} \\ h = a - h_{\text{cusp}} = a - \frac{c^2}{8r_\varepsilon} \\ w = \left\{\begin{array}{l} \frac{c}{2} + \frac{a - r_\varepsilon(1-\cos r_\varepsilon)}{\tan\kappa_r} + r_\varepsilon \sin\kappa_r \quad \rightarrow a > r_\varepsilon(1-\cos r_\varepsilon) \\ \\ \frac{c}{2} + \sqrt{r_\varepsilon^2 - (r_\varepsilon - a)^2} \rightarrow a \leq r_\varepsilon(1-\cos r_\varepsilon) \end{array}\right. \\ A = ca, \quad \theta = \tan^{-1}\frac{w}{h}, \quad L = \sqrt{h^2 + w^2} \end{array}\right\} \tag{4.46}$$

where A is the chip area, h and w are the projections of the approximate chord in the depth of cut and feed directions, respectively.

The cutting forces are projected in three Cartesian directions where the structural dynamics of the machine are defined as follows:

$$\left\{F_x\ F_y\ F_z\right\}_c^T = [C_{nm}]^T \left\{F_n\ F_r\ F_t\right\}_c^T = [C_{nm}]^T \left\{K_{nA}\ K_{rA}\ K_{tA}\right\}^T Lh_c. \tag{4.47}$$

By projecting the vibrations perpendicular to the chord, the equivalent regenerative chip thickness can be approximated as follows:

$$h_c(t) = c\sin\theta - \left\{\left[x(t)\cos\theta + y(t)\sin\theta\right] - \left[x(t-T)\cos\theta + y(t-T)\sin\theta\right]\right\}, \tag{4.48}$$

where T is the spindle rotation period. The vibrations in the direction of cutting speed (z) do not affect the chip thickness. The cutting forces contributed by the regenerative chip is reduced to

$$\begin{Bmatrix} F_x \\ F_y \\ F_z \end{Bmatrix}_c = [C_{nm}]^T \begin{Bmatrix} K_{nA} \\ K_{rA} \\ K_{tA} \end{Bmatrix} L \left[c\sin\theta - (1-e^{-sT})\left\{\cos\theta \ \sin\theta \ 0\right\} \begin{Bmatrix} x \\ y \\ z \end{Bmatrix} \right]. \quad (4.49)$$

The time-independent, static chip load ($c\sin\theta$) can be dropped from the equation because it does not affect the stability. By substituting $[C_{nm}]^T$ from (4.42), the dynamic cutting force from the regenerative cutting becomes

$$\begin{Bmatrix} F_x \\ F_y \\ F_z \end{Bmatrix}_c = -(1-e^{-sT})L \begin{bmatrix} \cos\theta & \sin\theta & 0 \\ -\sin\theta & \cos\theta & 0 \\ 0 & 0 & 1 \end{bmatrix}^T \begin{Bmatrix} K_{nA} \\ K_{rA} \\ K_{tA} \end{Bmatrix} \left\{\cos\theta \ \sin\theta \ 0\right\} \begin{Bmatrix} x \\ y \\ z \end{Bmatrix}, \quad (4.50)$$

which can be summarized as

$$\mathbf{F}_c(t) = -(1-e^{-sT})L\mathbf{DQ}(\mathbf{t}), \quad (4.51)$$

where the directional matrix $\mathbf{D}$ coefficients are

$$\mathbf{D} = \begin{bmatrix} K_{nA}\cos^2\theta - 0.5K_{rA}\sin 2\theta & 0.5K_{nA}\sin 2\theta - K_{rA}\sin^2\theta & 0 \\ K_{rA}\cos^2\theta + 0.5K_{nA}\sin 2\theta & K_{nA}\sin^2\theta + 0.5K_{rA}\sin 2\theta & 0 \\ K_{tA}\cos\theta & K_{tA}\sin\theta & 0 \end{bmatrix}. \quad (4.52)$$

4.4.2 Process Damping Gains Contributed by Flank Wear

The contact between the clearance face of the tool and the wavy surface contributes to the damping of the chatter during the dynamic cutting process. The process damping force is modeled by the compression of the volume of the work material under the flank face of the cutting tool [38, 39]. The normal (F_d) and friction (F_{dz}) forces caused by the contact are modeled (Fig. 4.10) as follows:

$$F_d = K_{sp}V_m, \quad F_{dz} = \mu_c F_d, \quad (4.53)$$

where K_{sp} is the experimentally identified contact force coefficient and μ_c is the coefficient of friction and assumed to be 0.3 for steel. The compressed volume of the material (V_m) under the tool flank is approximated as [38] follows:

$$V_m = -\frac{1}{2}L_c L_w^2 \frac{\dot{h}}{V_c}, \quad (4.54)$$

where $\dot{h}$ is the vibration velocity in the direction normal to the plane defined by the cutting edge and cutting speed (V_c), L_w is the wear land, and L_c is the total length of cutting edge in cut. For a differential element of a curved cutting edge segment with length dL, the vibration velocity is

$$\dot{h} = +\dot{x}\cos\chi_n + \dot{y}\sin\chi_n, \quad (4.55)$$

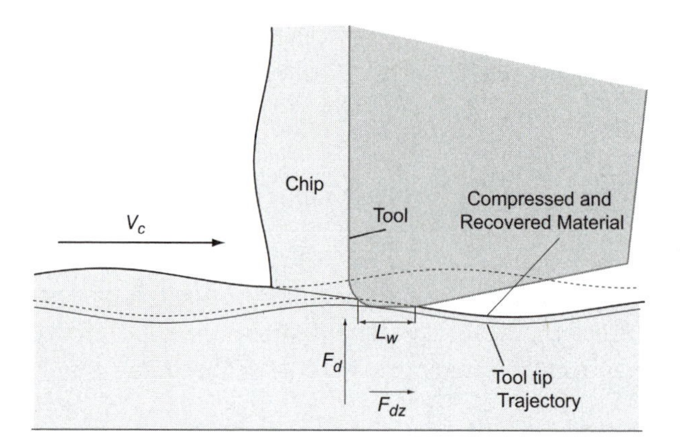

Figure 4.10: Model of process damping forces caused by tool wear.

where χ_n is the varying approach angle (4.9). Substituting Eqs. (4.54) and (4.55) into contact force (4.53),

$$dF_d = -\frac{dL \cdot L_w^2 K_{sp}}{2V_c}\dot{h} = -\frac{dL \cdot L_w^2 K_{sp}}{2V_c}(\dot{x}\cos\chi_n + \dot{y}\sin\chi_n). \tag{4.56}$$

and projecting it in x and y directions leads to differential contact forces in three directions as follows:

$$dF_{dx} = dF_d \cos\chi_n,\ dF_{dy} = dF_d \sin\chi_n,\ dF_{dz} = \mu_c dF_d \tag{4.57}$$

The process damping forces are organized in the matrix form as follows:

$$\begin{Bmatrix} dF_{dx} \\ dF_{dy} \\ dF_{dz} \end{Bmatrix} = -\frac{L_w^2 K_{sp}}{2V_c}\begin{bmatrix} \cos^2\chi_n & \cos\chi_n \sin\chi_n & 0 \\ \cos\chi_n \sin\chi_n & \sin^2\chi_n & 0 \\ \mu_c \cos\chi_n (\sin\chi_n + \cos\chi_n) & \mu_c \sin\chi_n (\sin\chi_n + \cos\chi_n) & 0 \end{bmatrix} dL \begin{Bmatrix} \dot{x} \\ \dot{y} \\ \dot{z} \end{Bmatrix}. \tag{4.58}$$

Assuming a constant flank wear (K_w), the velocity-dependent process damping force acting on the structure can be evaluated by integrating the differential forces along the cutting edge as follows:

$$\mathbf{F}_d = \int_0^{L_e} \begin{Bmatrix} dF_{dx} \\ dF_{dy} \\ dF_{dz} \end{Bmatrix} = \mathbf{J}_v \begin{Bmatrix} \dot{x} \\ \dot{y} \\ \dot{z} \end{Bmatrix}, \tag{4.59}$$

where the process damping matrix $\mathbf{J}_v$ is

$$\mathbf{J}_v = -\frac{L_w^2 K_{sp}}{2V_c}\int_0^{L_e}\left(\begin{bmatrix} \cos^2\chi_n & \cos\chi_n \sin\chi_n & 0 \\ \cos\chi_n \sin\chi_n & \sin^2\chi_n & 0 \\ \mu_c \cos\chi_n (\sin\chi_n + \cos\chi_n) & \mu_c \sin\chi_n (\sin\chi_n + \cos\chi_n) & 0 \end{bmatrix} dL\right). \tag{4.60}$$

By substituting $L_{\chi_r} = [a - r_\varepsilon (1 - \cos\chi_r)] / \sin\chi_r$, $dL = r_\varepsilon \cdot d\chi_n$ for the straight and curved sections of the cutting edge, the process damping matrix

($\mathbf{J}_v$) is defined as a function of depth of cut as follows:

$$\mathbf{J}_v = \begin{cases} -\frac{L_w^2 K_{sp}}{2V_c}\left(L_{\chi_r} \begin{bmatrix} \cos^2\chi_n & \cos\chi_n \sin\chi_n & 0 \\ \cos\chi_n \sin\chi_n & \sin^2\chi_n & 0 \\ \mu_c \cos\chi_n (\sin\chi_n + \cos\chi_n) & \mu_c \sin\chi_n (\sin\chi_n + \cos\chi_n) & 0 \end{bmatrix} + \int\limits_{-\tan^{-1} c/2r_\varepsilon}^{\chi_r} \begin{bmatrix} \cos^2\chi_n & \cos\chi_n \sin\chi_n & 0 \\ \cos\chi_n \sin\chi_n & \sin^2\chi_n & 0 \\ \mu_c \cos\chi_n (\sin\chi_n + \cos\chi_n) & \mu_c \sin\chi_n (\sin\chi_n + \cos\chi_n) & 0 \end{bmatrix} r_\varepsilon d\chi_n \right) & a > r_\varepsilon (1 - \cos\chi_r) \\ -\frac{L_w^2 K_{sp}}{2V_c}\left(\int\limits_{-\chi}^{\chi_r'} \begin{bmatrix} \cos^2\chi_n & \cos\chi_n \sin\chi_n & 0 \\ \cos\chi_n \sin\chi_n & \sin^2\chi_n & 0 \\ \mu_c \cos\chi_n (\sin\chi_n + \cos\chi_n) & \mu_c \sin\chi_n (\sin\chi_n + \cos\chi_n) & 0 \end{bmatrix} r_\varepsilon d\chi_n \right) & a \le r_\varepsilon (1 - \cos\chi_r) \end{cases}, \tag{4.61}$$

where $\chi_r' = \cos^{-1}(1 - a/r_\varepsilon)$, when $a \le r_\varepsilon (1 - \cos\chi_r)$. $\mathbf{J}_v$ relates the vibration velocities in (x, y) directions to the dynamic cutting forces.

4.4.3 Stability Analysis

The cutting forces are now expressed by the superposition of chip regeneration, i.e., cutting ($\mathbf{F}_c$) and process damping ($\mathbf{F}_d$) forces as follows:

$$\mathbf{F}_m(t) = \mathbf{F}_c(t) + \mathbf{F}_d(t) = -(1 - e^{-sT}) L\mathbf{DQ}(\mathbf{t}) + \mathbf{J}_v \mathbf{Q}(\mathbf{t}). \tag{4.62}$$

By substituting $\mathbf{Q(s)} = \Phi(\mathbf{s})\mathbf{F}_m(s)$ from Eq. (4.43) into Eq. (4.62) in Laplace domain,

$$\mathbf{F}_m(s) = -(1 - e^{-sT}) L\mathbf{DQ}(\mathbf{s}) + \mathbf{J}_v s\mathbf{Q}(\mathbf{s}) = \left[-(1 - e^{-sT}) L\mathbf{D} + s\mathbf{J}_v\right] \Phi(\mathbf{s})\mathbf{F}_m(s). \tag{4.63}$$

The characteristic equation of the dynamic cutting system becomes

$$[I] + \left[(1 - e^{-sT}) L\mathbf{D} - s\mathbf{J}_v\right] \Phi(\mathbf{s}) = \mathbf{0}. \tag{4.64}$$

For a known depth of cut (a), feed (c), and nose radius (r_ε), the critically stable equivalent width of cut (L) and spindle speed ($1/T$) are evaluated by applying the Nyquist stability criterion on the following characteristic equation in the frequency domain:

$$\left|\mathbf{I}_{[3\times3]} + \left[(1 - e^{-j\omega T}) L\mathbf{D}_{[3\times3]} - j\omega \mathbf{J}_{v[3\times3]}\right] \Phi(j\omega)_{[3\times3]}\right| = \mathbf{0}, \tag{4.65}$$

where $\mathbf{I}$ is the diagonal unit matrix; $\Phi(\mathbf{j}\omega)$ is the FRF matrix between the tool and workpiece (Eq. 4.43); $\mathbf{D}$ is the directional coefficient matrix for regenerative terms (Eq. 4.52); and $\mathbf{J}_v$ is the process damping gain matrix (Eq. 4.61). The stability is solved with the Nyquist criterion iteratively by assigning chord length (L) and spindle speed ($1/T$), and checking whether they lead to a stable or unstable turning operation. The Nyquist criterion is applied by finding the determinant of characteristic Eq. (4.65), and plotting its real and imaginary

TABLE 4.1. Modal Parameters of the Turning Tool Used in Chatter Tests

Modes	Directional Stiffness ($N/\mu m$)					
	XX	YY	ZZ	XY	ZX	YZ
$\omega_n = 242$ [Hz] $\zeta = 0.03$	125	91		−109		
$\omega_n = 340$ [Hz] $\zeta = 0.04$		59	185	735		133

parts on a complex plane. The process is unstable with chatter if the polar plot encircles the origin, as explained in Section 4.2.3. The critical stability (i.e. the limiting depth of cut) occurs when the polar plot passes from the origin. The stability inspection is repeated by scanning the range of L and T that are acceptable for the machine tool's operation limits and tool life. The radial depth of cut (a) can be evaluated from the critically stable chord length (L), tool geometry (r_ε, χ_r), and feed (c) as given in Eq. (4.46). The details of the presented model of turning stability with three-dimensional regeneration and process damping are given by Eynian and Altintas [45].

4.5 EXPERIMENTAL VALIDATION

A series of turning tests with various feeds, depths of cuts, and spindle speeds have been conducted on rigidly clamped, short AISI-1045 steel bars. The turret, which carries the tool, has dominant structural modes at 242 and 340 Hz with coupling terms that affect the regeneration in both feed (y) and radial depth of cut (x) directions. Experimentally identified modal parameters of the tool holder system are given in Table 4.1. The tool had a flank wear land of 0.13 mm. The cutting force coefficients are identified from chatter-free cutting tests, and the contact coefficient is identified from indentation tests and given in Figure 4.11.

The stability of the process is predicted by Model I presented here and a three-dimensional regenerative chip area (Model II) given by Eynian and Altintas [45]. The effect of nose radius and the depth of cut are considered by the use of equivalent chord length, but the dynamic changes in the chip area and chord length are neglected in model I. Model II considers cutting conditions (depth of cut, feed, and speed), tool geometry, and the regenerative displacements and their effects on the dynamic chip area and edge contact length. The onset of the chatter is detected by monitoring the sound pressure measured with a microphone and an accelerometer attached underneath the tool holder. When the frequency spectrum has significant strength around modal frequencies, but not at the spindle's rotational frequency, accompanied by high-pitch noise and poor surface finish, the presence of chatter is assumed. The stability limit at high speeds is predicted similarly by both models, and the influence of process damping at low speeds is predicted better by Model II. The chatter-free cutting conditions at higher speeds greater than 1,500 rev/min correspond to stability pockets. The experiments indicated an unanticipated decrease in the stability at lower speeds, where the cutting process exhibited poor shear and surface finish even without chatter. The vibration-free experimental evaluation of cutting force coefficients revealed an increase in magnitude at cutting speeds less than 100 m/min, which caused chatter during experiments in smaller depths of cut. Built-up edge and the process damping also becomes most effective at this zone, leading to increased stability. Since the proposed model evaluates the stability at each cutting condition, the speed-dependent cutting force coefficient is used similarly to the depth of cut and feed-dependent process gains.

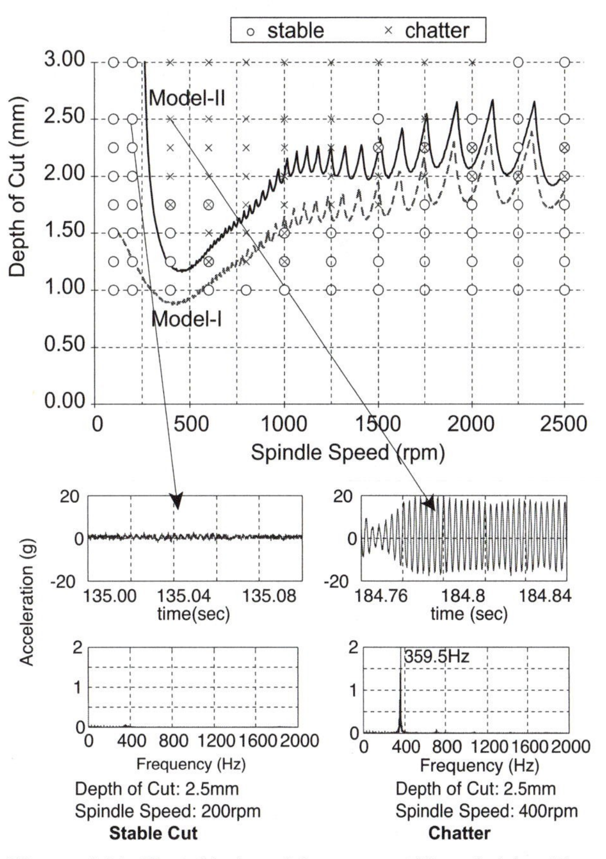

Figure 4.11: Unstable ($a = 2.5$ mm, $n = 400$ rev/min) cutting conditions. Feed rate $c = 0.1$ mm/rev and nose radius $r_\varepsilon = 0.8$ mm and approach angle $\kappa_r = 95$ degrees. Cutting force coefficients: $K_{nA} = 1,544, K_{rA} = -124, K_{tA} =$ 2,881 [N/mm] and contact coefficient $K_{sp} = 4.0 \times 10^{13}$ [N/m^3]. Flank wear width $L_w = 0.13$ [mm]. See Table 4.1 for the modal parameters.

4.6 ANALYTICAL PREDICTION OF CHATTER VIBRATIONS IN MILLING

The rotating cutting force and chip thickness directions and intermittent cutting periods complicate the application of orthogonal chatter theory to milling operations. The following analytical chatter prediction model was presented by Altintas and Budak [8, 34], who provide practical guidance to machine tool users for optimal process planning of depth of cuts and spindle speeds in milling operations [7].

4.6.1 Dynamic Milling Model

Milling cutters can be considered to have two orthogonal degrees of freedom as shown in Figure 4.12. The cutter is assumed to have N number of teeth with a zero helix angle. The cutting forces excite the structure in the

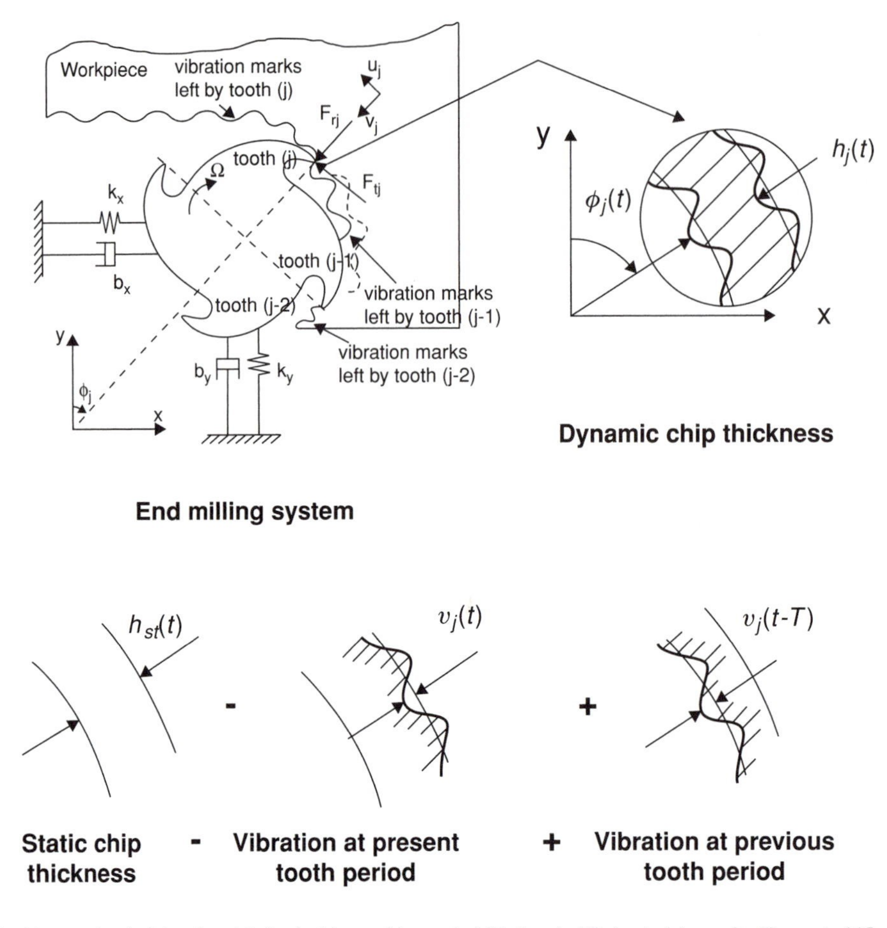

$$h_j(t) = s_j \sin\phi_j(t) - [-x(t)\sin\phi_j(t) - y(t)\cos\phi_j(t)] + [-x(t\text{-}T)\sin\phi_j(t) - y(t\text{-}T)\cos\phi_j(t)]$$

Figure 4.12: Self-excited vibrations in 2-DOF milling system.

feed (X) and normal (Y) directions, causing dynamic displacements x and y, respectively. The dynamic displacements are carried to rotating tooth number j in the radial or chip thickness direction with the coordinate transformation of $v_j = -x\sin\phi_j - y\cos\phi_j$, where ϕ_j is the instantaneous angular immersion of tooth j measured clockwise from the normal (Y) axis. If the spindle rotates at an angular speed of Ω(rad/s), the immersion angle varies with time as $\phi_j(t) = \Omega t$. The resulting chip thickness consists of a static part ($s_t \sin\phi_j$), attributed to rigid body motion of the cutter, and a dynamic component caused by the vibrations of the tool at the present and previous tooth periods. Because the chip thickness is measured in the radial direction (v_j), the total chip load can be expressed by

$$h(\phi_j) = [s_t \sin\phi_j + (v_{j,0} - v_j)]g(\phi_j), \tag{4.66}$$

where s_t is the feed rate per tooth and ($v_{j,0}, v_j$) are the dynamic displacements of the cutter at the previous and present tooth periods, respectively. The function

$g(\phi_j)$ is a unit step function that determines whether the tooth is in or out of cut, that is,

$$\left.\begin{aligned} g(\phi_j) &= 1 \leftarrow \phi_{\mathrm{st}} < \phi_j < \phi_{\mathrm{ex}}, \\ g(\phi_j) &= 0 \leftarrow \phi_j < \phi_{\mathrm{st}} \text{ or } \phi_j > \phi_{\mathrm{ex}}, \end{aligned}\right\}, \tag{4.67}$$

where ϕ_{st} and ϕ_{ex} are the start and exit immersion angles of the cutter to and from the cut, respectively. Henceforth, the static component of the chip thickness ($s_{\mathrm{t}} \sin \phi_j$) is dropped from the expressions because it does not contribute to the dynamic chip load regeneration mechanism. Substituting v_j into (4.66) yields

$$h(\phi_j) = [\Delta x \sin \phi_j + \Delta y \cos \phi_j] g(\phi_j), \tag{4.68}$$

where $\Delta x = x - x_0$ and $\Delta y = y - y_0$. Here, (x, y) and (x_0, y_0) represent the dynamic displacements of the cutter structure at the present and previous tooth periods, respectively. The tangential ($F_{\mathrm{t}j}$) and radial ($F_{\mathrm{r}j}$) cutting forces acting on the tooth j are proportional to the axial depth of cut (a) and chip thickness (h) as follows:

$$F_{\mathrm{t}j} = K_{\mathrm{t}} a h(\phi_j), \qquad F_{\mathrm{r}j} = K_{\mathrm{r}} F_{\mathrm{t}j}, \tag{4.69}$$

where cutting coefficients K_{t} and K_{r} are constant. Resolving the cutting forces in the x and y directions,

$$\begin{aligned} F_{xj} &= -F_{\mathrm{t}j} \cos \phi_j - F_{\mathrm{r}j} \sin \phi_j, \\ F_{yj} &= +F_{\mathrm{t}j} \sin \phi_j - F_{\mathrm{r}j} \cos \phi_j, \end{aligned} \tag{4.70}$$

and summing the cutting forces contributed by all teeth, we can write the total dynamic milling forces acting on the cutter as

$$F_x = \sum_{j=0}^{N-1} F_{x_j}(\phi_j); \qquad F_y = \sum_{j=0}^{N-1} F_{y_j}(\phi_j), \tag{4.71}$$

where $\phi_j = \phi + j\phi_{\mathrm{p}}$ and cutter pitch angle is $\phi_{\mathrm{p}} = 2\pi/N$. Substituting the chip thickness (4.68) and tooth forces (4.69) into (4.70) and rearranging the resulting expressions in matrix form yields

$$\begin{Bmatrix} F_x \\ F_y \end{Bmatrix} = \frac{1}{2} a K_{\mathrm{t}} \begin{bmatrix} a_{xx} & a_{xy} \\ a_{yx} & a_{yy} \end{bmatrix} \begin{Bmatrix} \Delta x \\ \Delta y \end{Bmatrix}, \tag{4.72}$$

where time-varying *directional dynamic milling force coefficients* are given by

$$a_{xx} = \sum_{j=0}^{N-1} -g_j[\sin 2\phi_j + K_r(1 - \cos 2\phi_j)],$$

$$a_{xy} = \sum_{j=0}^{N-1} -g_j[(1 + \cos 2\phi_j) + K_r \sin 2\phi_j],$$

$$a_{yx} = \sum_{j=0}^{N-1} g_j[(1 - \cos 2\phi_j) - K_r \sin 2\phi_j],$$

$$a_{yy} = \sum_{j=0}^{N-1} g_j[\sin 2\phi_j - K_r(1 + \cos 2\phi_j)].$$

Considering that the angular position of the parameters changes with time and angular velocity, we can express Eq. (4.72) in the time domain in a matrix form as [51, 79]

$$\{F(t)\} = \frac{1}{2} a K_t [A(t)]\{\Delta(t)\}. \tag{4.73}$$

The dynamic cutting force is converted from the time domain (4.73) to the frequency domain by taking the Fourier transform of Eq. (4.73) as follows:

$$\mathcal{F}\{F(t)\} = \frac{1}{2} a K_t \mathcal{F}\{[A(t)]\{\Delta(t)\}\} = \frac{1}{2} a K_t \mathcal{F}[A(t)]\} * \mathcal{F}[\{\Delta(t)\}]$$

$$\{F(\omega)\} = \frac{1}{2} a K_t \{[A(\omega)] * \{\Delta(\omega)\}\}. \tag{4.74}$$

where $*$ donates convolution integral. The vibration vectors at the present time (t) and previous tooth period $(t - T)$ are defined as

$$\{Q\} = \{x(t) \;\; y(t)\}^T; \qquad \{Q_0\} = \{x(t-T) \;\; y(t-T)\}^T, \tag{4.75}$$

or in the frequency domain,

$$\begin{aligned} \{Q(\omega)\} &= [\Phi(i\omega)]\{F(\omega)\}, \\ \{Q_0(\omega)\} &= e^{-i\omega T}\{Q(i\omega)\}, \end{aligned} \tag{4.76}$$

The frequency response function matrix ($[\Phi(i\omega)]$) of the structure at the cutter–workpiece contact zone is

$$[\Phi(i\omega)] = \begin{bmatrix} \Phi_{xx}(i\omega) & \Phi_{xy}(i\omega) \\ \Phi_{yx}(i\omega) & \Phi_{yy}(i\omega) \end{bmatrix}, \tag{4.77}$$

where $\Phi_{xx}(i\omega)$ and $\Phi_{yy}(i\omega)$ are the direct transfer functions in the x and y directions, and where $\Phi_{xy}(i\omega)$ and $\Phi_{yx}(i\omega)$ are the cross-transfer functions. Describing the vibrations at the vibration frequency ω in the frequency domain, substituting $\{\Delta\} = \{(x - x_0) \;\; (y - y_0)\}^T$ gives

$$\{\Delta(i\omega)\} = \{Q(i\omega)\} - \{Q_0(i\omega)\} = [1 - e^{-i\omega T}][\Phi(i\omega)]\{F(\omega)\}. \tag{4.78}$$

Substituting $\{\Delta(i\omega)\}$ into the dynamic milling force equation in frequency domain (4.74) gives

$$\{F(\omega)\} = \frac{1}{2}aK_t\left\{[A(\omega)] * [1 - e^{-i\omega T}][\Phi(i\omega)]\{F(\omega)\}\right\}. \tag{4.79}$$

As the cutter rotates, the directional factors vary with time, which is the fundamental difference between milling and operations such as turning where the direction of the force is constant. However, like the milling forces, $[A(t)]$ is periodic at tooth-passing frequency $\omega_T = N\Omega$ or tooth period $T = 2\pi/\omega_T$, i.e., $[A(t)] = [A(t+T)]$; thus, it can be expanded into the Fourier series as follows:

$$\left.\begin{array}{c} [A(\omega)] = \mathcal{F}[A(t)] = \sum\limits_{r=-\infty}^{+\infty}[A_r]\delta(\omega - r\omega_T) = \sum\limits_{r=-\infty}^{+\infty}[A_r]e^{ir\omega_T t} \\ [A_r] = \frac{1}{T}\int_0^T [A(t)]e^{-ir\omega_T t}\,dt \end{array}\right\}, \tag{4.80}$$

where δ and $\mathcal{F}$ denote the Dirac delta function and Fourier transformation, respectively. The directional matrix $[A(t)]$ is periodic at tooth-passing frequency ω_T or at pitch angle ϕ_p, and it has zero value when the tooth is out of cut, i.e., $\phi_{st} \le \phi \le \phi_{ex} \to [A(t)] \neq 0$. The Fourier coefficients are evaluated for N number of tooth as follows:

$$[A_r] = \frac{1}{T}\sum_{j=0}^{N-1}\int_0^T \begin{bmatrix} a_{xx,j} & a_{xy,j} \\ a_{yx,j} & a_{yy,j} \end{bmatrix} e^{-ir\omega_T t}dt. \tag{4.81}$$

By introducing a change of variable, $\phi_j(t) = \Omega(t + jT) = \Omega\tau_j$, where $\tau_j = t + jT$. The time-varying angular position can be transformed into a pure angular domain as $\omega_T t = N\Omega \cdot t = N \cdot \phi$, where $\phi = \Omega t$ is the angular rotation of the spindle with respect to the reference tooth ($j = 0$).

$$\begin{aligned} &\tau_j = t + jT, \quad d\tau_j = dt = \frac{d\phi}{\Omega} \\ &t = 0 \to \tau_{j_0} = jT, \quad \phi_j(0) = j\Omega T = j\phi_p \\ &t = T \to \tau_j = T + jT, \quad \phi_j(T) = (j+1)\Omega T = (j+1)\phi_p, \end{aligned} \tag{4.82}$$

where the pitch angle (ϕ_p) of the cutter $\phi_p = \Omega T = 2\pi/N$. By substituting $\omega_T t = N\phi$, the directional matrix becomes

$$\begin{aligned} [A_r] &= \frac{1}{\Omega T}\sum_{j=0}^{N-1}\int_{j\phi_p}^{(j+1)\phi_p} \begin{bmatrix} a_{xx,j} & a_{xy,j} \\ a_{yx,j} & a_{yy,j} \end{bmatrix} e^{-irN\phi}d\phi \\ &= \frac{1}{\phi_p}\left(\int_0^{\phi_p} \begin{bmatrix} a_{xx,0} & a_{xy,0} \\ a_{yx,0} & a_{yy,0} \end{bmatrix} e^{-irN\phi}d\phi + \int_{\phi_p}^{2\phi_p} \begin{bmatrix} a_{xx,1} & a_{xy,1} \\ a_{yx,1} & a_{yy,1} \end{bmatrix} e^{-irN\phi}d\phi + \ldots\right) \\ &= \frac{N}{2\pi}\int_0^{2\pi} \begin{bmatrix} a_{xx} & a_{xy} \\ a_{yx} & a_{yy} \end{bmatrix} e^{-irN\phi}d\phi. \end{aligned} \tag{4.83}$$

However, the periodic functions are nonzero only within the immersion interval $\langle \phi_{st}, \phi_{ex} \rangle$; therefore, the integral boundaries are modified as follows:

$$[A_r] = \frac{N}{2\pi} \int_{\phi_{st}}^{\phi_{ex}} \begin{bmatrix} a_{xx} & a_{xy} \\ a_{yx} & a_{yy} \end{bmatrix} e^{-irN\phi} d\phi = \frac{N}{2\pi} \begin{bmatrix} \alpha_{xx}^{(r)} & \alpha_{xy}^{(r)} \\ \alpha_{yx}^{(r)} & \alpha_{yy}^{(r)} \end{bmatrix}, \tag{4.84}$$

where each term depends on the harmonic counter (r) as follows:

$$\alpha_{xx}^{(r)} = \frac{i}{2} \left[-c_0 K_r e^{-irN\phi} + c_1 e^{-ip_1\phi} - c_2 e^{ip_2\phi} \right] \Big|_{\phi_{st}}^{\phi_{ex}}$$

$$\alpha_{xy}^{(r)} = \frac{i}{2} \left[-c_0 K_r e^{-irN\phi} + c_1 e^{-ip_1\phi} + c_2 e^{ip_2\phi} \right] \Big|_{\phi_{st}}^{\phi_{ex}}$$

$$\alpha_{yx}^{(r)} = \frac{i}{2} \left[c_0 K_r e^{-irN\phi} + c_1 e^{-ip_1\phi} + c_2 e^{ip_2\phi} \right] \Big|_{\phi_{st}}^{\phi_{ex}}$$

$$\alpha_{yx}^{(r)} = \frac{i}{2} \left[-c_0 K_r e^{-irN\phi} - c_1 e^{-ip_1\phi} + c_2 e^{ip_2\phi} \right] \Big|_{\phi_{st}}^{\phi_{ex}},$$

where $p_1 = 2 + Nr$, $p_2 = 2 - Nr$, $c_0 = 2/(Nr)$, $c_1 = (K_r - i)/p_1$, $c_2 = (K_r + i)/p_2$. Note that when $Nr = -2 \rightarrow p_1 = 0, c_1 = \infty$; $Nr = +2 \rightarrow p_2 = 0, c_2 = \infty$. Hence, for the special case of $Nr = \pm 2$, the integral is evaluated specially instead of by using the general parametric solutions. The number of harmonics (r) of the tooth-passing frequency (ω_T) to be considered for an accurate reconstruction of $[A(t)]$ depends on the immersion conditions and on the number of teeth in the cut. Altintas and Budak proposed zero-order and multifrequency solutions, where the number of harmonics is $r = 0$ and $r \geq 1$, respectively. Although the zero-order solution is solved directly and analytically, and proven to be practical in most milling operations, the multifrequency leads to improved accuracy when the radial immersion is small. If we consider a special situation where $r = 0, \pm 1$, the directional coefficient matrix becomes (Eq. 4.84) as follows:

$$\begin{aligned} [A(t)] &= \sum_{r=-1}^{+1} [A_r] e^{ir\omega_T t} \\ &= \begin{bmatrix} \alpha_{xx}^{(-1)} & \alpha_{xy}^{(-1)} \\ \alpha_{yx}^{(-1)} & \alpha_{yy}^{(-1)} \end{bmatrix} e^{-i\omega_T t} + \begin{bmatrix} \alpha_{xx}^{(0)} & \alpha_{xy}^{(0)} \\ \alpha_{yx}^{(0)} & \alpha_{yy}^{(0)} \end{bmatrix} + \begin{bmatrix} \alpha_{xx}^{(+1)} & \alpha_{xy}^{(+1)} \\ \alpha_{yx}^{(+1)} & \alpha_{yy}^{(+1)} \end{bmatrix} e^{i\omega_T t}. \end{aligned} \tag{4.85}$$

4.6.2 Zero-Order Solution of Chatter Stability in Milling

If the most simplistic approximation, the average component of the Fourier series expansion, is considered (i.e., $r = 0$), then

$$[A_0] = \frac{1}{T} \int_0^T [A(t)]\, dt = \frac{1}{\phi_p} \int_{\phi_{st}}^{\phi_{ex}} [A(\phi)]\, d\phi = \frac{N}{2\pi} \begin{bmatrix} \alpha_{xx} & \alpha_{xy} \\ \alpha_{yx} & \alpha_{yy} \end{bmatrix}, \tag{4.86}$$

where the integrated functions are given as

$$\alpha_{xx} = \tfrac{1}{2}\left[\cos 2\phi - 2K_r\phi + K_r \sin 2\phi\right]_{\phi_{st}}^{\phi_{ex}},$$

$$\alpha_{xy} = \tfrac{1}{2}\left[-\sin 2\phi - 2\phi + K_r \cos 2\phi\right]_{\phi_{st}}^{\phi_{ex}},$$

$$\alpha_{yx} = \tfrac{1}{2}\left[-\sin 2\phi + 2\phi + K_r \cos 2\phi\right]_{\phi_{st}}^{\phi_{ex}},$$

$$\alpha_{yy} = \tfrac{1}{2}\left[-\cos 2\phi - 2K_r\phi - K_r \sin 2\phi\right]_{\phi_{st}}^{\phi_{ex}}.$$

When the time-dependent terms are neglected, the system loses its periodic variation, and thus becomes a time-invariant system. The average directional factors depend on the radial cutting constant (K_r) and the width of cut bound by entry (ϕ_{st}) and exit (ϕ_{ex}) angles as illustrated in Figure 4.13. The dynamic milling expression (4.79) is reduced to the following:

$$\{F(\omega)\} = \frac{1}{2} a K_t \left\{[A_0][1 - e^{-i\omega T}][\Phi(i\omega)]\{F(\omega)\}\right\}, \tag{4.87}$$

where $[A_0]$ is the time-invariant, but immersion-dependent directional cutting coefficient matrix. Because the average cutting force per tooth period is independent of the helix angle, $[A_0]$ is valid for helical end mills as well. If the system is critically stable by vibrating at the chatter frequency ω_c, the roots of the characteristic equation are found from the determinant

$$\det\left[[I] - \frac{1}{2} K_t a (1 - e^{-i\omega_c T})[A_0][\Phi(i\omega_c)]\right] = 0.$$

The notation is further simplified by defining the oriented FRF matrix as

$$[\Phi_0(i\omega_c)] = \begin{bmatrix} \alpha_{xx}\Phi_{xx}(i\omega_c) + \alpha_{xy}\Phi_{yx}(i\omega_c) & \alpha_{xx}\Phi_{xy}(i\omega_c) + \alpha_{xy}\Phi_{yy}(i\omega_c) \\ \alpha_{yx}\Phi_{xx}(i\omega_c) + \alpha_{yy}\Phi_{yx}(i\omega_c) & \alpha_{yx}\Phi_{xy}(i\omega_c) + \alpha_{yy}\Phi_{yy}(i\omega_c) \end{bmatrix} \tag{4.88}$$

and the eigenvalue of the characteristic equation as

$$\Lambda = -\frac{N}{4\pi} a K_t (1 - e^{-i\omega_c T}). \tag{4.89}$$

The resulting characteristic equation becomes

$$\det|[I] + \Lambda[\Phi_o(i\omega_c)]| = 0. \tag{4.90}$$

The eigenvalue of Eq. (4.90) can easily be solved for a given chatter frequency ω_c, static cutting coefficient factors (K_t, K_r) (which can be stored as material-dependent quantities for any milling cutter geometry), radial immersion (ϕ_{st}, ϕ_{ex}), and the FRF of the structure (4.88). If two orthogonal degrees of freedom in feed (x) and normal (y) directions are considered (i.e., $\Phi_{xy} = \Phi_{yx} = 0.0$), the characteristic equation becomes just a quadratic function as follows:

$$a_0\Lambda^2 + a_1\Lambda + 1 = 0, \tag{4.91}$$

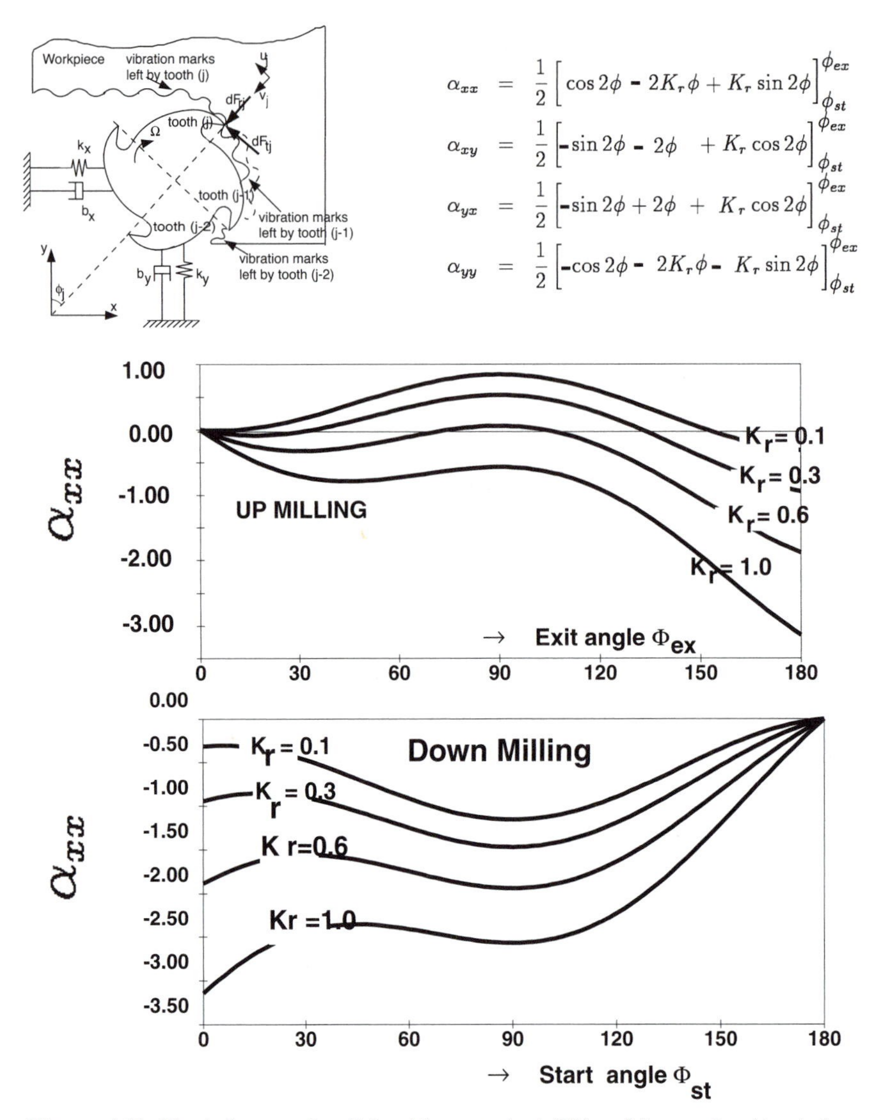

Figure 4.13: The influence of radial cutting constant (K_r) and immersion (ϕ_{st}, ϕ_{ex}) on average directional factors.

where

$$a_0 = \Phi_{xx}(i\omega_c)\Phi_{yy}(i\omega_c)(\alpha_{xx}\alpha_{yy} - \alpha_{xy}\alpha_{yx}),$$
$$a_1 = \alpha_{xx}\Phi_{xx}(i\omega_c) + \alpha_{yy}\Phi_{yy}(i\omega_c).$$

Then, the eigenvalue Λ is obtained as

$$\Lambda = -\frac{1}{2a_0}\left(a_1 \pm \sqrt{a_1^2 - 4a_0}\right). \tag{4.92}$$

As long as the plane of cut (x, y) is considered, the characteristic equation is still a simple quadratic function regardless of the number of modes considered in the machine tool structure. Because the FRFs are complex, the eigenvalue has a real and an imaginary part, $\Lambda = \Lambda_R + i\Lambda_I$. Substituting the eigenvalue and $e^{-i\omega_c T} = \cos\omega_c T - i\sin\omega_c T$ in Eq. (4.89) gives the critical axial depth of cut at chatter frequency ω_c as follows:

$$a_{\text{lim}} = -\frac{2\pi}{NK_t}\left[\frac{\Lambda_R(1-\cos\omega_c T) + \Lambda_I \sin\omega_c T}{(1-\cos\omega_c T)} + i\frac{\Lambda_I(1-\cos\omega_c T) - \Lambda_R\sin\omega_c T}{(1-\cos\omega_c T)}\right]. \tag{4.93}$$

Because a_{lim} is a real number, the imaginary part of Eq. (4.93) must vanish as follows:

$$\Lambda_I(1-\cos\omega_c T) - \Lambda_R \sin\omega_c T = 0. \tag{4.94}$$

By substituting

$$\kappa = \frac{\Lambda_I}{\Lambda_R} = \frac{\sin\omega_c T}{1-\cos\omega_c T} \tag{4.95}$$

into the real part of Eq. (4.93) (the imaginary part vanishes), the final expression for chatter-free axial depth of cut is found as

$$a_{\text{lim}} = -\frac{2\pi\Lambda_R}{NK_t}(1+\kappa^2). \tag{4.96}$$

Therefore, given the chatter frequency (ω_c), the chatter limit in terms of the axial depth of cut can be directly determined from Eq. (4.96).

The corresponding spindle speeds are also found in a manner similar to the chatter in orthogonal cutting presented in the previous section. From Eq. (4.95),

$$\kappa = \tan\psi = \frac{\cos(\omega_c T/2)}{\sin(\omega_c T/2)} = \tan[\pi/2 - (\omega_c T/2)] \tag{4.97}$$

and the phase shift of the eigenvalue is $\psi = \tan^{-1}\kappa$, and $\omega_c T = \pi - 2\psi + 2k\pi$ is the phase distance in one tooth period (T). If k is the integer number of full-vibration waves (i.e., lobes) imprinted on the cut arc, and $\epsilon = \pi - 2\psi$ is the phase shift between the inner and outer modulations (present and previous vibration marks), then

$$\omega_c T = \epsilon + 2k\pi. \tag{4.98}$$

Again, care must be taken in calculating the phase shift (ψ) from the real (Λ_R) and imaginary (Λ_I) parts of the eigenvalue. The spindle speed n(rev/min) is simply calculated by finding the tooth-passing period $T(s)$ as follows:

$$T = \frac{1}{\omega_c}(\epsilon + 2k\pi) \rightarrow n = \frac{60}{NT}. \tag{4.99}$$

In summary, the FRFs of the machine tool system are identified, and the directional coefficients are evaluated from the derived Eq. (4.86) for a specified cutter, workpiece material, and radial immersion of the cut. Then the stability lobes are calculated as follows [8]:

- Select a chatter frequency from transfer functions around a dominant mode.
- Solve the eigenvalue from Eq. (4.91).
- Calculate the critical depth of cut from Eq. (4.96).
- Calculate the spindle speed from Eq. (4.99) for each stability lobe $k = 0, 1, 2, \ldots$.
- Repeat the procedure by scanning the chatter frequencies around all dominant modes of the structure evident on the transfer functions.

The algorithm can be applied to milling with dynamics in three directions [6], or with three lateral and torsional flexibilities found in drilling [93] and plunge milling/boring [13] operations.

It may be argued that the use of the average Fourier coefficient $[A_0]$ may not be sufficient in the stability analysis. When the width of cut is small, the milling force waveforms are narrow and intermittent. Such waveforms have strong harmonic components in addition to the average value. When the author's group considered the high harmonics with the use of iterative solutions, they noticed that the stability results did not improve significantly even at very low immersion cuts. Physical analysis of the process illustrated in Figure 4.14 revealed that the high harmonics of the forcing function (i.e., milling forces) are low pass filtered within the dynamic cutting process, unless they ring another mode that is a tooth frequency away from the chatter frequency.

The process is explained by Figure 4.14 as follows. If chatter vibrations are present in a milling operation, the regenerative vibration spectra are dominated by the chatter frequency ω_c. The regenerative vibrations (i.e., *dynamic chip loads*) are multiplied by the immersion-dependent directional factors, which have an average value and additional strengths at the harmonics of tooth passing frequency ($\omega_T, 2\omega_T, \ldots$). The resulting dynamic milling forces have strengths at chatter frequency (ω_c) and at tooth passing harmonics away from the chatter frequency ($\omega_c + i\omega_T, i = \pm 1, \pm 2, \ldots$). The dynamic cutting force excites the structure whose dominant mode is very close to the chatter frequency ω_c. The frequency response function does not have any strength away from the natural mode that is causing the chatter; hence, it low-pass filters or attenuates the dynamic cutting force harmonics away from the chatter frequency. The chatter loop is closed when the structure leaves new chatter vibration marks on the present and previous surface cut during milling. However, if the structure has closely spaced modes, and if these modes happen to be distributed at tooth-passing frequency intervals, the use of the average Fourier coefficient $[A_0]$ may not be sufficient for accurate prediction of chatter stability. In such cases, higher harmonics of the directional factors must be considered, and the resulting stability expression shown in the next section must be solved with numerical iterations.

Example: Face Milling Cutter. A bull-nosed face milling cutter with two circular inserts is used in milling aluminum alloy Al-7075. The diameter of the cutter body is 31.75 mm, and the radius of circular inserts is 4.7625 mm. The

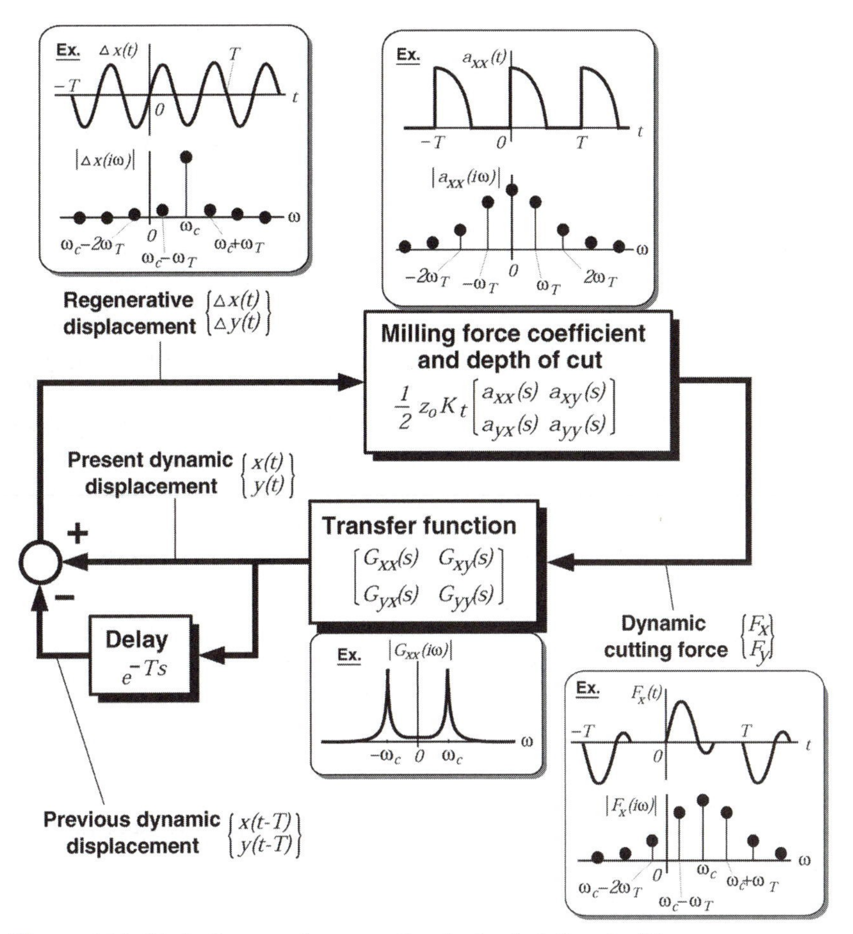

Figure 4.14: Block diagram of regenerative chatter in ball end milling process.

identified cutting constants are $K_t = 1{,}319.4$ [N/mm^2] and $K_r = 788.8$ [N/mm^2]. The FRF of the cutter mounted on the spindle is measured, and the modal parameters are identified in the feed (x) and normal (y) directions in Table 4.2.

TABLE 4.2. Identified Modal Parameters of a Bull-Nose Cutter Mounted on the Spindle of a Vertical Machining Center. The Transfer Function Units Are in [m/N]

Mode x	ω_{nx} [Hz]	ζ_x	$\sigma_x + j\nu_x$
1	452.77	0.123718	9.202966E−05−j1.862195E−04
2	1,448.53	0.01651	−4.181562E−05−j3.043618E−04
Mode y	ω_{ny} [Hz]	ζ_y	$\sigma_y + j\nu_y$
1	516.17	0.0243	−2.3929E−06−j1.721539E−04
2	1,407.64	0.0324	4.055052E−05−j3.618808E−04

The stability lobes of the cutter are shown in Figure 4.15 for a half-immersion down-milling operation. The stability lobes are predicted using time domain, numerical simulations, which consider all nonlinearities in the dynamic milling, process, such as tool jumping out, feed marks, and waves left by the previous cuts. The analytical and time domain solutions are in good agreement, although the analytical solution is linear and quite simple to implement in practice. Two cutting tests are conducted at $a = 4.7$ mm depth of cut. One is at the unstable spindle speed of $n = 9{,}500$ rev/min, where severe chatter vibrations were observed, which are evident from force measurements and their Fourier spectrum. The chatter occurred close to the second mode of the structure at 1,448 Hz. When the spindle speed is increased to $n = 14{,}000$ rev/min, the chatter vibrations disappeared, and the spectrum of forces indicate that the dominance of forced vibration or tooth passing frequency was at 467 Hz. Time domain simulations demonstrate the presence of poor surface finish when chatter vibrations are present during milling.

Example End Milling with a Flexible Cutter. A helical carbide end mill with four flutes, 19.05 mm diameter, and 10 degree rake is used in peripheral milling of aluminum alloy Al-7075. The identified cutting constants for the work material and cutter are found as $K_t = 796$ [N/mm^2] and $K_r = 0.212$. The transfer function of the cutter attached to the spindle is measured at the tool tip in both feed (x) and normal (y) directions with an instrumented hammer and accelerometer. The modal parameters of the structure converted to displacement domain are identified from modal analysis software and are given in Table 4.3. The experimentally measured and reconstructed transfer functions are compared in Figure 4.16, which indicates the accuracy of identified modal parameters. Later the modal parameters are used to simulate the stability lobes for a half-immersion down-milling and slotting of Al-7075 alloy. These results are shown in Figure 4.17. The minimum stable depth of cut seems to be about $a_{\mathrm{lim}} = 4$ mm in half-immersion up-milling and about $a_{\mathrm{lim}} = 1.5$ mm in slotting. The ideal stability pockets correspond to a spindle speed $n = 11{,}800$ rev/min, where the maximum material can be removed without chatter. Note that there is hardly any stability pocket at speeds less than $n = 5{,}000$ rev/min, which is due to the many tightly packed vibration waves at each tooth period. Because Al-7075 is not a hard material, it is clearly advisable to machine it at high speeds to enhance productivity. The stability lobes presented for this case have been experimentally proven on a machining center powered by a 15 kW spindle capable of 15,000 rev/min.

4.6.3 Multi-Frequency Solution of Chatter Stability in Milling

When the radial immersion of cut is small, the milling process will exhibit highly intermittent directional factors that will lead to force waveforms with high-frequency content. In such cases, the average directional factor $[A_0]$ may not be sufficient to predict the stability lobes at high spindle speeds with

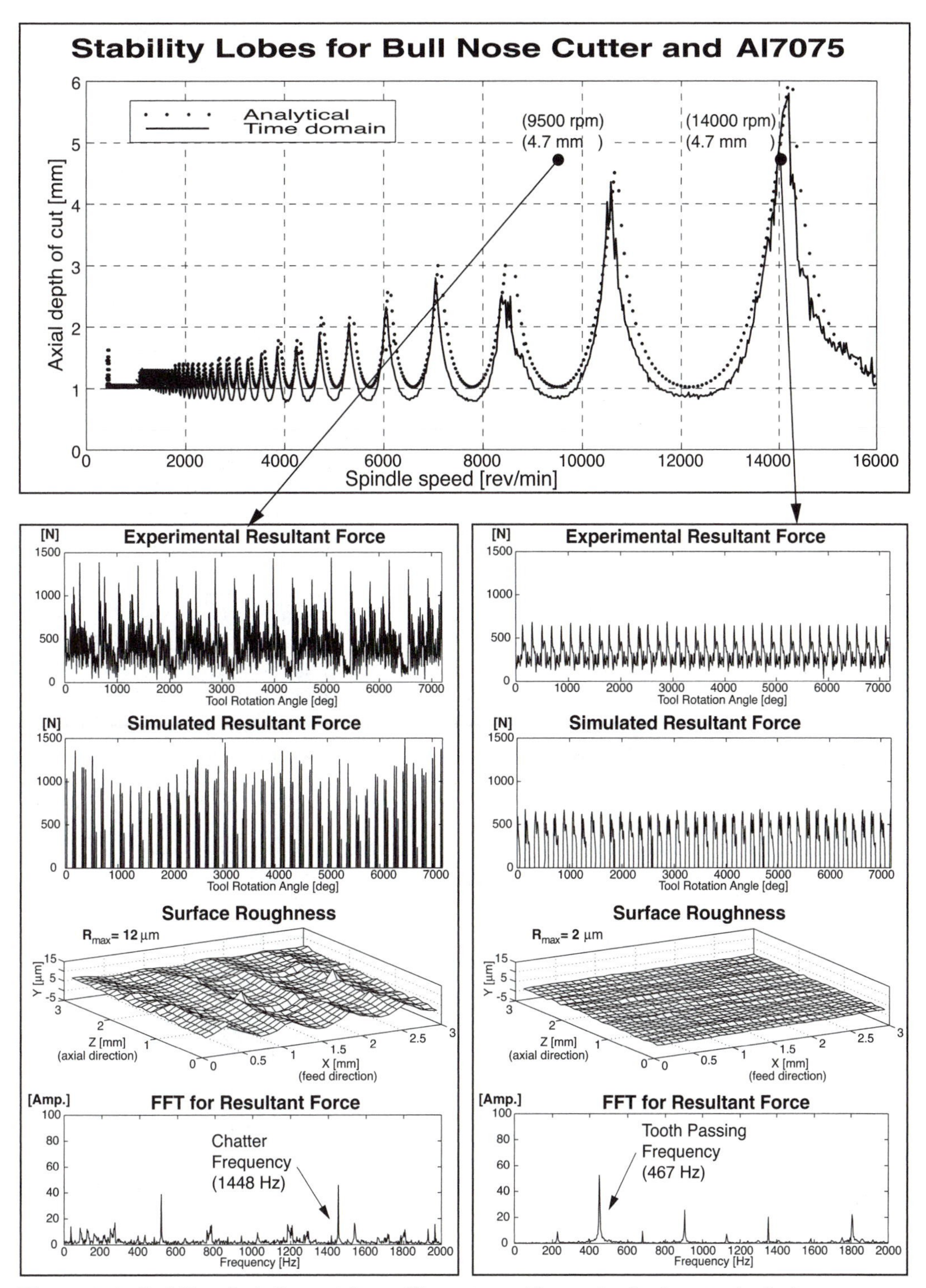

Figure 4.15: Stability of a bull-nosed face milling cutter with two circular inserts; the feed per tooth is $s_t = 0.05$ mm/tooth in sample measurenents and time domain simulations.

TABLE 4.3. Identified Modal Parameters of a Vertical Machining Center in x and y Directions. The transfer function units are in [m/N]

Mode x	ω_{nx} [Hz]	ζ_x	$\sigma_x + j\nu_x$
1	262.16	0.048	1.994606E−06−j 3.779091E−06
2	503.60	0.096	1.932445E−06−j 2.44616E−05
3	667.92	0.057	1.364547E−05−j 4.128093E−05
4	886.78	0.074	8.632813E−07−j 7.195716E−05
5	2,201.5	0.015	−5.011467E−06−j 4.623555E−05
6	2,799.4	0.027	−1.394105E−05−j 8.317728E−05
7	3,837.7	0.007	−6.874543E−06−j 6.560925E−05
8	4,923.6	0.019	−8.824935E−06−j 6.54481E−05
9	6,038.7	0.025	2.564520E−06−j 1.986276E−05
Mode y	ω_{ny} **[Hz]**	ζ_y	$\sigma_y + j\nu_y$
1	285.53	0.021	9.797268E−07−j 5.959362E−07
2	587.8	0.089	1.349004E−05−j 4.544067E−06
3	749.6	0.027	−3.106831E−05−j 3.816475E−05
4	804.9	0.075	3.033849E−05−j 8.813244E−05
5	1,573.7	0.027	1.283945E−06−j 8.678961E−06
6	2,038.1	0.016	1.298625E−06−j 1.270846E−05
7	2,303.3	0.220	9.518897E−07−j 3.750897E−05
8	2,681.0	0.019	1.190834E−05−j 2.825781E−05
9	2,870.5	0.014	−1.912932E−05−j 4.051860E−05
10	3,838.8	0.006	−1.119589E−05−j 8.475725E−05
11	4,928.9	0.017	−1.859880E−05−j 8.993226E−05
12	6,073.4	0.016	−7.608392E−06−j 2.022991E−05

small radial immersions. The frequency domain solution can be used to include higher-order harmonics at the expense of computational complexity as follows [34, 76].

The dynamic milling force equation (4.79) is revisited as a starting point as follows:

$$\{F(\omega)\} = \frac{1}{2} a K_{\mathrm{t}} \left\{ [A(\omega)] * [1 - e^{-i\omega T}][\Phi(i\omega)]\{F(\omega)\} \right\}. \tag{4.100}$$

Because of the periodicity of the directional matrix ($[A(\omega)]$) at tooth passing frequency (ω_T), the Floquet theory states that the periodic force has the following solution where the tool vibrates with an additional chatter frequency ω_c,:

$$\{F(t)\} = e^{i\omega_c t}\{P(t)\}, \quad \{P(t)\} = \sum_{l=-\infty}^{l=+\infty} \{P_l\} e^{il\omega_T t}, \tag{4.101}$$

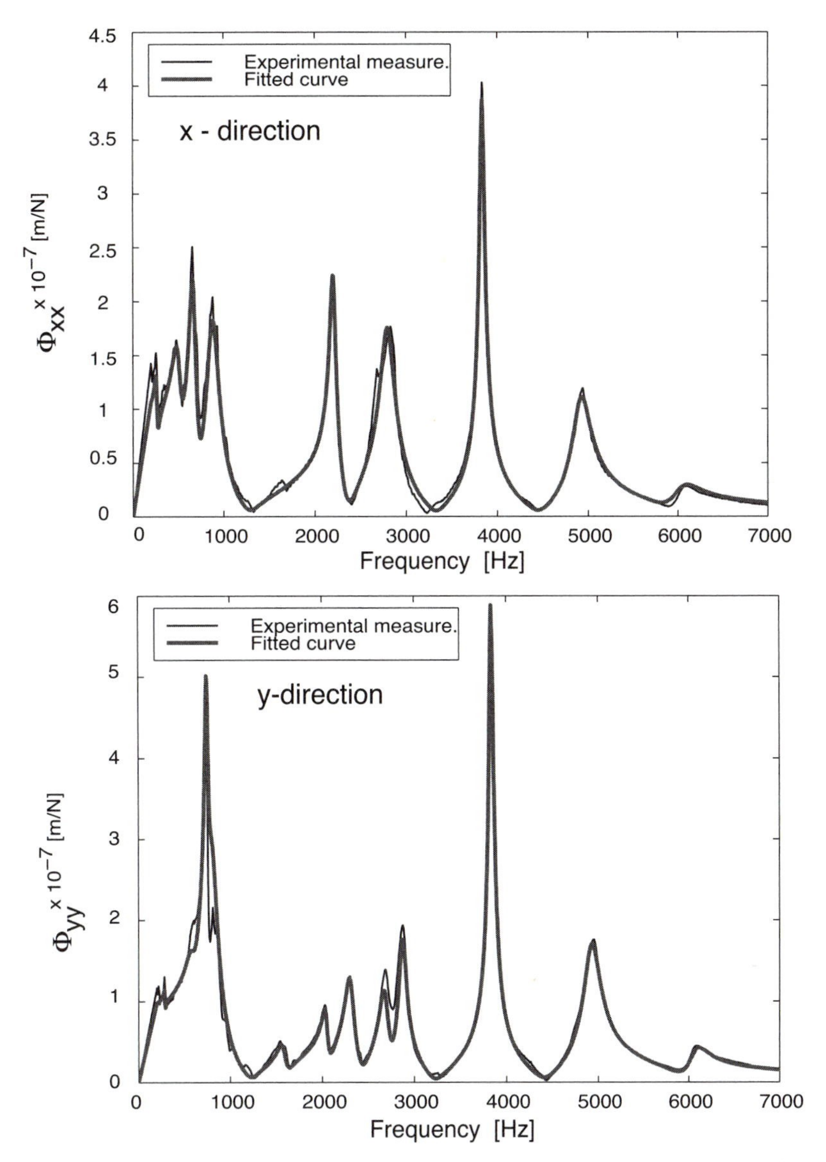

Figure 4.16: Experimentally measured and curve-fitted transfer functions of a vertical machining center at the tool tip.

where $\{P(t)\}$ is periodic at tooth passing frequency ω_T. The force has the following form in the frequency domain according to the modulation theorem:

$$\{F(\omega)\} = \mathcal{F}\left[\{F(t)\}\right] = \mathcal{F}\left[e^{i\omega_c t}\{P(t)\}\right] = \mathcal{F}\left[e^{i\omega_c t}\sum_{l=-\infty}^{l=+\infty}\{P_l\}e^{il\omega_T t}\right]$$

$$= \mathcal{F}\left[\sum_{l=-\infty}^{l=+\infty}\{P_l\}e^{i(\omega_c + l\omega_T)t}\right] = \sum_{l=-\infty}^{l=+\infty}\{P_l\}\delta\left[\omega - \left(\omega_c + l\omega_T\right)\right], \qquad (4.102)$$

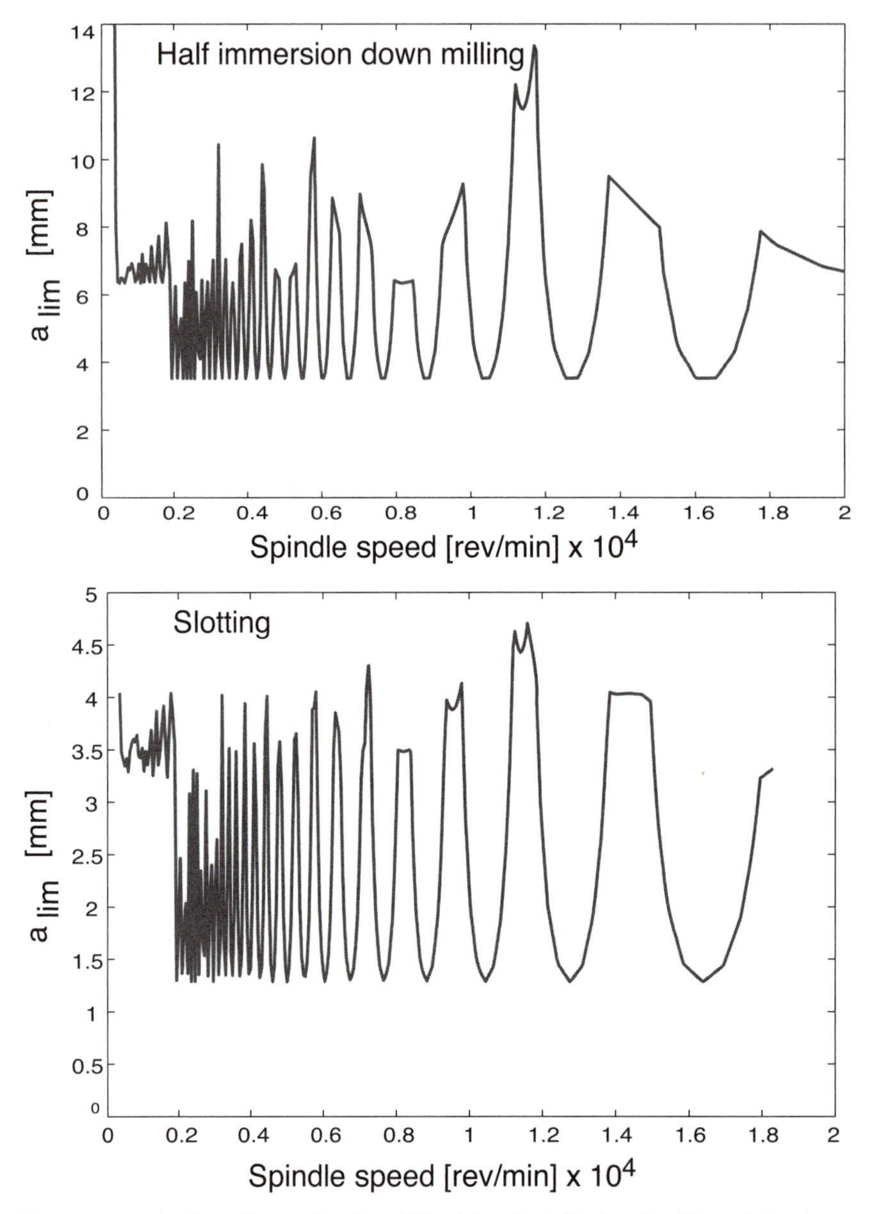

Figure 4.17: Analytically predicted stability lobes for helical end milling of aluminum alloy Al-7075.

where $\delta\left[\omega-\left(\omega_c+l\omega_T\right)\right]$ is Dirac delta function. By substituting $\{F(\omega)\}$ into dynamic milling Eq. (4.100),

$$\{F(\omega)\}=\frac{1}{2}aK_{\mathrm{t}}[A(\omega)]*\left\{\left(1-e^{-i\omega T}\right)[\Phi(i\omega)]\sum_{l=-\infty}^{l=+\infty}\{P_l\}\delta\left[\omega-\left(\omega_c+l\omega_T\right)\right]\right\}. \tag{4.103}$$

The Diract delta function δ samples the terms in the functions when $\omega = \omega_c + l\omega_T$ as follows:

$$\{F(\omega)\} = \frac{1}{2} a K_t [A(\omega)] * \left(\sum_{l=-\infty}^{l=+\infty} \left\{ \left(1 - e^{-i(\omega_c + l\omega_T)T}\right) [\Phi(\omega_c + l\omega_T)]\{P_l\}\delta\left[\omega - \left(\omega_c + l\omega_T\right)\right] \right\} \right). \tag{4.104}$$

Because $l\omega_T T = l\omega_T \frac{2\pi}{\omega_T} = l2\pi$ and $e^{-il\omega_T T} = 1$, $\{F(\omega)\}$ is reduced to

$$\begin{aligned}\{F(\omega)\} &= \frac{1}{2} a K_t [A(\omega)] * \left(\sum_{l=-\infty}^{l=+\infty} \{(1 - e^{-i\omega_c T}) [\Phi(\omega_c + l\omega_T)]\{P_l\}\delta\left[\omega - \left(\omega_c + l\omega_T\right)\right]\} \right) \\ &= \frac{1}{2} a K_t \left(1 - e^{-i\omega_c T}\right) [A(\omega)] * \left(\sum_{l=-\infty}^{l=+\infty} \{[\Phi(\omega_c + l\omega_T)]\{P_l\}\delta\left[\omega - \left(\omega_c + l\omega_T\right)\right]\} \right).\end{aligned} \tag{4.105}$$

By substituting the Fourier expansion of directional factor matrix $[A(\omega)]$ from Eq. (4.80),

$$\{F(\omega)\} = \frac{1}{2} a K_t \left(1 - e^{-i\omega_c T}\right) \left(\sum_{r=-\infty}^{+\infty} [A_r]\delta(\omega - r\omega_T) \right) * \left(\sum_{l=-\infty}^{l=+\infty} \{[\Phi(\omega_c + l\omega_T)]\{P_l\}\delta\left[\omega - \left(\omega_c + l\omega_T\right)\right]\} \right). \tag{4.106}$$

By applying the theory of Cauchy for the products of infinite but converging series ($\sum_{n=0}^{\infty} a_n \cdot \sum_{n=0}^{\infty} b_n = \sum_{n=0}^{\infty} c_n \rightarrow c_n = \sum_{k=1}^{n} a_{n-k} b_k$),

$$\{F(\omega)\} = \frac{1}{2} a K_t \left(1 - e^{-i\omega_c T}\right) \times \left(\sum_{r=-\infty}^{+\infty} \sum_{l=-\infty}^{l=+\infty} [A_{r-l}][\Phi(\omega_c + l\omega_T)]\{P_l\}\delta\left[\omega - \left(r - l\right)\omega_T\right] * \delta\left[\omega - \left(\omega_c + l\omega_T\right)\right] \right) \tag{4.107}$$

and applying the shifting theorem on Dirac delta functions ($\delta(\omega - a) * \delta(\omega - b) = \delta[\omega - (a + b)]$),

$$\begin{aligned}\delta\left[\omega - (r - l)\omega_T\right] * \delta\left[\omega - \left(\omega_c + l\omega_T\right)\right] &= \delta\left\{\omega - \left[(r - l)\,\omega_T + \omega_c + l\omega_T\right]\right\} \\ &= \delta\left\{\omega - [\omega_c + r\omega_T]\right\}\end{aligned} \tag{4.108}$$

on the Fourier expansion of dynamic force $\{F(\omega)\}$ (Eq. 4.106) leads to the following:

$$\{F(\omega)\} = \sum_{r=-\infty}^{+\infty} \left[\frac{1}{2} a K_t \left(1 - e^{-i\omega_c T}\right) \left(\sum_{l=-\infty}^{l=+\infty} [A_{r-l}][\Phi(\omega_c + l\omega_T)]\{P_l\} \right) \right.$$
$$\left. \times\, \delta\left[\omega - (\omega_c + r\omega_T)\right] \right]. \quad (4.109)$$

By expressing the force with its harmonics from Eq. (4.102) $\{F(\omega)\} = \sum_{l=-\infty}^{l=+\infty} \{P_l\} \delta\left[\omega - \left(\omega_c + l\omega_T\right)\right]$, the Fourier coefficients of the force can be expressed as

$$\{P_r\} = \Lambda \left(\sum_{l=-\infty}^{+\infty} [A_{r-l}][\Phi(\omega_c + l\omega_T)]\{P_l\} \right), \ldots (r = 0, \pm 1, \pm 2, .. \pm \infty)$$
$$\Lambda = \frac{1}{2} a K_t \left(1 - e^{-i\omega_c T}\right). \quad (4.110)$$

If only one term of the harmonics is considered, i.e., $(r, l) = \epsilon(-1, 0, +1)$,

$$\{P_0\} = \Lambda\left([A_0][\Phi(\omega_c)]\{P_0\} + [A_1][\Phi(\omega_c - \omega_T)]\{P_{-1}\}\right.$$
$$\left. + [A_{-1}][\Phi(\omega_c + \omega_T)]\{P_1\} + \cdots\right) \rightarrow r = 0; l = 0, -1, +1$$
$$\{P_{-1}\} = \Lambda\left([A_{-1}][\Phi(\omega_c)]\{P_0\} + [A_0][\Phi(\omega_c - \omega_T)]\{P_{-1}\}\right.$$
$$\left. + [A_{-2}][\Phi(\omega_c + \omega_T)]\{P_1\} + \cdots\right) \rightarrow r = -1; l = 0, -1, +1$$
$$\{P_{+1}\} = \Lambda\left([A_1][\Phi(\omega_c)]\{P_0\} + [A_2][\Phi(\omega_c - \omega_T)]\{P_{-1}\}\right.$$
$$\left. + [A_0][\Phi(\omega_c + \omega_T)]\{P_1\} + \cdots\right) \rightarrow r = +1; l = 0, -1, +1$$

.

.

By collecting the terms in matrix form,

$$\begin{Bmatrix} \{P_0\} \\ \{P_{-1}\} \\ \{P_1\} \end{Bmatrix} = \Lambda \left(\overbrace{\begin{bmatrix} [A_0] & [A_1] & [A_{-1}] \\ [A_{-1}] & [A_0] & [A_{-2}] \\ [A_1] & [A_2] & [A_0] \end{bmatrix}}^{l=0,-1,+1;\, p=0,-1,+1} \begin{Bmatrix} [\Phi(\omega_c)] \\ [\Phi(\omega_c - \omega_T)] \\ [\Phi(\omega_c + \omega_T)] \end{Bmatrix} \right) \begin{Bmatrix} \{P_0\} \\ \{P_{-1}\} \\ \{P_1\} \end{Bmatrix}$$

Note that if $[\Phi]$ is $\left[2 \times 2\right]$ matrix, then each $\{P\} = \{2 \times 1\}$ vector and $[A] = \left[2 \times 2\right]$ matrix. By collecting the terms $\{P\}$, the stability of the system is defined by the following eigenvalue problem:

$$\left\{[I] - \Lambda [A_{r-l}][\Phi(\omega_c + l\omega_T]]\right\}\{P\} = \{0\}, \quad (4.111)$$

where $\{P\}$ is the eigenvector and Λ is the eigenvalue. Note that for each eigenvalue (Λ), there will be one eigenvector $\{P\}$. If we continue to use the same example with one harmonic of the periodic function,

$$\det\left|\begin{bmatrix}[I]_{2x2} & 0 & 0\\ 0 & [I] & 0\\ 0 & 0 & [I]\end{bmatrix} - \Lambda\left(\overbrace{\begin{bmatrix}[A_0]_{2x2} & [A_1] & [A_{-1}]\\ [A_{-1}] & [A_0] & [A_{-2}]\\ [A_1] & [A_2] & [A_0]\end{bmatrix}\begin{Bmatrix}[\Phi(\omega_c)]_{2x2}\\ [\Phi(\omega_c-\omega_T)]\\ [\Phi(\omega_c+\omega_T)]\end{Bmatrix}}^{k=0,-1,+1;r=0,-1,+1}\right)\right| = 0, \tag{4.112}$$

where the matrix $[I]$, $[A]$, $[\Phi]$ have 2×2 dimensions if the system is flexible in two orthogonal directions (i.e., x, y).The determinant of the system equation would lead to eigenvalues (Λ) as follows:

$$\det\left|[I] - \Lambda[A_{r-l}][\Phi(\omega_c + l\omega_T]\right| = 0, \quad (r, l) = 0, \pm1, \pm2, \ldots \tag{4.113}$$

where (r, l) represents column and row indices of the the directional matrix, respectively. The directional and frequency response function matrix have (2×2) dimensions for 2-DOF system as follows:

$$[A_{r-l}][\Phi(\omega_c + r\omega_T] = \begin{bmatrix}\alpha_{xx}^{(r-l)} & \alpha_{xy}^{(r-l)}\\ \alpha_{yx}^{(r-l)} & \alpha_{yy}^{(r-l)}\end{bmatrix}\begin{bmatrix}\Phi_{xx}(\omega_c + l\omega_T) & \Phi_{xx}(\omega_c + l\omega_T\\ \Phi_{xy}(\omega_c + l\omega_T) & \Phi_{yy}(\omega_c + l\omega_T\end{bmatrix} \tag{4.114}$$

If we consider r number of tooth passing frequency harmonics, the dimension of the eigenvalue matrix becomes $D = n_{\mathrm{dof}}(2r + 1)$, where n_{dof} is the number of orthogonal flexibility directions considered in chatter. For example, if $r = 1$, and flexibility in (x, y) directions is considered ($n_{\mathrm{dof}} = 2$), the matrix dimension becomes $D = 6$. If we consider that the system is flexible only in the x or y direction (i.e., $n_{\mathrm{dof}} = 1$) , $D = 3$. The number of eigenvalues obtained from the solution would be equal to the size of the matrix (D). By noting that $e^{-i\omega_c T} = \cos\omega_c T - i\sin\omega_c T$, the eigenvalue number q can be represented by

$$\Lambda_q = \Lambda_{R,q} + i\Lambda_{I,q} = \frac{1}{2}aK_t[1 - e^{-i\omega_c T}] = \frac{1}{2}aK_t[1 - \cos\omega_c T + i\sin\omega_c T]$$

$$a = \frac{2(\Lambda_{R,q} + i\Lambda_{I,q})}{K_t[1 - \cos\omega_c T + i\sin\omega_c T]} = \frac{2(\Lambda_{R,q} + i\Lambda_{I,q})[(1 - \cos\omega_c T) - i\sin\omega_c T]}{K_t[(1 - \cos\omega_c T)^2 + (\sin\omega_c T)^2]}$$

$$a = \frac{[\Lambda_{R,q}(1 - \cos\omega_c T) + \Lambda_{I,q}\sin\omega_c T]}{K_t(1 - \cos\omega_c T)} + i\frac{[\Lambda_{I,q}(1 - \cos\omega_c T) - \Lambda_{R,q}\sin\omega_c T]}{K_t(1 - \cos\omega_c T)}.$$

Because the depth of cut is a physical quantity, the imaginary part of a must be zero.

$$\Lambda_{I,q}(1-\cos\omega_c T) = \Lambda_{R,q}\sin\omega_c T \rightarrow \frac{\Lambda_{I,q}}{\Lambda_{R,q}} = \frac{\sin\omega_c T}{1-\cos\omega_c T}$$

$$\frac{\Lambda_{I,q}}{\Lambda_{R,q}} = \frac{2\sin\frac{\omega_c T}{2}\cos\frac{\omega_c T}{2}}{1-(\cos^2\frac{\omega_c T}{2}-\sin^2\frac{\omega_c T}{2})}$$

$$= \frac{2\sin\frac{\omega_c T}{2}\cos\frac{\omega_c T}{2}}{\sin^2\frac{\omega_c T}{2}+\cos^2\frac{\omega_c T}{2}-\cos^2\frac{\omega_c T}{2}+\sin^2\frac{\omega_c T}{2}}$$

$$= \frac{2\sin\frac{\omega_c T}{2}\cos\frac{\omega_c T}{2}}{2\sin^2\frac{\omega_c T}{2}}$$

$$\tan\psi = \frac{\Lambda_{I,q}}{\Lambda_{R,q}} = \frac{\cos\frac{\omega_c T}{2}}{\sin\frac{\omega_c T}{2}} = \tan\left(\frac{\pi}{2}-\frac{\omega_c T}{2}+k\pi\right), \text{ where } k = 0, 1, 2, \ldots$$

$$\psi = \frac{\pi-\omega_c T+2k\pi}{2}, \rightarrow \omega_c T = \pi - 2\psi + 2k\pi$$

$$\omega_c T = \varepsilon + 2k\pi, \quad \varepsilon = \pi - 2\psi, \quad \psi = \tan^{-1}\frac{\Lambda_{I,q}}{\Lambda_{R,q}}$$

Spindle speed n [rev/ min) $= \dfrac{60}{NT} = \dfrac{60\omega_c}{N(\varepsilon+2k\pi)}$.

The axial depth of cut is as follows:

$$a_q = \frac{[\Lambda_{R,q}(1-\cos\omega_c T)+\Lambda_{I,q}\sin\omega_c T]}{K_t(1-\cos\omega_c T)}$$

$$= \frac{\Lambda_{R,q}}{K_t}\left[\frac{1-\cos\omega_c T+\frac{\Lambda_{I,q}}{\Lambda_{R,q}}\sin\omega_c T}{1-\cos\omega_c T}\right]$$

$$= \frac{\Lambda_{R,q}}{K_t}\left[1+\frac{\Lambda_{I,q}}{\Lambda_{R,q}}\frac{\sin\omega_c T}{1-\cos\omega_c T}\right] = \frac{\Lambda_{R,q}}{K_t}\left[1+\frac{\Lambda_{I,q}}{\Lambda_{R,q}}\frac{\Lambda_{I,q}}{\Lambda_{R,q}}\right]$$

$$a_q = \frac{\Lambda_{R,q}}{K_t}\left[1+\left(\frac{\Lambda_{I,q}}{\Lambda_{R,q}}\right)^2\right].$$

The critically stable spindle speed and axial depth of cut are summarized as

$$n = \frac{60}{NT} = \frac{60\omega_c}{N(\varepsilon+2k\pi)}$$

$$a_q = \frac{\Lambda_{R,q}}{K_t}\left[1+\left(\frac{\Lambda_{I,q}}{\Lambda_{R,q}}\right)^2\right]. \tag{4.115}$$

However, the evaluation of phase shift (ε) between the waves requires the solution of eigenvalues Λ_q, which depends on the evaluation of the FRF at the harmonics of tooth-passing frequencies ($[\Phi(\omega_c \pm l\omega_T)]$). Hence, unlike in the zero-order solution, there is no direct solution to the depth of cut. Instead, a range of spindle speeds must be scanned for each chatter frequency to find the eigenvalues. For a given speed, we can calculate the eigenvalues and the corresponding value of depth of cut by using Eq. (4.115). The computational load is significantly higher in the multi-frequency solution in comparison with the zero-frequency solution. For example, if we scan a frequency range of 1,000Hz at 1-Hz intervals, speed range of 0 to 15,000 at 100 rev/min intervals, take 2 harmonics of the tooth-passing frequency for a two-dimensional flexibility system ($D = n_{\text{dof}}(2h_r + 1) = 2(2 \times 2 + 1) = 10$, we need to have $1{,}000 \times 15{,}000/100 = 1{,}500{,}000$ iterations. If each iteration takes 0.1 sec, the computation time is 150,000 sec = 150,000/60 = 2,500 min = 41.63 [hr]. There will be D number of eigenvalues, and at each iteration the most conservative and positive depth of cut must be considered as a final solution. However, the zero-order solution, which considers only the average directional factors $[A_0]$, gives the stability lobes directly within a second or two. The zero-order and multifrequency solutions give almost identical results within the speed range, which is less than or equal to the natural frequencies. If the speed is greater than the natural frequency, and the immersion is very low, the process is highly intermittent and the multifrequency solution would yield more accurate lobes.

Additional difficulties exist with the iterative solution of the depth of cut given in Eq. (4.115). Because there are $D = n_{\text{dof}}(2r + 1)$ number of eigenvalues, there are also $D = n_{\text{dof}}(2r + 1)$ number of possible axial depths of cuts. Taking the minimum of all depths of cuts does not necessarily lead to the correct solution. The harmonics ($\omega_c \pm l\omega_T$) may create artificial flexibility at $\Phi_{xx}(\omega_c \pm l\omega_T)$. Only one of the solutions, which must be identified by checking the eigenvalues, must be valid. The following procedure ensures the correct identification of acceptable eigenvalues that lead to the accurate prediction of a critically stable depth of cut for a given set of spindle speed and chatter frequency.

The complex eigenvalue $\Lambda_q = \Lambda_{R,q} + i\Lambda_{I,q}$ can be expressesd by its magnitude and phase as follows:

$$|\Lambda_q| = \sqrt{\Lambda_{R,q}^2 + \Lambda_{I,q}^2}, \quad \tan\psi_q = \frac{\Lambda_{I,q}}{\Lambda_{R,q}}$$

$$\sin\psi_q = \frac{\Lambda_{I,q}}{|\Lambda_q|}, \qquad \cos\psi_q = \frac{\Lambda_{R,q}}{|\Lambda_q|}. \tag{4.116}$$

Noting that the imaginary part of the axial depth of cut is zero, which led to the following eigenvalue root condition,

$$\tan\psi = \frac{\Lambda_{I,q}}{\Lambda_{R,q}} = \frac{\cos\frac{\omega_c T}{2}}{\sin\frac{\omega_c T}{2}} \rightarrow \Lambda_{R,q}\cos\frac{\omega_c T}{2} - \Lambda_{I,q}\sin\frac{\omega_c T}{2} = 0. \tag{4.117}$$

By normalizing it with the magnitude of the eigenvalue,

$$\frac{\Lambda_{R,q}\cos\frac{\omega_c T}{2} - \Lambda_{I,q}\sin\frac{\omega_c T}{2}}{\sqrt{\Lambda_{R,q}^2 + \Lambda_{I,q}^2}} = 0$$

$$\cos\psi_q \cos\frac{\omega_c T}{2} - \sin\psi_q \sin\frac{\omega_c T}{2} = 0$$

$$\cos(\frac{2\psi_q + \omega_c T}{2}) = 0, \;\; \psi_q = \tan^{-1}\frac{\Lambda_{I,q}}{\Lambda_{R,q}}. \tag{4.118}$$

For each eigenvalue q, the phase condition ($\cos(\frac{2\psi_q+\omega_c T}{2}) = 0$) must be satisfied. Because the eigenvalue solution is iterative, the convergence and identification of roots are numerically challenging, and require a well-tuned and efficient computer algorithm. Further inspection is needed to eliminate the false eigenvalue solution created by the FRFs that are $\pm l\omega_T$ away from the chatter frequency ω_c.

Once chatter occurs at frequency ω_c, the dynamic force exhibits frequencies at $\omega_c \pm l\omega_T$ as follows:

$$\{F(t)\} = \begin{Bmatrix} F_x(t) \\ F_y(t) \end{Bmatrix} = \sum_{l=-\infty}^{+\infty} \{P_l\} e^{i(\omega_c + l\omega_T)t} = \sum_{l=-\infty}^{+\infty} \begin{Bmatrix} P_{lx} \\ P_{ly} \end{Bmatrix} e^{i(\omega_c + l\omega_T)t}$$
$$= \cdots + \{P_{-2}\}e^{i(\omega_c - 2\omega_T)t} + \{P_{-1}\}e^{i(\omega_c - \omega_T)t} + \{P_0\}e^{i\omega_c t} + \{P_1\}e^{i(\omega_c + \omega_T)t}$$
$$+ \{P_2\}e^{i(\omega_c + 2\omega_T)t} + \cdots \tag{4.119}$$

where $\{\cdots P_{-2x}\ P_{-1x}\ P_{0x}\ P_{1x}\ P_{2x}\ \cdots\}^T$ and $\{\cdots P_{-2y}\ P_{-1y}\ P_{0y}\ P_{1y}\ P_{2y}\ \cdots\}^T$ are the eigenvectors in flexible x and y directions, respectively. Each term in the eigenvector represents the amplitude of the spectrum at the corresponding frequency ($\omega_c \pm l\omega_T$). If the system chatters at the frequency ω_c, the force and vibration caused by the unstable, regenerative dynamic chip will lead to largest force and vibration at the same frequency ω_c. As a result, the term $P_0 e^{i\omega_c t}$ in Eq. (4.119) will have the highest spectrum amplitude P_0, i.e., $P_0 > \in (\cdots P_{-2}, P_{-1}, P_1, P_2, ...)$, which can be checked more efficiently by normalizing the eigenvector by P_0 in each flexible direction as follows:

$$|\overline{p}| = \left\{\cdots \left|\frac{P_{-2}}{P_0}\right| \left|\frac{P_{-1}}{P_0}\right| 1 \left|\frac{P_1}{P_0}\right| \left|\frac{P_2}{P_0}\right| \cdots\right\}^T. \tag{4.120}$$

If any of the normalized eigenvector terms, i.e., $\left|\frac{P_l}{P_0}\right| > 1$, the solution gives a false depth of cut and must be neglected during the iterative solution. If the false solution is not neglected, the multifrequency solution may always converge to the zero-order solution where the harmonics are not considered.

Example 5. Low Immersion Milling. The stability of a milling system with a low radial depth of cut is predicted by considering average directional factors $[A_0]$ and up to three harmonics ($r = 3$) of tooth-passing frequency according to the zero and multi-frequency solution, respectively. The cutting conditions and structural parameters are given in Figure 4.18. Whereas both solutions predicted chatter at point A, zero-frequency and multifrequency solutions predicted contradictory stability results in operating points B and C. The cutting forces and vibrations are predicted by using the time domain simulations as shown in Figure 4.18.

At point A ($n = 30{,}000$ rev/min, $a = 2$ mm), both zero- and multi-frequency solutions give chatter, and both solutions, i.e., depths of cut, are almost equal to each other. The force spectrum exhibits dominant tooth-passing frequency harmonics plus the spread of chatter frequency at the integer multiples of tooth-passing frequency. However, there is only one dominant frequency in the spectrum of vibrations in y direction, which is the chatter frequency. The chatter frequency is obtained as $\omega_{c,\mathrm{TDS}} = 947.23$ Hz, $\omega_{c,\mathrm{MFS}} = 946.9$ Hz from time domain simulation (TDS) and multi-frequency solution (MFS), respectively. This type of regular chatter vibration is called Hopf Bifurcation in the literature.

Point B ($n = 34{,}000$ rev/min, $a = 3$ mm) is located in the stable, added lobe predicted by the multi-frequency solution, but predicted as an unstable zone by the zero-frequency solution. Time domain simulation, presented in Figure 4.18b, shows that the process is indeed stable. The dominant frequencies of both the tool vibration and cutting force are only at the tooth-passing frequency

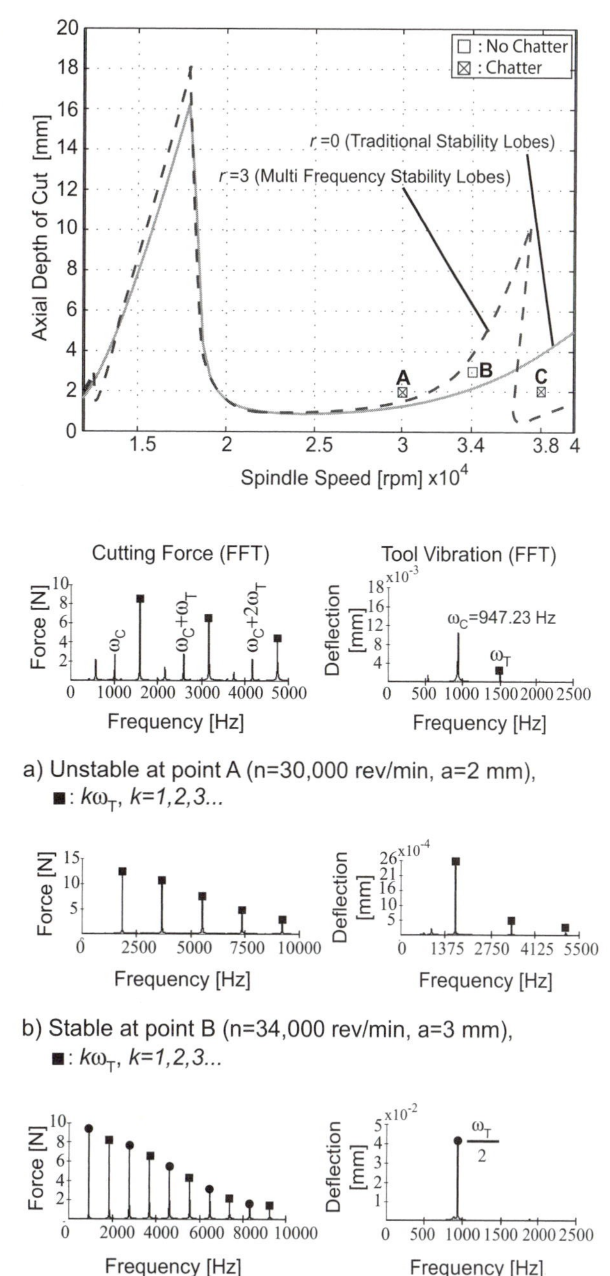

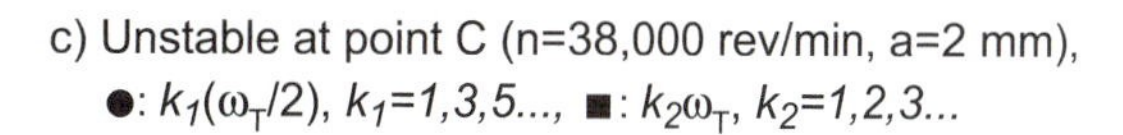

Figure 4.18: Stability lobes for low-immersion down-milling, structural dynamic parameters in normal (y) direction: $\omega_{ny} = 907[Hz]$, $k_y = 1.4 \times 10^6[N/m]$, $\zeta_y = 0.013$, and the system is rigid in x direction. The radial depth of cut: $1.256[mm]$, 3 fluted end mill with $23.6[mm]$ diameter. Feedrate=0.12 mm/rev/tooth. Cutting coefficients (Al-6061): $K_t = 500[MPa]$ and $K_r = 0.2$

(ω_T) plus its integer harmonics because of the periodic behavior of the stable milling process.

Point C ($n = 38{,}000$ rev/min, $a = 2$ mm) is predicted to be stable by the zero-frequency solution, but is indicated as an unstable zone by the multi-frequency solution. The simulated cutting forces and tool vibrations indicate clear instability because of growing amplitudes, and their spectra are given in Figure 4.18c. It can be seen that only tooth-passing frequency and its harmonics ($k\omega_T$), as well as half-tooth-passing frequency and its odd harmonics ($(2k+1)(\omega_T/2)$), are dominant, where k is a positive integer number. This type of chatter vibration is called flip bifurcation in the literature. The multi-frequency solution predicts the most dominant chatter vibration frequency as $\omega_{c,\text{MFS}} = 950.5$ Hz which is almost equal to half of the tooth-passing frequency $\omega_{c,\text{TDS}} = \omega_T/2 = 950$ Hz. Note that such a spectrum, which is dominated at the integer multiples of half-tooth-passing frequency in cutting forces, must not be confused with the forced vibrations that occur at the integer multiples of tooth passing frequency. The cutting conditions at point C clearly exhibit chatter as proven by the time domain simulations of the process. Most common milling conditions experienced in industry exhibit regular chatter, which can be accurately predicted by the zero frequency solution.

4.7 CHATTER STABILITY OF DRILLING OPERATIONS

The general equations of motion for the dynamic drilling system can be formulated in the stationary frame as follows:

$$[M]\begin{Bmatrix} \ddot{x}_c(t) \\ \ddot{y}_c(t) \\ \ddot{z}_c(t) \\ \ddot{\theta}_c(t) \end{Bmatrix} + [C]\begin{Bmatrix} \dot{x}_c(t) \\ \dot{y}_c(t) \\ \dot{z}_c(t) \\ \dot{\theta}_c(t) \end{Bmatrix} + [K]\begin{Bmatrix} x_c(t) \\ y_c(t) \\ z_c(t) \\ \theta_c(t) \end{Bmatrix} = \begin{Bmatrix} F_x(t) \\ F_y(t) \\ F_z(t) \\ T_c(t) \end{Bmatrix},$$

Figure 4.19: Dynamic model of drill vibrations.

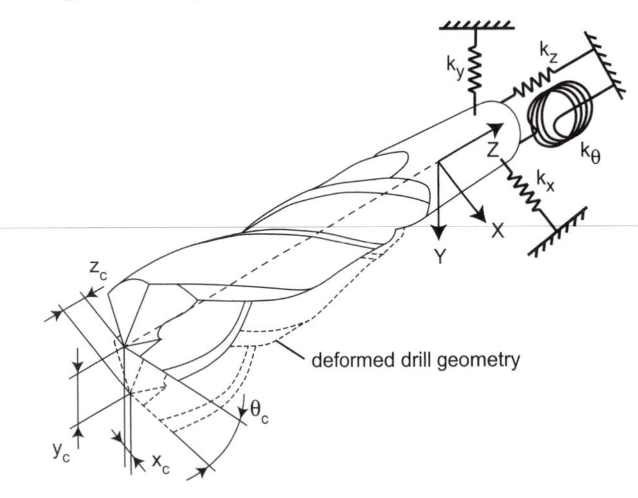

where (x_c, y_c) denote the lateral, (z_c) the axial deflections of the drill in the global coordinate system, as illustrated in Figure 4.19. (θ_c) is the torsional deflection of the drill bit itself, with respect to the rigid body spindle motion. The rotation speed of the tool is Ω in [rad/s]. The matrices $[M]$, $[C]$, and $[K]$ contain the lumped mass, damping, and stiffness characteristics reflected at the drill tip, respectively. The external cutting loads acting on the drill include two lateral forces (F_x, F_y), thrust force (F_z), and torque (T_c). The dynamic properties of the drill are either predicted from the finite element method or measured experimentally. The drill end is in contact with the material during the drilling process, which significantly alters the dynamic stiffness of the drill structure. Thrust compresses the drill, whereas torque extends the drill by untwisting it. This sign is taken into account in the proposed frequency domain solution [92, 93].

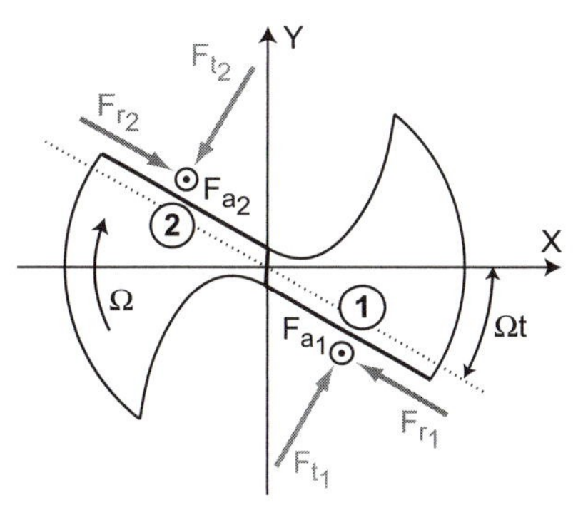

Figure 4.20: Elemental cutting forces acting on the edges of a two-fluted drill.

4.7.1 Dynamic Drilling Force Model

The (positive) force components acting on each flute are tangential (t), radial (r), and axial (a), as illustrated in Figure 4.20 and defined as follows:

$$\begin{aligned} F_{t1} &= k_{tc} b h_1 , \quad F_{r1} = k_{rc} F_{t1} , \quad F_{a1} = k_{ac} F_{t1} \\ F_{t2} &= k_{tc} b h_2 , \quad F_{r2} = k_{rc} F_{t2} , \quad F_{a2} = k_{ac} F_{t2}, \end{aligned} \tag{4.121}$$

where h_1 is the uncut chip thickness for flute 1 measured in the spindle axis direction, b is the radial depth of cut, defined by the difference of tool radius and pilot hole radius: $b = R - R_p$. The radial and axial forces are expressed to be proportional to the tangential force. The total cutting forces acting in the X, Y and Z directions at the tool tip are (see 4.20) as follows:

$$\left. \begin{aligned} F_x(t) &= (F_{t1} - F_{t2}) \sin \Omega t - (F_{r1} - F_{r2}) \cos \Omega t \\ F_y(t) &= (F_{t1} - F_{t2}) \cos \Omega t + (F_{r1} - F_{r2}) \sin \Omega t \\ F_z(t) &= F_{a1} + F_{a2} \\ T_c(t) &\simeq R_t \left(F_{t1} + F_{t2}\right) \end{aligned} \right\} \tag{4.122}$$

where R_t is the torque arm for calculating the cutting torque T_c from tangential and radial forces. The dynamic chip thickness is influenced by vibrations in three orthogonal directions and one torsional direction. The static chip thickness equals the feed per revolution (f_r) divided by the number of flutes (N), which is two for a twisted drill as follows:

$$h_s = \frac{f_r}{N}. \tag{4.123}$$

The change in chip thickness due to regenerative displacements dx, dy on each flute are:

$$\left. \begin{aligned} dh_1 &= \tfrac{1}{\tan \varkappa_t} \left(dx \cos \Omega t - dy \sin \Omega t\right) \\ dh_2 &= \tfrac{-1}{\tan \varkappa_t} \left(dx \cos \Omega t - dy \sin \Omega t\right) \end{aligned} \right\}, \tag{4.124}$$

where $2\varkappa_t$ is the tip angle of the drill and Ωt is the tool rotation angle. An increase in chip thickness on flute 1 is accompanied by an equal-sized decrease in chip thickness on flute 2. The chip thickness change is illustrated in Figure 4.21, where $du = dx \cos \Omega t - dy \sin \Omega t$ is the tool deflection in the direction of the cutting lips.

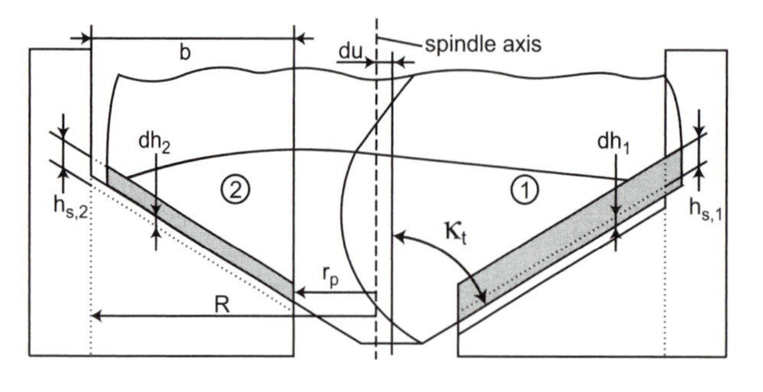

Figure 4.21: Structural dynamic model of drill with radial, axial, and torsional vibrations.

The regenerative displacements are

$$\{\Delta r\} = \begin{Bmatrix} dx \\ dy \\ dz \\ d\theta \end{Bmatrix} = \begin{Bmatrix} x_c(t) - x_c(t-T) \\ y_c(t) - y_c(t-T) \\ z_c(t) - z_c(t-T) \\ \theta_c(t) - \theta_c(t-T) \end{Bmatrix}, \tag{4.125}$$

where $T = 2\pi/N\Omega$ is the tooth period. The dynamic chip thickness due to torsional vibrations (positive in the direction of tool rotation) is expressed for two teeth as follows:

$$dh_1 = dh_2 = -\frac{f_r}{2\pi} d\theta, \tag{4.126}$$

and depends on the feed per revolution f_r. The axial vibrations influence the chip thickness directly as follows:

$$dh_1 = dh_2 = -dz \tag{4.127}$$

The total change in chip thickness becomes

$$\begin{Bmatrix} dh_1 \\ dh_2 \end{Bmatrix} = \begin{Bmatrix} \frac{dx \cos \Omega t - dy \sin \Omega t}{\tan \varkappa_t} - dz - \frac{f_r}{2\pi} d\theta \\ \frac{-(dx \cos \Omega t - dy \sin \Omega t)}{\tan \varkappa_t} - dz - \frac{f_r}{2\pi} d\theta \end{Bmatrix}. \tag{4.128}$$

The static chip thickness (h_s) is neglected because it does not contribute to the stability. The dynamic forces depend on the dynamic chip thicknesses dh_1 and dh_2 as follows:

$$\begin{Bmatrix} F_x \\ F_y \\ F_z \\ T_c \end{Bmatrix} = k_{tc} b \begin{Bmatrix} (dh_1 - dh_2) \sin \Omega t - k_{rc}(dh_1 - dh_2) \cos \Omega t \\ (dh_1 - dh_2) \cos \Omega t + k_{rc}(dh_1 - dh_2) \sin \Omega t \\ k_{ac}(dh_1 + dh_2) \\ (dh_1 + dh_2)(1 - k_{rc}) R_t \end{Bmatrix}, \tag{4.129}$$

where the sum of dynamic chip thicknesses $(dh_1 + dh_2)$ and difference $(dh_1 - dh_2)$ are

$$\begin{Bmatrix} dh_1 + dh_2 \\ dh_1 - dh_2 \end{Bmatrix} = \begin{Bmatrix} -2dz - 2\frac{f_r}{2\pi} d\theta \\ \frac{2}{\tan \varkappa_t} (dx \cos \Omega t - dy \sin \Omega t) \end{Bmatrix}, \tag{4.130}$$

which are substituted in Eq. (4.129) with $\phi = \Omega t$ to obtain the dynamic cutting forces acting on the tool, are as follows:

$$\begin{Bmatrix} F_x \\ F_y \\ F_z \\ T_c \end{Bmatrix} = k_{tc} b \begin{Bmatrix} \frac{2}{\tan \varkappa_t} (dx \cos\phi - dy \sin\phi)(\sin\phi - k_{rc}\cos\phi) \\ \frac{2}{\tan \varkappa_t} (dx \cos\phi - dy \sin\phi)(\cos\phi + k_{rc}\sin\phi) \\ k_{ac}(-2dz - 2\frac{f_r}{2\pi} d\theta) \\ (-2dz - 2\frac{f_r}{2\pi} d\theta)(1 - k_{rc}) R_t \end{Bmatrix}$$

$$= -2k_{tc} b \left[B(\phi)\right] \begin{Bmatrix} dx \\ dy \\ dz \\ d\theta \end{Bmatrix},$$

where time-varying directional matrix $\left[B(\phi)\right]$ is obtained by considering the identities ($\cos^2\phi = (1 + \cos 2\phi)/2$, $\sin^2\phi = (1 - \cos 2\phi)/2$, $\sin\phi\cos\phi = (\sin 2\phi)/2$) as follows:

$$\left[B(\phi)\right] = \begin{bmatrix} \frac{-1}{\tan \varkappa_t}\left(\frac{\sin 2\phi}{2} - k_{rc}\cos^2\phi\right) & \frac{1}{\tan \varkappa_t}\left(-\sin^2\phi + k_{rc}\frac{\sin 2\phi}{2}\right) & 0 & 0 \\ \frac{-1}{\tan \varkappa_t}\left(\cos^2\phi - k_{rc}\frac{\sin 2\phi}{2}\right) & \frac{1}{\tan \varkappa_t}\left(-\frac{\sin 2\phi}{2} - k_{rc}\sin^2\phi\right) & 0 & 0 \\ 0 & 0 & k_{ac} & \frac{f_r}{2\pi} k_{ac} \\ 0 & 0 & (1 - k_{rc}) R_t & \frac{f_r}{2\pi}(1 - k_{rc}) R_t \end{bmatrix} \quad (4.131)$$

$$= \begin{bmatrix} \frac{-1}{2\tan \varkappa_t}\left[\sin 2\phi - k_{rc}(1 + \cos 2\phi)\right] & \frac{1}{2\tan \varkappa_t}\left[-(1 - \cos 2\phi) + k_{rc}\sin 2\phi\right] & 0 & 0 \\ \frac{-1}{2\tan \varkappa_t}\left[(1 + \cos 2\phi) + k_{rc}\sin 2\phi\right] & \frac{1}{2\tan \varkappa_t}\left[-\sin 2\phi - k_{rc}(1 - \cos 2\phi)\right] & 0 & 0 \\ 0 & 0 & k_{ac} & \frac{f_r}{2\pi} k_{ac} \\ 0 & 0 & (1 - k_{rc}) R_t & \frac{f_r}{2\pi}(1 - k_{rc}) R_t \end{bmatrix}. \quad (4.132)$$

The dynamic drilling coefficient matrix $[B(\phi)]$ depends on time t, spindle speed Ω, cutting coefficients k_{rc} and k_{ac}, tool tip angle $\varkappa_t$, torque arm R_t, and feed per revolution f_r. Because the forces in the x and y directions depend on the chip thickness difference $(dh_1 - dh_2)$, axial and torsional deflections do not affect the lateral cutting forces (F_x, F_y). Similarly, as the torque and thrust depend on the chip thickness sum $(dh_1 + dh_2)$, lateral tool deflections do not affect the dynamic thrust force F_z or torque T_c. The dynamic cutting forces can be summarized as follows:

$$\{F(t)\} = -2k_{tc} b \left[B(\phi)\right] \{\Delta r\}, \quad (4.133)$$

where $\{\Delta r\}$ contains the regenerative displacements from Eq. (4.125). The time-dependent matrix $[B(\phi)]$ is periodic at tooth-passing frequency $N\Omega$ or tooth period $T = 2\pi/N\Omega$. By considering only the mean of the directional factors as similar to milling stability, the time variation in lateral directions are removed.

$$\left[B_0\right] = \frac{1}{T}\int_0^T \left[B(t)\right] dt = \frac{1}{\phi_p}\int_{\phi_{st}}^{\phi_{ex}} \left[B(\phi)\right] d\phi, \quad (4.134)$$

where the pitch angle of the tool is $\phi_p = 2\pi/N$. The mean directional matrix $[B_0]$ is valid only between entry (ϕ_{st}) and exit (ϕ_{ex}) angles of the tool. Although the entry and exit angles can vary in plunge milling operations, which have the identical process dynamics as drilling, the twist drills and cylinder boring heads have $\phi_{st} = 0$ and $\phi_{ex} = \pi$.

$$[B_0] = \frac{1}{\phi_p}\int_{\phi_{st}}^{\phi_{ex}} [B(\phi)]\, d\phi = \begin{bmatrix} \beta_{xx} & \beta_{xy} & 0 & 0 \\ \beta_{yx} & \beta_{yy} & 0 & 0 \\ 0 & 0 & \beta_{zz} & \beta_{z\theta} \\ 0 & 0 & \beta_{\theta z} & \beta_{\theta\theta} \end{bmatrix}. \tag{4.135}$$

The average directional factors for the two-fluted drill are ($\phi_p = \pi$, $\phi_{st} = 0$ and $\phi_{ex} = \pi$) as follows:

$$[B_0] = \frac{1}{\pi}\int_0^{\pi} [B(\phi)]\, d\phi = \begin{bmatrix} \frac{+k_{rc}}{2\tan\chi_t} & \frac{+1}{2\tan\chi_t} & 0 & 0 \\ \frac{-1}{2\tan\chi} & \frac{k_{rc}}{2\tan\chi_t} & 0 & 0 \\ 0 & 0 & k_{ac} & \frac{f_r}{2\pi}k_{ac} \\ 0 & 0 & (1-k_{rc})R_t & \frac{f_r}{2\pi}(1-k_{rc})R_t \end{bmatrix}, \tag{4.136}$$

which depend only on the cutting force coefficients and drill geometry. By substituting $[B_0]$ into Eq. (4.133), the dynamic cutting force becomes time invariant as

$$\{F(t)\} = -2k_{tc}b\,[B_0]\{\Delta r\}. \tag{4.137}$$

Because $[B_0]$ has zero off-diagonal terms, the lateral dynamic cutting forces (F_x, F_y) and axial force/cutting torque (F_z, T_c) are decoupled from the corresponding lateral vibrations (dx, dy) and axial/torsional $(dz, d\theta)$ vibrations.

4.8 FREQUENCY DOMAIN SOLUTION OF DRILLING STABILITY

The structural dynamics of the drill at its cutting lips can be represented by the following FRF matrix:

$$[\Phi(i\omega)] = \begin{bmatrix} \Phi_{xx} & 0 & 0 & 0 \\ 0 & \Phi_{yy} & 0 & 0 \\ 0 & 0 & \Phi_{zz} & \Phi_{z\theta} \\ 0 & 0 & \Phi_{\theta z} & \Phi_{\theta\theta} \end{bmatrix}. \tag{4.138}$$

It is assumed that there is no coupling between the lateral directions, and no cross-talk from lateral directions into axial and torsional directions either. The drill acts like a coil spring under torsional vibrations. It shrinks axially when twisted in torsion, and extends when it unwinds. The cross FRFs $(\Phi_{z\theta}, \Phi_{\theta z})$ between the axial force (F_z) and torque (T_c) represent the coupling. The drilling system is critically stable when the harmonic, regenerative displacements $\{\Delta r\}$ occur at the chatter frequency ω_c with a constant amplitude as follows:

$$\{\Delta r(i\omega_c)\} = \left(1 - e^{-i\omega_c T}\right)\{r(i\omega_c)\} = \left(1 - e^{-i\omega_c T}\right)[\Phi(i\omega)]\{F\}\,\mathrm{e}^{i\omega_c t}, \tag{4.139}$$

where $\omega_c T$ is the regenerative phase delay between the vibrations at successive tooth periods T. By substituting $\{\Delta r\,(i\omega_c)\}$ into Eq. (4.137), the dynamic drilling force is converted into the following eigenvalue problem:

$$\{F\}\,\mathrm{e}^{i\omega_c t} = -2\mathrm{k}_{tc} b_{\lim}\left(1 - e^{-i\omega_c T}\right)\left[B_0\right]\left[\Phi\,(i\omega)\right]\{F\}\,\mathrm{e}^{i\omega_c t}, \tag{4.140}$$

where $b_{\lim}$ is the critically stable depth of cut. The determinant of the system leads to eigenvalue solution with critically stable depth of cut and speed as follows:

$$\det\left(\left[I\right] + \Lambda\left[B_0\right]\left[\Phi\,(i\omega)\right]\right) = 0, \tag{4.141}$$

where the eigenvalue is

$$\Lambda = 2k_{tc} b_{\lim}\left(1 - e^{-i\omega_c T}\right) = 2k_{tc} b_{\lim}\left[(1 - \cos\omega_c T) + i\sin\omega_c T\right] \tag{4.142}$$

The determinant (Eq. 4.141) can be partitioned into lateral and torsional/axial components due to zero off-diagonal elements of mean directional matrix $[B_0]$.

$$\left.\begin{array}{l} \left|\begin{bmatrix}1 & 0\\ 0 & 1\end{bmatrix} + \Lambda_{xy}\begin{bmatrix}\beta_{xx} & \beta_{xy}\\ \beta_{yx} & \beta_{yy}\end{bmatrix}\begin{bmatrix}\Phi_{xx} & 0\\ 0 & \Phi_{yy}\end{bmatrix}\right| = 0 \\ \left|\begin{bmatrix}1 & 0\\ 0 & 1\end{bmatrix} + \Lambda_{z\theta}\begin{bmatrix}\beta_{zz} & \beta_{z\theta}\\ \beta_{\theta z} & \beta_{\theta\theta}\end{bmatrix}\begin{bmatrix}\Phi_{zz} & \Phi_{z\theta}\\ \Phi_{\theta z} & \Phi_{\theta\theta}\end{bmatrix}\right| = 0 \end{array}\right\}. \tag{4.143}$$

The eigenvalues Λ_{xy} and $\Lambda_{z\theta}$ lead to critical depth of cut and speed because of the lateral and torsional/axial structural dynamics of the drill. The critically stable depth of cut (b_{im}) and spindle speed (n) can be solved using the same procedure shown for milling. Each eigenvalue will have real and imaginary parts as follows:

$$\Lambda = \Lambda_R + i\Lambda_I = 2k_{tc} b_{\lim}\left[(1 - \cos\omega_c T) + i\sin\omega_c T\right]$$

$$b_{\lim} = \frac{1}{4k_{tc}}\left[\frac{\Lambda_R\,(1 - \cos\omega_c T) + \Lambda_I \sin\omega_c T}{1 - \cos\omega_c T} + i\frac{\Lambda_I\,(1 - \cos\omega_c T) - \Lambda_R \sin\omega_c T}{1 - \cos\omega_c T}\right]. \tag{4.144}$$

Because the imaginary part of the depth of cut needs to be zero,

$$\Lambda_I\,(1 - \cos\omega_c T) - \Lambda_R \sin\omega_c T = 0 \rightarrow \chi = \frac{\Lambda_I}{\Lambda_R} = \frac{\sin\omega_c T}{1 - \cos\omega_c T}. \tag{4.145}$$

By substituting χ into the real part of the critical depth of cut (Eq. 4.144),

$$b_{\lim} = \frac{\Lambda_R\left[1 + \chi^2\right]}{4k_{tc}}. \tag{4.146}$$

The corresponding spindle speed is obtained similar to milling as

$$T = \frac{\epsilon + 2k\pi}{\omega_c}, \epsilon = \pi - 2\tan^{-1}\frac{\Lambda_I}{\Lambda_R} \rightarrow n(\text{rev/min}) = \frac{60}{2T}. \tag{4.147}$$

The stability lobes of the drilling can be plotted by using the identical procedure explained for milling. However, the stability lobes must be separately evaluated for lateral vibrations by using eigenvalues Λ_{xy} and torsional/axial vibrations considering eigenvalues $\Lambda_{z\theta}$. It is possible to see which part of the structural dynamics affect the productivity most, which leads to hints for improved drill geometry design and spindle speed selection. It must be noted that the mechanics of drilling at its chisel edge are complex, and the drill wanders in the hole, which changes the stiffness and damping of the process during drilling operations. As a result, the stability of drilling is difficult to predict accurately in practice. Instead, the trend and sources of chatter needs to be interpreted from the stability analysis presented here.

4.9 SEMIDISCRETE TIME DOMAIN SOLUTION OF CHATTER STABILITY

Insperger and Stépán [53] presented an analytical solution of chatter stability in a discrete time domain. The delayed differential equation is discretized at discrete time intervals, which allow linear, time domain simulation of forces and vibrations while predicting the stability of the cutting system for a given set of cutting conditions. The application of the semidiscrete time domain method on orthogonal cutting and milling is presented in the following sections.

4.9.1 Orthogonal Cutting

Let the dynamics of the orthogonal cutting system be represented by the following delayed differential equation:

$$\frac{d^2x(t)}{dt^2} + 2\zeta\omega_n\frac{dx(t)}{dt} + \omega_n^2 x(t) = F(t) = \frac{\omega_n^2}{k_x}K_f a\left[-x(t) + x(t-T)\right]. \quad (4.148)$$

By organizing the differential equation with two state variables,

$$x_1(t) = x(t), \quad x_2(t) = \frac{dx(t)}{dt} \quad (4.149)$$

$$\frac{d^2x(t)}{dt^2} = \frac{dx_2(t)}{dt} = -\omega_n^2\left(1 + \frac{K_f a}{k_x}\right)x_1(t) - 2\zeta\omega_n x_2(t) + \frac{\omega_n^2}{k_x}K_f a x_1(t-T).$$

It can be expressed as first-order equations as follows:

$$\underbrace{\begin{Bmatrix}\dot{x}_1(t)\\ \dot{x}_2(t)\end{Bmatrix}}_{\dot{y}(t)} = \underbrace{\begin{bmatrix}0 & 1\\ -\omega_n^2\left(1+\frac{K_f a}{k_x}\right) & -2\zeta\omega_n\end{bmatrix}}_{L}\underbrace{\begin{Bmatrix}x_1(t)\\ x_2(t)\end{Bmatrix}}_{y(t)} + \underbrace{\begin{bmatrix}0 & 0\\ \frac{\omega_n^2}{k_x}K_f a & 0\end{bmatrix}}_{R}\underbrace{\begin{Bmatrix}x_1(t-T)\\ 0\end{Bmatrix}}_{y(t-T)} \quad (4.150)$$

$$\dot{y}(t) = Ly(t) + Ry(t-T) \quad (4.151)$$

The delay period T is divided into m number of discrete time intervals Δt, i.e., $T = m \cdot \Delta t$. Let the value of $\{y(t_i)\}$ at the current time t_i be expressed as y_i and at time $t_i - T \to y(t_i - T) = y\left[(i-m)\Delta t\right] = y_{i-m}$. When the sampling interval Δt is very small, the value of $y(t-T)$ can be approximated by averaging the values at two consecutive sampling intervals as follows:

$$y(t-T) \approx \frac{y(t_i - T + \Delta t) + y(t_i - T)}{2} = \frac{y_{i-m+1} + y_{i-m}}{2} \to t \in \left[t_i, t_{i+1}\right]. \quad (4.152)$$

The differential equation $\dot{y}$ has homogenous $(y_{ih}(t))$ and particular $(y_{ip}(t))$ solutions in small time intervals Δt as

$$y_i(t) = y_{ih}(t) + y_{ip}(t). \quad (4.153)$$

The homogenous solution is obtained as follows:

$$\dot{y}_{ih}(t) = L y_{ih}(t) \to y_{ih}(t) = C_0 e^{L(t-t_i)}, \quad (4.154)$$

where C_0 depends on the initial conditions. The particular solution is given by

$$\dot{y}_{ip}(t) = L y_{ip}(t) + R\left(\frac{y_{i-m+1} + y_{i-m}}{2}\right)$$

$$y_{ip}(t) = u(t)e^{L(t-t_i)} \to \frac{d}{dt} y_{ip}(t) = e^{L(t-t_i)} \frac{du(t)}{dt} + L e^{L(t-t_i)} u(t). \quad (4.155)$$

By substituting $y_{ip}(t) = u(t)e^{L(t-t_i)}$ and forcing the derivatives $(\dot{y}_{ip}(t) = dy_{ip}(t)/dt)$ to be equal,

$$e^{L(t-t_i)}\dot{u}(t) + L e^{L(t-t_i)} u(t) = L e^{L(t-t_i)} u(t) + R\left(\frac{y_{i-m+1} + y_{i-m}}{2}\right), \quad (4.156)$$

which leads to

$$\dot{u}(t) = \frac{1}{2} e^{-L(t-t_i)} R\left(y_{i-m+1} + y_{i-m}\right)$$

$$u(t) = -\frac{1}{2} L^{-1} e^{-L(t-t_i)} R\left(y_{i-m+1} + y_{i-m}\right). \quad (4.157)$$

Substituting $u(t)$ into $y_{ip}(t) = e^{L(t-t_i)} u(t)$,

$$y_{ip}(t) = e^{L(t-t_i)} u(t) = -\frac{1}{2} e^{L(t-t_i)} L^{-1} e^{-L(t-t_i)} R\left(y_{i-m+1} + y_{i-m}\right).$$

By using the exponential matrix property of $Y e^{X} Y^{-1} = e^{YXY^{-1}}$,

$$L y_{ip}(t) = -\frac{1}{2}\left[L e^{L(t-t_i)} L^{-1}\right] e^{-L(t-t_i)} R\left(y_{i-m+1} + y_{i-m}\right)$$

$$\begin{aligned} y_{ip}(t) &= -\frac{1}{2} L^{-1}\left[e^{LLL^{-1}(t-t_i)}\right] e^{-L(t-t_i)} R\left(y_{i-m+1} + y_{i-m}\right) \\ &= -\frac{1}{2} L^{-1} e^{L(t-t_i)} e^{-L(t-t_i)} R\left(y_{i-m+1} + y_{i-m}\right) \\ &= -\frac{1}{2} L^{-1} R\left(y_{i-m+1} + y_{i-m}\right). \end{aligned} \quad (4.158)$$

The complete solution of the system $y_i(t)$ is as follows:

$$y_i(t) = y_{ih}(t) + y_{ip}(t) = C_0 e^{L(t-t_i)} + e^{L(t-t_i)} u(t)$$

$$y_i(t) = C_0 e^{L(t-t_i)} - \frac{1}{2} L^{-1} R \left(y_{i-m+1} + y_{i-m}\right). \tag{4.159}$$

When the system is at $t = t_i$,

$$y_i = y_{ih}(t_i) + y_{ip}(t_i) = C_0 - \frac{1}{2} L^{-1} R \left(y_{i-m+1} + y_{i-m}\right)$$

$$C_0 = y_i + \frac{1}{2} L^{-1} R \left(y_{i-m+1} + y_{i-m}\right). \tag{4.160}$$

Because the solution is valid at discrete time intervals Δt, the time difference between two successive intervals is replaced by $t_{i+1} - t_i = \Delta t$ in Eq. (4.159) as follows:

$$y_{i+1} = e^{L\Delta t} C_0 - \frac{1}{2} L^{-1} R \left(y_{i-m+1} + y_{i-m}\right). \tag{4.161}$$

By substituting C_0 from Eq. (4.160), the linear solution of the delayed difference equation is expressed at discrete time intervals Δt as follows:

$$y_{i+1} = e^{L\Delta t} y_i + e^{L\Delta t} \frac{1}{2} L^{-1} R \left(y_{i-m+1} + y_{i-m}\right) - \frac{1}{2} L^{-1} R \left(y_{i-m+1} + y_{i-m}\right).$$

$$= e^{L\Delta t} y_i + \frac{1}{2} \left(e^{L\Delta t} - I\right) L^{-1} R \left(y_{i-(m-1)} + y_{i-m}\right). \tag{4.162}$$

The solution requires the previous value y_i and values a delay before (y_{i-m}, y_{i-m+1}). The discrete time values of the states can be expressed in matrix form as

$$\underbrace{\begin{bmatrix} y_i \\ y_{i-1} \\ \cdot \\ \cdot \\ y_{i-(m-1)} \\ y_{i-m} \end{bmatrix}}_{Y_i} = \underbrace{\begin{bmatrix} e^{L\Delta t} & 0 & 0 & 0 & 0 \\ I & 0 & 0 & 0 & 0 \\ \cdot & I & 0 & 0 & 0 \\ \cdot & \cdot & \cdot & \cdot & \cdot \\ 0 & 0 & I & 0 & 0 \\ 0 & 0 & 0 & I & 0 \end{bmatrix}}_{B_1} \underbrace{\begin{bmatrix} y_{i-1} \\ y_{i-2} \\ \cdot \\ \cdot \\ y_{i-m} \\ y_{i-(m+1)} \end{bmatrix}}_{Y_{i-1}} \tag{4.163}$$

$$+ \underbrace{\frac{1}{2} \left(e^{L\Delta t} - I\right) L^{-1} R \begin{bmatrix} 0 & 0 & \cdot & \cdot & I & I \\ 0 & 0 & \cdot & \cdot & 0 & 0 \\ \cdot & & \cdot & \cdot & \cdot & \cdot \\ \cdot & & \cdot & \cdot & \cdot & \cdot \\ 0 & 0 & 0 & 0 & 0 & 0 \\ 0 & 0 & 0 & 0 & 0 & 0 \end{bmatrix}}_{B_2} \underbrace{\begin{bmatrix} y_{i-1} \\ y_{i-2} \\ \cdot \\ \cdot \\ y_{i-m} \\ y_{i-(m+1)} \end{bmatrix}}_{Y_{i-1}}$$

$$\{Y_i\} = [B]\{Y_{i-1}\} \to [B] = [B_1] + [B_2],$$

where the first row is given by Eq. (4.162), and the rest of them are simply equal to themselves, i.e., $y_{i-1} = y_{i-1}, y_{i-2} = y_{i-2}, \ldots, y_{i-m} = y_{i-m}$. The delayed differential equation, which describes the chatter dynamics for the orthogonal cutting condition, is discretized with m time segments and represented by Eq. (4.163). The prediction of each vibration (y_i) at time segment $t = i\Delta t$ requires the previous value of vibration (y_{i-1}), the vibration generated one spindle revolution before $(y_{i-m} \rightarrow t_{i-m} = t_i - m\Delta t = t_i - T)$, and one just before that $(y_{i-(m+1)} \rightarrow t_{i-(m+1)} = t_i - (m+1)\Delta t = t_t - T - \Delta t)$. In other words, the system needs to remember the vibrations during the previous revolutions of the spindle. The solution therefore requires m number of initial values of the vibrations generated during the previous spindle period, which are not available. However, because the transient vibrations eventually disappear, one can assume a nonzero, and constant steady-state static deflection (i.e., $\{y\} = \{x \;\; \dot{x}\} \rightarrow x_{-1} = x_{-2} = \ldots = x_{-m} = \frac{\omega_n^2}{k_x} K_f a,\ \dot{x} = 0)$ as initial conditions. The system will converge to true steady-state vibrations after one or two revolutions of the spindle. When the depth of cut and speed are defined, Eq. (4.163) allows prediction of vibration history, as well as the dynamic cutting forces, by substituting the vibrations in Eq. (4.148) ($F_i = \frac{\omega_n^2}{k_x} K_f a \left[-x_i + x_{i-m}\right]$) when the system is stable. The accuracy of the solution depends on the number of time intervals within one spindle period. It must be noted that the sampling interval Δt must be able to capture the vibration frequency; hence, $\Delta t \leq \pi/\omega_c$ or $m \geq 2$.

Because the transition matrix $[B]$ is time invariant and constant for a specified speed and depth of cut, the discretized equation (Eq. 4.163) can be used to assess the stability of the turning system, as well, by simply checking its eigenvalues.

$$\{Y_i\} = [B]\{Y_{i-1}\} \rightarrow |\lambda[I] - [B]| = 0. \tag{4.164}$$

If any eigenvalue λ of transition matrix $[B]$ is outside the unit circle, i.e., $|\lambda| > 1$, it corresponds to having a positive real pole in the continuous system $(\lambda = e^{+\sigma_i T} > 1)$, the orthogonal system will be unstable and chatter will occur. The stability lobes can also be constructed by scanning the spindle speed n (rev/s) $(T[s] = 1/n)$ at the operating range of the machine at increments, and increasing the axial depth of cut at acceptable increments of Δa until the chatter limit is reached. Although the frequency domain stability predicts the lobes directly, the semidiscrete method requires a trial of cutting conditions; hence, it is a computationally more costly process. Furthermore, the accuracy of the semidiscrete method is highly dependent on the discrete sampling time (Δt), spindle speed interval along the lobes (Δn), and axial depth of cut increments (Δa). On the other hand, the semidiscrete method allows direct prediction of steady-state vibrations, velocities, and cutting forces at any stable cutting condition, because it is based on the analytical, time domain solution of the dynamic cutting equation in discrete time intervals.

4.9.2 Discrete Time Domain Stability Solution in Milling

Stepan et al. [49, 52] extended the semidiscretization method to predict the stability of milling. The dynamic milling equation was previously derived as Eq. (4.73).

$$\{F(t)\} = \frac{1}{2} a K_t [A(t)]\{\Delta(t)\}. \tag{4.165}$$

As the cutter rotates, the directional factors vary with time, and $[A(t)]$ is periodic at tooth passing frequency $\omega_T = N\Omega$ or tooth period $T = 2\pi/\omega_T$. For the simplicity of mathematical illustration, let the machine have the following two orthogonal flexibilities reflected at the cutter:

$$\Phi_{xx}(s) = \frac{\omega_{nx}^2/k_x}{s^2 + 2\varsigma_x\omega_{nx}s + \omega_{nx}^2} \rightarrow \ddot{x}(t) + 2\varsigma_x\omega_{nx}\dot{x}(t) + \omega_{nx}^2 x(t) = \frac{\omega_{nx}^2}{k_x} F_x(t) \tag{4.166}$$

$$\Phi_{yy}(s) = \frac{\omega_{ny}^2/k_y}{s^2 + 2\varsigma_y\omega_{ny}s + \omega_{ny}^2} \rightarrow \ddot{y}(t) + 2\varsigma_x\omega_{nx}\dot{y}(t) + \omega_{nx}^2 y(t) = \frac{\omega_{ny}^2}{k_y} F_y(t). \tag{4.167}$$

The dynamic cutting forces are

$$\begin{Bmatrix} F_x(t) \\ F_y(t) \end{Bmatrix} = \frac{1}{2} a K_t \begin{bmatrix} a_{xx} & a_{xy} \\ a_{yx} & a_{yy} \end{bmatrix} \left(\begin{Bmatrix} x(t) \\ y(t) \end{Bmatrix} - \begin{Bmatrix} x(t-T) \\ y(t-T) \end{Bmatrix} \right).$$

The resulting dynamics of the milling process is expressed by the following coupled, delayed differential equations:

$$\begin{Bmatrix} \ddot{x}(t) + 2\varsigma_x\omega_{nx}\dot{x}(t) + \omega_{nx}^2 x(t) \\ \ddot{y}(t) + 2\varsigma_y\omega_{ny}\dot{y}(t) + \omega_{ny}^2 y(t) \end{Bmatrix} = \frac{1}{2} a K_t \begin{bmatrix} \omega_{nx}^2/k_x & 0 \\ 0 & \omega_{ny}^2/k_y \end{bmatrix} \begin{bmatrix} a_{xx} & a_{xy} \\ a_{yx} & a_{yy} \end{bmatrix} \left(\begin{Bmatrix} x(t) \\ y(t) \end{Bmatrix} - \begin{Bmatrix} x(t-T) \\ y(t-T) \end{Bmatrix} \right). \tag{4.168}$$

By rearranging the equation of motion in matrix form,

$$\begin{bmatrix} 1 & 0 \\ 0 & 1 \end{bmatrix} \begin{Bmatrix} \ddot{x} \\ \ddot{y} \end{Bmatrix} + \begin{bmatrix} 2\varsigma_x\omega_{nx} & 0 \\ 0 & 2\varsigma_y\omega_{ny} \end{bmatrix} \begin{Bmatrix} \dot{x} \\ \dot{y} \end{Bmatrix} + \begin{bmatrix} \omega_{nx}^2 & 0 \\ 0 & \omega_{ny}^2 \end{bmatrix} \begin{Bmatrix} x \\ y \end{Bmatrix} = \frac{1}{2} a K_t \begin{bmatrix} \omega_{nx}^2/k_x & 0 \\ 0 & \omega_{ny}^2/k_y \end{bmatrix} \begin{bmatrix} a_{xx} & a_{xy} \\ a_{yx} & a_{yy} \end{bmatrix} \left(\begin{Bmatrix} x(t) \\ y(t) \end{Bmatrix} - \begin{Bmatrix} x(t-T) \\ y(t-T) \end{Bmatrix} \right).$$

By defining the state variables as

$$x_1 = x(t),\ x_2 = \frac{dx(t)}{dt}, \dot{x}_1 = x_2\ ; y_1 = y(t),\ \ y_2 = \frac{dy(t)}{dt} = \dot{y}_1,$$

the periodic milling dynamics with time-varying, self-excitation and delay terms expressed in Eq. (4.168) is organized as first-order equations as follows:

$$\{\dot{q}\} = [L]\{q\} + [R]\{q(t-T)\}, \tag{4.169}$$

where

$$\{z_i\} = \begin{Bmatrix} \{q_i\} \\ \{q_{i-1}\} \\ \cdot \\ \cdot \\ \{q_{i-m+1}\} \\ \{q_{i-m}\} \end{Bmatrix}_{(m+1)x1}$$

$$[B_i] = \begin{bmatrix} e^{[L_i]\Delta t} & [0] & \cdot & \cdot & \frac{1}{2}\left(e^{[L_i]\Delta t} - [I]\right)[L_i]^{-1}[R_i] & \frac{1}{2}\left(e^{[L_i]\Delta t} - [I]\right)[L_i]^{-1}[R_i] \\ [I] & & & & [0] & [0] \\ & [I] & & & [0] & \cdot \\ \cdot & & \cdot & \cdot & \cdot & \cdot \\ \cdot & & \cdot & [I] & \cdot & [0] \\ & & & & \cdot & [0] \\ & & & & [I] & [0] \end{bmatrix}_{4[m+1]\times 4[m+1]}$$

The state matrices $[L(t)]$ and $[R(t)]$ depend on time-varying directional matrix $[A(t)]$ that can be evaluated at each time interval $(t = i\Delta t)$ from Eq. (4.170). The time-varying milling process can be simulated by solving the discrete set of recursive equations (4.179) at Δt time intervals. Because the process is periodic at tooth passing interval T, it is sufficient to solve the equations at m number of time intervals.

The stability of the system can be evaluated by expressing Eq. (4.179) at m number of intervals within the tooth period T as follows:

$$\{z_{i+m}\} = [\Phi]\{z_i\} = [B]_m \cdot \cdot [B]_2[B]_1\{z_i\}. \tag{4.180}$$

According to the Floquet theory, the linear periodic system (Eq. 4.180) will be unstable if any of the eigenvalues of the transition matrix $[\Phi]$ have a modulus greater than one, critically stable if the modulus is unity, and stable if the modulus is less than unity. Unlike the zero-order frequency domain solution that gives the critical stability borders, the lobes directly, the stability must be searched iteratively by checking trial spindle speeds and depths of cuts when a semidiscrete time domain solution is used. The semidiscrete time domain solution considers the time-varying, periodic coefficients $[A(t)]$ described at discrete time intervals Δt; hence, the accuracy of the stability is expected to be higher, especially when the process is highly interrupted at small radial cutting depths.

Example 6. Low Immersion, Highly Intermittent Milling. Altintas et al., compared the frequency and discrete time domain stability solutions in [20] by using a highly interrupted, low-immersion milling operation. When the cutting process is highly intermittent at high spindle speeds, i.e., low radial immersion and a small number of teeth, the zero-order solution cannot predict the lobes as accurately as the multi–frequency and semi-discrete solutions that consider the time–varying directional factors. The problem will be most evident at speeds that start from the lobe where the tooth passing and natural frequency of the structure matches. A specific example is created to illustrate the problem as shown in Figure 4.23. The cutter has three flutes with a zero helix, cutting aluminium with a half-immersion down-milling mode. The structural modes are at 510 [Hz] and 802 [Hz], which corresponds to the highest stability pockets around 10,200 [rev/min] and 16,040 [rev/min], respectively. The process is highly intermittent because of a low immersion and small number of teeth. It can be seen that the zero-order, multifrequency and semidiscretization methods give almost the same stability lobes until the stability pocket zone around 10,200 [rev/min], but deviate afterward where the zero-order solution cannot capture the influence of time-varying directional factors on the stability. The process is stable at point A where forced vibrations occur at the tooth-passing frequency ω_T[Hz]. The process chatters around the natural mode of 802 [Hz] at point B. However, after 16,040 [rev/min], the zero-order solution cannot predict the added lobes that occur at 26,000 [rev/min] and 35,000 [rev/min]. However, both multifrequency and semidiscrete solutions predict the added lobes in perfect agreement. The multifrequency solution used three harmonics of the directional factors, and the semi-discrete solution used a frequency scale of 40 times the highest natural frequency (i.e., 40 × 802 Hz). The physics of added lobes have been explained by Merdol and Altintas [76], and are briefly discussed here. The higher harmonics of directional factors flip the FRF from left to right, bringing it to the zone of modal frequency on the right. This creates additional stability lobes. This phenomenon is illustrated at cutting condition C. For example, at $a = 30$ mm depth of cut and $n = 26{,}000$ [rev/min], the process is stable. The tooth-passing frequency is 1300 Hz, which is considered by the multifrequency solution. Both results flip the frequency response functions and bring them to the zones of 510 Hz and 802 Hz, creating a stability pocket at 26,000 [rev/min]. At spindle speed 38,000 [rev/min] (point D), the process chatters exactly at the half of the tooth-passing frequency. However, as the depth of the cut increases from 30 mm toward 60 mm, the added stability lobe appears due to the same phenomenon. Readers must be cautioned, however, that this phenomenon rarely occurs in practice, because the machine is usually not operated at tooth-passing frequencies beyond the natural modes. If operated at a higher mode, the machine would always distort the added lobes. If there is no higher mode, the machine can operate at very high speeds without unbalance issues.

4.10 PROBLEMS

1. A three-fluted end mill is used in machining Al-7075 alloy. The tangential and radial cutting constants are given as $K_t = 600$ MPa and $K_r = 0.07$, respectively. The end mill is represented by two orthogonal modes, which are evaluated as follows:

$$\omega_{nx} = 593.75 \text{ Hz}, \quad \zeta_x = 3.9\%, \quad K_x = 5.59 \times 10^6 \text{ [N/m]},$$
$$\omega_{ny} = 675 \text{ Hz}, \quad \zeta_y = 3.5\%, \quad K_y = 5.71 \times 10^6 \text{ [N/m]}.$$

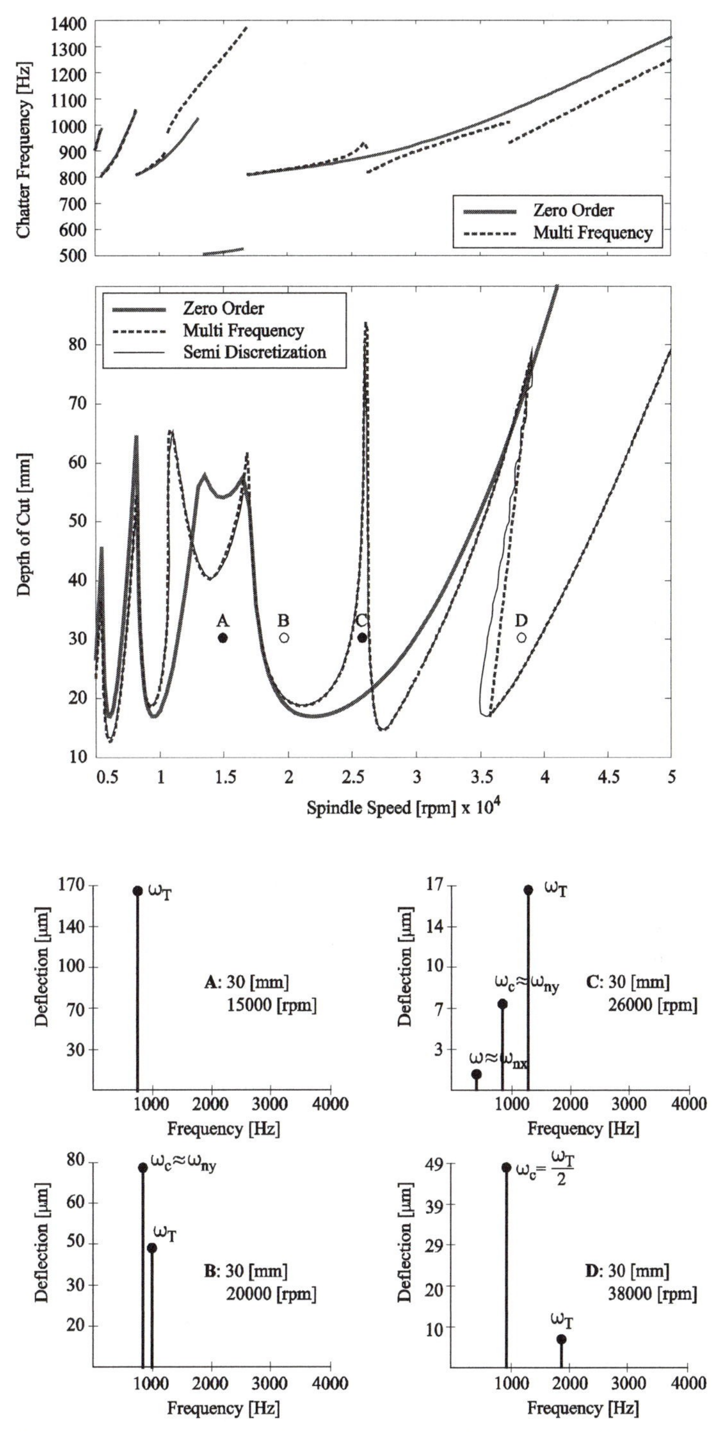

Figure 4.23: Comparison of the zero-order, multifrequency solution with $r = 3$ harmonics, and semidiscretization-based solutions. Cutter: three flutes with zero helix. Cutting condition: half-immersion down-milling. Cutting coefficients: $K_t = 900$ [N/mm^2], $K_r = 270$ [N/mm^2]. Structural dynamic parameters: $\omega_{nx} = 510$ [Hz], $\omega_{nx} = 802$ [Hz]; $\zeta_x = 0.04$, $\zeta_x = 0.05$, $k_x = 96.2 \times 10^6$ [N/m], $k_y = 47.5 \times 10^6$ [N/m].

TABLE 4.4. Modal Parameters of an End Mill Attached to a Spindle

Model	X		Y	
Parameters	Model 1	Model 2	Model 1	Model 2
ω_n(Hz)	479.4	588.7	493.3	637.9
ζ	0.51	0.029	0.068	0.027
σ_{12} (1/kg)	$1.304945.10^{-5}$	$-2.725243.10^{-6}$	$1.532536.10^{-5}$	$7.659394.10^{-6}$
$j\nu_{12}$ (1/kg)	$-2.467218.10^{-5}$	$-6.02617.10^{-5}$	$-7.680401.10^{-5}$	$-1.260438.10^{-5}$
σ_{22} (1/kg)	$6.117676.10^{-6}$	$-2.225907.10^{-6}$	$9.606246.10^{-6}$	$4.553267.10^{-6}$
$j\nu_{22}$ (1/kg)	$-1.408889.10^{-5}$	$-4.572173.10^{-5}$	$-5.5804.10^{-5}$	$-7.4116.10^{-6}$
σ_{32} (1/kg)	$1.415152.10^{-6}$	$-2.237442.10^{-6}$	$3.953402.10^{-6}$	$1.35349.10^{-6}$
$j\nu_{32}$ (1/kg)	$-1.094268.10^{-5}$	$-3.452623.10^{-5}$	$-3.913159.10^{-5}$	$-6.116283.10^{-6}$

Plot the stability lobes of the milling system for full- and half-immersion cuts.

2. The transfer functions of a 127-mm-long end mill with 100-mm diameter attached to the spindle are measured at three points in both feed (x) and normal (y) directions (see Fig. 4.24). The accelerometer was attached to point 2, and the hammer instrumented with an impact force sensor is applied to all three points. The measured transfer functions are curve fitted with a modal analysis software, and the modal parameters are given in Table 4.4.

Figure 4.24: A four-fluted long helical end mill attached to a spindle; the cutter has diameter $d = 100$ mm and length $l = 127$ mm.

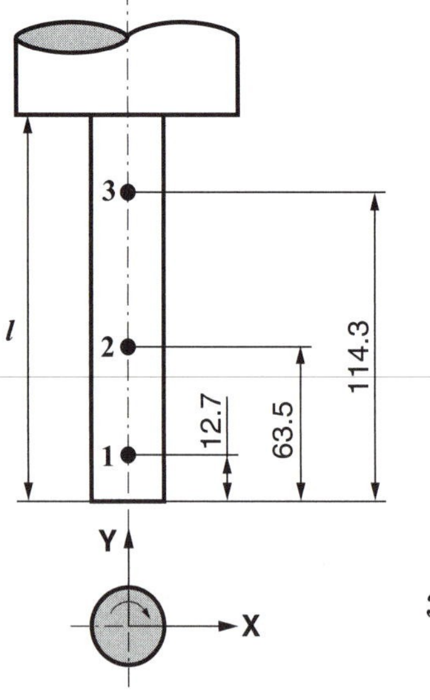

   a. Identify the real modal matrix in both (x, y) directions.
   b. Plot the mode shapes in the y direction.
   c. Evaluate the direct transfer function at point 1 in directions x and y (i.e., $\Phi_{xx} =$?, $\Phi_{yy} =$?).
   d. An aircraft Al-7075 wing spar is down-milled with a width of cut $b = 10$-mm using the same cutter and material. The cutting constants are given as $K_t = 752$ MPa and $K_r = 0.3$. Evaluate the stability lobes by assuming the dynamics only at the tool tip (point 1). The cross-transfer functions are assumed to be negligible. Plot the stability lobes and provide a short recommendation for selecting the most optimal cutting conditions.

3. Consider a slender end mill with the measured transfer functions at its tip (Φ_{11}) and middle (Φ_{12}) as shown in Figure 4.25.

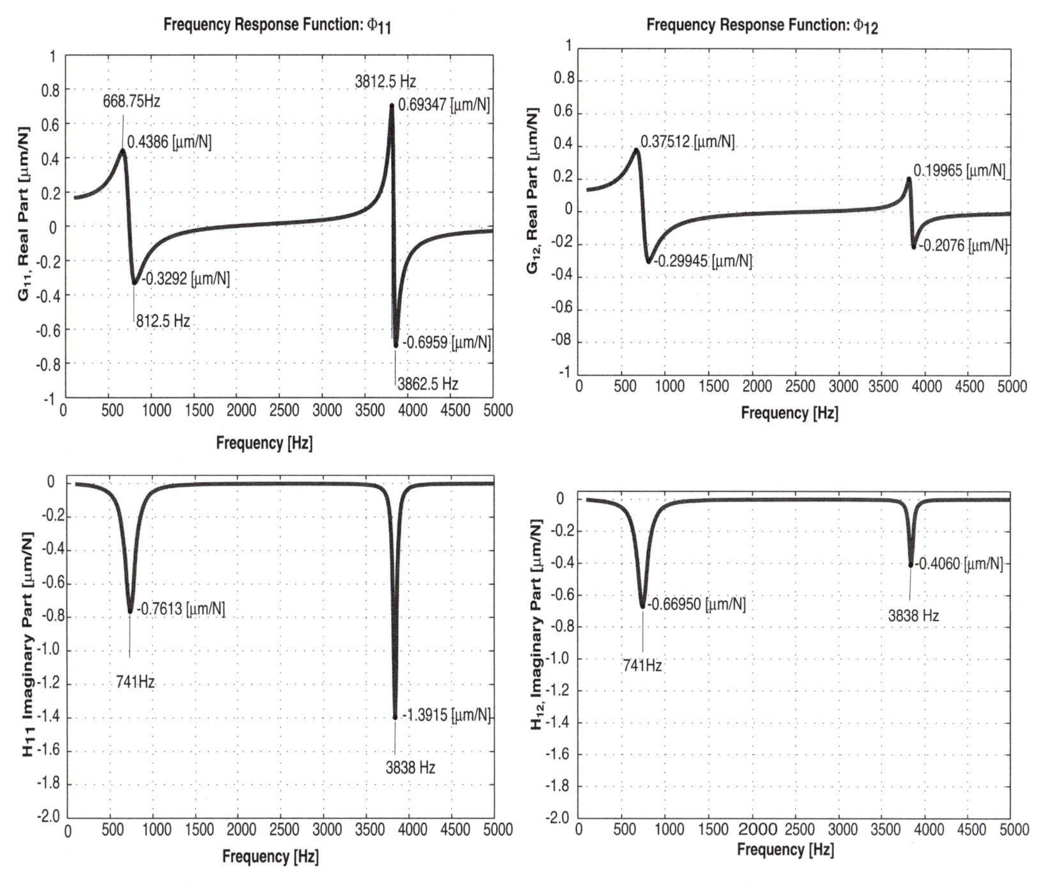

Figure 4.25: FRF measurements of a tool at the tool tip (Φ_{11}) and in the middle of stick out (Φ_{12}).

a. Calculate and plot the mode shapes.
b. Express the direct transfer functions of the beam at the two measurement points.
c. Assume that the cutting coefficients for the material is $K_t = 750$ MPa, $K_r = 0.3$. The number of teeth on the cutter is 6, and the machining is in half-immersion up-milling. Calculate the absolute stability limit in the frequency domain ? (Hints: Use the property of rigidity in x direction, $\Phi_{xx} = 0$. and avoid calculating unnecessary parts of the stability.)

4. An aluminium Al-7050-T6 block with 1,000 mm width and 30 mm depth needs to be roughed with a 40 mm diameter cutter on a machine with maximum 20,000 rev/min spindle speed capacity. The spindle has 20 kW power and 30 Nm torque capacity above 8,000 rev/min spindle speed. The tool manufacturer recommends 0.2 mm/tooth/rev chip load for Al-7050. The structural dynamics of the machine measured at the tool tip is given in Table 4.5. Identify the number of teeth, spindle speed, radial and axial depth of cut to achieve maximum material removal rate without violating

TABLE 4.5. Modal Parameters of a 40-mm-Diameter Cutter Mounted on a Spindle

	ω_n [Hz]	k [N/μm]	ζ[%]	$2k\zeta$[N/μm]
Feed direction (xx)				
Mode 1	808	11	3.4	0.748
Mode 2	1, 347	27	2.44	1.32
Normal direction (yy)				
Mode 1	788	18	3.01	1.08
Mode 2	1, 273	37	2.73	2.02

TABLE 4.6. Modal Parameters of a Two-Fluted 16-mm Diameter Twist Drill

Modes	Frequency -$\omega_n[Hz]$	Stiffness -k	Damping (%)
Φ_{xx}	363	16 [N/μm]	2
Φ_{yy}	338	16 [N/μm]	2
Φ_{zz}	3,358	105 [N/μm]	2
$\Phi_{\theta\theta}$	3,358	778 [Nm/μ rad]	2
$\Phi_{z\theta}$	3,358	0.43 [Nm/μm]	2
$\Phi_{\theta z}$	3,358	492,400 [N/μ rad]	2

torque, power, and chatter limits of the machine? The recommendations are used by the numerically controlled programmer.

5. A two-fluted twist drill with 16-mm diameter is used to open holes on Al-7050 aluminum alloys with 4-mm-diameter predrilled pilot holes. The feed rate was 0.3 mm/flute with a half-tip angle of $\chi_t = 45$ deg. The cutting force coefficients are identified as $k_{tc} =$ 1,200 MPa, $k_{rc} = 0.3$, $k_{ac} = 0.23$ with a torque arm of $R_t = 6.62$ mm. The identified modal parameters of the drill bit is given in Table 4.6. Predict the stability of the drill.

CHAPTER FIVE

TECHNOLOGY OF MANUFACTURING AUTOMATION

5.1 INTRODUCTION

Numerically controlled (NC) machine tools were developed to fulfill the contour-machining requirements of complex aircraft parts and forming dies. The first NC machine tool was developed by Parsons Company and MIT in 1952 [63]. The first-generation NC units used digital electronic circuits and did not contain any actual central processing unit; therefore, they were called NC or *hard-wired* NC machine tools. In the 1970s, *computer numerically controlled* (CNC) machine tools were developed with minicomputers used as control units. With the advances in electronics and computer technology, current CNC systems use several high-performance microprocessors and programmable logical controllers that work in a parallel and coordinated fashion. Current CNC systems allow simultaneous servoposition and velocity control of all the axis monitoring of controller and machine tool performance, online part programming with graphical assistance, in-process cutting process monitoring, and in-process part gauging for completely unmanned machining operations. Manufacturers offer most of these features as options.

5.2 COMPUTER NUMERICALLY CONTROLLED UNIT

A typical CNC machine tool has three fundamental units: the mechanical machine tool unit, power units (motors and power amplifiers), and the CNC unit. Here, a brief introduction of a CNC system from the user's point of view is presented.

5.2.1 Organization of a CNC Unit

A CNC unit of a machine tool consists of one or more central processing units (CPUs), input/output devices, operator interface devices, and programmable logical controllers.

The CNC unit may contain several CPUs, or microprocessors, depending on the tasks required by the machine tool. A very basic three-axis CNC milling machine requires the fine coordinated feeding velocity and position control of all three axes and the spindle speed simultaneously. Current CNC systems tend to use multiple CPUs depending on the number of computation tasks.

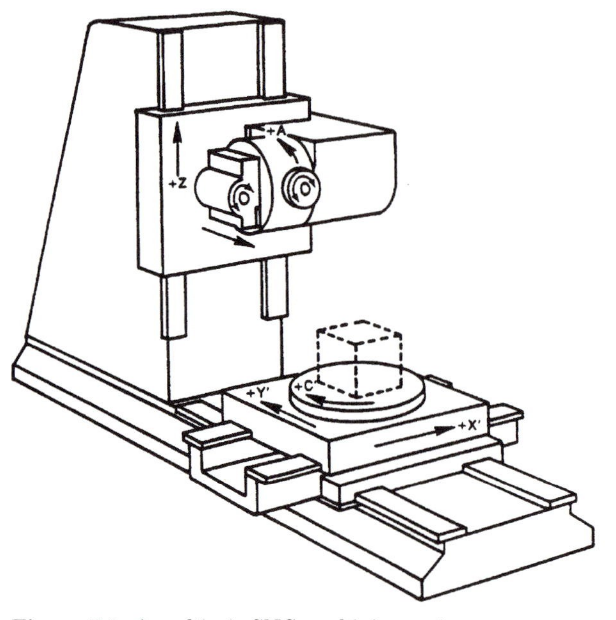

Figure 5.1: A multiaxis CNC machining center.

In such a multiprocessor-based CNC system, additional CPU modules can be added to the CNC to extend its intelligence and functions. For example, in a machining center, such as the one shown in Figure 5.1, contouring control, pallet control, coordination of all eight axes, cutting process monitoring, and graphically assisted part programming can take place simultaneously owing to the use of several computers on its CNC system.

Each CNC unit has a keyboard and monitor for the operator, as well as portable storage units and high-speed communication ports for loading NC programs and linking to computer networks. In a fairly computerized factory floor, it is common to have a direct communication link between the master computer of the factory and individual CNC units. NC programs, production schedules, and recording of production times and operation cycles are continuously electronically communicated between the CNC units and the master computer. Such systems are called *distributed numerically controlled (DNC)* systems. CNC systems, equipped with a monitor and a keyboard, allow operators to edit NC programs on site. However, in an organized production floor, this is not recommended. The capital and operating costs of CNC machine tools are rather high, and their use for other than continuous metal removal may not be justified. The process planners in the engineering office test NC programs with the aid of graphically animated tool paths and machine tool motions before they are released to the production floor. The methods of NC programming and tool path verification are explained in the following sections.

Miscellaneous logic functions have to be carried out routinely by the CNC system. For example, activating the cutting lubricant pump, turning the spindle clockwise (CW) or counterclockwise (CCW), touching travel limits of the guides, and tool-changing functions require only coordinated ON/OFF type logic signals, which are carried out by the programmable logical control (PLC) units of the CNC systems.

Each CNC unit is equipped with emergency stop, shutdown, feed rate, and spindle speed override buttons and a manual jog and machine homing switches for easy use of the machine tool by the operators. The number of operator interface units depends very much on the type and the functions of the machine tool, and on the richness of the CNC software library, as well.

5.2.2 CNC Executive

NC programs are written in an internationally recognized standard language. Each commercial CNC unit is expected to accept standard NC codes. The NC codes are transferred to the CNC unit in ASCII format by the use of input devices. The CNC executive is the main system software that decodes the NC codes block by block and sends appropriate commands to physical control, computation, and PLC units of the CNC system. For example, a 10-mm distance to be traveled at a 200-mm/s feed velocity command can be translated as follows: The real-time clock is set to generate 10,000 pulses at a rate of 200,000 pulses/s (1 pulse = 0.001 mm position). The position pulses (i.e., the discrete velocity commands) are directed to the indicated machine tool axis position control units by converting to their analog voltage equivalent (i.e., typically within ± 10 V range). The analog voltage is amplified by the power unit and fed to the axis drive motors to deliver the desired motion. There are also digital drives with onboard computers, and they accept the discrete velocity commands directly. Miscellaneous functions, such as spindle ON and tool change commands, are translated as Boolean logic signals (+5 V or −5 V) for PLC units.

The CNC executive software is expected to carry out NC functions in a logical order. For example, it must turn the spindle before a controlled feeding or machining action takes place. It must load the tool on the spindle before positioning the tool for machining, although such commands may be given in the same NC block.

5.2.3 CNC Machine Tool Axis Conventions

The Electronic Industries Association, in its standard RS-367-A, lists 14 different axis designations or types of motion. Whereas common machine tools may have up to five axes, gear-shaping and cutter-grinding machine tools may have up to fourteen axes of controlled motions.

The machine tools are programmed in a Cartesian coordinate system. The Z axis is always aligned with the direction of spindle. The primary X motion is normally parallel to the longest dimension of the primary machine tool table. The Y axis is normally parallel to the shortest dimension of the primary machine tool table. Characters A, B, and C designate angular motions around the X, Y, and Z axes, respectively. Axis conventions of various machine tool configurations are shown in Figure 5.2.

5.2.4 NC Part Program Structure

An NC part program represents the machining sequences, or *blocks*, used to produce a desired component shape. Each block starts with the letter N followed by the block sequence number. One block of a typical NC program would look like

$$\text{N0040 G91 X25.00 Y10.00 Z} - \text{12.55 F150 S1100 T06 M03 M07.}$$

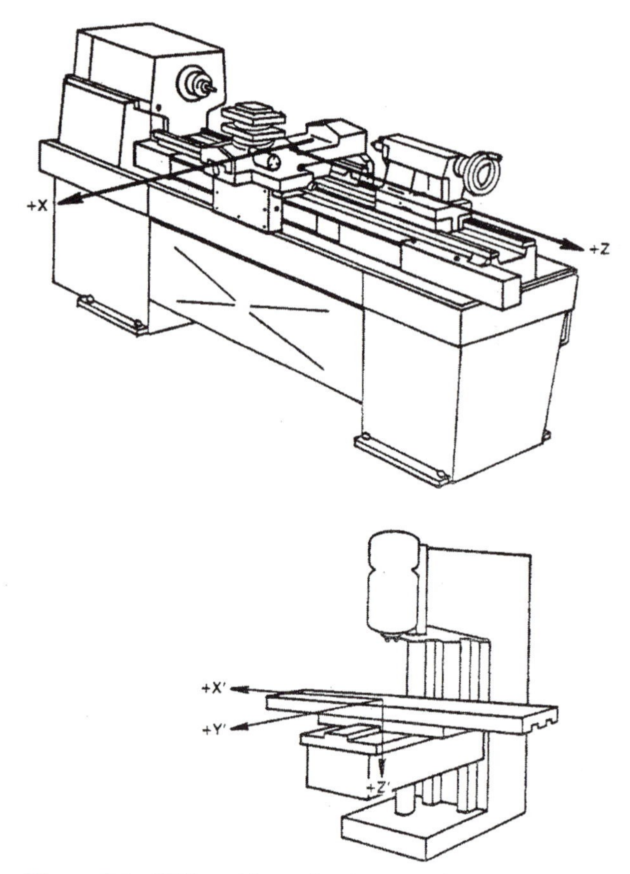

Figure 5.2: CNC machine tool axis conventions.

The NC program consists of *blocks*, and each block consists of several *words*. A word starts with a character followed by numbers that represent a specific command for the machine tool. The word starting with G represents the *preparatory* functions, and the M words are the *miscellaneous* functions in the NC program. The words starting with F and S represent the feed and spindle speeds, respectively. T represents the tool number. x, y, and z letters followed by the scalars represent the motion lengths in the designated axes. The particular NC block given above is interpreted as follows by the CNC executive software: Load tool number 6 on the spindle (T06), rotate the spindle CW at 1,100 rev/min (S1100, M03) and turn the cutting fluid on (M07) before the motion starts. Move the machine tool in 25-mm, 10-mm, and 12.55-mm increments (G91) in the x, y, and z directions with the resultant feeding velocity of 150 mm/min along the tool path. N040 represents the 40th block sequence of the entire part program.

The list of commonly used NC words are as follows:

N...: Block sequence number.
G...: Preparatory function.
X...: Primary X motion dimension.
Y...: Primary Y motion dimension.
Z...: Primary Z motion dimension.
U...: Secondary motion parallel to X axis.
V...: Secondary motion parallel to Y axis.
W...: Secondary motion parallel to Z axis.
A...: Angular dimension about X axis.
B...: Angular dimension about Y axis.
C...: Angular dimension about Z axis.
I...: Interpolation parameter or thread lead parallel to X.

$J\ldots$: Interpolation parameter or thread lead parallel to Y.
$K\ldots$: Interpolation parameter or thread lead parallel to Z.
$F\ldots$: Feed word.
$M\ldots$: Miscellaneous function.
$S\ldots$: Spindle speed word.
$T\ldots$: Tool number word.
$R\ldots$: Rapid traverse dimension in Z axis.

Digital codes in a word may have to be written in a specific given format for some CNC units. However, in general, most control systems allow free formatted word fields. All NC words and their functions are given in the following text according to ISO 1056 international standards.

Preparatory (G) Functions

- *G00:* Point-to-point positioning. This provides point-to-point positioning of the feed drives at a rapid traverse rate of the machine tool along an uncontrolled random path. It must be noted that each drive traverses at the rapid feeding velocity independently; therefore, the programmer must be careful to avoid collision of the tool with the clamps or the workpiece.
- *G01:* Linear interpolation. The tool path velocity is kept constant at the given feed along the indicated straight line. The feed drives' velocities are coordinated to keep the tool along the straight line.
- *G02, G03:* Circular interpolation CW (*G*02) or CCW (*G*03). An arc generated by the coordinated motion of the two axes in which the tool moves CW CCW, when viewing the plane of motion in the negative direction of the axis perpendicular to the plane of motion.
- *G04:* A timed delay of a programmed motion.
- *G07:* A controlled velocity increase and decrease to programmed rate starting immediately.
- *G17–G19:* Used to identify the plane for such functions as circular interpolation, cutter compensation, and other functions as required.
- *G21–G23:* Precision cornering mode selection for linear and circular interpolation.
- *G33:* Constant lead thread cutting.
- *G34, G35:* Increasing lead (*G*34) or decreasing lead (*G*35) thread cutting.
- *G40:* Command that will discontinue any active cutter compensation or offset.
- *G41, G42:* Cutter compensation left (*G*41), cutter compensation right (*G*42). Cutter on left (or right) side of work surface looking from cutter in the direction of relative cutter motion with displacement normal to the cutter path to adjust for the difference between actual and programmed cutter radius or diameters.

- *G70:* Mode for programming in imperial (inch) units.
- *G71:* Mode for programming in metric (mm) units.
- *G80:* Cancel the active fixed cycle.
- *G81–G89:* A preset series of operations that direct machine axis movement and/or cause spindle operation to complete fixed machining cycles such as boring, drilling, tapping, or combinations thereof. Fixed cycles are also called *canned cycles* by machine tool manufacturers.
- *G90:* Coordinate inputs are given in *absolute coordinates* from a fixed Cartesian coordinate center.
- *G91:* Coordinate inputs are given *incrementally* from a previous tool location.
- *G94:* The feeds are given in inches (or millimeters)/minute.
- *G95:* The feed rate is given in inches (or millimeters)/revolution.

Miscellaneous Functions

- *M00:* Program stop. Terminates further program execution after the completion of other commands in the block.
- *M01:* Optional stop if it is enabled by the operator. The program continues after the execution of a *continue* command by the operator.
- *M02:* End of program indicating completion of machining cycle. Stops spindle, coolant, and feed after the completion of all commands in the last NC block.
- *M03, M04:* Start spindle CW (*M*03) or CCW (*M*04).
- *M05:* Spindle off.
- *M06:* Tool change.
- *M07, M08:* Cutting fluid ON (*M*07), OFF (*M*08).
- *M19:* Oriented spindle stop at a predetermined angular position.
- *M30:* End of program. It stops feed, spindle, and cutting fluid and rewinds the NC program to the beginning.
- *M49:* Prevents operator from overriding spindle and feed speeds.

Note that any G or M codes used in the NC program are active until their opposites are used, which overrides the previous ones.

5.2.5 Main Preparatory Functions

G00 – Point-to-Point Positioning

The table (or cutter) is positioned from one point to another without having to coordinate the velocities of any of the moving axes. This mode is usually used at a rapid traverse rate for drilling and tapping and for rapid positioning of the cutter without cutting.

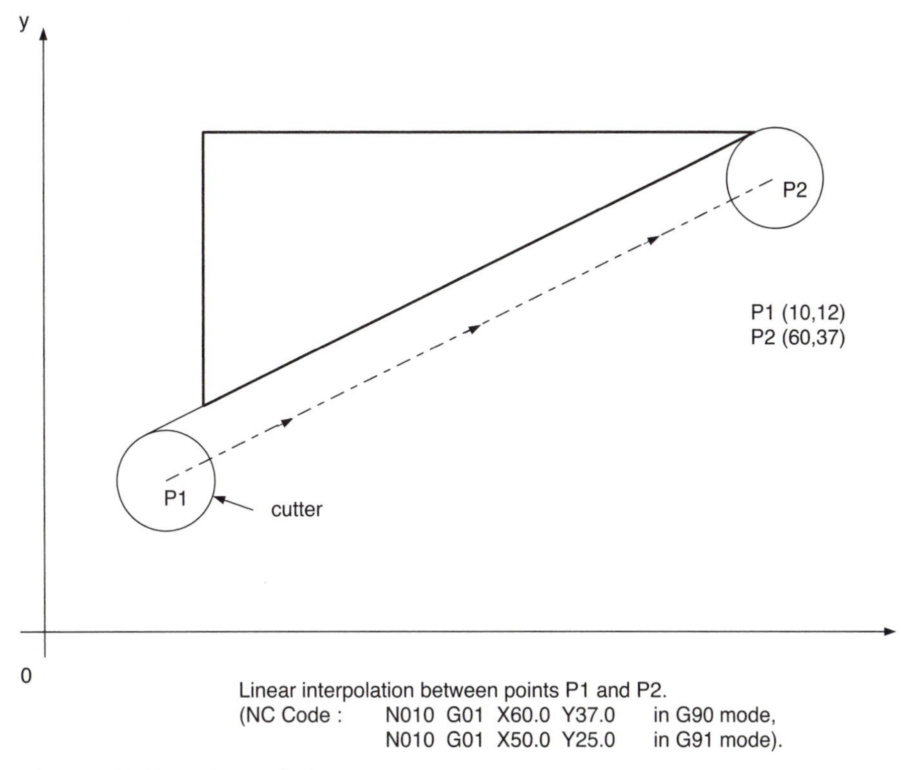

Figure 5.3: Linear interpolation.

Continuous Path Contouring

The table's position is continuously controlled to keep the cutter on a desired trajectory profile (i.e., slanted line, arc, or spline segment) during machining. This involves continuous manipulation of velocities of each axis involved during contour machining. Real-time digital interpolation methods, which are explained in detail in the next chapter, are used for continuous path machining. The two fundamental interpolation commands, which are provided on every commercial CNC unit, are explained below.

G01 – Linear Interpolation Code. The velocities of two axes are controlled to keep the tool on a straight path in a plane of motion. Figure 5.3 shows a line segment to be machined by a milling cutter. The coordinates of the end mill center for the beginning point P_1(10 mm, 12 mm) and end point P_2(60 mm, 37 mm) are given in the XY plane. To keep the end mill following the straight line (P_1P_2) at a given vector feed velocity, the linear interpolation command G01 must be used. The corresponding NC programs in absolute (G90) and incremental (G91) programming modes, respectively, are as follows:

$$N0010\,G90\,G01X60.0Y37.0F300$$

or

$$N0010\,G91\,G01X50.0Y25.0F300.$$

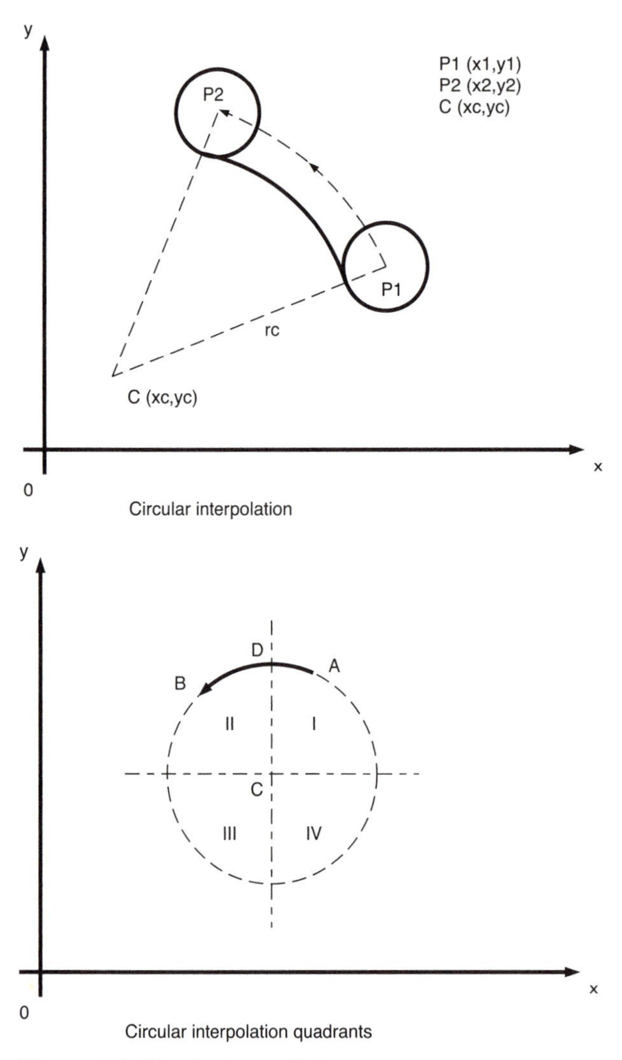

Figure 5.4: Circular interpolation.

G02, G03 – Circular Interpolation Codes. The velocities of two axes on a plane of motion are varied to keep the tool following the given arc at the specified feed velocity. Two types of circular interpolation commands are used in CNC systems. Some CNC systems require the coordinates of the arc center and arc's end point, whereas others need the radius of the arc and its end point. Figure 5.4 shows a sample arc segment to be contoured by an end mill. CNC assumes that the tool is located at the beginning point of the arc. Looking down the plane, or with respect to the previous tool motion, the arc contouring can be either CW (G02) or CCW (G03). In Figure 5.4, the tool is required to move in a CCW direction (G03) at a constant contour feed velocity of f. If the CNC requires the radius (r_c) and the end point of the arc (P_2), the following NC program is used for the contour milling in absolute coordinates:

$$\text{N010 G90 G03 } X\, x_2\ Y\, y_2\ R\, r_c\ F\, f.$$

If a CNC requires the coordinates of the center point of the arc and the arc's end point, the NC program is

$$\text{N010 G90 G03 } X\, x_2\ \ Y\, y_2\ \ I\, i_c\ \ J\, j_c\ \ F\, f,$$

where the values of i_c and j_c are used to define the center of the arc with respect to the starting point. The following algorithm can be used to calculate the arc center offset:

$$i_c = x_c - x_1, \quad j_c = y_c - y_1.$$

The interpolation parameter K is used when one of the axes is Z. Some CNC systems allow circular arcs to be programmed only in one quadrant at a time. Each arc must be programmed in a separate NC block. For example, the circular arc shown in Figure 5.4 must be programmed in two NC blocks as follows:

N010 G90 G03 $X\ x_D\ Y\ y_D\ I\ (x_C - x_A)\ J\ (y_C - y_A)$ F200

N020 G03 $X\ x_B\ Y\ y_B\ I\ (x_C - x_D)\ J\ (y_C - y_D)$

Note that new-generation CNC systems allow programming of more than one arc quadrant in one block.

G81 to G89 – Fixed Cycles

The fixed cycle (or canned cycle) commands set up the CNC unit for different types of automatic operations. Use of fixed cycles permits the reading of X–Y location, a rapid Z plane plunge (programmed with R word), a final Z feed point and a feed rate into the control. The values of the words are stored until they are replaced with new numbers or canceled by G80. By the sole reading of a new X and Y location, the stored sequence of events is performed at the new location. The following NC codes illustrate the use of a canned drill cycle:

```
N050  G81  X125.0  Y237.5  Z-112.5  R2.5  F300
N060       X325.0
N070               Y137.5
N080  G80
```

In block N050, the cutter is positioned at rapid traverse to 125.0 mm in X and 237.5 mm in Y. The spindle would then speed to 2.5 mm in Z and feed at 300 mm/min to a Z depth of −112.5 mm. The final motion for the block would be rapid retraction of the spindle to the R plane of 2.5 mm in Z. The subsequent blocks N060 and N070 would drill an additional two holes at the indicated new locations. The G80 in block N080 cancels the fixed drill cycle G81. It must be noted that some machine tool control units may use different formats and codes from the listed standard codes.

An NC part program listing for the sample workpiece profile shown in Figure 5.5 is given as follows:

N01G90	Absolute coordinates
N02G71	Metric units (mm)
N03G92X-12.5 Y-12.5 Z50.0	Cutter starts from here with respect to the part zero
N04G00Z2.5 M03 S800	Spindle on CW, move rapidly to 2.5 mm above the part
N05G01Z-7.5 F25.0 M08	Coolant on, plunge with feed in Z (P1)
N06 X162.5 F125	Move in x with 125 mm/min feed to P2
N07 Y0.0	Move to P3 at the same feed (125 mm/min)
N08G02X220 Y57.5 I57.5 J0	CW circular interpolation (P4)

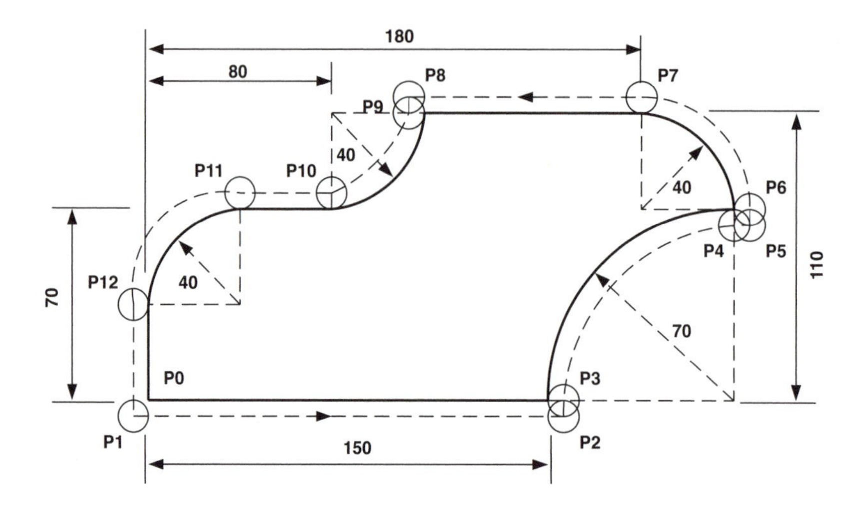

Points	X	Y	Circle center
P0	0	0	
P1	-12.5	-12.5	
P2	-162.5	-12.5	
P3	-162.5	0.0	
P4	220	57.5	c(220,0)
P5	232.5	57.5	
P6	232.5	70	
P7	180	122.5	c(180,70)
P8	107.5	122.5	
P9	107.5	110	
P10	80	82.5	c(80,110)
P11	40	82.5	
P12	-12.5	30	c(40,30)
P1	-12.5	-12.5	

Figure 5.5: Sample workpiece for NC programming.

```
N09G01X232.5                        Move one cutter radius in x
N10      Y70.0                      Move in y (P6)
N11G03X180 Y122.5 I-52.5 J0         CCW relative to the tool
                                       motion in block N10 (P7)
N12G01X107.5                        Move left (P8)
N13      Y110                       Move to (P9)
N14G02X80.0 Y82.5 I-27.5 J0         CW circular interpolation (P10)
N15G01X40.0                         Move left (P11)
N16G03X-12.5 Y30.0 I0 J-52.5        CCW circular interpolation (P12)
N17G01Y-12.5                        Return to starting point (P1)
N18      Z3.8                       Move up to z = 3.8 mm position
N19G00Z50 M09 M05                   Move to Z = 50 rapidly,
                                       coolant and spindle are off
N20M30                              Program ends, move the cutter
                                       to starting point.
```

5.3 COMPUTER-ASSISTED NC PART PROGRAMMING

It is rather tedious and unproductive to generate NC part programs manually, as demonstrated in the previous section. In computer-assisted NC part programming, the computer automatically generates the codes. The philosophy behind the computer-assisted programming is rather straightforward to describe. If a part geometry can be parametrically defined in a computer code, the tool path can also be generated for given dimensions of a tool according to the desired sequences of machining. The programmer must, of course, select tools, feeds, and speeds and generate the tool paths according to the process-planning strategy for the machining operation.

Computer-assisted part programming can be done with and without the aid of computer graphics. Both methods require mathematical representation of geometric entities in a computer program. Before introducing the computer-aided NC programming techniques used in industry, the fundamental principles of analytical geometry used in computer-aided design and computer-aided manufacturing (CAD/CAM) systems are briefly presented in the next section.

5.3.1 Basics of Analytical Geometry

The scientific principles or even the technology of CAD/CAM are too widespread to cover in this text. However, a basic introduction to geometric entities and their manipulation in both CAD and CNC machines is essential for leading readers to more advanced topics related to CAD/CAM. As briefly presented in NC programming with auto automatically programmed tools (APT) and advanced computer graphic tools, the geometry of the workpiece is represented by points, lines, circles, splines, surfaces, and solids. The geometric entities are manipulated by translation, rotation, intersection, and trimming operations. The objects are either represented by their wire frame or solid models. Using shading and hidden line-surface removal techniques, the user can visualize the designed part on the graphics terminal. Because the detailed presentation of CAD techniques and computer graphics are well covered in dedicated texts [43, 118], instead, we will briefly introduce the mathematical formulation of the very basic geometric entities used in both CAD and real-time CNC interpolation algorithms.

Vectors and Lines

A vector in a Cartesian coordinate system is expressed in Figure 5.6 as

$$\vec{r} = r_x\vec{i} + r_y\vec{j} + r_y\vec{k}, \tag{5.1}$$

where $r_x = r\cos\alpha$, $r_y = r\cos\beta$, and $r_z = r\cos\gamma$ are the directional factors. The magnitude of the vector is $|\,\vec{r}\,| = (r_x^2 + r_y^2 + r_y^2)^{1/2}$. The unit vector, which has a magnitude of unity, is $\vec{r}/\,|\,\vec{r}\,| = \cos\alpha\vec{i} + \cos\beta\vec{j} + \cos\gamma\vec{k}$. A vector between two points $(P_1(x_1, y_1, z_1), P_2(x_2, y_2, z_2))$ can be constructed as

$$\overrightarrow{P_1P_2} = (x_2 - x_1)\vec{i} + (y_2 - y_1)\vec{j} + (z_2 - z_1)\vec{k}. \tag{5.2}$$

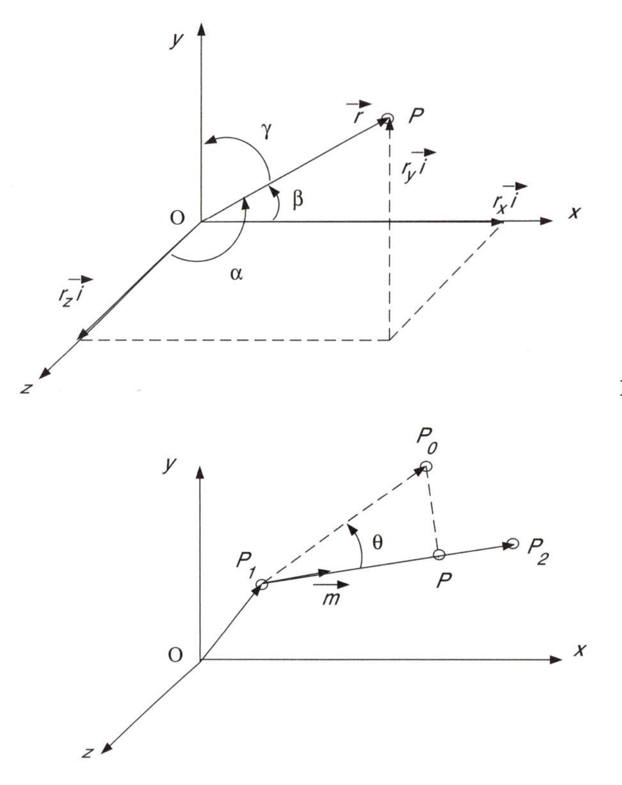

Figure 5.6: Vectors and lines.

A unit vector along $\overrightarrow{P_1P_2}$ is

$$\vec{m} = \frac{\overrightarrow{P_1P_2}}{|\overrightarrow{P_1P_2}|} = \frac{x_2 - x_1}{L}\vec{i} + \frac{y_2 - y_1}{L}\vec{j} + \frac{z_2 - z_1}{L}\vec{k}, \tag{5.3}$$

where $L = |\overrightarrow{P_1P_2}| = [(x_2 - x_1)^2 + (y_2 - y_1)^2 + (z_2 - z_1)^2]^{1/2}$ is the length of the line connecting the two points.

The products of vectors provide angles, projections, and distances between various geometric entities. Consider two vectors $\vec{A} = \overrightarrow{P_1P_0} = A_x\vec{i} + A_y\vec{j} + A_y\vec{k}$ and $\vec{B} = \overrightarrow{P_1P_2} = B_x\vec{i} + B_y\vec{j} + B_y\vec{k}$. The dot product of the two vectors is a scalar as follows:

$$\vec{A} \cdot \vec{B} = A_xB_x + A_yB_y + A_zB_z = |\vec{A}||\vec{B}|\cos\theta, \qquad 0 \le \theta \le \pi. \tag{5.4}$$

Note that $\vec{i}\cdot\vec{i} = \vec{j}\cdot\vec{j} = \vec{k}\cdot\vec{k} = 1$ and $\vec{i}\cdot\vec{j} = \vec{j}\cdot\vec{k} = \vec{k}\cdot\vec{i} = 0$. Hence, the angle between the two vectors can be evaluated by $\theta = \arccos[\vec{A}\cdot\vec{B}]/[|\vec{A}||\vec{B}|]$.

The vector product of $(\vec{A}, \vec{B})$ is a vector as follows:

$$\vec{A} \times \vec{B} = (A_yB_z - A_zB_y)\vec{i} + (A_{zy}B_x - A_xB_z)\vec{j} + (A_xB_y - A_yB_x)\vec{k}$$

$$= (|\vec{A}||\vec{B}|\sin\theta)\vec{n}, \qquad 0 \le \theta \le \pi, \tag{5.5}$$

where $\vec{n}$ is a unit vector perpendicular to the plane formed by vectors $\vec{A}$ and $\vec{B}$ and θ is the angle between the two vectors. The distance between a point and a line can be evaluated by using vector algebra. The unit vector of line P_1P_2 is $\vec{m} = \overrightarrow{P_1P_2} / \mid \overrightarrow{P_1P_2} \mid$. The distance between point P_0 and line P_1P_2 is (Fig. 5.6)

$$\mid \overrightarrow{P_0P} \mid = \mid \overrightarrow{P_1P_0} \times \vec{m} \mid = \mid \overrightarrow{P_1P_0} \mid \cdot \mid \vec{m} \mid \sin\theta.$$

Translation and Rotation of Objects

Translating an object from one location to another requires that every point and entity describing it experience the same amount of displacement. If the displacement vector is $\vec{d}$, and the object to be translated is $\vec{P}$, then the translated position of the object becomes

$$vecP^* = \vec{P} + \vec{d}. \tag{5.6}$$

The objects can be scaled by simply multiplying their coordinates as follows:

$$\vec{P^*} = [S]\vec{P}, \tag{5.7}$$

where the scale matrix $[S]$ is

$$[S] = \begin{bmatrix} s_x & 0 & 0 \\ 0 & s_y & 0 \\ 0 & 0 & s_z \end{bmatrix}.$$

An object can be rotated about the z axis, and the relevant equation is

$$\vec{P^*} = [R_z]\vec{P}, \tag{5.8}$$

where the rotation matrix $[R_z]$ is

$$[R_z] = \begin{bmatrix} \cos\theta & -\sin\theta & 0 \\ \sin\theta & \cos\theta & 0 \\ 0 & 0 & 1 \end{bmatrix},$$

where θ is the CCW rotation angle. Similarly, the rotation matrices for the x and y axes are given by

$$[R_x] = \begin{bmatrix} 1 & 0 & 0 \\ 0 & \cos\theta & -\sin\theta \\ 0 & \sin\theta & \cos\theta \end{bmatrix}, \quad [R_z] = \begin{bmatrix} \cos\theta & 0 & \sin\theta \\ 0 & 1 & 0 \\ -\sin\theta & 0 & \cos\theta \end{bmatrix}.$$

Rotation with respect to an arbitrary axis can be derived using vector algebra [43].

Circles

A point $P(x, y, z)$ on a circle (Fig. 5.7) can be defined by the radius and angular distance,

$$P(x, y, z) = \begin{Bmatrix} x \\ y \\ z \end{Bmatrix} = \begin{Bmatrix} x_c + R\cos\theta \\ y_c + R\sin\theta \\ z_c \end{Bmatrix}, \quad 0 \le \theta \le 2\pi. \tag{5.9}$$

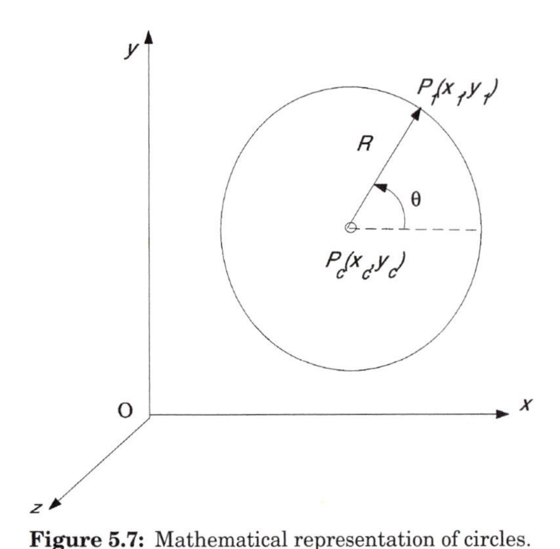

Figure 5.7: Mathematical representation of circles.

The following equation can be used to identify the center (x_c, y_c) and radius r of a circle passing through three points as follows:

$$r^2 = (x - x_c)^2 + (y - y_c)^2. \tag{5.10}$$

Alternatively, the three points can be used to evaluate the coefficients (c_1, c_2, c_3) of a nonparametric equation defining the circle in the (x, y) plane as follows:

$$y = c_1 + c_2 x + c_3 x^2.$$

Cubic Splines

Parts such as dies, molds, and gas turbine blades have sculptured surfaces that are designed by fitting smooth curves to a set of design points. Although various spline-fitting techniques are available in CAD systems, they can all be derived by understanding the algorithm for a cubic spline fit. A series of sequential design points $P_0, P_1, \ldots, P_{i-1}, P_i, P_{i+1}, \ldots, P_n$ are shown in Figure 5.8. The points must be connected by series of cubic spline segments. Consider an arbitrary point $P_i(u)$ on the cubic spline segment S_i as follows:

$$P_i(u) = A_i u^3 + B_i u^2 + C_i u + D_i, \qquad 0 \le u \le 1. \tag{5.11}$$

Note that the end points of the spline segment are $P_i(u = 0) \equiv P_{i-1}, P_i(u = 1) \equiv P_i$. By increasing the parameter u from zero toward unity with small increments, the spline segment can be generated from a series of points provided that the spline parameters A_i, B_i, C_i, and D_i are known. The derivative of Eq. (5.11) gives the tangent vectors

$$P_i'(u) = 3A_i u^2 + 2B_i u + C_i. \tag{5.12}$$

With the use of the boundary conditions at both ends of the spline segment, the parameters A_i, B_i, C_i, and D_i are evaluated. The resulting expressions for the cubic spline and its tangent and curvature expressions are as follows:

$$\begin{aligned}
P_i(u) &= [2(P_{i-1} - P_i) + P_{i-1}' + P_i']u^3 \\
&\quad + [3(P_i - P_{i-1}) - 2P_{i-1}' - P_i']u^2 \\
&\quad + P_{i-1}'u + P_{i-1}, \\
P_i'(u) &= [6(P_{i-1} - P_i) + 3(P_{i-1}' + P_i')]u^2 \\
&\quad + [6(P_i - P_{i-1}) - 4P_{i-1}' - 2P_i']u + P_{i-1}', \\
P_i''(u) &= [12(P_{i-1} - P_i) + 6(P_{i-1}' + P_i')]u \\
&\quad + [6(P_i - P_{i-1}) - 4P_{i-1}' - 2P_i'].
\end{aligned} \tag{5.13}$$

Similar expressions can be written for the spline segment S_{i+1} bounded by points P_i and P_{i+1}. Like an elastic beam, curvature continuity is imposed at the knots where the spline segments S_i and S_{i+1} meet [i.e., $P_i''(u = 1) = P_{i+1}''(u = 0)$]. The resulting expression provides a relationship between the unknown tangent vectors and known points as follows:

$$P_{i-1}' + 4P_i' + P_{i+1}' = 3P_{i+1} - 3P_{i-1}. \qquad (5.14)$$

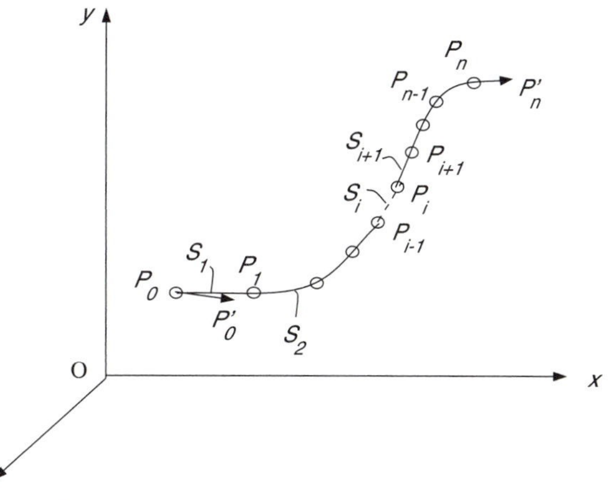

Figure 5.8: Cubic spline curve fit.

To solve all tangent vectors at every knot, the tangents at the two end points (P_0, P_n) must be provided. One can either impose a tangent depending on the design criteria or adjacent geometric entities, or one can approximate them. For example, one approach is to use the lines connecting the first and last knots with their nearest neighbors, and use their slope as follows:

$$P_0' = \frac{\overrightarrow{P_0P_1}}{|\overrightarrow{P_0P_1}|}, \qquad P_n' = \frac{\overrightarrow{P_{n-1}P_n}}{|\overrightarrow{P_nP_{n-1}}|}. \qquad (5.15)$$

Alternatively, the two ends can be relaxed like a beam with moment-free ends (i.e., $P_0'' = P_n'' = 0$), which leads to two more required expressions as follows:

$$\begin{aligned} 2P_0' + P_1' &= 3P_1 - 3P_0, \\ 2P_n' + P_{n-1}' &= 3P_n - 3P_{n-1}. \end{aligned} \qquad (5.16)$$

Equations (5.14) and (5.16) can be combined in the following matrix:

$$\begin{bmatrix} 2 & 1 & 0 & 0 & . & . & . & 0 & 0 & 0 \\ 1 & 4 & 1 & 0 & . & . & . & 0 & 0 & 0 \\ . & . & . & . & & & & . & . & . \\ . & . & . & . & & & & . & . & . \\ . & . & . & . & & & & . & . & . \\ 0 & 0 & 0 & . & . & . & 1 & 4 & 1 & 0 \\ 0 & 0 & 0 & . & . & . & 0 & 1 & 4 & 1 \\ 0 & 0 & 0 & . & . & . & 0 & 0 & 1 & 2 \end{bmatrix} \begin{bmatrix} P_0' \\ P_1' \\ P_2' \\ . \\ . \\ . \\ P_{n-2}' \\ P_{n-1}' \\ P_n' \end{bmatrix} = 3 \begin{bmatrix} P_1 - P_0 \\ P_2 - P_0 \\ P_3 - P_1 \\ . \\ . \\ . \\ P_{n-2} - P_{n-4} \\ P_{n-1} - P_{n-3} \\ P_n - P_{n-1} \end{bmatrix}. \qquad (5.17)$$

Hence, the tangents at all knots can be evaluated by solving matrix Eq. (5.17). Substituting the tangents of each spline segment into the cubic Eq. (5.13) yields to a spline fit equation:

$$\begin{aligned} P_i(u) &= [A_i \ B_i \ C_i \ D_i][u^3 \ u^2 \ u \ 1]^T \\ &= [-2u^3 + 3u^2]P_{i+1} + [2u^3 - 3u^2 + 1]P_i \\ &\quad + (u^3 - u^2)P'_{i+1} + (u^3 - 2u^2 + u)P'_i, \qquad 0 \le u \le 1. \end{aligned} \tag{5.18}$$

There are better curve-fitting techniques for smoother design and NC tool path generation. Some of the best-known methods include Bezier curves [118] and bicubic splines. The splines are used to model the geometry and NC tool paths of parts with sculptured surfaces. Higher-order splines are also used to generate the splined tool paths in real-time CNC applications.

5.3.2 APT Part Programming Language

Owing to difficulties in manual NC programming, a computer language APT was developed at Massachusetts Institute of Technology (1956) to automate part programming. Like any other high-level language, APT allows arithmetic operations, subroutines, macros, looping logic, and so on. However, the main difference is that it allows parametric representations of geometric entities ranging from a single point in space to very complex, three-dimensional sculptured surfaces. The tool path is generated by commanding the given tool geometry to travel in a selected tool path on the defined part geometry. The computed tool path coordinates and cutting conditions are then stored in a computer file with a general standard format, which is also automatically converted to the specific manual code of the selected CNC machine tool.

APT was the most common and standard code used in the industry. APT has been replaced by graphics-based, user-friendly interactive CAD/CAM software systems in industry. However, because CAD/CAM systems use APT standards in processing NC tool path data, it is useful to understand at least the very basic structure of the APT programming system.

Geometric Statements

APT is no longer used to define the part geometry since the adoption of integrated CAD/CAM systems in industry. Only a brief logic behind the geometric statements is explained here. The part geometry is defined by using points, lines, arcs, space curves, and surfaces. Each geometric entity is mathematically represented and stored as variables by APT. The user must define each geometric entity according to the format required by APT. The formats and the use of geometric statements can be found in the APT dictionary generated by Computer Aided Manufacturing-International (CAM-I) [36] Each geometry is defined according to the following format:

(Geometry's symbol) = **GEOMETRY**/(dimensions and the parameters of the geometry).

Some of the fundamental geometric statements are illustrated in the following. Note that the lowercase letters represent labels or dimensions to be entered by the programmer; capital letters and operator symbols (, / = $) must be used as they are shown.

- **Pname = POINT/ x, y, z**
  Here, x, y, and z are the coordinates of the point. Zero value is assumed when the z dimension is omitted.

  Example: P1 = POINT/2.0,3.2,2.0

- **Lx = LINE/P1, P2**
  Line between points P1 and P2.
- Circlex = CIRCLE/CENTER, P1, RADIUS, 40
  Circle with a center at point P1 and 40-mm radius.

Planes are generally used to define the planes of motion for the tool movements. Its surface normal vector and its amplitude define a plane as follows:

Planex = PLANE/a,b,c,d

where a, b, and c are the cosines of the angles between the surface normal vector and the plane, and d is the length of the vector.

Example: PL1=PLANE/0, 0, 1, 7.5

Plane PL1 is parallel to the xy plane, and the height is 7.5 along the z axis.

Tool Motion Statements

There are two types of tool motion commands in APT: point-to-point (PTP) and continuous tool motion commands.

PTP Tool Motion Commands. If the tool is required to move to a specified point from its current position, PTP commands are used. There are two types of PTP motion commands,

GOTO/point, feed rate

and

GODLTA/point, feed rate

Both commands take the tool to the point indicated to the right of the slash with an optional feed rate provided. The point coordinates are assumed to be given in absolute coordinates (G90) by a GOTO command and incremental (G91) coordinates by a GODLTA command. If only one coordinate is given in the GODLTA/ command, incremental z axis motion is assumed.

Example: GOTO/10,2,1

The tool is commanded to move to the point that has absolute coordinates of $x = 10\,\text{mm}, y = 2\,\text{mm}$, and $z = 1\,\text{mm}$.

Example: GODLTA/10,2,1

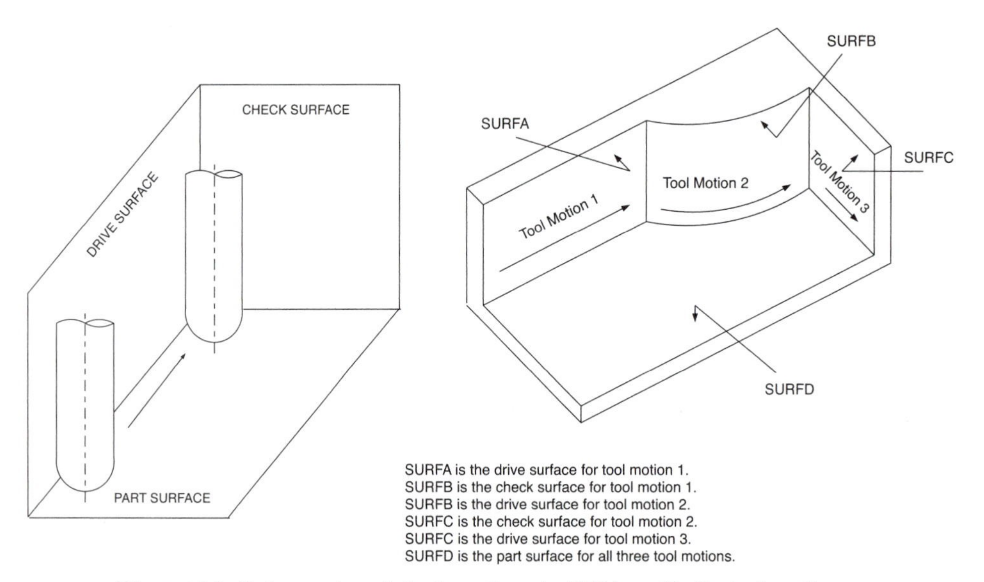

Figure 5.9: Drive, part, and check surfaces in APT to guide the tool motion.

The tool is commanded to move incrementally 10 mm, 2 mm, and 1 mm in the x, y, and z axis, respectively.

Example: GODLTA/-10

The tool is commanded to move 10 mm in the negative z axis direction.

Continuous Tool Motion Commands. The tool is commanded to move along a controlled path defined by the *part, drive*, and *check* surfaces.

If we consider a three-axis machining process with a vertical CNC milling machine tool, the tool's end stays on the part surface, the tool's axis stays parallel to the drive surface, and the tool's motion is bounded by the check surface (see Fig. 5.9). If the part surface is a three-dimensional sculptured surface (i.e., an aircraft wing or a turbine blade surface), a three-axis machine tool is required. If two of the surfaces are sculptured then five or more simultaneously controlled axes are required on the CNC machining center.

Before the tool can be commanded to move along a given profile, it must be brought from a specific point. Usually this point corresponds to tool-change location coordinates with respect to the part program zero. This is achieved by a

FROM/point

command, followed by

GO/{TO,ON,PAST}, drive surface, {TO, ON, PAST}, part surface, {TO, ON, PAST}, check surface

where one of the three modifiers (TO, ON, PAST) is selected to define the position of the tool with respect to the surface. In direction vector (INDIRV/vector) or in direction point (INDIRP/point) commands may be used ahead of

GO/... statements to direct the tool motion in a desired direction. These three statements are often called *startup* commands in APT vocabulary. The startup commands are typically followed by continuous motion statements:

$$\left\{\begin{array}{c}\text{GOFWD}\\ \text{GOBACK}\\ \text{GOLFT}\\ \text{GORGT}\end{array}\right\} \text{/drive surface, } \left\{\begin{array}{c}\text{TO}\\ \text{ON}\\ \text{PAST}\\ \text{TANTO}\end{array}\right\} \text{, check surface}$$

where GOFWD, GOBACK, GOLFT, and GORGT motion statements correspond to continuous tool motion movement in the *forward, backward, left*, or *right* direction relative to the preceding motion (see Figs. 5.10 and 5.11).

Example: A typical motion command sequence might be

```
GO      / TO, L1, ON, PL1, TO, C1
GOFWD   / C1, TO, L1
GOLFT   / L1, ON, L1

GOTO    /P1
```

Cutter Location File and Postprocessing

APT processor or CAD/CAM systems produce two cutter location (CL) files . The first is in readable ASCII, and the second is in binary format. The ASCII printout can be obtained by inserting a CLPRNT/ON command before the motion statements. The center coordinates of the tool are printed for the programmer to debug the APT program. The binary file has a standard format. Each record in the file may have two or more fields depending on the type of command. The length of each field is again several bytes long and is determined by the command's function. For example,

Program	Class	Code	Field
PARTNO/A0010	2000	1045	A0010
TOOL/14	2000	1025	10
COOLNT/ON	2000	1030	71
FEDRAT/30.0	2000	1009	30
FROM/0,0, 3	5000	3	0.000 0.000 0.000
GODLTA/0.3	5000	4	0.300
GOTO/5.0, 3.7	5000	5	5.000 3.700
FINI	2000	1	

Class word defines whether the statement produces a miscellaneous function (M function), a PTP motion, or a continuous tool motion. The second word gives the specific code for the statement. The third field, which may be several words long, gives the coordinates of the motion or logic value for the miscellaneous functions. For example, PARTNO/ , TOOL/ , and COOLNT/ FINI are called postprocessor commands, and all are classified under 2000. COOLNT has a

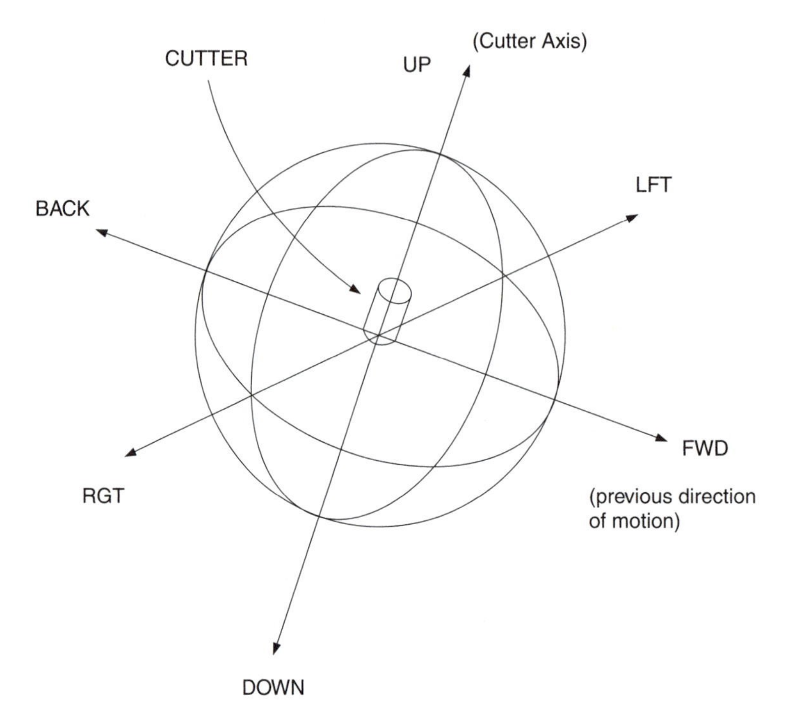

Figure 5.10: Feed motion directions in APT.

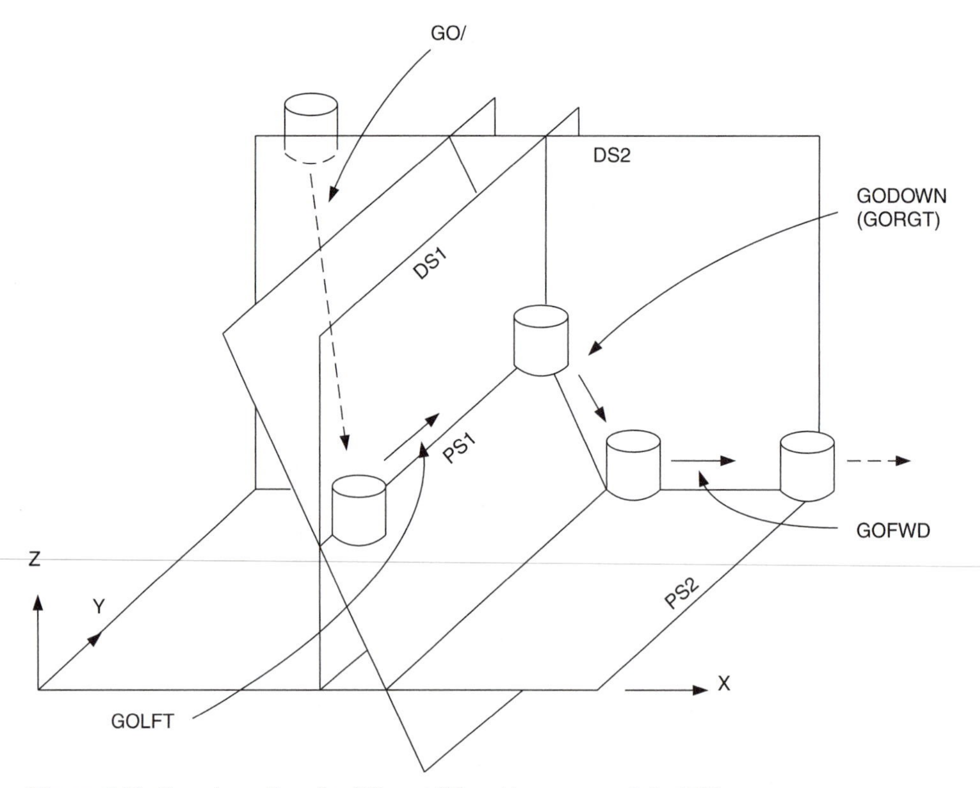

Figure 5.11: Sample motions for different GO motion commands in APT.

code value of 1030 and 71 means ON. GOTO/ and GODLTA/ commands belong to the point motion class and are grouped under 5000. GOTO has a code of 5, followed by three double-precision floating point long field words that give the x, y, and z coordinates of the target point.

The CL file has an internationally standard format. Each CNC machine tool builder provides a postprocessor for its specific CNC machine tool. After generating the CL file, the programmer *postprocesses* the CL file and obtains the final manual equivalent NC code automatically. Note that if any dimensions change in the workpiece, the geometric statements can be updated, processed, and postprocessed without having to program the part again.

NC Part Programming with CAD Systems

CAD systems have already replaced APT. Essentially, most CAD systems have started with the APT processor subroutines but have added interactive graphical displays. Each geometric statement can be made interactively using a mouse. The logic of geometry generation is similar to that of the APT commands. CAD has the advantage of allowing visual inspection of the part geometry on the computer workstation. Most current CAD systems allow three-dimensional construction of the part geometry while simultaneously providing top, side, front, and isometric views. They incorporate solid modeling technology that helps the designer to visualize the actual shape of the part more easily.

Once the geometry is designed on the CAD system, the process planner can pull the part to a computer workstation for a tool path generation. The tool paths are also interactively generated on the graphics station. The process planner defines the tool, feed rate, spindle speed, coolant, and program zero in almost identical fashion to APT. The tool path can be seen on the screen for visual interference checking. Once the tool path is accepted, a CL file is generated by the CAD system. The CL file has an APT standard format, and thus postprocessors developed for APT can be used to generate the final manual equivalent NC code on CAD/CAM systems. Figure 5.12 shows some of the sample CAD-generated tool paths and the machined sculptured part using the IDEAS™ CAD/CAM system.

5.4 TRAJECTORY GENERATION FOR COMPUTER-CONTROLLED MACHINES

The CNC motion control computer receives NC blocks that contain the mode of interpolation (i.e., linear, G01; circular, G02 or G03; or spline, G05), the coordinates of end point ($P_e(x_e, y_e)$), arc radius ($R = \sqrt{I_c^2 + J_c^2}$), and vector feed (f). The coordinates and derivatives of a series of points along a spline trajectory are provided for quintic spline interpolation.

The acceleration (A), deceleration (D), and jerk (J) values are either set to default values within the CNC or given by the NC programmer within the NC part program. The units are converted into *counts* in the CNC system. The tool

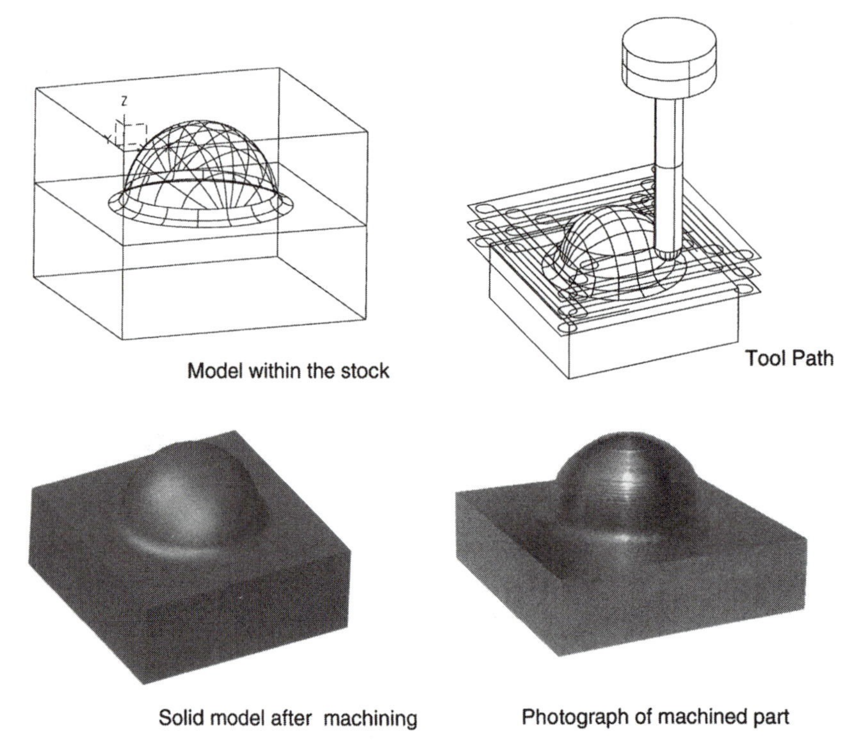

Figure 5.12: Sample tool paths and three-axis ball end milling of a sculptured surface.

path segment (i.e., line, arc, or spline in an NC block) is broken into N number of small segments in axis directions at interpolation time intervals of T_i.

The minimum interpolation time has to be equal to or an integer multiple of the axis position control loop closure time (T). The feed f is provided by the NC part program, and the minimum interpolation period $T_{\min}$ is set within the CNC control software. The interpolation step size is calculated as

$$\Delta u = f T_{\min} \tag{5.19}$$

for set values of f and $T_{\min}$. Either step size Δu or interpolation period T_i is kept constant during velocity command generation. The step size Δu can be equal to or greater than the position feedback resolution, and the interpolation period must be equal to or greater than the discrete servocontrol interval ($T_i \geq T$).

5.4.1 Interpolation with Constant Displacement

First, let us consider a design approach where the step size Δu is set until $T_{\min}$ or f is changed. When the feed (f) is changed during machining by a feed-override switch or a sensor-based machining process control module, Δu is kept constant, but the interpolation time T_i

is updated as

$$T_i = \frac{\Delta u}{f}. \tag{5.20}$$

The selection of the minimum interpolation time $T_{\min}$ is limited by the computation load and speed of the CNC motion control computer where the algorithm is executed. By using a varying interpolation period, many axes are synchronized from a one-dimensional calculation. The addition of other axes has no effect on the velocity profiling; therefore, the position and velocity interpolation are decoupled. Assuming that the total displacement along an arbitrary path is L, the interpolator task is executed N times at interpolation time intervals of T_i as follows:

$$N = \frac{L}{\Delta u}. \tag{5.21}$$

N is always rounded to the next higher even integer for computational efficiency. The feed is reduced accordingly for correction. The total number of iterations (N) is divided into a number of stages depending on the type of velocity profile used for trajectory generation. A simple to implement trapezoidal velocity profile and a complex but accurate parabolic velocity profile for high-speed drives are presented here.

Trapezoidal Velocity Profile

The acceleration and deceleration of the axis is controlled by imposing a trapezoidal velocity profile shown in Figure 5.13 on the position command generation algorithm. The trapezoidal velocity profile is simple to implement, computationally advantageous, and suitable for most low-speed, low-cost machines. The total number of interpolation steps (N) is divided into an acceleration (N_1), constant velocity (N_2), and deceleration (N_3) zones according to the trapezoidal velocity profile shown in Figure 5.13; that is, $N = N_1 + N_2 + N_3$. The counters for acceleration and deceleration distances (N_1, N_3) are calculated from the target feed f [count/s], acceleration A [count/s^2], deceleration D [count/s^2], and displacement step Δu using the trapezoidal velocity profile. If the initial feed is zero, the total tool path length (l_1) traveled during the acceleration period ($0 < t < t_1$) can be found from Figure 5.13a as

$$l_1 = \int_0^{t_1} At\,dt = \frac{At_1^2}{2}. \tag{5.22}$$

Because $t_1 = f/A$ for a constant acceleration, the number of interpolation intervals during acceleration is

$$N_1 = \frac{l_1}{\Delta u} = \frac{f^2}{2A\Delta u}. \tag{5.23}$$

Similarly, the deceleration counter is found as

$$N_3 = \frac{l_3}{\Delta u} = \frac{f^2}{2D\Delta u}. \tag{5.24}$$

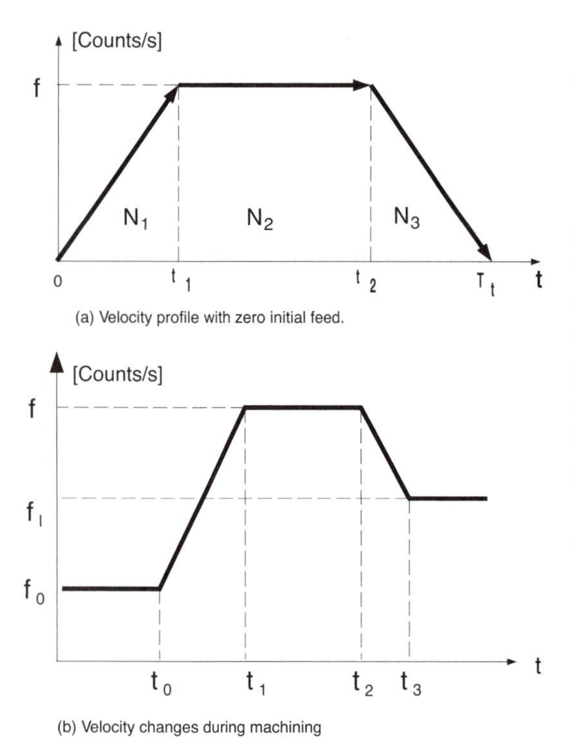

Figure 5.13: Trapezoidal velocity profile of feed.

The constant velocity zone counter (N_2) is the remaining period as follows:

$$N_2 = N - (N_1 + N_2) \tag{5.25}$$

In some NC machining applications, it is not desirable to bring the machine to a full stop before switching to the next tool path segment. Alternatively, the operator or sensor-based adaptive process control algorithms may manipulate the feed within an NC block. Let the CNC system accelerate from feed f_0 to the new feed command f. The acceleration counter can be found as (Figure 5.13b) follows:

$$\begin{aligned} l_1 &= \int_{t_0}^{t_1} [f_0 + A(t - t_0)]dt \\ &= \int_0^{\tau_a} [f_0 + A\tau]d\tau = f_0\tau_a + \frac{A\tau_a^2}{2}, \end{aligned} \tag{5.26}$$

where $\tau_a = t_1 - t_0 = (f - f_0)/A$ and

$$l_1 = \frac{f^2 - f_0^2}{2A}, \tag{5.27}$$

which leads to

$$N_1 = \frac{l_1}{\Delta u} = \frac{f^2 - f_0^2}{2A \cdot \Delta u}. \tag{5.28}$$

Similarly, if the system decelerates from feed f to f_l, then

$$l_3 = \int_{t_2}^{t_3} [f - D(t - t_2)]dt = f\tau_d - \frac{D\tau_d^2}{2} = \frac{f^2 - f_l^2}{2D}, \tag{5.29}$$

which leads to the following number of interpolation periods during deceleration:

$$N_3 = \frac{l_3}{\Delta u} = \frac{f^2 - f_l^2}{2D \cdot \Delta u}. \tag{5.30}$$

The counters N, N_1, N_2, and N_3 are rounded integers. Note that if the desired feed is not reached because of a short tool path, the acceleration and deceleration periods are made equal ($A = D$); i.e., (if $N_2 < 0 \rightarrow N_2 = 0, N_1 = N_3 = N/2$).

The interpolation period T_i must be changed at each interval during acceleration and deceleration. Because the traveled tool path segment Δu is kept constant, the following expression can be written between two interpolation periods:

$$\Delta u = \int_{t_{k-1}}^{t_k} At dt = \frac{A}{2}\left(t_k^2 - t_{k-1}^2\right) = \frac{A}{2}\left(t_k - t_{k-1}\right)\left(t_k + t_{k-1}\right).$$

By substituting $T_i(k) = t_k - t_{k-1}$ and $t_k = f(k)/A$, $t_{k-1} = f(k-1)/A$, the interpolation period $(T_i(k))$ during acceleration and deceleration where the velocity changes is found at each increment as

$$T_i(k) = \frac{2\Delta u}{f(k) + f(k-1)}. \tag{5.31}$$

For small acceleration and deceleration values, this equation can be approximated as $T_i(k) = \Delta(u)/f(k)$ with caution. The approximation reduces the real-time computation load, but it produces jerks at large velocity increments. The following pseudocode algorithm is given to calculate interpolation time intervals for acceleration, constant feed, and deceleration phases:

for	$k = 1, N_1$	; Iterate during acceleration period.
	$f(k) = \sqrt{f_0^2 + 2kA\Delta u}$	; Next f calculated from initial f_0.
	$T_i(k) = \frac{2\Delta u}{f(k)+f(k-1)}$	; Next interpolation period time.
next	k	;
for	$k = 1, N_2$	; Iterate during constant feed period.
	$T_i = \Delta u/f$	; Period is always the same.
next	k	;
for	$k = 1, N_3$	; Iterate during deceleration period.
	$f(k) = \sqrt{f^2 - 2kD\Delta u}$	; Next f calculated from previous f_0.
	$T_i(k) = \frac{2\Delta u}{f(k)+f(k-1)}$	; Next interpolation period time.
next	k	;

Depending on the *on-the-fly* increase or decrease in the feed, the new values of N_1, N_2, N_3, and T_i are recalculated and updated. Note that acceleration, constant feed, and deceleration phases are coded into different functions and are executed when they are required. Because most of the terms are constants, the interpolation rate is constrained by the square root.

Note that the feed, acceleration, and deceleration regulated for vector displacement guarantees synchronized and correct velocity values in all active drives contributing to the vectorial motion in the space. If we take a two-axis

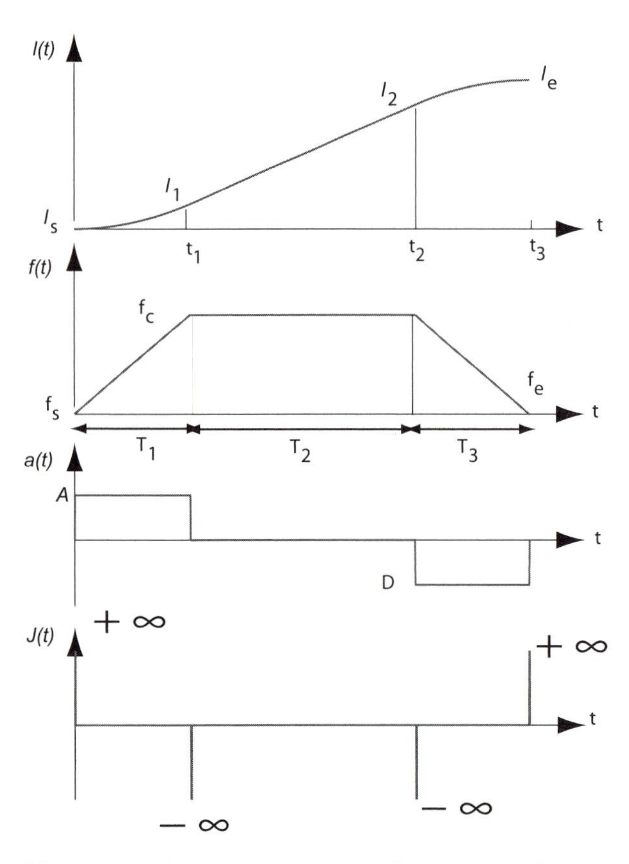

Figure 5.14: Trajectory generation with constant acceleration and fixed interpolation period.

motion in the x and y directions, the differential vector displacement Δu can be written as

$$\vec{\Delta u} = \Delta x\vec{i} + \Delta y\vec{j}, \tag{5.32}$$

where i and j are unit vectors in the x and y directions, respectively. If we divide both sides with interpolation time interval T_i, we get

$$\frac{\vec{\Delta u}}{T_i} = \frac{\Delta x}{T_i}\vec{i} + \frac{\Delta y}{T_i}\vec{j},$$

$$\vec{f} = f_x\vec{i} + f_y\vec{j},$$

where the feed amplitude is $f = \sqrt{f_x^2 + f_y^2}$, and f_x and f_y are the resulting velocities of the x and y drives. Hence, once Δu, interpolation time T_i, and counters N_1, N_2, and N_3 are calculated, the velocities and incremental positions in the x and y drives are automatically defined by the algorithm. The interpolation algorithms presented in the next section clarify the engineering implementation of the method.

5.4.2 Acceleration-Limited Velocity Profile Generation with Constant Interpolation Period

Let us consider a design approach where the interpolation period T_i is constant. Typically T_i is selected equal to control interval if the motion control computer has sufficient computation power in real time, or integer multiples of it.

The acceleration, velocity, and displacement profiles can be expressed as a function of time (t) as follows:

$$a(t) = \begin{Bmatrix} A & 0 \le t < t_1 & A > 0 \\ 0 & t_1 \le t < t_2 & \\ D & t_2 \le t < t_3 & D < 0 \end{Bmatrix}, \tag{5.33}$$

where A and D are the acceleration and deceleration, respectively. In most applications $D = -A$. The velocity profile is expressed as

$$f(t) = \begin{Bmatrix} f_s + \int Adt = f_s + At & 0 \le t < t_1 & f_1 = f_s + AT_1 = f_c \\ f_c & t_1 \le t < t_2 & f_2 = f_c \\ f_c + \int Ddt = f_c + Dt & t_2 \le t < t_3 & f_3 = f_c + DT_3 = f_e \end{Bmatrix}, \quad (5.34)$$

where f_s, f_c, f_e are the start, steady, and end feed values of the motion segments, respectively. The time segments are $T_1 = t_1$, $T_2 = t_2 - t_1$, $T_3 = t_3 - t_2 = T - (T_1 + T_2)$, where T is the total travel time along the path segment. The integration of feed leads to displacement traveled at each motion segment as

$$l(t) = \begin{Bmatrix} l_s + \int (f_s + At)dt = l_s + f_s t + \frac{1}{2}At^2 & 0 \le t < t_1 & l_1 = l_s + f_s T_1 + \frac{1}{2}AT_1^2 \\ l_1 + \int f_c dt = l_1 + f_c t & t_1 \le t < t_2 & l_2 = l_1 + f_c T_2 \\ l_2 + \int (f_c + Dt)dt = l_2 + f_c t + \frac{1}{2}Dt^2 & t_2 \le t < t_3 & l_3 = l_2 + f_c T_3 + \frac{1}{2}DT_3^2 = l_s + L \end{Bmatrix}, \quad (5.35)$$

where L is the total path segment length. By substituting $l_1$and l_2 into the l_3, the total path length L can be expressed as a function of motion time zones, feed, acceleration, and deceleration.

$$\begin{aligned} L &= l_3 - l_s = l_2 + f_c T_3 + \frac{1}{2}DT_3^2 - l_s \\ &= l_1 + f_c T_2 + f_c T_3 + \frac{1}{2}DT_3^2 - l_s \\ &= l_s + f_s T_1 + \frac{1}{2}AT_1^2 + f_c T_2 + f_c T_3 + \frac{1}{2}DT_3^2 - l_s \\ L &= f_s T_1 + \frac{1}{2}AT_1^2 + f_c T_2 + f_c T_3 + \frac{1}{2}DT_3^2 \end{aligned} \quad (5.36)$$

The time segments T_1, T_2, and T_3 must be identified to generate trajectory commands at each interpolation period. The acceleration (T_1) and deceleration (T_3) periods can be evaluated as

$$\begin{aligned} f_s + AT_1 = f_c &\rightarrow T_1 = \tfrac{f_c - f_s}{A} \\ f_c + DT_3 = f_e &\rightarrow T_3 = \tfrac{f_e - f_c}{D} \end{aligned} \quad (5.37)$$

By substituting T_1, T_3 into the total displacement length (L), the unknown T_2 can be found as

$$\begin{aligned} L &= f_s T_1 + \frac{1}{2}AT_1^2 + f_c T_2 + f_c T_3 + \frac{1}{2}DT_3^2 \\ L &= f_s\left(\frac{f_c - f_s}{A}\right) + \frac{1}{2}A\left(\frac{f_c - f_s}{A}\right)^2 + f_c T_2 + f_c\left(\frac{f_e - f_c}{D}\right) + \frac{1}{2}D\left(\frac{f_e - f_c}{D}\right)^2 \\ T_2 &= \frac{L}{f_c} - \left[\left(\frac{1}{2A} - \frac{1}{2D}\right)f_c + \left(\frac{-f_s^2}{2A} + \frac{f_e^2}{2D}\right)\frac{1}{f_c}\right]. \end{aligned} \quad (5.38)$$

Solution 7. *Details of derivation for T_2*

$$T_2 = \left\{L - \left[f_s\left(\frac{f_c - f_s}{A}\right) + \frac{1}{2}A\left(\frac{f_c - f_s}{A}\right)^2 + f_c\left(\frac{f_e - f_c}{D}\right) + \frac{1}{2}D\left(\frac{f_e - f_c}{D}\right)^2\right]\right\}\frac{1}{f_c}$$

$$= \left\{L - \left[\frac{f_s f_c - f_s^2}{A} + \frac{f_c^2 - 2f_s f_c + f_s^2}{2A} + \frac{f_c f_e - f_c^2}{D} + \frac{f_e^2 - 2f_e f_c + f_c^2}{2D}\right]\right\}\frac{1}{f_c}$$

$$= \left\{L - \left[\frac{2f_s f_c - 2f_s^2 + f_c^2 - 2f_s f_c + f_s^2}{2A} + \frac{2f_e f_c - 2f_c^2 + f_e^2 - 2f_e f_c + f_c^2}{2D}\right]\right\}\frac{1}{f_c}$$

$$= \left\{L - \left[\frac{-f_s^2 + f_c^2}{2A} + \frac{-f_c^2 + f_e^2}{2D}\right]\right\}\frac{1}{f_c}$$

$$= \left\{L - \left[\left(\frac{1}{2A} - \frac{1}{2D}\right)f_c^2 + \left(\frac{-f_s^2}{2A} + \frac{f_e^2}{2D}\right)\right]\right\}\frac{1}{f_c}.$$

When the acceleration and deceleration values are set equal ($D = -A$), the steady-state velocity time T_2 becomes

$$T_2 = \frac{L}{f_c} + \left[-\frac{f_c}{A} + \left(\frac{f_e^2 + f_s^2}{2A}\right)\frac{1}{f_c}\right]. \tag{5.39}$$

The total motion duration becomes $T_t = T_1 + T_2 + T_3$.

When the path segments are too small, the machine may not be able to reach to the command feed (f_c), and the steady-state velocity zone may be zero or negative, i.e., $T_2 \leq 0$. If the limited acceleration leads to such an incompatibility with the trajectory generation, the command feed must be reduced to a compatible value that can be achieved without saturating the drives.

If $T_2 \leq 0$ in Eq. (5.38), then it is forced to become zero to identify the possible feed from a given set of acceleration, deceleration, and the start and end feed values as follows:

$$T_2 = \frac{L}{f_{cm}} - \left[\left(\frac{1}{2A} - \frac{1}{2D}\right)f_{cm} + \left(\frac{-f_s^2}{2A} + \frac{f_e^2}{2D}\right)\frac{1}{f_{cm}}\right] = 0 \tag{5.40}$$

$$f_{cm} = \sqrt{\frac{2ADL - (f_e^2 A - f_s^2 D)}{D - A}}.$$

Note that if $D = -A$, the the allowable feed becomes

$$f_{cm} = \sqrt{AL + \frac{(f_e^2 + f_s^2)}{2}}. \tag{5.41}$$

The new acceleration and deceleration periods are computed as ($T_2 = 0$) follows:

$$\begin{aligned} f_s + AT_1 = f_{cm} \quad &\rightarrow \quad T_1 = \tfrac{f_{cm} - f_s}{A} \\ f_{cm} + DT_3 = f_e \quad &\rightarrow \quad T_3 = \tfrac{f_e - f_{cm}}{D}. \end{aligned} \tag{5.42}$$

In addition to considering the zero steady-state travel zone ($T_2 = 0$), it is important to consider numerical roundoff errors in trajectory generation. If the interpolation period T_i is fixed in the CNC system, the total number of interpolation steps must be an integer number as follows:

$$N = ceil\left(\frac{T_t}{T_i}\right)$$

However, if the T_t/T_i is not a perfect integer, which is always possible, then the fraction must be considered to correct the duration of motion according to trajectory generation. Because the interpolation period (T_i) is constant, the algorithm leads to actual total travel time $T_t' = NT_i$, which can be different than T_t if T_t/T_i is not an integer. The acceleration, steady-state, and deceleration time lengths (T_1, T_2, T_3) are adjusted to guarantee that the total time is equal to T_t as follows:

$$T_j' = \left(\frac{T_t'}{T_t}\right).T_j, \quad \text{where } j = 1, 2, 3.$$

Because T_1', T_2', and T_3' are now different than previously estimated travel periods (T_1, T_2, and T_3), it is necessary to remodify the nominal feed, acceleration, and deceleration. From Eq. (5.38), the new feed f_{cn} is calculated as follows:

$$f_{cns} = \frac{2L - f_s T_1' - f_e T_3'}{T_1' + 2T_2' + T_3'}$$

$$A_n = \frac{f_{cn} - f_s}{T_1'}, \quad D_n = \frac{f_e - f_{cns}}{T_3'}.$$

The new kinematic profiles must be regenerated by following the new feed (f_{cn}), acceleration (A_n), and deceleration (B_n) values by substituting them in Eqs. (5.33), (5.34), (5.35). The trajectory generation can be rearranged as follows:

$$l(t) = \begin{Bmatrix} l_s + f_s\tau + \frac{1}{2}A_n\tau^2 & 0 \le \tau < T_1' & l_{1n} = l_s + f_s T_1' + \frac{1}{2}A_n T_1'^2 \\ l_{1n} + f_{cn}\tau & 0 \le \tau < T_2' & l_{2n} = l_{1n} + f_{cn}T_2' \\ l_{2n} + f_{cn}\tau + \frac{1}{2}D_n\tau^2 & 0 \le \tau < T_3' & = l_s + L \end{Bmatrix}. \tag{5.43}$$

The real-time computation can be optimized by converting Eq. (5.43) into difference equations by substituting $\tau = kT_i$, where T_i is the constant interpolation period.

$$l(k) = \begin{Bmatrix} l_s + f_s(kT_i) + \frac{1}{2}A_n(kT_i)^2 & k = 1, 2, \ldots, (N_1 = T_1'/T_i) & \begin{matrix} l(k-1) + f_s T_i + \Delta_1(k - \frac{1}{2}) \\ \Delta_1 = A_n T_i^2,\ l_{1n} = l(N_1) \end{matrix} \\ l_{1n} + f_{cn}(kT_i) & k = 1, 2, \ldots, (N_2 = T_2'/T_i) & \begin{matrix} l(k-1) + \Delta_2, \\ \Delta_2 = f_{cn}T_i,\ l_{2n} = l(N_2) \end{matrix} \\ l_{2n} + f_{cn}(kT_i) + \frac{1}{2}D_n(kT_i)^2 & k = 1, 2, \ldots, (N_3 = T_3'/T_i) & \begin{matrix} l(k-1) + f_{cn}T_i + \Delta_3(k - \frac{1}{2}), \\ \Delta_3 = D_n T_i^2,\ l_{3n} = l(N_3) \end{matrix} \end{Bmatrix}. \tag{5.44}$$

The constant displacements (Δ_1, Δ_2, Δ_3) can be calculated during initialization and simply added recursively during real-time trajectory generation. Note that $l(k)$ is a discrete displacement command in space and must be decoupled into individual axes using the kinematics of the machine.

Solution 8. *Details of derivation for the acceleration phase of trajectory generation.*

$$\begin{aligned} l_s + f_s(kT_i) + \frac{1}{2}A_n(kT_i)^2 &= l_s + f_s(k-1)T_i + f_sT_i + \frac{1}{2}A_n\left[(k-1)\,T_i + T_i\right]^2 \\ &= l_s + f_s(k-1)T_i + f_sT_i + \frac{1}{2}A_n[(k-1)\,T_i]^2 + A_n\,(k-1)\,T_i^2 + \frac{1}{2}A_nT_i^2 \\ &= l_s + f_s(k-1)T_i + \frac{1}{2}A_n[(k-1)\,T_i]^2 + f_sT_i + A_nT_i^2(k-\frac{1}{2}) \\ &= l(k-1) + f_sT_i + A_nT_i^2(k-\frac{1}{2}). \end{aligned}$$

5.4.3 Jerk-Limited Velocity Profile Generation

The trapezoidal velocity profile introduced in the previous section is simple to implement and suitable for most machines. However, because it uses a constant acceleration, the jerk or the derivative of the acceleration is zero, which leads to various oscillations and noise on the feed and acceleration when interpolating along complex tool paths. The acceleration times inertia or mass acts as a dynamic torque or force on the feed drive structure. If the feed drive acceleration command produced by the trajectory generator is not smooth, the resulting acceleration torque for the ball screw and force for the linear motor drives contain high-frequency components that excite the structural dynamics of the feed drives and cause undesired vibrations. To obtain smooth velocity and acceleration profiles, jerk-limited trajectory generation algorithms are used, and these are presented in this section.

Kinematic Profiles

The kinematic time profiles of jerk (J), acceleration (a), feed rate (f), and trajectory command position (l) are shown in Figure 5.15. Before the NC block motion is started, the initial and final values of position (l_s, l_e) and feed rate (f_s, f_e), maximum acceleration (A), deceleration (D), and jerk (J) limits are defined. The maximum acceleration/deceleration limits are identified from the maximum torque and force limits of the drive motors. The acceleration time is set depending on the peak torque/force delivery periods of the amplifier. The jerk limit is set by the maximum acceleration divided by the acceleration time. From Figure 5.15, the jerk ($J(\tau)$), acceleration ($a(\tau)$), feed ($f(\tau)$), and displacement ($l(\tau)$) along the tool path can be expressed. The jerk is constant when the acceleration increases linearly as follows:

$$J(\tau) = \left\{ \begin{array}{ll} J_1 & 0 \le t < t_1 \\ 0 & t_1 \le t < t_2 \\ -J & t_2 \le t < t_3 \\ 0 & t_3 \le t < t_4 \\ -J_5 & t_4 \le t < t_5 \\ -0 & t_5 \le t < t_6 \\ J_7 & t_6 \le t < t_7 \end{array} \right\} \tag{5.45}$$

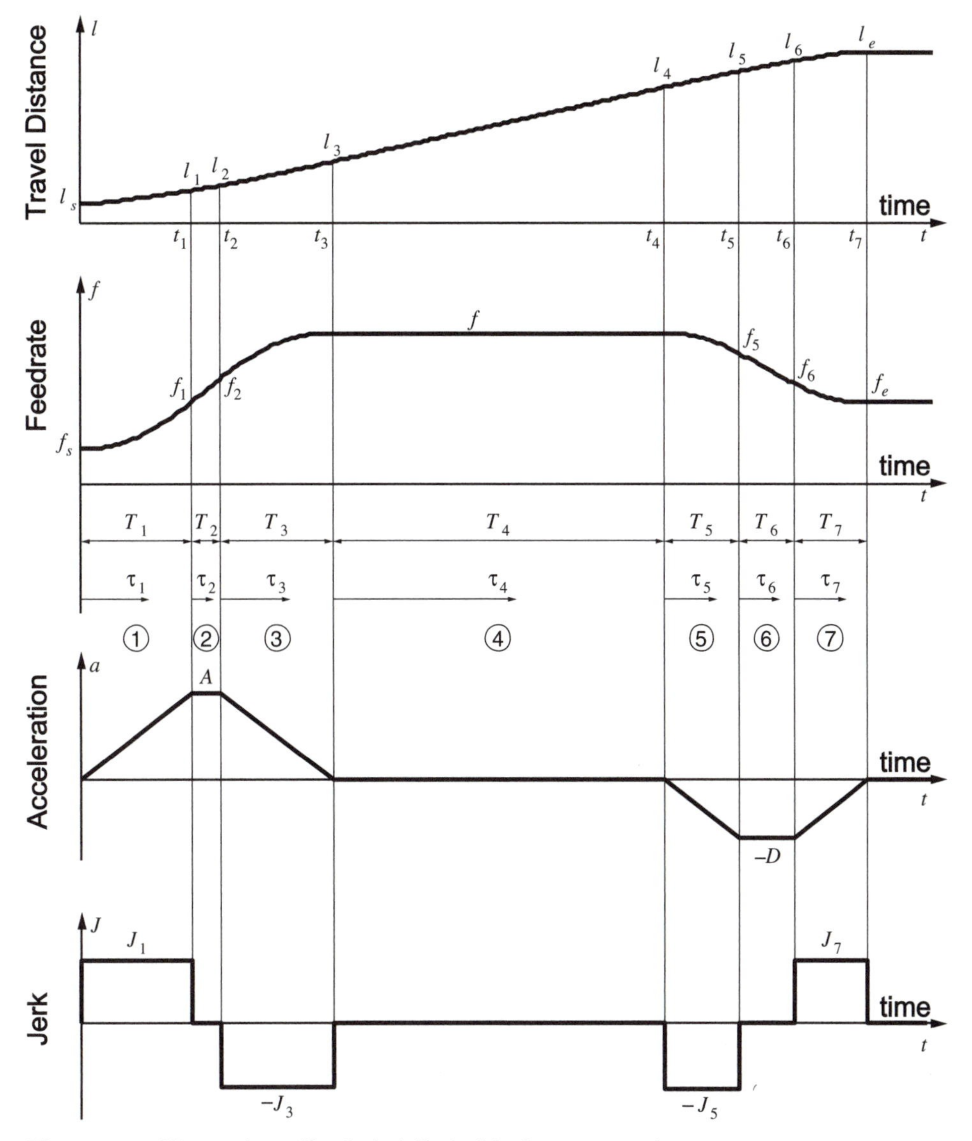

Figure 5.15: Kinematic profiles for jerk-limited feed rate generation.

The acceleration is found by integrating the jerk while considering its initial conditions as follows:

$$
\begin{aligned}
a(t) &= a(t_i) + \int_{t_i}^{t} J(\tau)d\tau, \\
f(t) &= f(t_i) + \int_{t_i}^{t} a(\tau)d\tau, \qquad (5.46) \\
l(t) &= l(t_i) + \int_{t_i}^{t} f(\tau)d\tau.
\end{aligned}
$$

The jerk is constant during acceleration (T_1, T_3) and deceleration (T_5, T_7) periods and is zero during constant acceleration (T_2, T_4, T_6). By integrating the jerk at each period shown in Figure 5.15, the acceleration profiles are obtained as follows:

$$a(\tau) = \begin{Bmatrix} J_1\tau_1, & 0 \le t < t_1 \\ A, & t_1 \le t < t_2 \\ A - J_3\tau_3, & t_2 \le t < t_3 \\ 0, & t_3 \le t < t_4 \\ -J_5\tau_5, & t_4 \le t < t_5 \\ -D, & t_5 \le t < t_6 \\ -D + J_7\tau_7, & t_6 \le t < t_7 \end{Bmatrix}, \tag{5.47}$$

where τ_k $(k = 1, 2, \ldots, 7)$ denotes relative time (i.e., $\tau_k = t - t_{k-1}$) at each each motion zone. By integrating the acceleration (Eq. 5.47) at each period, the feed velocity profile for each phase in Figure 5.15 is found to be

$$f(\tau) = \begin{Bmatrix} f_s + \frac{1}{2}J_1\tau_1^2, & f_s : \text{initial feed}, & 0 \le t < t_1 \\ f_1 + A\tau_2, & f_1 = f_s + \frac{1}{2}J_1T_1^2, & t_1 \le t < t_2 \\ f_2 + A\tau_3 - \frac{1}{2}J_3\tau_3^2, & f_2 = f_1 + AT_2, & t_2 \le t < t_3 \\ f_3, & f = f_3 = f_2 + AT_3 - \frac{1}{2}J_3T_3^2, & t_3 \le t < t_4 \\ f_4 - \frac{1}{2}J_5\tau_5^2, & f = f_4 = f_3, & t_4 \le t < t_5 \\ f_5 - D\tau_6, & f_5 = f_4 - \frac{1}{2}J_5T_5^2, & t_5 \le t < t_6 \\ f_6 - D\tau_7 + \frac{1}{2}J_7\tau_7^2, & f_6 = f_5 - DT_6, & t_6 \le t < t_7 \end{Bmatrix}, \tag{5.48}$$

where T_k is the period of phase k. Integrating (5.48) once more yields the displacement profile as

$$l(\tau) = \begin{Bmatrix} l_s + f_s\tau_1 + \frac{1}{6}J_1\tau_1^3, & l_s : \text{initial position}, & 0 \le t < t_1 \\ l_1 + f_1\tau_2 + \frac{1}{2}A\tau_2^2, & l_1 = l_s + f_sT_1 + \frac{1}{6}J_1T_1^3, & t_1 \le t < t_2 \\ l_2 + f_2\tau_3 + \frac{1}{2}A\tau_3^2 - \frac{1}{6}J_3\tau_3^3, & l_2 = l_1 + f_1T_2 + \frac{1}{2}AT_2^2, & t_2 \le t < t_3 \\ l_3 + f_3\tau_4, & l_3 = l_2 + f_2T_3 + \frac{1}{2}AT_3^2 - \frac{1}{6}J_3T_3^3, & t_3 \le t < t_4 \\ l_4 + f_4\tau_5 - \frac{1}{6}J_5\tau_5^3, & l_4 = l_3 + f_3T_4, & t_4 \le t < t_5 \\ l_5 + f_5\tau_6 - \frac{1}{2}D\tau_6^2, & l_5 = l_4 + f_4T_5 - \frac{1}{6}J_5T_5^3, & t_5 \le t < t_6 \\ l_6 + f_6\tau_7 - \frac{1}{2}D\tau_7^2 + \frac{1}{6}J_7\tau_7^3, & l_6 = l_5 + f_5T_6 - \frac{1}{2}DT_6^2, & t_6 \le t < t_7 \end{Bmatrix}, \tag{5.49}$$

where l_k is the total displacement reached at the end of phase k.

The incremental distance (L_k) traveled at each phase (k) is

$$L_k = l_k - l_{k-1}, \tag{5.50}$$

where the initial displacement is $l_0 = l_s$. The sum of the distances traveled at each step should be equal to the total distance to be traveled (Eqs. 5.49 and 5.50) as follows:

$$L = l_e - l_s = \sum_{k=1}^{7} L_k, \tag{5.51}$$

where L is the total distance traveled within the NC tool path segment. From the trapezoidal acceleration/deceleration profiles,

$$A = J_1 T_1 = J_3 T_3, \quad D = J_5 T_5 = J_7 T_7 \tag{5.52}$$

should hold, although this may require readjustment after the first initialization step. Considering that the desired feed (f) is reached at the end of phase 3, we have

$$f_3 = f \rightarrow T_2 = \frac{1}{A}\left[f - f_s - \frac{1}{2}J_1 T_1^2 - AT_3 + \frac{1}{2}J_3 T_3^2\right], \tag{5.53}$$

and similarly, considering that the final feed (f_e) is reached at the end of phase 7, we have

$$\begin{aligned} f_7 &= f_e = f_6 - DT_7 + \tfrac{1}{2}J_7 T_7^2, \\ T_6 &= \tfrac{1}{D}\left[f - f_e - \tfrac{1}{2}J_5 T_5^2 - DT_7 + \tfrac{1}{2}J_7 T_7^2\right]. \end{aligned} \tag{5.54}$$

Initialization

Before generating incremental position commands in the interpolation stage, the number of interpolation steps and the time interval between each step must be predetermined to implement the jerk-limited kinematic profile given in Figure 5.15. To design a general algorithm, the following conditions are evaluated at the initialization step.

The total number of interpolation steps (N) is checked. If $2 < N \leq 4$, then $N = 4$ is selected to guarantee at least the presence of the acceleration and deceleration phases ($1, 3, 5,$ and 7) in Figure 5.15. If $N \leq 2$, then $N = 2$ is selected to allow an acceleration and deceleration. Note that these conditions would only occur when the motion is very small, such as in high-speed spline interpolation applications or in precision positioning. If the acceleration stage exists, the desired feed (f) must be reached within the first three phases, which implies that $T_2 \geq 0$. If the jerk values are equal ($J_1 = J_3$) (Eq. 5.52), the acceleration condition requires that

$$T_1 = T_3 = A/J_1, \quad T_2 = \frac{f - f_s}{A} - \frac{A}{J_1} \geq 0. \tag{5.55}$$

If Eq. (5.55) does not hold, then the magnitude of the acceleration must be reduced to its maximum possible limit as

$$A = \text{sgn}(A) \cdot \sqrt{J_1(f - f_s)} \tag{5.56}$$

and T_2 is set to zero. Similarly, if the deceleration stage exists,

$$T_5 = T_7 = D/J_5, \quad T_6 = \frac{f - f_e}{D} - \frac{D}{J_5} \geq 0. \tag{5.57}$$

If the deceleration stage does not exist, the deceleration limit is reduced as

$$D = \text{sgn}(D) \cdot \sqrt{J_5(f - f_e)}, \tag{5.58}$$

and T_6 is set to zero. If the displacement length is large enough to encompass the constant feed stage, $T_4 \geq 0$ and must be identified. The total travel length is calculated from Eqs. (5.49) and (5.51) by substituting T_2 and T_6 from Eqs. (5.54) and (5.57) as

$$L = \left(\frac{1}{2A} + \frac{1}{2D}\right) f^2 + \left(\frac{A}{2J_1} + \frac{D}{2J_5} + T_4\right) f + \left(\frac{Af_s}{2J_1} + \frac{Df_e}{2J_5} - \frac{f_s^2}{2A} - \frac{f_e^2}{2D}\right). \tag{5.59}$$

If the tool path distance is too short to reach the target feed (f), its corresponding terms must not be considered in Eq. (5.59). To have a constant feed (f) phase, $T_4 \geq 0$ should hold as follows:

$$T_4 = \frac{1}{f}\left[L - \left\{\left(\frac{1}{2A} + \frac{1}{2D}\right) f^2 + \left(\frac{A}{2J_1} + \frac{D}{2J_5}\right) f + \left(\frac{Af_s}{2J_1} + \frac{Df_e}{2J_5} - \frac{f_s^2}{2A} - \frac{f_e^2}{2D}\right)\right\}\right] \geq 0. \tag{5.60}$$

If Eq. (5.60) is not satisfied, then $T_4 = 0$ and the magnitude of the feed rate is reduced to its maximum possible value by solving Eq. (5.60). If the quadratic equation yields complex roots, the initial and final feeds are set to zero ($f_s = f_e = 0$), and the path reinitialization is repeated by adjusting the initial parameters (A, D, J).

Number of Interpolation Steps in Each Phase

The incremental displacement Δu is decided by the linear, circular, or spline interpolation algorithms. Since the displacement in each phase is defined in Eq. (5.50), the number of interpolation steps in phases $(1, 3, 5,$ and $7)$ can be expressed as

$$\begin{aligned} N_1 &= \text{round}\,(L_1/\Delta u), \quad N_3 = \text{round}\,(L_3/\Delta u), \\ N_5 &= \text{round}\,(L_5/\Delta u), \quad N_7 = \text{round}\,(L_7/\Delta u). \end{aligned} \tag{5.61}$$

If any of (N_1, N_3, N_5, N_7) are zero for nonzero L_1, L_3, L_5, L_7 because of rounded integers, they are set to one. The total number of steps for the acceleration (N_{ac}) and deceleration (N_{dec}) stages are calculated as

$$N_{ac} = \text{round}\,\frac{(L_1 + L_2 + L_3)}{\Delta u}, \quad N_{dec} = \text{round}\,\frac{(L_5 + L_6 + L_7)}{\Delta u}, \tag{5.62}$$

from which the number of steps for constant acceleration (phase 2) and constant deceleration (phase 6) are found as follows:

$$N_2 = N_{\mathrm{ac}} - (N_1 + N_3), \quad N_6 = N_{\mathrm{dec}} - (N_5 + N_7). \tag{5.63}$$

The number of interpolation steps for the constant feed (phase 4) becomes

$$N_4 = N - (N_{\mathrm{ac}} + N_{\mathrm{dec}}). \tag{5.64}$$

The number of interpolation steps ($N_1, \ldots, N_7$) for each phase of the jerk-limited velocity profile generation (Fig. 5.15) is now defined. Although the total number of interpolation steps is still N, the displacement at each phase may be changed because of constraints, that is,

$$L'_k = N_k \cdot \Delta u, \quad k = 1, 2, \ldots, 7, \tag{5.65}$$

where L'_k is the new displacement value, which may be different than the value given in Eq. (5.50). This requires the readjustment of all acceleration, deceleration, jerk, and time periods.

For the acceleration stage (phases 1, 2, and 3) when $T_2 > 0$, Eq. (5.53) for T_2 and the new displacements L'_1, L'_2, L'_3 are substituted in Eqs. (5.49) and (5.50), resulting in the following equations:

$$\left.\begin{array}{l} f_s T_1 + \frac{1}{6} A T_1^2 - L'_1 = 0 \\ -\frac{1}{8} A T_1^2 + \frac{1}{8} A T_3^2 - \frac{1}{2} f_s T_1 - \frac{1}{2} f T_3 + \frac{f^2 - f_s^2}{2A} - L'_2 = 0 \\ f T_3 - \frac{1}{6} A T_3^2 - L'_3 = 0 \end{array}\right\} \rightarrow \text{for } T_2 > 0. \tag{5.66}$$

If $T_2 = 0$, Eq. (5.50) can not be used. Instead, the feed rate condition for the end of the acceleration stage (f_3 in Eq. 5.48) is used, which leads to

$$\left.\begin{array}{l} f_s T_1 + \frac{1}{6} A T_1^2 - L'_1 = 0 \\ \frac{1}{2} A T_1 + \frac{1}{2} A T_3 + f_s - f = 0 \\ \frac{1}{3} A T_3^2 + \frac{1}{2} A T_1 T_3 + f_s T_3 - L'_3 = 0 \end{array}\right\} \rightarrow \text{for } T_2 = 0. \tag{5.67}$$

A similar approach is applied to the deceleration stage (phases 5, 6, and 7). By substituting T_6 and the new displacements L'_5, L'_6, L'_7 in Eqs. (5.49) and (5.50), we get

$$\left.\begin{array}{l} f T_5 - \frac{1}{6} D T_5^2 - L'_5 = 0 \\ \frac{1}{8} D T_5^2 - \frac{1}{8} D T_7^2 - \frac{1}{2} f T_5 - \frac{1}{2} f_e T_7 + \frac{f^2 - f_e^2}{2D} - L'_6 = 0 \\ f_e T_7 + \frac{1}{6} D T_7^2 - L'_7 = 0 \end{array}\right\} \rightarrow \text{for } T_6 > 0. \tag{5.68}$$

For the case $T_6 = 0$ the feed rate condition at the end of deceleration is used as follows:

$$\left.\begin{array}{l} fT_5 - \frac{1}{6}DT_5^2 - L_5' = 0 \\ \frac{1}{2}DT_5 + \frac{1}{2}DT_7 + f_e - f = 0 \\ -\frac{1}{3}DT_7^2 - \frac{1}{2}DT_5T_7 + fT_7 - L_7' = 0 \end{array}\right\} \rightarrow \text{for } T_6 = 0. \tag{5.69}$$

For the acceleration stage, either Eq. (5.66) for $T_2 > 0$ or Eq. (5.67) for $T_2 > 0$ is used to find the new values of T_1, T_3, A, and T_2. For the deceleration stage, either Eq. (5.68) for $T_6 > 0$ or Eq. (5.69) for $T_6 = 0$ is used to find the updated values of T_5, T_7, D, and T_6. These equations are nonlinear and can be solved iteratively by using the Newton–Raphson numerical algorithm. T_4 is found in Eq. (5.60). Once the values of $T_1, T_2, \ldots, T_7$ and A, D are found, the jerk values are updated by using Eqs. (5.52) and (5.57), the final feed rate values reached at the end of each phase are recalculated from Eq. (5.48).

In short, the number of interpolation steps $(N_1, \ldots, N_7)$, time intervals $(T_1, \ldots, T_7)$, feed values $(f_1, \ldots, f_7)$, displacements $(L_1', \ldots, L_7')$, and possible accelerations and decelerations (A, D, J) are found for each phase of the motion block before the real-time interpolation starts.

Recursively Executed Real-Time Part

Here the interpolation period at each step must be identified. After the initialization part is completed as explained earlier, the continuously executed real-time part of the jerk-limited feed rate generation algorithm is invoked at each interpolation step. For any of the seven phases in Figure 5.15, the following general displacement formula can be used:

$$l(\tau_k) = \frac{1}{6}J_{0,k}\tau_k^3 + \frac{1}{2}a_{0,k}\tau_k^2 + f_{0,k}\tau_k + l_{0,k}, \tag{5.70}$$

where $J_{0,k}$, $a_{0,k}$, $f_{0,k}$, and $l_{0,k}$ are jerk, acceleration, feed rate, and displacement values, respectively, and they are calculated at the initialization stage. τ_k is the relative time parameter at the beginning of each phase k. Because the incremental displacement step Δu is predetermined at the interpolation, the displacement at interpolation step number m can be found as

$$l(\tau_{k,m}) = m \cdot \Delta u = \frac{1}{6}J_{0,k}\tau_{k,m}^3 + \frac{1}{2}a_{0,k}\tau_{k,m}^2 + f_{0,k}\tau_{k,m} + l_{0,k}, \tag{5.71}$$

which leads to the solution of accumulated time $(\tau_{k,m})$ within the phase k. Although Eq. (5.71) can be solved analytically, it is more efficient to solve it in real time by using the Newton–Raphson iterative algorithm. The interpolation period for within phase k at interpolation step m can be evaluated by

$$T_{i(k,m)} = \tau_{k,m} - \tau_{k,m-1}. \tag{5.72}$$

Alternatively, by replacing the initial conditions at each step, the interpolation period can be solved from Eq. (5.71) as well, but at the expense of numerical roundoff errors.

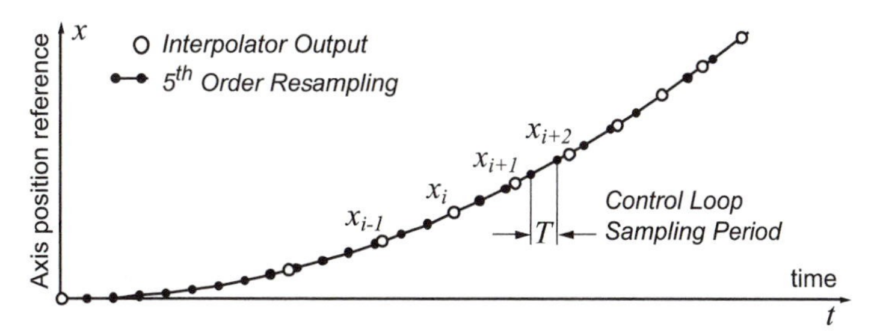

Figure 5.16: Resampling of interpolator output trajectory at control loop frequency.

Reconstruction of Reference Trajectory at Servoloop Control Frequency

Reference trajectories generated with varying interpolation periods may cause sudden changes in the discrete position commands at each axis and, hence, cause undesirable jerks on the drives. To smooth the generated trajectory, the discrete position commands of each drive at varying interpolation periods must be resampled at the servoloop frequency. Note that the resampling is done after the trajectory is generated by the interpolation stage presented in the following section, although it is presented here because of its relevance to jerk-limited trajectory generation.

A sample trajectory is shown in Figure 5.16, where the interpolator and resampled trajectory at servocontrol interval T are shown. If we consider one of the drives, the x axis, two consecutive displacements of the drive are denoted by x_i, x_{i+1}, where i is the interpolation counter. A fifth-order polynomial is fit between the two points as

$$\tilde{x}(\tau) = A_r\tau^5 + B_r\tau^4 + C_r\tau^3 + D_r\tau^2 + E_r\tau + F_r, \tag{5.73}$$

where the time varies as $0 \le \tau \le (t_{i+1} - t_i)$. The coefficients $(A_r, B_r, C_r, D_r, E_r, F_r)$ are identified from the boundary conditions as

$$\begin{aligned}
&\tilde{x}(0) = x_i, \quad \tilde{x}(t_{i+1} - t_i) = x_{i+1},\\
&\frac{d\tilde{x}(0)}{d\tau} = \dot{x}_i, \quad \frac{d\tilde{x}(t_{i+1} - t_i)}{d\tau} = \dot{x}_{i+1},\\
&\frac{d^2\tilde{x}(0)}{d\tau^2} = \ddot{x}_i, \quad \frac{d^2\tilde{x}(t_{i+1} - t_i)}{d\tau^2} = \ddot{x}_{i+1},
\end{aligned} \tag{5.74}$$

where $(\dot{x}_i, \ddot{x}_i; \dot{x}_{i+1}, \ddot{x}_{i+1})$ are the feed rate and acceleration estimates at the beginning and end of the interpolated position command segment, respectively. They can be estimated using a third-order cubic polynomial similar to the method described in the quintic or fifth-order spline interpolation as explained in Section 5.5.3. For each new reference point coming out of the interpolator algorithm, the polynomial coefficients are recalculated, and the reference axis position commands are generated at servocontrol loop interval T from Eq. (5.73) (i.e., $\tau = 0, T, 2T, \ldots (t_{i+1} - t_i)$).

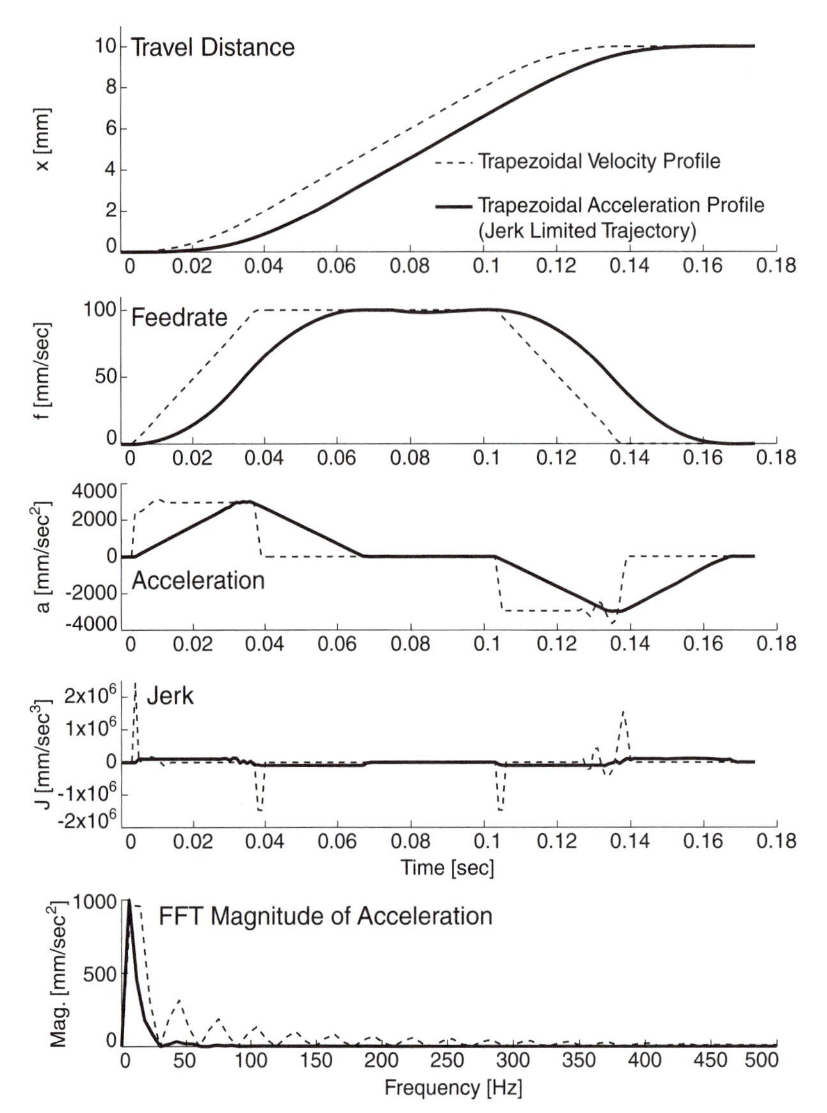

Figure 5.17: Comparison trajectories generated with jerk-limited and trapezoidal velocity profiles.

Example. The trajectory generation with simple trapezoidal velocity and jerk-limited trapezoidal acceleration profiles are compared in Figure 5.17 during a single axis (x) motion. Transient oscillations are present in the acceleration and jerk when the trapezoidal velocity profile is used. However, it can be seen that both acceleration and jerk become very smooth when the trapezoidal acceleration is used instead. The presence of many harmonics in the acceleration indicates that the trapezoidal velocity profile may excite structural modes of the feed drive; hence, it is not recommended for high-speed machine tool drives. Further smoothing is achieved on the acceleration and jerk when the

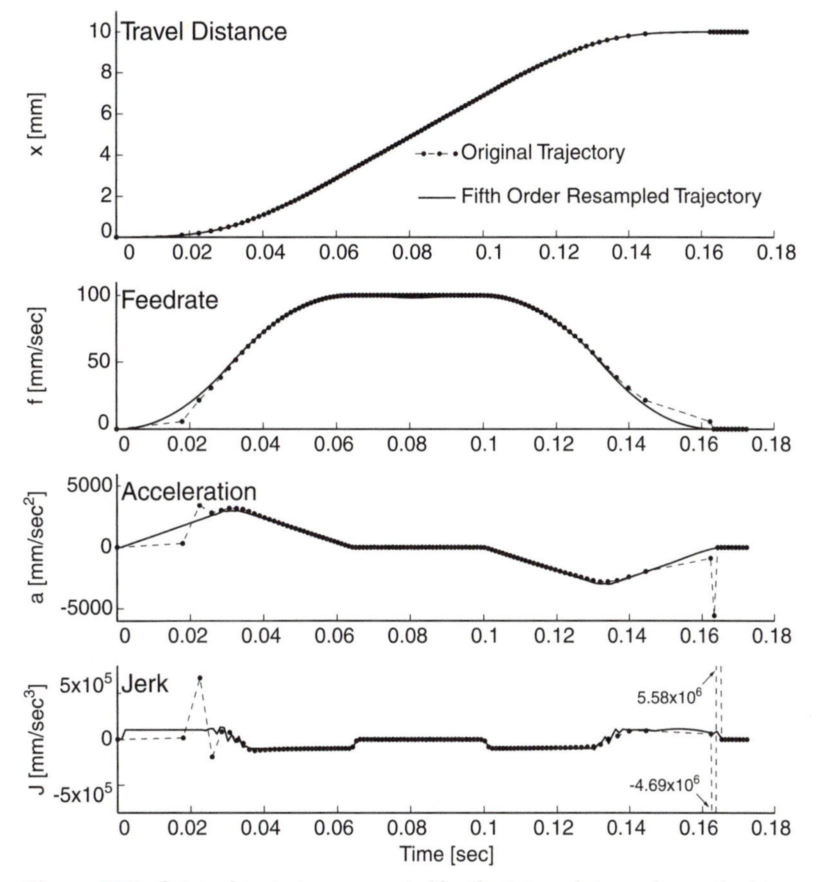

Figure 5.18: Original trajectory generated by the interpolator and smoothed trajectory with fifth-order resampling at the axis level.

interpolated axis displacements are resampled using a fifth-order polynomial (see Fig. 5.18).

5.5 REAL-TIME INTERPOLATION METHODS

CNC systems must be capable of following the complex trajectories that may be present on the parts to be machined. Most of the common geometries can be constructed from linear and circular segments. However, high-speed machining of dies, molds, and aerospace parts with sculptured surfaces requires real-time spline interpolation as well.

In this section, common design methods for real-time two-axis linear, circular, and quintic spline interpolation methods are presented. The algorithms are general and can easily be extended to motions containing more than two active axes.

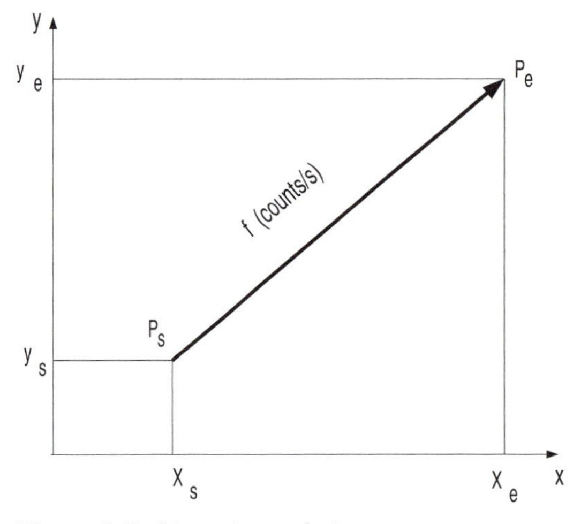

Figure 5.19: Linear interpolation.

5.5.1 Linear Interpolation Algorithm

Method I: Constant Displacement-Varying Interpolation Period

This interpolation method is based on the digital integration of the velocity components in two axes.

Let us assume that a cutting tool center is to follow the linear path shown in Figure 5.19. The starting point of the tool is $P_s(x_s, y_s)$ and the end point is $P_e(x_e, y_e)$. At time t the position of the axes will be at

$$\begin{aligned} x(t) &= x(k \cdot \Delta t) = x_s + \textstyle\int_0^t f_x(t)dt, \\ y(t) &= y(k \cdot \Delta t) = y_s + \textstyle\int_0^t f_y(t)dt, \end{aligned} \tag{5.75}$$

where the axis velocities f_x and f_y and interpolation time interval T_i vary with time during acceleration and deceleration periods, but remain constant during the constant feed zone as explained in the previous section. Because the interpolation algorithm is executed N times at time intervals T_i, Eq. (5.75) can be expressed in discrete form as

$$x(k) = x_s + \textstyle\sum_{j=1}^{k} f_x(j)T_i(j) = x_s + \sum_{j=1}^{k-1} f_x(j)T_i(j) + f_x(k)T_i(k),$$

$$y(k) = y_s + \textstyle\sum_{j=1}^{k} f_y(j)T_i(j) = y_s + \sum_{j=1}^{k-1} f_y(j)T_i(j) + f_y(k)T_i(k),$$

or

$$\begin{aligned} x(k) &= x(k-1) + f_x(k)T_i(k), \\ y(k) &= y(k-1) + f_y(k)T_i(k). \end{aligned} \tag{5.76}$$

Note that the axis velocities at time interval k are

$$f_x(k) = \frac{\Delta x}{T_i(k)}, \quad f_y(k) = \frac{\Delta y}{T_i(k)}. \tag{5.77}$$

Manipulation of interpolation time interval T_i results in manipulation of the axis feeds according to the vector feed and the acceleration/deceleration velocity profile. However, the incremental displacements in both axes are constant and are given by

$$\Delta x = \frac{x_e - x_s}{N}, \quad \Delta y = \frac{y_e - y_s}{N}. \tag{5.78}$$

Substituting Eqs. (5.77) and (5.78) into (5.76) gives the recursive digital linear interpolation equations as follows:

$$\begin{aligned} x(k \cdot \Delta t) &= x_s + k \cdot \Delta x = x(k-1) + \Delta x, \\ y(k \cdot \Delta t) &= y_s + k \cdot \Delta y = y(k-1) + \Delta y. \end{aligned} \tag{5.79}$$

The integration increments Δx and Δy are constant and calculated once at the beginning of the interpolation routine. A real-time computer implementation of the linear interpolation is given below.

T_i	The interpolation period
f	The requested feed rate
x_s, y_s	Start position
x_e, y_e	End position
δx	Total distance in x axis
δy	Total distance in y axis
$\text{sign}(x)$	Direction of x movement
$\text{sign}(y)$	Direction of y movement
N	The number of interpolation iterations
dx, dy	Step size for each axis
x_{rem}	Remainder for x axis step size
y_{rem}	Remainder for y axis step size

(All values are integer except T_i and f)

The initialization calculations in the CNC are

$\delta x = \text{abs}(x_e - x_s)$
$\delta y = \text{abs}(y_e - y_s)$
$\text{sign}(x) = \text{sign}(x_e - x_s)$
$\text{sign}(y) = \text{sign}(y_e - y_s)$
$\Delta u = fT_i$
$N = \sqrt{\delta x^2 + \delta y^2}/\Delta u$
$dx = \text{fix}(\delta x/N)$
$dy = \text{fix}(\delta y/N)$
$x_{rem} = \delta x - (dx \cdot N)$
$y_{rem} = \delta y - (dy \cdot N)$
$\text{line}(x_s, dx, x_{rem}, \text{sign}(x), N)$; sent to x axis position controller
$\text{line}(y_s, dy, y_{rem}, \text{sign}(y), N)$; sent to y axis position controller

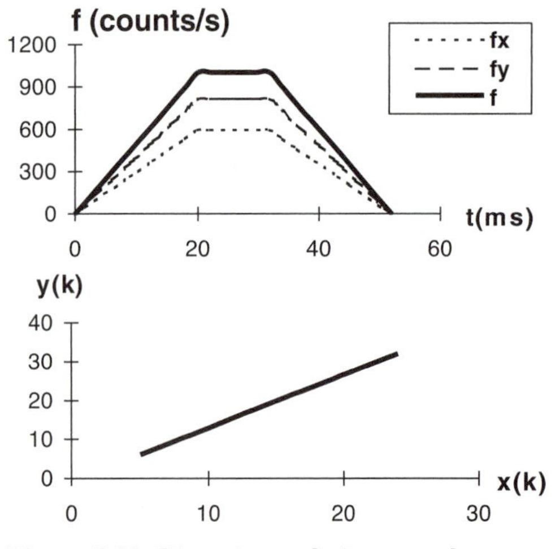

Figure 5.20: Linear interpolation example.

After initialization of as many axes as are needed, each can be calculated at every period T_i with the following:

function line $(x_s, dx, x_{rem}, \text{sign}(x), N)$
$x(1) = x_s$
$x_{error} = 0$
for $i = 2, N+1$
  $x(i) = x(i-1) + \text{sign}(x) \cdot dx$
  $x_{error} = x_{error} + x_{rem}$
  if $(x_{error} >= N)$
    $x(i) = x(i) + \text{sign}(x)$
    $x_{error} = x_{error} - N$
  end
end

The generated position commands at each interpolation time interval T_i are then sent to the digital servo control algorithm of each axis in the CNC.

Example. The following NC block is entered to the CNC in *counts*:

N010 G01 G90 X24 Y32 F1000

This NC program provides $P_e(24, 32)$ and $f = 1{,}000$ [counts/s]. The initial coordinates were given as $P_s(5, 6)$, the acceleration and deceleration values of the CNC are set as $A = D = 50{,}000$ [counts/s^2], and the trapezoidal velocity profile is used. The minimum interpolation time is $T_{min} = 0.002$ s in the CNC. The interpolation algorithm is set as follows:

$$\delta x = x_e - x_s = 19, \quad \delta y = y_e - y_s = 26, \quad L = \sqrt{\delta x^2 + \delta y^2} = 32,$$
$$\Delta u = f T_{min} = 2, \quad N = L/\Delta u = 16, \quad N_1 = N_3 = f^2/(2A\Delta u) = 5,$$
$$\Delta x = \delta x/N = 1.1875, \quad \Delta y = \delta y/N = 1.625.$$

The numerical interpolation results are shown in Table 5.1 and Figure 5.20.

Method II: Varying Displacement–Constant Interpolation Period

The acceleration limited trajectory generation was derived previously as follows:

$$l(k) = \begin{cases} l_s + f_s(kT_i) + \frac{1}{2}A(kT_i)^2 & k = 1, 2, .., (N_1 = T_1/T_i) & l(k-1) + f_s T_i + \Delta_1(k - \frac{1}{2}), \; \Delta_l = AT_i^2, \; l_1 = l(N_1) \\ l_1 + f_c \tau & k = 1, 2, .., (N_2 = T_2/T_i) & l(k-1) + \Delta_2, \; \Delta_2 = f_c T_i, \; l_2 = l(N_2) \\ l_2 + f_c(kT_i) + \frac{1}{2}D(kT_i)^2 & k = 1, 2, ..(N_3 = T_3/T_i) & l(k-1) + f_c T_i + \Delta_3(k - \frac{1}{2}), \; \Delta_3 = DT_i^2, \; l_1 = l(N_3) \end{cases} \tag{5.80}$$

where the displacement $l(k)$ is calculated at each time interval k at constant interpolation period T_i. If the tool travels from point $P_s(x_s, y_s)$ to point $P_e(x_e, y_e)$

TABLE 5.1. Linear Interpolation Steps between Points $P_s(5, 6)$ and $P_e(24, 32)$

Clock Pulses k	Feed $f(k)$ [counts/s]	Interpolation Interval T_i (ms)	Time t [ms]	$x(k)$ [counts]	$y(k)$ [counts]
0	0.00	0.00000	0.0	5.00	6.00
1	447.21	0.00894	8.9	6.19	7.63
2	632.46	0.00370	12.6	7.38	9.25
3	774.60	0.00284	15.5	8.56	10.88
4	894.43	0.00240	17.9	9.75	12.50
5	1,000.00	0.00211	20.0	10.94	14.13
6	1,000.00	0.00200	22.0	12.13	15.75
7	1,000.00	0.00200	24.0	13.31	17.38
8	1,000.00	0.00200	26.0	14.50	19.00
9	1,000.00	0.00200	28.0	15.69	20.63
10	1,000.00	0.00200	30.0	16.88	22.25
11	1,000.00	0.00200	32.0	18.06	23.88
12	894.43	0.00211	34.1	19.25	25.50
13	774.60	0.00240	36.5	20.44	27.13
14	632.46	0.00284	39.4	21.63	28.75
15	447.21	0.00370	43.1	22.81	30.38
16	0.00	0.00894	52.0	24.00	32.00

along the linear path, the direction is described by the following unit vector:

$$\frac{\overrightarrow{P_1P_2}}{\left|\overrightarrow{P_1P_2}\right|} = \frac{(x_e - x_s)}{L}\vec{i} + \frac{(y_e - y_s)}{L}\vec{j}. \tag{5.81}$$

The tool position at interpolation period k can be evaluated as follows:

$$\begin{aligned}\overrightarrow{l(k)} &= \left(\frac{(x_e - x_s)}{L}\vec{i} + \frac{(y_e - y_s)}{L}\vec{j}\right) l(k) \\ &= \frac{(x_e - x_s)}{L} l(k)\vec{i} + \frac{(y_e - y_s)}{L} l(k)\vec{j} \\ \overrightarrow{l(k)} &= \Delta x \cdot l(k)\vec{i} + \Delta y \cdot l(k)\vec{j} \\ \overrightarrow{l(k)} &= \delta x(k)\vec{i} + \delta y(k)\vec{i}.\end{aligned} \tag{5.82}$$

Substituting $l(k)$ into the linear interpolation leads to following recursive difference equations that need to be executed at constant time interval T_i.

$$l(k) = \left\{\begin{array}{ll} \Delta x\left[l(k-1) + f_sT_i + \Delta_1(k-\frac{1}{2})\right] = x(k-1) + \Delta x\left[f_sT_i + \Delta_1(k-\frac{1}{2})\right] & \Delta_l = AT_i^2 \\ \Delta y\left[l(k-1) + f_sT_i + \Delta_1(k-\frac{1}{2})\right] = y(k-1) + \Delta y\left[f_sT_i + \Delta_1(k-\frac{1}{2})\right] & k = 1, 2, .., (N_1 = T_1'/T_i) \\ \Delta x\left[l(k-1) + \Delta_2\right] = x(k-1) + \Delta x\Delta_2 & \Delta_2 = f_cT_i \\ \Delta y\left[l(k-1) + \Delta_2\right] = y(k-1) + \Delta y\Delta_2 & k = 1, 2, .., (N_2 = T_2'/T_i) \\ \Delta x\left[l(k-1) + f_cT_i + \Delta_3(k-\frac{1}{2})\right] = x(k-1) + \Delta x\left[f_cT_i + \Delta_3(k-\frac{1}{2})\right] & \Delta_3 = DT_i^2 \\ \Delta y\left[l(k-1) + f_cT_i + \Delta_3(k-\frac{1}{2})\right] = y(k-1) + \Delta y\left[f_cT_i + \Delta_3(k-\frac{1}{2})\right] & k = 1, 2, ..(N_3 = T_3'/T_i) \end{array}\right\} \tag{5.83}$$

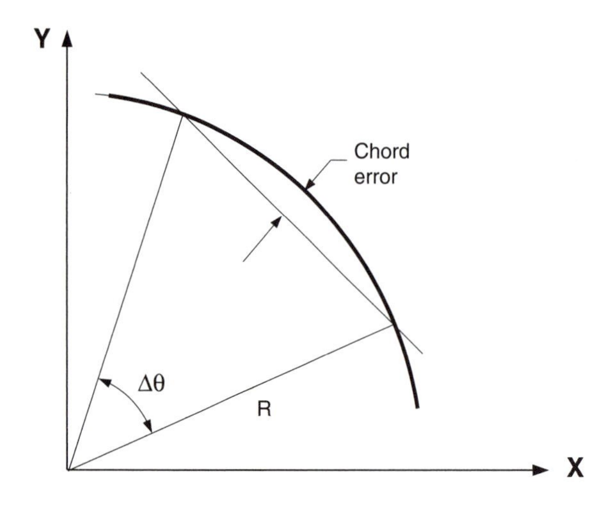

The algorithm is first initialized by identifying $(\Delta x, \Delta y)$, $(\Delta_l = AT_i^2, N_1 = T_1/T_i)$, $(\Delta_2 = f_c T_i, N_2 = T_2/T_i)$, $(\Delta_3 = DT_i^2, N_3 = T_3/T_i)$, and followed by the recursive calculations listed above at each time interval T_i. Because the displacement $l(k)$ is found by integrating the feed, which is in turn calculated by integrating the acceleration, each displacement command at time interval k inherently contains projected velocities in x and y directions.

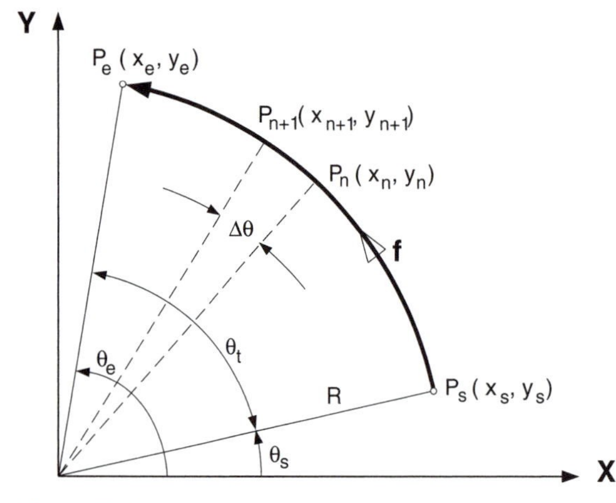

Figure 5.21: Circular interpolation.

5.5.2 Circular Interpolation Algorithm

Method I: Constant Displacement-Varying Interpolation Period

Consider a circular segment whose center is at the origin of the CNC coordinate system (see Fig. 5.21). The circular tool path has a length of

$$L = R(\theta_e - \theta_s) = R\theta_t. \tag{5.84}$$

The arc must be broken into N small segments for digital interpolation. The length of each segment is Δu, with a corresponding angular segment of $\Delta\theta$ as shown in Figure 5.21. The chord error (chord_error) must be kept less than the resolution of position sensing systems (i.e., 1 count) as follows:

$$\text{chord\_error} = R\left(1 - \cos\frac{\Delta\theta}{2}\right) \leq 1, \tag{5.85}$$

which is satisfied when the angular segment $\Delta\theta$ is selected as follows:

$$\Delta\theta \leq 2\cos^{-1}\left(\frac{R-1}{R}\right). \tag{5.86}$$

The selection of

$$\Delta\theta = \cos^{-1}\left(\frac{R-1}{R}\right), \quad \cos\Delta\theta = \frac{R-1}{R} \tag{5.87}$$

guarantees half a count chord error and is more advantageous for computation. The corresponding chord segment is found as

$$\Delta u = R\Delta\theta. \tag{5.88}$$

The circular tool path is broken into N_1, N_2, N_3 segments for trapezoidal velocity profiling as explained earlier. The tool has a feed speed of f [counts/s] on the arc, and the arc segment Δu is traveled in one interpolation period T_i. The interpolation period T_i varies during acceleration and deceleration zones (N_1, N_3), but it remains constant during the steady-state feed zone (N_2) as shown in Figure 5.13. The angular feed velocity $\omega(t)$ and instantaneous angular position $(\theta(t))$ of the cutter at time t are

$$\omega = \frac{f}{R}, \quad \theta(t) = \omega t = \frac{f}{R}t, \tag{5.89}$$

respectively. Coordinates of a point on the arc can be expressed as

$$x(t) = R\ \cos\theta(t) = R\ \cos\left(\frac{f}{R}t\right),$$

$$y(t) = R\ \sin\theta(t) = R\ \sin\left(\frac{f}{R}t\right). \tag{5.90}$$

The velocities of the x and y axis feed drives are

$$f_x = \frac{dx}{dt} = -\frac{f}{R}R\sin\left(\frac{f}{R}t\right) = -\frac{f}{R}\cdot y(t),$$

$$f_y = \frac{dy}{dt} = \frac{f}{R}R\cos\left(\frac{f}{R}t\right) = \frac{f}{R}\cdot x(t). \tag{5.91}$$

Because the velocities are coupled with the positions, digital integration of these equations always produces some errors. A decoupled recursive circular interpolation explained in the following can be easily implemented on CNC systems that are capable of handling floating point arithmetic operations.

Consider two points, P_n and $P_{\mathrm{n}+1}$ on the arc, which are $\Delta\theta$ away from each other (Figure 5.21) as follows:

$$\begin{aligned} x_\mathrm{n} &= R\ \cos(\theta_\mathrm{s} + n\Delta\theta), \quad y_\mathrm{n} = R\ \sin(\theta_\mathrm{s} + n\Delta\theta), \\ x_{\mathrm{n}+1} &= R\ \cos[\theta_\mathrm{s} + (n+1)\Delta\theta], \quad y_{\mathrm{n}+1} = R\ \sin[\theta_\mathrm{s} + (n+1)\Delta\theta]. \end{aligned} \tag{5.92}$$

The coordinates of $P_{\mathrm{n}+1}$ can be expanded as

$$x_{\mathrm{n}+1} = R\ \cos(\theta_\mathrm{s} + n\Delta\theta + \Delta\theta),$$

$$y_{\mathrm{n}+1} = R\ \sin(\theta_\mathrm{s} + n\Delta\theta + \Delta\theta),$$

or

$$x_{\mathrm{n}+1} = R\cos(\theta_\mathrm{s} + n\Delta\theta)\cos\Delta\theta - R\sin(\theta_\mathrm{s} + n\Delta\theta)\sin\Delta\theta,$$

$$y_{\mathrm{n}+1} = R\sin(\theta_\mathrm{s} + n\Delta\theta)\cos\Delta\theta + R\cos(\theta_\mathrm{s} + n\Delta\theta)\sin\Delta\theta.$$

The products of trigonometric functions can be expanded as

$$\sin(\theta_s + n\Delta\theta)\sin\Delta\theta = \frac{1}{2}\{\cos[\theta_s + (n-1)\Delta\theta] - \cos[\theta_s + (n+1)\Delta\theta]\},$$

$$\cos(\theta_s + n\Delta\theta)\sin\Delta\theta = \frac{1}{2}\{-\sin[\theta_s + (n-1)\Delta\theta] + \sin[\theta_s + (n+1)\Delta\theta]\}.$$

Substituting the trigonometric expansion into the coordinates of P_{n+1}, we get

$$x_{n+1} = R\cos(\theta_s + n\Delta\theta)\cos\Delta\theta - \frac{R}{2}\cos[\theta_s + (n-1)\Delta\theta] + \frac{R}{2}\cos[\theta_s + (n+1)\Delta\theta],$$

$$y_{n+1} = R\sin(\theta_s + n\Delta\theta)\cos\Delta\theta - \frac{R}{2}\sin[\theta_s + (n-1)\Delta\theta] + \frac{R}{2}\sin[\theta_s + (n+1)\Delta\theta].$$

When the coordinates of x_n and y_n are substituted in Eq. (5.92), the following difference equations are obtained:

$$x_{n+1} = x_n\cos\Delta\theta - \frac{1}{2}x_{n-1} + \frac{1}{2}x_{n+1},$$

$$y_{n+1} = y_n\cos\Delta\theta - \frac{1}{2}y_{n-1} + \frac{1}{2}y_{n+1}.$$

The discrete position equations can be reduced to a set of decoupled recursive axis equations as follows:

$$\begin{aligned} x_{n+1} &= 2x_n\cos\Delta\theta - x_{n-1}, \\ y_{n+1} &= 2y_n\cos\Delta\theta - y_{n-1}. \end{aligned} \tag{5.93}$$

Note that the integration interval $\Delta\theta$ is selected according to the maximum chord error criterion, and $\cos\Delta\theta$ is given by Eq. (5.87). The interpolation time intervals are calculated according to the velocity profile as explained in Section 5.4. The numerical value of $\Delta\theta$ is precomputed and stored in the memory before the real time, recursive interpolation algorithm starts.

The algorithm has been tested on the research milling machine controlled by an in-house–developed CNC system. A two-axis 117-count (0.148 mm) linear move, followed by an 80-count (0.102 mm) move in the x direction and a full circular path with an 80-count (0.102 mm) radius were executed on the machine tool at an intended feed of 4,000 counts/s. The command and measured tool paths, actual feed along the tool path, and the corresponding axis velocities are shown in Figure 5.22. With A = 50,000 counts/s^2 acceleration, the target feed is never reached because of the short travel distances. However, the path is still followed with a satisfactory accuracy.

Example. The following NC block is entered to the CNC in counts:

N010 G90 G03 X90 Y-90 I- 90 J0 F1000

This NC program provides $P_e(90, -90)$, $f = 1{,}000$ [counts/s], and $R = \sqrt{I^2 + J^2} = 90$ [counts]. The initial coordinates were given as $P_s(180, 0)$, and the acceleration (A) and deceleration (D) values of the CNC are set to

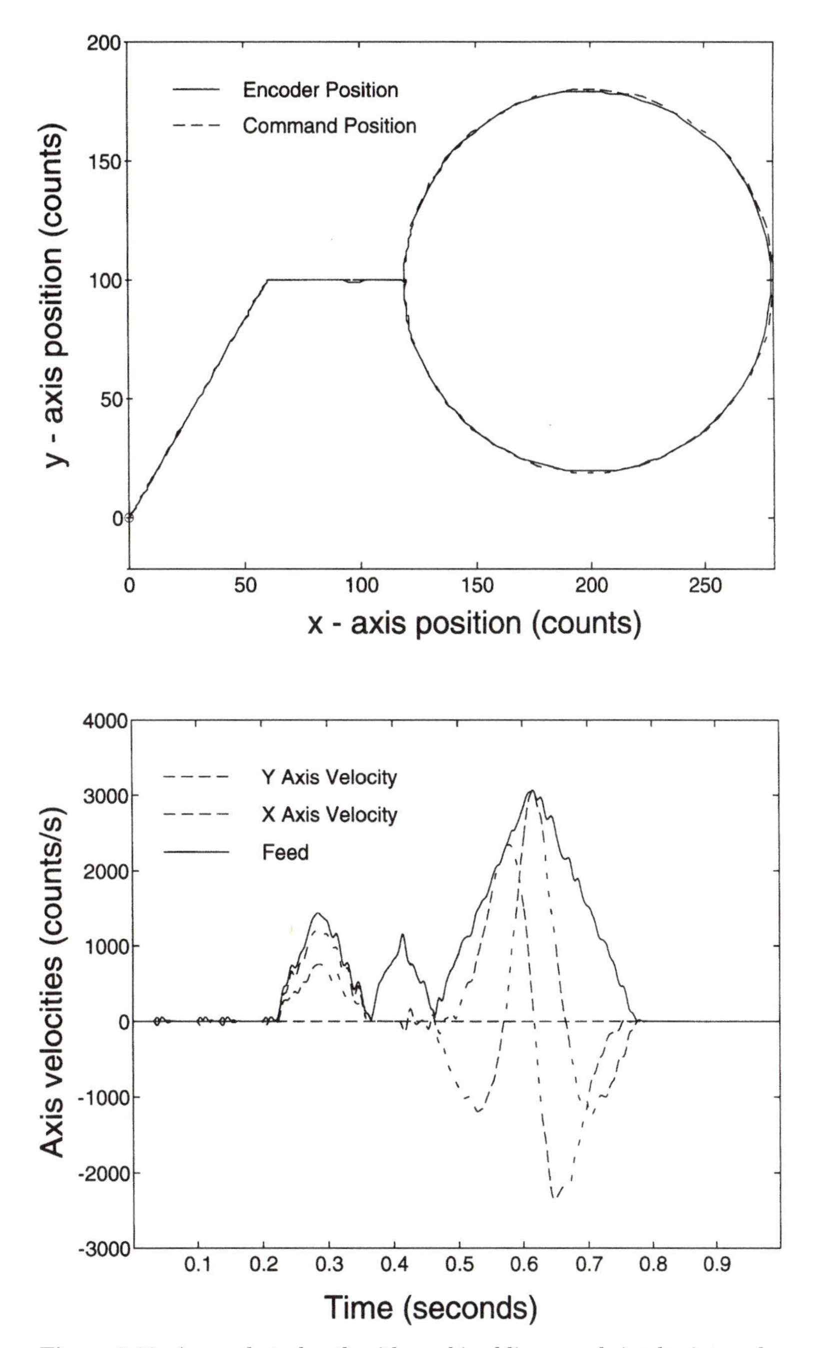

Figure 5.22: A sample tool path with combined linear and circular interpolation segments. The actual position and velocities were evaluated from encoder measurements. 1 count = 0.00127 mm.

TABLE 5.2. Circular Interpolation Steps Between Points $P_s(180, 0)$ and $P_e(90, -90)$

Clock Pulses *k*	Feed *f*(*k*) [counts/s]	Interpolation Interval T_i (ms)	Time *t* [ms]	*x*(*k*) [counts]	*y*(*k*) [counts]
0	0.00	0.0000	0	180.00	0.00
1	463.53	0.1158	116	176.02	26.46
2	655.53	0.0480	164	164.44	50.58
3	802.86	0.0368	201	146.28	70.23
4	927.06	0.0310	232	123.14	83.68
5	1,000.0	0.0278	260	97.08	89.72
6	1,000.0	0.0268	286	70.38	87.84
7	1,000.0	0.0268	313	45.43	78.19
8	1,000.0	0.0268	340	24.41	61.63
9	1,000.0	0.0268	367	9.19	39.62
10	1,000.0	0.0268	394	1.11	14.11
11	1,000.0	0.0268	421	0.89	−12.65
12	886.08	0.0284	449	8.55	−38.29
13	755.17	0.0327	482	23.41	−60.54
14	596.17	0.0397	522	44.15	−77.45
15	374.90	0.0553	577	68.95	−87.50
16	0.00	0.1432	720	90.00	−90.00

$A = D = 4{,}000$ [counts/s^2] with a trapezoidal velocity profile. The minimum interpolation time is $T_{\min} = 0.002$ s in the CNC. The interpolation algorithm is set as follows:

$$\theta_s = 0, \quad \theta_e = \theta_t = 3\pi/2 = 4.71239, \quad \Delta\theta = 0.298419,$$
$$\Delta u = R\Delta\theta = 26.857723, \quad N_1 = N_3 = f^2/(2A\Delta u) = 5, \quad N = \theta_t/\Delta\theta = 16.$$

The numerical interpolation results are shown in Table 5.2 and Figure 5.23.

Method II: Varying Displacement–Constant Interpolation Period

The instantaneous angular position of the tool along the circular tool path is

$$\theta(k) = \frac{l(k)}{R}. \tag{5.94}$$

The circular interpolation equation derived for the constant displacement method can be used here as well as follows:

$$\begin{aligned} x_{k+1} &= 2x_k \cos\Delta\theta(k) - x_{k-1}, \\ y_{k+1} &= 2y_k \cos\Delta\theta(k) - y_{k-1}, \end{aligned} \tag{5.95}$$

where

$$\Delta\theta(k) = \frac{l(k) - l(k-1)}{R} = \frac{f_s T_i + \Delta_1(k - \frac{1}{2})}{R} \rightarrow \quad n = 1, 2, \ldots, \tag{5.96}$$

Because k is increasing, $\Delta\theta(k)$ becomes time varying at each interpolation interval. It may be worthwhile to see whether the equation can be reduced further for computational efficiency.

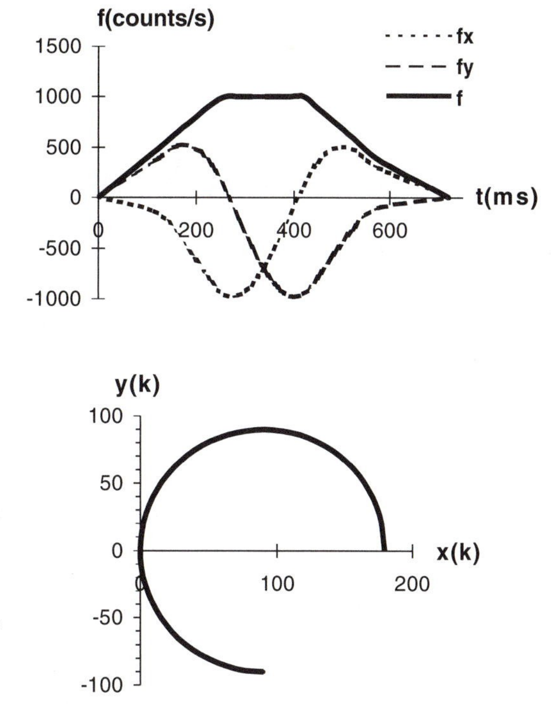

Figure 5.23: Circular interpolation example.

5.5.3 Quintic Spline Interpolation within CNC Systems

In addition to linear and circular interpolation, modern CNCs offer parabolic, helical, and spline interpolations. Instead of breaking sculptured paths into small linear and circular segments, it is preferred to move the cutting tool along the complex path by the use of the path geometry itself [101]. This reduces the length of the NC program and improves the smooth generation of velocity and acceleration in addition to the contouring along the path. Real-time quintic spline interpolation is most useful in high-speed machining of dies and molds [116], and its implementation in the CNC is presented here.

A series of n knots $(P_1, P_2, \ldots, P_n)$ along the tool path are to be connected by quintic spline segments with smooth transitions (see Fig. 5.24). A spline segment S_i connecting two knots (P_i, P_{i+1}) is expressed by a fifth-order polynomial as follows:

$$\{S_{qi}\} = \{A_{qi}\}u^5 + \{B_{qi}\}u^4 + \{C_{qi}\}u^3 + \{D_{qi}\}u^2 + \{E_{qi}\}u + \{F_{qi}\}, \qquad (5.97)$$

where $q : \{x, y, z\}$, $u \in [0, l]$ and $\{A_{qi}\}$, $\{B_{qi}\}$, $\{D_{qi}\}$, $\{E_{qi}\}$, $\{F_{qi}\}$ are vectors containing the coefficients for each coordinate (i.e., $q : x, y, z$ for a three-axis machine), and u is the spline parameter that changes between zero and spline segment length l_i. The number of elements in each vector is therefore equal to the number of axes in motion. The real-time interpolation algorithm requires the values of all spline coefficients (A, B, C, D, E, F), the spline length (l_i), and the feed, acceleration, and jerk limits of the machine tool drives as input parameters. The quintic spline interpolation has both off-line and on-line mathematical steps. First, a quintic spline is fit to a series of knots along a sculptured tool path as explained in Section 5.3.1, and the coefficients are identified. The fitting can either be done on the CAD/CAM system during tool path generation, and the coefficients of the spline can be transferredto CNC via NC codes, or

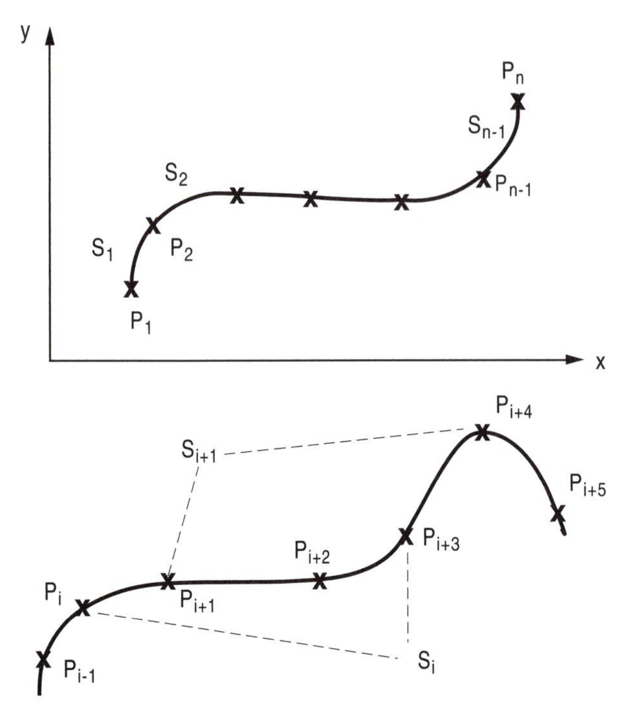

Figure 5.24: Spline passing through knots and the use of cubic polynomials in estimating the first and second derivatives.

they can be identified within the preprocessing stage of CNC by using the raw knot coordinates, supplied within the NC program. The length of the spline segment is estimated by approximating its chord length. The coefficients, knot coordinates, and approximated chord length are used to generate individual axis coordinates in real time by considering the feed, acceleration, and jerk limits of the machine tool drives. Each step in the implementation is shown in Figure 5.25 and explained in the following sections.

Evaluation of Quintic Spline Parameters

Because there are six coefficients, we need six boundary conditions to evaluate them. These are the coordinates (P_i, P_{i+1}), first derivatives $(\frac{dP_i}{du}, \frac{dP_{i+1}}{du})$, and second derivatives $(\frac{d^2P_i}{du^2}, \frac{d^2P_{i+1}}{du^2})$ at the two knots (P_i, P_{i+1}). The first and second derivatives of the quintic spline (5.97) with respect to u are

$$\begin{aligned} \left\{\frac{dS_i}{du}\right\} &= 5\{A_i\}u^4 + 4\{B_i\}u^3 \\ &\quad + 3\{C_i\}u^2 + 2\{D_i\}u + \{E_i\}, \\ \left\{\frac{d^2S_i}{du^2}\right\} &= 20\{A_i\}u^3 + 12\{B_i\}u^2 \\ &\quad + 6\{C_i\}u + 2\{D_i\}. \end{aligned} \tag{5.98}$$

The coefficients are evaluated from the following matrix:

$$\begin{bmatrix} u_0^5 & u_0^4 & u_0^3 & u_0^2 & u_0 & 1 \\ u_1^5 & u_1^4 & u_1^3 & u_1^2 & u_1 & 1 \\ 5u_0^4 & 4u_0^3 & 3u_0^2 & 2u_0 & 1 & 0 \\ 5u_1^4 & 4u_1^3 & 3u_1^2 & 2u_1 & 1 & 0 \\ 20u_0^3 & 12u_0^2 & 6u_0 & 2 & 0 & 0 \\ 20u_1^3 & 12u_1^2 & 6u_1 & 2 & 0 & 0 \end{bmatrix} \begin{Bmatrix} A_i \\ B_i \\ C_i \\ D_i \\ E_i \\ F_i \end{Bmatrix} = \begin{Bmatrix} P_i \\ P_{i+1} \\ \frac{dP_i}{du} \\ \frac{dP_{i+1}}{du} \\ \frac{d^2P_i}{du^2} \\ \frac{d^2P_{i+1}}{du^2} \end{Bmatrix}, \tag{5.99}$$

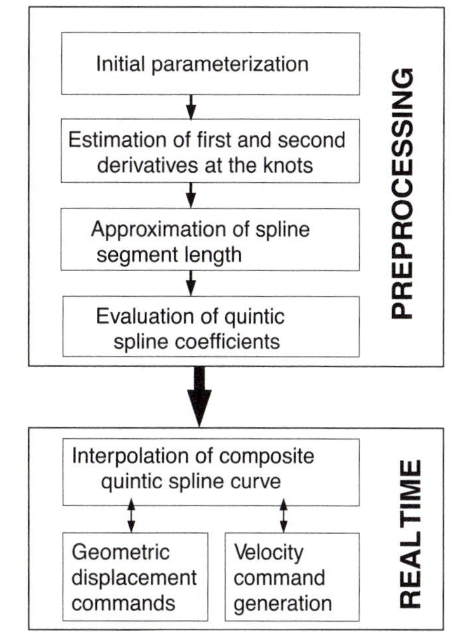

Figure 5.25: Preprocessing and real-time processing steps of quintic spline interpolation.

which leads to the solution of six quintic spline coefficients with $u_0 = 0, u_1 = l_i$ as follows:

$$\begin{aligned}
\{A_i\} &= \frac{1}{u_1^5}\left[6\left(P_{i+1} - P_i\right) - 3\left(\frac{dP_{i+1}}{du} + \frac{dP_i}{du}\right)u_1 + \frac{1}{2}\left(\frac{d^2P_{i+1}}{du^2} - \frac{d^2P_i}{du^2}\right)u_1^2\right],\\
\{B_i\} &= \frac{1}{u_1^4}\left[-15\left(P_{i+1} - P_i\right) + \left(7\frac{dP_{i+1}}{du} + 8\frac{dP_i}{du}\right)u_1 - \left(\frac{d^2P_{i+1}}{du^2} - \frac{3}{2}\frac{d^2P_i}{du^2}\right)u_1^2\right],\\
\left\{C_i\right\} &= \frac{1}{2u_1^3}\left[20\left(P_{i+1} - P_i\right) - 4\left(2\frac{dP_{i+1}}{du} + 3\frac{dP_i}{du}\right)u_1 + \left(\frac{d^2P_{i+1}}{du^2} - 3\frac{d^2P_i}{du^2}\right)u_1^2\right],\\
\{D_i\} &= \frac{1}{2}\frac{d^2P_i}{du^2},\\
\{E_i\} &= \frac{dP_i}{du},\\
\{F_i\} &= P_i.
\end{aligned}$$

To evaluate the coefficients (5.100), the first and second derivatives at the two ends (P_i, P_{i+1}) of the spline segment must be known.

Identification of Derivatives Using Parabolic Approximation

The derivatives at the knots are estimated by approximating a parabola between three consecutive knots. Consider the following parabola:

$$\{S_i(u)\} = \{a_i\}u^2 + \left\{b_i\right\}u + \{c_i\}, \quad u \in [0, L_i, L_{i+1}]. \tag{5.100}$$

passing through three consecutive knots P_i, P_{i+1}, P_{i+2}, that is,

$$\begin{bmatrix} u_0^2 & u_0 & 1 \\ u_1^2 & u_1 & 1 \\ u_2^2 & u_2 & 1 \end{bmatrix} \begin{Bmatrix} \{a_i\} \\ \{b_i\} \\ \{c_i\} \end{Bmatrix} = \begin{Bmatrix} \{P_i\} \\ \{P_{i+1}\} \\ \{P_{i+2}\} \end{Bmatrix}, \tag{5.101}$$

where $u_0 = 0, u_1 = L_i$, and $u_2 = L_i + L_{i+1}$. The distance between each knot is approximated as a line as follows:

$$L_i = |P_{i+1} - P_i| = \sqrt{(x_{i+1} - x_i)^2 + (y_{i+1} - y_i)^2 + (z_{i+1} - z_i)^2} \tag{5.102}$$

for a three-axis path. The parameters of the parabola can be identified from the matrix expression as Eq. (5.101) as

$$\{a_i\} = \frac{(P_{i+2} - P_i)(u_1 - u_0) - (P_{i+1} - P_i)(u_2 - u_0)}{(u_2^2 - u_0^2)(u_1 - u_0) - (u_1^2 - u_0^2)(u_2 - u_0)},$$

$$\{b_i\} = \frac{(P_{i+1} - P_i) - (u_1^2 - u_0^2)\{a_i\}}{(u_1 - u_0)},$$

$$\{c_i\} = P_i - u_0^2\{a_i\} - u_0\{b_i\}. \tag{5.103}$$

The first and second derivatives of the parabola at knot P_i are

$$\left\{\frac{dS_i(u)}{du}\right\} = 2\{a_i\}u + \{b_i\},$$

$$\left\{\frac{d^2S_i(u)}{du^2}\right\} = 2\{a_i\}. \tag{5.104}$$

Note that the second derivative has a unique solution, and the first derivative has three solutions using $u_0 = 0, u_1 = L_i$, and $u_2 = L_i + L_{i+1}$ at subsequent knots. An average value of the first derivative can be used for better accuracy at the mid knots. The solution is unique at the first and last knot.

Identification of Derivatives Using Cubic Spline Approximation

Experience has indicated that the parabolic approximation of the spline with a linear chord length assumption does not lead to smooth feed rate changes along the quintic spline path. Although computationally more demanding, a cubic spline approximation for the derivatives at the knots and a near–arc length estimation for the chord length between the knots are found to generate smoother feed and acceleration during quintic spline interpolation.

Consider a cubic spline passing through the knots $P_i, P_{i+1}, P_{i+2}, P_{i+3}$ shown in Figure 5.24:

$$\{S_i(u)\} = \{a_i\}u^3 + \{b_i\}u^2 + \{c_i\}u + \{d_i\}, \quad u \in [0, L_i + L_{i+1} + L_{i+2}], \tag{5.105}$$

where L_i is the length of the chord segment between knots (P_i, P_{i+1}) and is to be estimated using near–arc length parameterization. The number of elements in each vector is equal to the number of active axes on the machine. The boundary conditions at each knot are

$$\{S_i(u)\} = \begin{Bmatrix} P_i, & u_0 = 0, \\ P_{i+1}, & u_1 = L_i, \\ P_{i+2}, & u_2 = L_i + L_{i+1}, \\ P_{i+3}, & u_3 = L_i + L_{i+1} + L_{i+2}. \end{Bmatrix}. \tag{5.106}$$

By substituting the boundary conditions at four knots (5.106) into the cubic spline equation (5.105) the spline parameters can be evaluated as follows:

$$\begin{bmatrix} u_0^3 & u_0^2 & u_0 & 1 \\ u_1^3 & u_1^2 & u_1 & 1 \\ u_2^3 & u_2^3 & u_2^3 & 1 \\ u_3^3 & u_3^3 & u_3^3 & 1 \end{bmatrix} \begin{Bmatrix} \{a_i\} \\ \{b_i\} \\ \{c_i\} \\ \{d_i\} \end{Bmatrix} = \begin{Bmatrix} P_i \\ P_{i+1} \\ P_{i+2} \\ P_{i+3} \end{Bmatrix}, \tag{5.107}$$

which leads to the following solution:

$$\{a_i\} = \left[\frac{(P_{i+1}-P_i)u_2u_3(u_2-u_3)+(P_{i+2}-P_i)u_3u_1(u_3-u_1)+(P_{i+3}-P_i)u_1u_2(u_1-u_2)}{\Delta_c}\right],$$

$$\{b_i\} = \left[\frac{(P_{i+1}-P_i)u_2u_3(u_3^2-u_2^2)+(P_{i+2}-P_i)u_3u_1(u_1^2-u_3^2)+(P_{i+3}-P_i)u_1u_2(u_2^2-u_1^2)}{\Delta_c}\right],$$

$$\{c_i\} = \left[\frac{(P_{i+1}-P_i)u_2^2u_3^2(u_2-u_3)+(P_{i+2}-P_i)u_3^2u_1^2(u_3-u_1)+(P_{i+3}-P_i)u_1^2u_2^2(u_1-u_2)}{\Delta_c}\right],$$

$$\{d_i\} = \{P_i\}, \tag{5.108}$$

where

$$\Delta_c = u_1u_2u_3\left[u_1^2(u_2-u_3)+u_2^2(u_3-u_1)+u_3^2(u_1-u_2)\right]. \tag{5.109}$$

The derivatives of the spline at each knot are

$$\left.\begin{aligned} \frac{dS_i}{du} &= 3\{a_i\}u^2 + 2\{b_i\}u + \{c_i\} \\ \frac{d^2S_i}{du^2} &= 6\{a_i\}u + 2\{b_i\} \end{aligned}\right\} \quad u \in [0,\ u_3]. \tag{5.110}$$

If there are n knots on the spline, the derivatives at the first two knots (P_1, P_2) and the last two knots are evaluated by the use of the coordinates of the first four (P_1, P_2, P_3, P_4) and last four $(P_{n-3}, P_{n-2}, P_{n-1}, P_n)$. The derivatives at the remaining knots are evaluated for two mid knots of each spline segment that spans over four points. For example, spline segment S_i is used to evaluate the derivatives at knots P_{i+1}, P_{i+2} in Figure 5.24. This will lead to two values for each derivative, which are obtained from two consecutive spline segments. An average of the two derivative values is used in the quintic spline equation (5.100).

Near–Arc Length Parameterization

The length of a spline segment between the two consecutive knots (P_i, P_{i+1}) is better approximated by assuming an arc length instead of a straight line. When the spline (dS_i) is parameterized by its arc length, it must have a unit tangent along the spline segment. A closed arc length can be found when the deviation of (dS/du) is minimized. The least-squares solution is formulated as

$$\frac{d}{du}\int_0^l \left(\frac{dS}{du} - 1\right)^2 du = 0. \tag{5.111}$$

An analytical solution of the length (l) from this equation is difficult. A reasonable solution can be obtained by considering the midpoint of the chord as follows:

$$\left|\frac{dS}{du}\left(u=\frac{l}{2}\right)\right| = 1. \tag{5.112}$$

By substituting $u = l/2$ and quintic spline coefficients (5.100) into the first derivative of quintic spline (Eq. 5.98), we get

$$\left|\frac{15}{8l_i}(P_{i+1}-P_i) - \frac{7}{16}\left(\frac{dP_{i+1}}{du}+\frac{dP_i}{du}\right) + \frac{1}{32}\left(\frac{d^2P_{i+1}}{du^2}-\frac{d^2P_i}{du^2}\right)l_i\right| = 1. \tag{5.113}$$

Squaring both sides of the equation leads to

$$\begin{aligned}&\left(\frac{15}{8l_i}\Delta P\right)^2 + \left(\frac{-7}{16}\left(\frac{dP_{i+1}}{du}+\frac{dP_i}{du}\right)\right)^2 + \left(\frac{l_i}{32}\Delta^2 P\right)^2 \\ &\quad - \frac{105}{64l_i}\Delta P\left(\frac{dP_{i+1}}{du}+\frac{dP_i}{du}\right)^2 - \frac{7}{256}\Delta^2 P\left(\frac{dP_{i+1}}{du}+\frac{dP_i}{du}\right)l_i \\ &\quad + \frac{15}{128}\Delta P\cdot\Delta^2 P = 1,\end{aligned}$$

where $\Delta P = P_{i+1} - P_i$ and $\Delta^2 P = \frac{d^2P_{i+1}}{d^2u} - \frac{d^2P_i}{d^2u}$. Equation (5.114) is rearranged as a fourth-order polynomial as follows:

$$al_i^4 + bl_i^3 + cl_i^2 + dl_i + e = 0, \tag{5.114}$$

where the coefficients are

$$a = \left(\frac{d^2P_{i+1}}{d^2u} - \frac{d^2P_i}{d^2u}\right)^2,$$

$$b = -28\left(\frac{d^2P_{i+1}}{d^2u} - \frac{d^2P_i}{d^2u}\right)\left(\frac{dP_{i+1}}{du}+\frac{dP_i}{du}\right),$$

$$c = -1024 + 196\left(\frac{dP_{i+1}}{du}+\frac{dP_i}{du}\right)^2 + 120\,(P_{i+1}-P_i)\left(\frac{d^2P_{i+1}}{d^2u} - \frac{d^2P_i}{d^2u}\right),$$

$$d = -1680\,(P_{i+1}-P_i)\left(\frac{dP_{i+1}}{du}+\frac{dP_i}{du}\right),$$

$$e = 3600\,(P_{i+1}-P_i)^2.$$

Equation (5.114) can be solved by an iteration scheme such as the Newton–Raphson method [86]. Estimation of spline segment length (l_i) by using the near–arc length approximation (Eq. 5.114) minimizes the deviation of the unit tangent vector much better than the linear chord length parameterization; hence it leads to a smoother feed rate and acceleration during quintic spline interpolation.

Spline interpolation is summarized as follows (Fig. 5.25):

Estimate the derivatives at the knots from Eq. (5.110).

Estimate the chord length (l_i) from Eq. (5.114).

Evaluate the quintic spline coefficients from Eq. (5.100) by substituting the chord length (l_i) and derivatives.

Generate incremental motion commands from the quintic spline expression (5.97) by varying $u = 0, \delta u, 2\delta u, \ldots, k\delta u, \ldots, N\delta u$, where $\delta u = l_i/N$. Adjust the number of interpolation steps and the time intervals depending on the feed, acceleration, and position feedback measurement resolution as explained in the linear and circular interpolation algorithms.

Examples. The quintic spline interpolation and jerk-limited trajectory generation are implemented within an open CNC developed in house [11]. A spiral tool path has been generated in real time on a two-axis high-speed XY table. The actual spiral is generated by a pen attached to the stationary spindle, and resulting feed, acceleration, and jerk on each drive and along the tool path are shown in Figure 5.26. The beginning of the tool path corresponds to the center of the spiral, where some transients can be seen on the acceleration and jerk. A second example is shown in Figure 5.27, where a complex sculptured surface was represented with a series of quintic splines in a preprocessing stage. The part is machined on a three-axis milling machine controlled by the open CNC. The machined part is also shown in Figure 5.27. Both examples use a quintic spline with jerk-limited trajectory generation.

5.6 PROBLEMS

1. Interpret the following NC program block by block. Indicate G and M functions for each block, and plot the complete tool center path with the corresponding coordinates. Mark the corresponding NC blocks on the tool path segments.

```
N010 G90 G70 M03 S1200 T05
N020 G00 X0.375 Y0.875
N030 Z0.1
N040 G01 Z-0.1 F10.0 M08
N050 X2.0 F20.0
N060 G02 X2.375 Y0.5 I0.0 J-0.375
N070 G01 Y0.375
N080 X3.625
N090 Y1.25
N100 X3.5
N110 G02 Y1.875 I0.0 J0.375
N120 G01 X3.625
N130 Y2.625
N140 X2.375
```

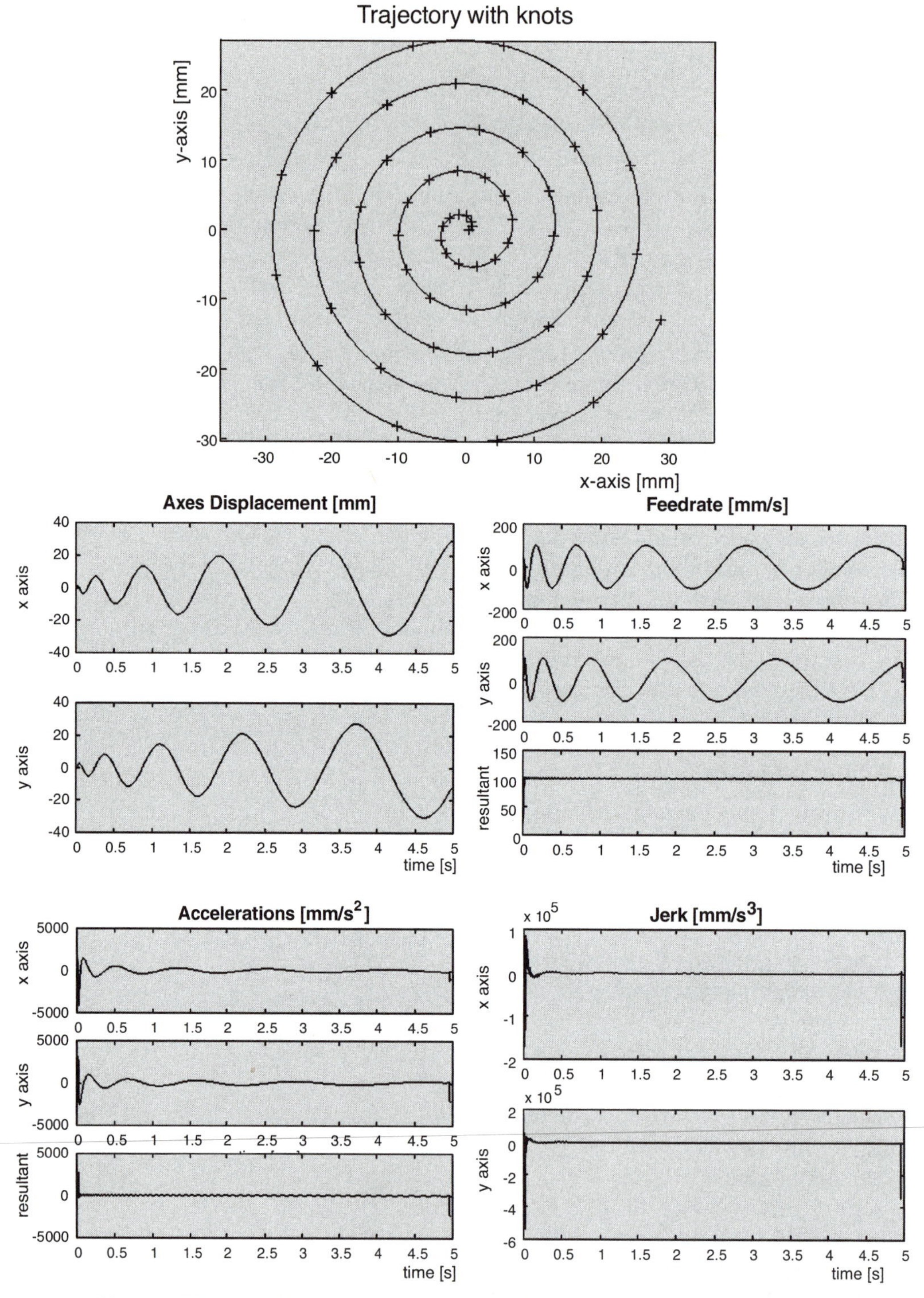

Figure 5.26: A spiral tool path generated by real-time quintic spline interpolation and corresponding feed rate, acceleration, and jerk values produced by jerk-limited trajectory generation.

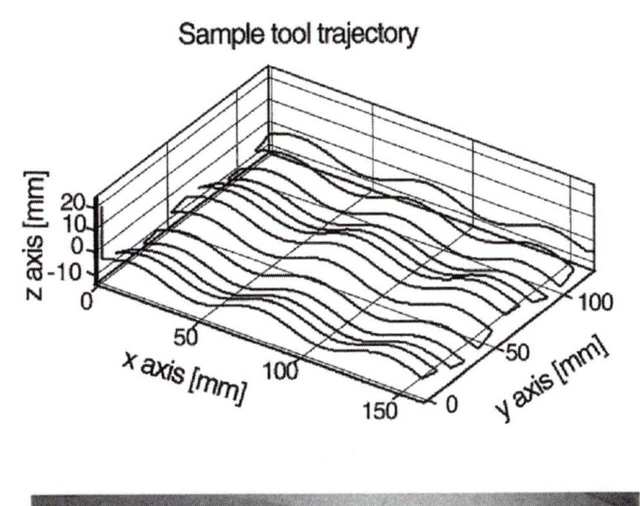

Figure 5.27: A sample sculptured surface tool path and machined part using quintic spline interpolator and jerk-limited trajectory generation.

```
N150 Y2.5
N160 G02 X2.0 Y2.125 I-0.375 J0.0
N170 G01 X0.375
N180 Y0.875
N190 Z0.15
N200 G00 Z2.0 M09 M05
N210 M30
```

2. Interpret the following NC program block by block. Indicate G and M functions for each block, and plot the complete NC tool path. Mark the corresponding NC blocks on the tool path segments.

```
N010 G90 G71 M03 S1200 T01
N020 G00 X40.0
N030 G01 X155.0 F360 M08
N040 Y26.0
```

```
N050 G03 X125.0 Y56.0 I-30 J0
N060 G03 X110.0 Y52.0 I 0 J-30.0
N070 G01 X40.0 Y12.0
N080 Y0
N090 X0
N100 M30
```

3. Interpret the following NC program block by block. Indicate G and M functions for each block, and plot the complete NC tool center path with corresponding coordinates. Mark the corresponding NC blocks on the tool path segments.

```
N010 G90 G71 M03 S1200 T01
N020 G00 Z7
N030 G01 Z0 F100 M08
N040 X30
N050 X90 Y10
N060 Y40
N070 G02 G91 X-30 Y30 I0 J30
N080 G01 X-40
N090 G03 G90 X0 Y50 I+10 J-30
N100 G01 Y0
N110 Z7
N120 G00 X30 Y40.0
N140 G01 Z-2.5
N150 Z7
N160 G00 X0 Y0 Z20 M05 M02
```

4. Sample Interpolation Problems: Assume a machine tool system with equal acceleration and deceleration rates of $A = D = 20{,}000$ count/s^2, and a minimum interpolation time of $T_{min} = 0.002$ s. The position and feed units are in *counts* and *counts/s*, respectively.

   a. Illustrate the working principle of a linear interpolator for an NC block:

      N010 G91 G01 X5 Y – 3 F6000

      Calculate the interpolation parameters and show the resulting tool path on a graph and in a table. (Note: All units are in *counts*, and feed is in *counts/s*.)

   b. Illustrate the working principle of a circular interpolator for an NC block

      N010 G91 G02 X60 Y – 30 I – 60 J – 30 F10000

      Calculate the interpolation parameters and show the resulting tool path on a graph and in a table. (Note: All units are in *counts*, and feed is in *counts/s*.)

5. A CNC interpolator has been designed to generate trapezoidal feed rate profiles at a constant sampling period of $T_s = 0.01$ [s]. The acceleration and deceleration magnitudes have been set to $A = D = 250$ mm/s^2 and the desired feed rate is $f = 10$ [mm/s] with starting and end feed rates of $f_{st} = 0$, $f_{end} = 0$, respectively.
   a. Using this information, interpolate a circular trajectory with a radius of $R = 0.5[mm]$ where the starting point is $x_s = y_s = 0$. Calculate the acceleration (T_1), constant feed (T_2), and deceleration (T_3) time lengths of the trajectory; process them with respect to the interpolation period; compute the number of interpolation steps for each section, and interpolate x_r and y_r axis reference commands in each section. Please fill in the table with your results. You need to calculate only one sample per acceleration, constant velocity, and deceleration phases. The center of the circle is given by ($x_c = 0.5$ mm, $y_c = 0$) and the feed is in CW direction.
   b. Depending on the feed rate (f) and the radius of the circle (R), this circular tool path is also generating sinusoidal reference commands at different frequencies to x and y axis. Calculate the approximate excitation frequency (Hz) of this circular tool path (Hint: Do not consider any acceleration/deceleration section and assume that the feed rate is achieved immediately as in a step velocity change).

Sampling	**Time**	**Arc Displacement**	**Feed Rate**	***x* axis**	***y* axis**
Interval (k)	t(s)	s [mm]	$\frac{ds}{dt}$	x_r [mm]	y_r [mm]

CHAPTER SIX

DESIGN AND ANALYSIS OF CNC SYSTEMS

6.1 INTRODUCTION

A diagram of a typical three-axis computer numerically controlled (CNC) machining center is shown in Figure 6.1. The CNC machining center consists of mechanical, power electronic, and CNC units. The mechanical unit consists of beds, columns, spindle assembly, and feed drive mechanisms. Spindle and feed drive motors and their servoamplifiers, high-voltage power supply unit, and limit switches are part of the power electronics group. The CNC consists of a computer unit and position and velocity sensors for each drive mechanism. The operator enters the numerically controlled (NC) program to the CNC unit. The CNC computer processes the data and generates discrete numerical position commands for each feed drive and velocity command for the spindle drive. The numerical commands are converted into signal voltage (± 5 V or ± 10 V) and sent to servoamplifiers of analog drives, or sent numerically to digital drives that process and amplify them to the high-voltage levels required by the motors. As the drives move, sensors measure their velocity and position. The CNC periodically executes digital control laws at fixed sampling intervals that maintain the feed speed and tool path at programmed rates by using sensor feedback measurements.

The fundamental principles of designing CNC systems are covered in this chapter. First, the sizing and selection of drive motors are presented, followed by physical structure and modeling of a servodrive control system. The mathematical modeling and analysis of drive systems are covered both in the time and frequency domain. The chapter includes sample CNC design examples from real life.

6.2 MACHINE TOOL DRIVES

The drives in machine tools are classified as spindle and feed drive mechanisms. Spindle drives rotate over a wide velocity range (i.e., up to 35,000 rev/min), whereas the feed drives usually convert angular motions of the motors to linear traverse speeds, which can range up to 30,000 mm/min. In this text, the servocontrol of feed drives is covered only, although the material can be easily extended to spindle drives because the fundamental design and analysis methods are quite similar.

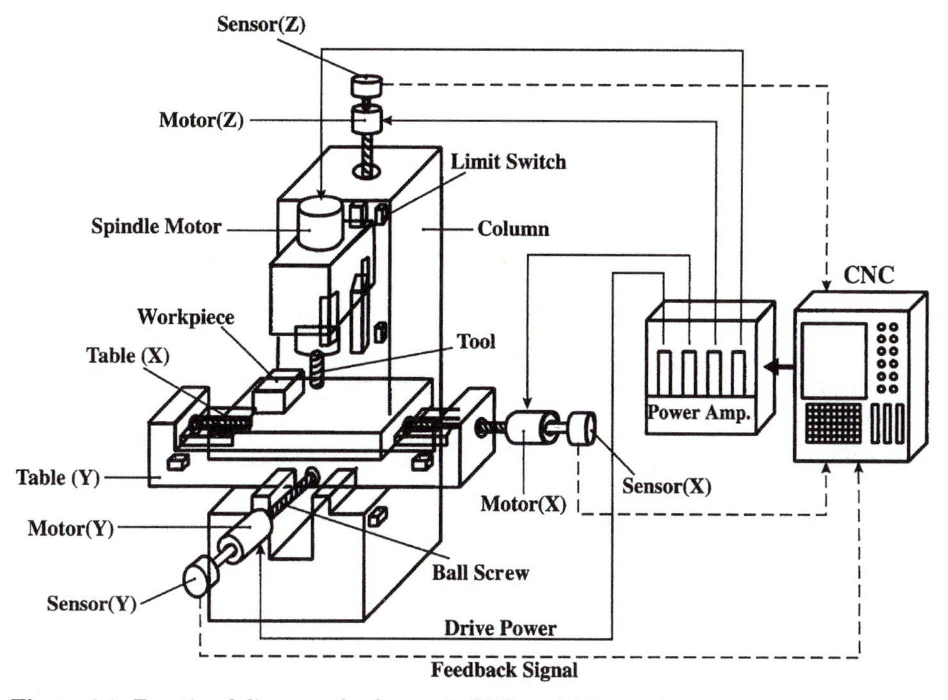

Figure 6.1: Functional diagram of a three-axis CNC machining center.

Let us take one of the feed drives in a machine tool as an example. The feed drive has the following mechanical components: the table with a workpiece, the nut, the ball lead screw, the torque reduction gear set, and the servomotor (see Fig. 6.2). Because of their efficient torque delivery capability at various speeds, the most common servomotors used in the feed drives are direct current (dc) motors. However, alternating current (ac) servomotors have also gained popularity because of their improved performance. The electrical components of a servomotor system comprise the servomotor amplifier, velocity and position feedback transducers, a digital computer, and a digital to analog converter circuit.

6.2.1 Mechanical Components and Torque Requirements

The feed drive motor has to overcome both the static and dynamic loads in the machine tool. The sources of the static loads are the friction losses in the guideways and bearings and the cutting forces acting in the feeding direction of the table. The motor must deliver a high enough dynamic torque to accelerate the table, workpiece, and leadscrew assembly for a short period of time until the drive reaches the desired steady-state speed. The dynamic torque is given as *peak torque* or *peak current* delivery with a period of 2 to 3 seconds by the servomotor manufacturers. The motors must have a sufficiently high continuous torque delivery range and a sufficient peak torque and

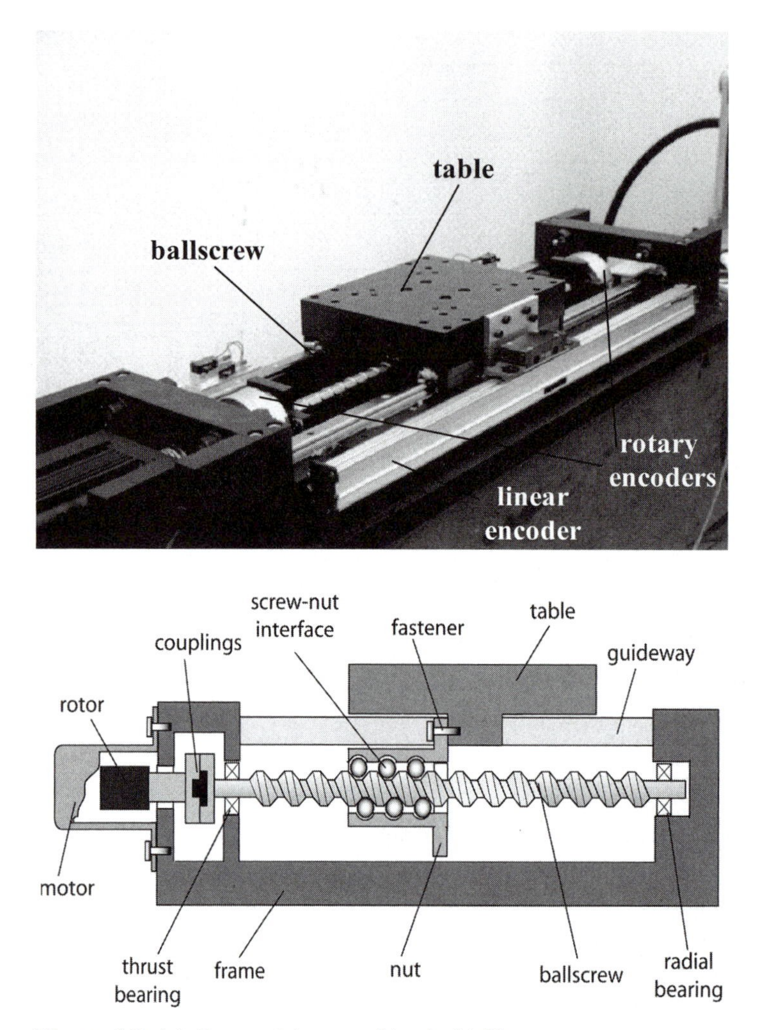

Figure 6.2: A ballscrew-driven machine tool table.

delivery period to overcome the static and dynamic loads, respectively. Estimation of static and dynamic motor loads are briefly introduced in the following paragraphs.

Static Loads

There are three sources of static loads: the friction in the slideways, frictional losses in the feed drive bearings, and cutting forces.

The friction in the guideways depends on the type of contact between the sliding table and the stationary guideway. High friction coefficients are found at plain lubricated guideways where the metal-to-metal table–guideway surface contact area is large. The metal contact surface area is reduced in hydrostatic and hydrodynamic guideways where pressurized lubricant is injected between the table and the guideway. The friction coefficient is probably smallest in the guideway designs where roller bearings are used at the guideway–table assembly (see Fig. 6.2). The torque reflected on the feed drive motor due to friction (T_{gf}) in the guideways can be estimated as

$$T_{gf} = \frac{h_p}{2\pi} \mu_{gf} [(m_t + m_w)g + F_z], \tag{6.1}$$

where μ_{gf} is the friction coefficient on the guideways, m_t is the table mass, m_w is the workpiece mass, F_z is the normal cutting force on the table, h_p is the leadscrew pitch length, and g is the gravitational acceleration (9.81 m/s^2).

The friction coefficient for plain guideways typically ranges from 0.05 to 0.1 and the vertical cutting force (F_z) can be taken as 10 percent of the maximum resultant cutting force in a typical vertical milling machine tool [105].

Axial thrust bearings are used at both ends of the leadscrew to absorb the feed forces and also to guide the screw radially at the same time [109]. Axial thrust bearings are preloaded in tension to offset the backlash produced by the thermal expansion of the leadscrew because of friction in the feed drive assembly. In addition to the preload, the axial thrust bearings are loaded by the feed forces. The feed forces can be estimated by using the cutting mechanics relationships given in the metal cutting chapter. The torque lost in the bearings and preload is estimated as

$$T_{lf} = \mu_b \frac{d_b}{2} (F_f + F_p), \tag{6.2}$$

where μ_b is the friction coefficient on bearings (typically approximately 0.005), d_b is the mean bearing diameter, F_f is the maximum feed force on the table, and F_p is the preload force.

The torque reflected on the leadscrew shaft due to the cutting forces in the feed direction is given as

$$T_f = \frac{h_p}{2\pi} F_f. \tag{6.3}$$

The total static disturbance load reflected on the leadscrew shaft (T_s) is found by summing the three torque values calculated in Eqs. (6.1), (6.2), and (6.3) as follows:

$$T_s = T_{gf} + T_{lf} + T_f. \tag{6.4}$$

In the cases where the static torque (T_s) is too large, a gear reduction can be applied between the motor shaft and the leadscrew. The gear reduction ratio (r_g) is defined as

$$r_g = \frac{z_l}{z_m} = \frac{n_m}{n_l}, \tag{6.5}$$

where z_m is the number of teeth on the motor's gear, z_l is the number of teeth on the feedscrew's gear, n_m is the motor's angular velocity (rev/min), and n_l is the feedscrew's angular velocity (rev/min).

To reduce the speed, we must have $z_l > z_m$, which gives a gear reduction ratio larger than one (i.e., $r_g > 1$). The reduced reflected torque on the motor's shaft (T_{sr}) is found as

$$T_{sr} = \frac{T_s}{r_g}. \tag{6.6}$$

The CNC designer must select a dc motor that has a larger continuous torque delivery capacity than the static torque reflected on the motor's shaft.

Dynamic Loads

Machine tools require high-acceleration torque during speed changes. The reflected inertia on the motor's shaft consists of the inertia of the table, the workpiece, the leadscrew, the gears, and the motor's shaft. The moment of inertia of the table and workpiece reflected on the leadscrew shaft is

$$J_{tw} = (m_t + m_w)\left(\frac{h_p}{2\pi}\right)^2. \tag{6.7}$$

The moment of inertia of the leadscrew with a pitch diameter of d_p is

$$J_l = \frac{1}{2} m_l \left(\frac{d_p}{2}\right)^2, \tag{6.8}$$

where m_l is the mass of the leadscrew shaft. The total inertia reflected on the motor's shaft is

$$J_e = \frac{J_{tw} + J_l}{r_g^2} + J_m, \tag{6.9}$$

where J_m is the inertia of the motor's shaft, and $r_g \geq 1$ is the gear reduction ratio between the feedscrew and motor speeds.

There is another friction torque in the drive system that is proportional to the velocity, namely, the viscous friction torque. The total dynamic torque required to accelerate the inertia J_e and to overcome viscous friction and the static loads is given as

$$T_d = J_e \frac{d\omega}{dt} + B\omega + T_{sr}, \tag{6.10}$$

where ω is the angular velocity of the motor and B is the viscous friction coefficient. Note that, because the cutting is performed at low feeds, the contribution of cutting forces to the static torque (T_s) does not have to be considered in Eq. (6.10). The peak torque value delivered by the motor must be larger than the dynamic torque calculated from Eq. (6.10). If a gear reduction is used between the motor's shaft and leadscrew, the dynamic torque reflected on the motor's shaft is reduced; see Eqs. (6.9) and (6.10).

Example. A vertical milling machine is to be retrofitted with three identical dc servomotors. Because the largest load is applied in the longitudinal axis, the motors are selected according to the torque requirements of this axis. The following parameters are given for the feed drive axis:

$m_t = 20$ kg – table mass,
$m_w = 30$ kg – maximum mass for the workpiece,
$m_l = 2$ kg – leadscrew mass,
$h_p = 0.020$ m/rev. – pitch of the feedscrew,
$d_p = 0.020$ m – feedscrew diameter,
$J_m = 2.875 \times 10^{-4}$ kg m^2 – motor's shaft, coupling, encoder and tachonerator inertia,
$r_g = 1.$ – gear reduction ratio,
$\mu_{gf} = 0.1$ – friction coefficient in the guides,
$\mu_b = 0.005$ – friction coefficient of bearings,
$F_z = 1{,}000$ N – maximum vertical force,
$B = 0.005$ Nm/(rad/s) – viscous damping coefficient,
$F_f = 5{,}000$ N – maximum feeding force,
$F_p = 2{,}000$ N – preload force in thrust bearings,
$a_l = 5$ m/s^2 – desired acceleration of the table.

Static Torque

The static torque contributed by the friction in the guideways (Eq. 6.1) is

$$T_{gf} = 0.1\frac{0.020}{2\pi}[(20 + 30)9.81 + 1{,}000] = 0.4744\,\text{Nm}.$$

The torque lost in the bearings due to friction (Eq. 6.2) is

$$T_{lf} = 0.005\frac{0.02}{2}(5{,}000 + 2{,}000) = 0.3500\,\text{Nm}.$$

The torque required to overcome feed forces (Eq. 6.3) is

$$T_f = \frac{0.020}{2\pi}5{,}000 = 15.90\,\text{Nm}.$$

Thus, the total required continuous torque from the dc motor (Eq. 6.4) is

$$T_{sr} = \frac{0.4744 + 0.35 + 15.90}{1.0} = 16.72\,\text{Nm}.$$

Dynamic Load

The moment of inertia of the table and workpiece reflected on the feedscrew shaft (Eq. 6.7) is

$$J_{tw} = (20 + 30)\left(\frac{0.020}{2\pi}\right)^2 = 5.066 \cdot 10^{-4}\ \text{kg m}^2.$$

The leadscrew's inertia (Eq. 6.8) is

$$J_l = \frac{1}{2}2.0\left(\frac{0.02}{2}\right)^2 = 1 \cdot 10^{-4}\ \text{kg m}^2.$$

Because the motor is directly connected to the leadscrew of the machine (i.e., $r_g = 1.0$), the total inertia reflected on the motor's shaft is found (Eq. 6.9) to be

$$J_e = 5.066 \cdot 10^{-4} + 1 \cdot 10^{-4} + 2.875 \cdot 10^{-3} = 8.9411 \cdot 10^{-4}\,\text{kg m}^2.$$

The angular acceleration of the motor's shaft is

$$\frac{d\omega}{dt} = \frac{a_l}{(h_p/2\pi)} = \frac{5}{0.020}2\pi = 1{,}570\,\text{rad/s}^2.$$

The dynamic torque required is found from Eq. (6.10) as follows:

$$T_d = 8.9411 \cdot 10^{-4}\,\text{kg m}^2 \times 1{,}570\,\text{rad/s}^2 + 0.005\,\text{Nm/(rad/s)} \times \frac{0.5}{0.020}2\pi(\text{rad/s}) + 16.72\,\text{Nm} = 18.90\,\text{Nm}.$$

Thus the selected servomotor must be able to attain 18.90 Nm dynamic torque for a period of acceleration (0.1 s).

6.2.2 Feedback Devices

There are two basic feedback devices in the feed drive control system: position and velocity feedback transducers. Tachometers are used as the velocity transducers and encoders are used as the position feedback transducers in general [63].

Tachometers

A tachometer is a small permanent magnet dc motor mounted directly on the rear of the servomotor's shaft. The tachometer produces a voltage proportional to the actual velocity of the motor shaft. It has a factory set constant and an adjustable gain that enables tuning of the velocity feedback loop. The transfer function between the actual motor velocity and the tachometer circuit output is given as

$$\frac{V_t(s)}{\omega(s)} = T_g \cdot H_g, \tag{6.11}$$

where $V_t(s)$ is the output voltage of the tachometer circuit, $\omega(s)$ is the actual angular velocity of the motor shaft, H_g is the tachometer constant, T_g is the adjustable tachometer gain, and s is the Laplace operator.

Encoders

Encoders are used as digital position measurement transducers in servodrives. The encoder is based on the principle of emitting light with photodiodes. The encoder, which can be a disk or linear scale type, contains bands of dark and transparent segments. Light is transmitted from one side of the transparent band to the other side where there is a photodiode receiver. The photodiode gives a logic signal (i.e., a binary code) depending on the number of dark and transparent bands detected at incremental positions of the encoder. Linear encoders are used to measure the actual position of the table, whereas rotary shaft encoders are used to measure the angular position of the motor's shaft. In precision machines, a linear encoder may be used for more accurate measurement of the actual position of the table. However, if there is an unmodeled backlash in the feed drive system, the table-mounted linear encoder may produce a *limit cycle* in the position control servo. Because disk encoders are mounted directly on the motor shaft, they do not feel the backlash and, therefore, they do not produce a nonlinear limit cycle. Manufacturers of servomotors usually provide tachometer and shaft encoders installed on the rear of the motor's shaft in the factory. Encoder is provided with a *line density* and the type of sensing decoder used. For example, a shaft encoder with 1,000 line density and a quadrature sensing decoder circuit gives 4,000 *counts* or impulses per revolution of the shaft. Encoders are simply represented as a gain K_e (counts/rad or counts/mm) in position control loop analysis.

Example. A 1,000 line encoder with a quadrature sensing decoder is used as a position feedback transducer in a feed drive control system. The motor is directly connected to the leadscrew, which has a pitch length of 5.08 mm. The encoder gain is calculated as

$$K_e = \frac{4 \cdot 1{,}000}{2\pi} \quad \text{(counts/rad)}$$

or

$$K_e = \frac{5.08}{4{,}000} = 0.00127\,\text{mm}\ (0.00005\,\text{in})/\text{count}.$$

Thus, one count signal sent by the encoder corresponds to 0.00127 mm linear movement of the table. Henceforth, *count* is used to represent the basic length unit of the position control servosystem.

6.2.3 Electrical Drives

The feed drives can be powered by either electrical motors (step, dc, or ac motors) or hydraulic motors. The type of motor used depends very much on the torque delivery and time response requirements of the machine tool or robot drive.

Hydraulic motors are used when a wide torque range and rapid response are required from a drive system. Rotary hydraulic motors are used in heavy-duty industrial robots, lathes, and milling machines. Linear hydraulic motors with piston-displacement units are used in grinding machines, planers, and shapers where reciprocating motion is required. The disadvantages of hydraulic drives include leakage, low efficiency, sensitivity to dirt in the supply oil, and high maintenance costs.

Electrical stepping motors are not common in feed drive systems where the cutting load is high. They are used without feedback devices. The step motor is controlled with angular step motions that are sent in the form of control pulses from a computer. At startups and braking, if the required dynamic torque is large, the step motor may slip some of the position control pulses.

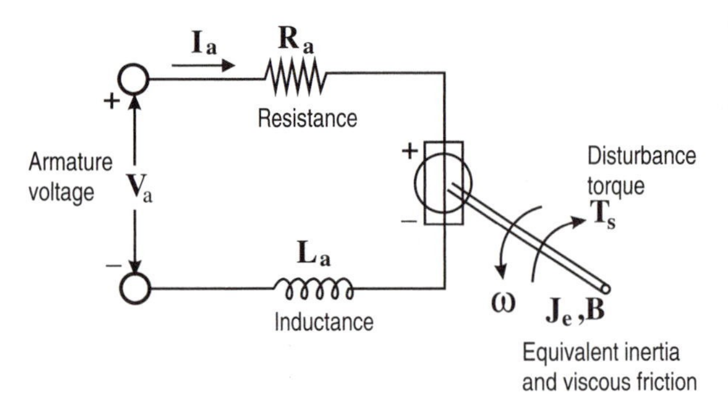

Figure 6.3: Electrical diagram of a dc motor.

Because there is no feedback device, the accuracy in machining is sacrificed. Light demonstration CNC machines and material handling units usually use electrical stepping motors.

The use of alternating current (ac) motors is also common in CNC machines. The speed of the ac motor is controlled by varying the frequency of the supply voltage. The fundamental problem has been to design the low-cost inverters used to vary the frequency of the supply voltage. However, recent microprocessor technology has enabled the calculation of firing frequencies used in the inverter.

The most common motors used in the feed drives are direct current (dc) motors because they allow a wide range of operating speeds with the sufficiently large torque delivery required by machine tools and robots. In the next section, analysis of a permanent magnet dc motor is explained to provide a detailed understanding of the control of dc motors. However, the analysis and modeling procedures are similar for both dc and ac servomotors.

6.2.4 Permanent Magnet Armature-Controlled dc Motors

An electrical diagram of a permanent magnet dc motor is shown in Figure 6.3. The speed of the dc motor is controlled by feeding a dc voltage V_a to the armature (rotor) of the motor. The dc voltage produces a variable dc current I_a on the armature, which in turn creates a magnetic field between the rotor and stationary stator. Note that the current drawn by the armature can not exceed the maximum current supply capacity of the power amplifier. The current limit is treated as a nonlinearity in the dc motor control system. It is evident that peak current will be drawn for a short period during acceleration and braking of the motor. Peak current and its duration time are given by the amplifier manufacturers. Magnetic flux is constant, whereas the armature voltage is variable in armature-controlled dc motors. The magnetic field produces a torque, which is used for rotating the rotor connected to the motor's shaft. In dc motors, a back electromotor voltage (e.m.v.) V_b, which is proportional to the rotor speed, is produced in the rotor circuit.

The following fundamental dynamic equations govern the motion of dc motors.

The applied armature voltage V_a is derived by applying Kirchoff's law to the motor's circuit as follows:

$$V_a(t) = R_a I_a(t) + L_a \frac{dI_a(t)}{dt} + K_b \omega(t), \tag{6.12}$$

where ω is the angular velocity (rad/s), I_a is the armature current (A), R_a is the armature resistance (Ω), L_a is the armature inductance (H), K_b is the motor's voltage (e.m.v.) constant (V/(rad/s)), and t is the time (s).

The magnetic field produces a useful motor torque T_m, which is proportional to the armature current I_a as follows:

$$T_m(t) = K_t\, I_a(t). \tag{6.13}$$

The useful torque produced by the motor is spent in accelerating the reflected inertia on the motor's shaft, overcoming the friction in the motor's bearings and guideways, and resisting against feed cutting forces and friction loads reflected as a disturbance torque on the motor's shaft. Thus,

$$T_m(t) = J_e \frac{d\omega(t)}{dt} + B\,\omega(t) + T_s(t), \tag{6.14}$$

where J_e is the reflected inertia on the motor shaft, B is the equivalent friction coefficient (viscous damping), and T_s is the static disturbance torque reflected on the motor shaft.

Note that the disturbance torque consists of cutting and friction components. The Coulomb friction torque opposes the velocity (ω). The direction of the velocity is considered in the block diagram (Fig. 6.4) using the sign function sgn [4]. In Eq. (6.14) the viscous friction is assumed to be proportional to the speed. The behavior of the friction torque varies depending on the type of guideways used in the machine tool. The viscous friction torque is linearly proportional to the velocity, and the proportionality constant is B.

Note that a Coulomb-type sticking friction is usually dominant in feed drives. Coulomb friction demands constant friction torque (or current) and is independent of the feeding velocity variation. The Coulomb friction torque and viscous friction coefficient can be identified from Eq. (6.14) by measuring the armature current at various steady-state feeding velocities (i.e., $d\omega/dt = 0$). The dc component of the resulting curve gives the constant current drawn by Coulomb friction, and the slope of the linear component is the viscous damping coefficient (B). To avoid nonlinearity in the system, Coulomb friction is disregarded and an average viscous damping value is assumed.

Taking the Laplace transforms of Eqs. (6.12), (6.13), and (6.14) gives

$$\begin{aligned} I_a(s) &= \frac{V_a(s) - K_b\,\omega(s)}{L_a\, s + R_a}, \\ T_m(s) &= K_t\, I_a(s), \\ \omega(s) &= \frac{T_m(s) - T_s(s)}{J_e\, s + B}. \end{aligned} \tag{6.15}$$

These equations have physical interpretations. The transfer function of the armature current is derived using the error voltage as an input. The error voltage is the difference between the supplied reference armature voltage V_a and back e.m.v., which acts like a feedback signal. The resulting current produces

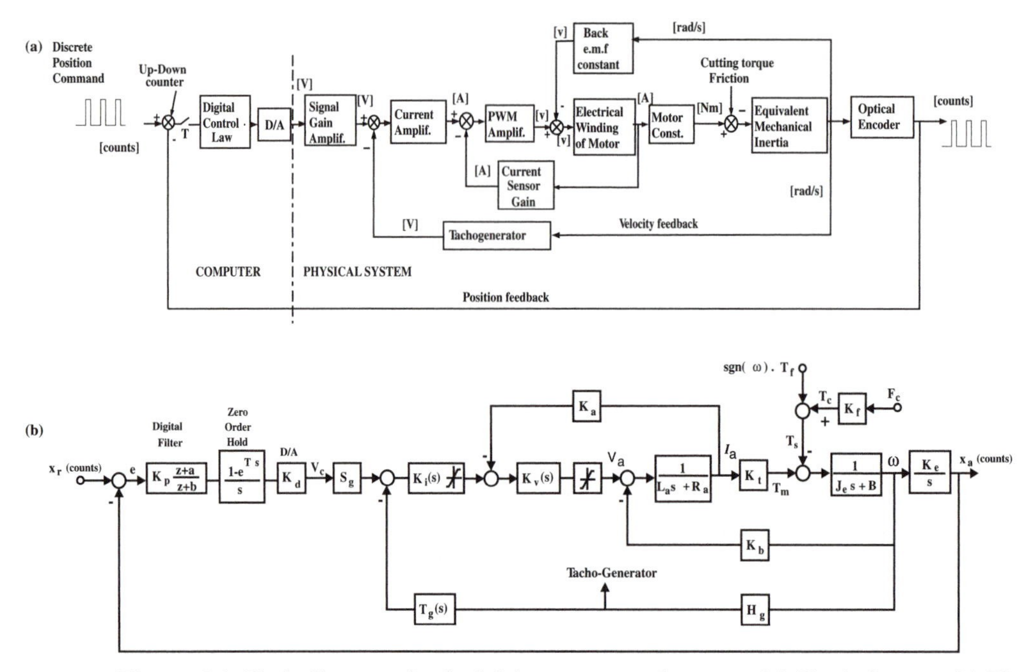

Figure 6.4: Block diagram of a feed drive servocontrol system. (a) Physical parts. (b) Block diagram.

a useful torque $T_m(s)$. A portion of this torque is spent in overcoming the disturbance torque $T_s(s)$. The remaining torque is used to accelerate the inertia and overcome the viscous friction torque, which is proportional to the velocity. The two important design parameters are

$$\tau_e = \frac{L_a}{R_a}, \qquad \tau_m = \frac{J_e}{B}, \tag{6.16}$$

where τ_m is the mechanical time constant and τ_e is the electrical time constant of the dc motor. Both time constants are provided by the motor's manufacturer by considering only the inertia of the motor's shaft and the friction in the motor's shaft bearings. The mechanical time constant of the dc motor obviously increases when the reflected inertia of the table and leadscrew assembly are taken into account.

dc Motor Power Amplifiers and Velocity Control Loop

The armature voltage is supplied by a power amplifier. The power amplifier receives a large constant dc voltage from a transformer, which converts ac line voltage to the desired dc voltage level. The power amplifier illustrated in this text is a pulse width modulated (PWM), current-controlled amplifier. However, the operation of other types of power amplifiers are quite similar.

The block diagram of a complete velocity control loop is shown in Figure 6.4. The power amplifier receives a velocity command signal V_c from the digital to analog converter output of the digital controller. The velocity command signal is first buffered with an adjustable differential preamplifier represented by a gain S_g. The output of S_g is compared with an actual velocity signal measured by the tachometer feedback unit. The resulting velocity error signal (in volts) is converted to a demand current by the current amplifier, which has a gain of K_I. Most current amplifiers use an armature current feedback signal to improve the dynamic response of the motor. The feedback current signal is pulled from a current sense coupler and compared with the demand current. The PWM circuit generates a varying sawtooth shape dc voltage at certain frequency. The PWM frequency is usually higher than 10 kHz. Low-frequency PWM switching signals (up to 6 kHz) generate irritating audible noise. The current error signal is modulated by the PWM circuit, which is modeled as a gain K_v. The resulting dc voltage becomes an ON–OFF type rectangular waveform. The average voltage level of the waveform (dc value) is used as armature voltage V_a in the calculations. The complete block diagram of the amplifier, motor, and tachometer feedback unit are shown in Figure 6.4a.

The block diagram is organized with temporary state variables to illustrate the derivation of the velocity loop transfer function. The following relationships can be expressed from the block diagram by the use of temporary states, V_1, V_2, and V_3 as follows:

$$
\begin{aligned}
V_1(s) &= S_g V_c(s) - T_g H_g \omega(s), \\
V_2(s) &= K_I V_1(s) - K_a I_a(s) \\
&= K_I S_g V_c(s) - K_I T_g H_g \omega(s) - K_a I_a(s), \\
V_3(s) &= K_v V_2(s) - K_b \omega(s) \\
&= K_v K_I S_g V_c(s) - (K_v K_I T_g H_g + K_b)\omega(s) - K_v K_a I_a(s).
\end{aligned}
$$

The transfer function between the current and state V_3 is

$$I_a(s) = \frac{V_3(s)}{L_a s + R_a}.$$

Substituting the value of V_3 into the current expression yields

$$I_a(s) = \frac{K_v K_I S_g}{L_a s + R_a + K_v K_a} V_c(s) - \frac{K_v K_I T_g H_g + K_b}{L_a s + R_a + K_v K_a} \omega(s). \tag{6.17}$$

The motor's mechanical transfer function (see Eq. 6.15) is

$$\omega(s) = \frac{T_m(s) - T_s(s)}{J_e s + B}$$

or

$$\omega(s) = \frac{K_t}{J_e s + B} I_a(s) - \frac{1}{J_e s + B} T_s(s). \tag{6.18}$$

Substituting the current expression (6.17) into Eq. (6.18) yields the transfer function between the output velocity ω and the velocity command input voltage

V_c and the disturbance torque T_s as follows:

$$\omega(s) = \frac{K_1}{s^2 + K_2\, s + K_3} V_c(s) - \frac{(1/J_e)[s + (R_a + K_v K_a)/L_a]}{s^2 + K_2\, s + K_3} T_s(s), \tag{6.19}$$

where

$$K_1 = \frac{K_t S_g K_I K_v}{L_a J_e},$$

$$K_2 = \frac{B}{J_E} + \frac{R_a + K_v K_a}{L_a},$$

$$K_3 = \frac{B\,(R_a + K_v K_a) + K_t(K_b + H_g T_g K_v K_I)}{J_e\, L_a}.$$

The feed drive servovelocity controller is designed to have a fast rise time with zero overshoot at step changes in the velocity. Let us analyze the velocity loop as a function of the velocity command input voltage V_c. The transfer function (Eq. 6.19) can be expressed as

$$\frac{\omega(s)}{V_c(s)} = \frac{K_1}{s^2 + 2\,\xi\,\omega_n\, s + \omega_n^2}. \tag{6.20}$$

Here, the natural frequency (w_n) and damping ratio (ξ) of the velocity loop are defined as follows:

$$\omega_n = \sqrt{K_3} \quad [\text{rad/s}],$$
$$\xi = \frac{K_2}{2\sqrt{K_3}} < 1, \tag{6.21}$$

where $K_1, K_2 > 0$. The time domain step response of this underdamped velocity servo is expressed as

$$\omega(t) = V_c \cdot \frac{K_1}{K_3}\left[1 - \frac{e^{-\xi\omega_n t}}{\sqrt{1-\xi^2}} \cdot \sin(\omega_d\, t + \phi)\right], \tag{6.22}$$

where damped natural frequency ω_d and phase shift ϕ are defined as

$$\omega_d = \omega_n\sqrt{1-\xi^2},$$

$$\phi = \tan^{-1}\left(\frac{\sqrt{1-\xi^2}}{\xi}\right).$$

The variable gains of the amplifier (i.e., S_g, T_g, K_I, K_v) are tuned to have a desired velocity loop gain and step response characteristics as shown in Figure 6.5. When a unit step input (i.e., $V_c = 1\,\text{V}$) is applied on the amplifier input port, the maximum response of the velocity loop occurs at time t_p where the derivative of the velocity is zero (i.e., $d\omega(t)/dt = 0$). At the first overshoot, the following expression can be obtained from the time derivative of Eq. (6.22):

$$t_p = \frac{\pi}{\omega_d}. \tag{6.23}$$

Typical design values for a feed drive servo may be a damping ratio of $\xi = 0.707$ and a peak time of $t_p = 10\,\text{ms}$. The natural frequency ω_n can be estimated from Eq. (6.23) as follows:

$$\omega_n = \frac{\pi}{t_p\sqrt{1-\xi^2}} = 444 \text{ rad/s} = 70 \text{ Hz}.$$

The corresponding servoparameters are identified by substituting the values of ξ and ω_n into Eq. (6.21). It is fairly obvious that one can not demand a higher natural frequency than the maximum capacities of the motor and amplifier gains can provide.

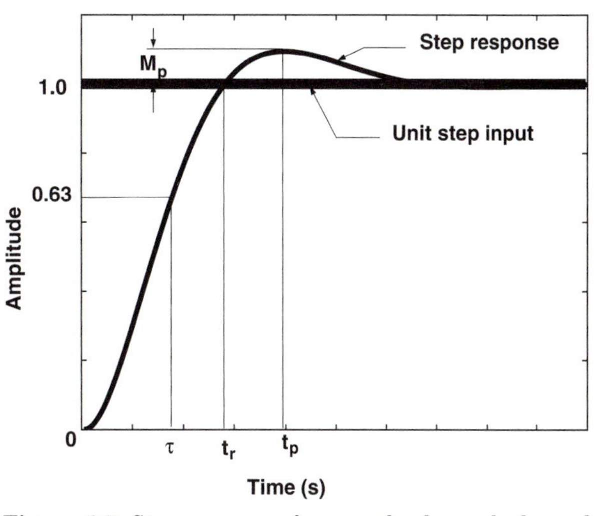

Figure 6.5: Step response of a second-order underdamped system (τ = time constant, t_p = peak time, t_r = rise time, M_p = overshoot).

6.2.5 Position Control Loop

The position loop consists of an up–down counter, an encoder, a digital compensation filter, and a digital to analog converter.

Up–Down Counter

The up–down counter register receives both command and measured position counts. The position commands increment while the encoder feedback counts decrement the contents of the up–down counter. The instantaneous content of the counter represents the accumulated or integrated position error within digital servocontrol interval T. The block diagram representation of the up–down counter is

$$\frac{X_a(s)}{\omega(s)} = \frac{K_e}{s}. \tag{6.24}$$

Digital Compensation Filter

The contents of the up–down counter, the position error, is sampled every T seconds. The discrete position error $E(k)$ is the difference between the reference and actual position of the table as follows:

$$E(k) = X_r(k) - X_a(k), \tag{6.25}$$

where $X_r(k)$ is the discrete reference position, $X_a(k)$ is the discrete actual position, and z is the discrete forward time shift operator.

The error is passed from a digital filter, which has a typical transfer function form of

$$D(z) = K_{\mathrm{p}} \frac{z+a}{z+b}, \tag{6.26}$$

where K_{p} is the position control filter gain, a is the filter's zero, and b is the filter's pole.

The digital filter is programmable and resides in the servomotion control computer. The filter's parameters are tuned to provide a desired transient response of the position control system.

Digital to Analog (D/A) Converter

The motion control computer sends the digital filter's output to a D/A converter circuit of the motion control computer. The D/A converter is modeled as a zero-order hold (ZOH) and a gain K_{d}. The D/A gain is found as

$$K_{\mathrm{d}} = \frac{\text{voltage range of the D/A chip}}{2^{nb}}, \tag{6.27}$$

where nb is the number of bits used by the D/A chip in converting a binary number to analog voltage. For example, a 12-bit D/A converter chip with $\pm 10\,\mathrm{V}$ voltage range has a gain of

$$K_{\mathrm{d}} = \frac{20\,\mathrm{V}}{2^{12}} = 0.00488\ (\mathrm{V/count}).$$

6.3 TRANSFER FUNCTION OF THE POSITION LOOP

The block diagram of the complete position control system can be organized as shown in Figure 6.4. The system has continuous and discrete components. The continuous part of the system's transfer function is represented in the Laplace domain as

$$G_{\mathrm{c}}(s) = \frac{K_1}{s^2 + K_2\, s + K_3} \frac{K_{\mathrm{e}}}{s}. \tag{6.28}$$

The velocity control signal V_{c} of the digital motion control unit is applied to the power amplifier at T second intervals via a D/A converter with a gain of K_{d}. The ZOH equivalent of the $G_{\mathrm{c}}(s)$ for a 1-ms sampling interval is [81] as follows:

$$G_{\mathrm{c}}(z) = K_{\mathrm{d}}(1 - z^{-1})\, \mathcal{Z}\left[\frac{G_{\mathrm{c}}(s)}{s}\right],$$

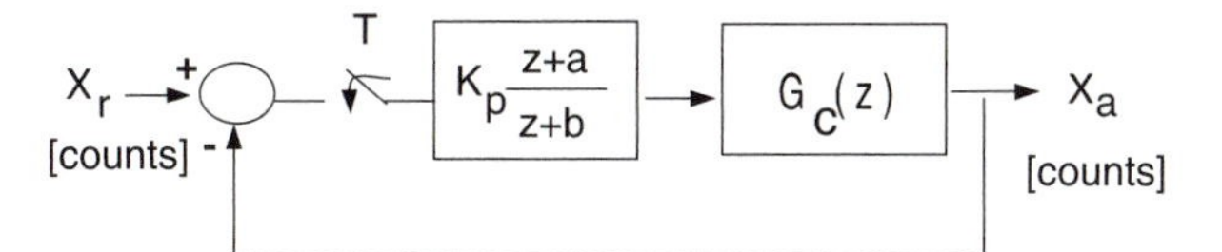

Figure 6.6: Discrete time position control loop.

which, after taking the z transform, becomes

$$G_c(z) = \frac{K_d K_1 K_e}{K_3} \frac{b_2 z^2 + b_1 z + b_0}{(z-1)(z^2 + a_1 z + a_0)}, \tag{6.29}$$

where the forward operator is equivalent to $z = e^{sT}$. The derived expressions for the parameters of $G_c(z)$ are given as

$$\begin{aligned}
b_2 &= T - \frac{1}{\omega_d} e^{-\xi\omega_n T} \sin(\omega_d T) \\
&\quad - \frac{K_2}{K_3} \left\{ 1 - e^{-\xi\omega_n T} \left[\frac{\xi\omega_n}{\omega_d} \sin(\omega_d T) + \cos(\omega_d T) \right] \right\}, \\
b_1 &= 2\, e^{-\xi\omega_n T} \left[\frac{\sin(\omega_d T)}{\omega_d} - T \cos(\omega_d T) \right] \\
&\quad + \frac{K_2}{K_3} (1 - e^{-2\xi\omega_n T}) - 2\, e^{-\xi\omega_n T} \sin(\omega_d T) \frac{K_2}{K_3} \frac{\xi\omega_n}{\omega_d}, \\
b_0 &= T e^{-2\xi\omega_n T} - \frac{1}{\omega_d} e^{-\xi\omega_n T} \sin(\omega_d T) \\
&\quad + \frac{K_2}{K_3} \left\{ e^{-2\xi\omega_n T} + e^{-\xi\omega_n T} \left[\frac{\xi\omega_n}{\omega_d} \sin(\omega_d T) - \cos(\omega_d T) \right] \right\}, \\
a_1 &= -2\, e^{-\xi\omega_n T} \cos(\omega_d T), \\
a_0 &= e^{-2\xi\omega_n T}.
\end{aligned}$$

The reduced block diagram of the equivalent discrete position control system is shown in Figure 6.6. The closed-loop transfer function of the complete feed drive control system is derived as follows:

$$G_{cl}(z) = \frac{X_a(k)}{X_k(z)} = \frac{D(z)\, G_c(z)}{1 + D(z) G_c(z)},$$

or

$$G_{cl}(z) = K_{cl} \frac{z^3 + \beta_2 z^2 + \beta_1 z + \beta_0}{z^4 + \alpha_3 z^3 + \alpha_2 z^2 + \alpha_1 z + \alpha_0}. \tag{6.30}$$

The parameters of the closed-loop transfer function $G_{\mathrm{cl}}(z)$ are given as follows:

$$\beta_2 = \frac{b_1 + a\, b_2}{b_2},$$

$$\beta_1 = \frac{b_0 + a\, b_1}{b_2},$$

$$\beta_0 = \frac{a\, b_0}{b_2},$$

$$\alpha_3 = \frac{K_1\, K_{\mathrm{e}}\, K_{\mathrm{d}}\, K_{\mathrm{p}}}{K_3}\, b_2 + b + a_1 - 1,$$

$$\alpha_2 = \frac{K_1\, K_{\mathrm{e}}\, K_{\mathrm{d}}\, K_{\mathrm{p}}}{K_3}\, (b_1 + a\, b_2) + b(a_1 - 1) + a_0 - a_1,$$

$$\alpha_1 = \frac{K_1\, K_{\mathrm{e}}\, K_{\mathrm{d}}\, K_{\mathrm{p}}}{K_3} (b_0 + a\, b_1) + b(a_0 - a_1) - a_0,$$

$$\alpha_0 = \frac{K_1\, K_{\mathrm{e}}\, K_{\mathrm{d}}\, K_{\mathrm{p}}}{K_3} \cdot a \cdot b_0 - b a_0,$$

$$K_{\mathrm{cl}} = \frac{K_1\, K_{\mathrm{e}}\, K_{\mathrm{d}}\, K_{\mathrm{p}}\, b_2}{K_3}.$$

The transfer function (Eq. 6.30) can be expressed by the use of the backward time shift operator z^{-1} by multiplying both numerator and denominator by z^{-4}. This gives

$$G_{\mathrm{cl}}(z^{-1}) = K_{\mathrm{cl}}\, \frac{z^{-1} + \beta_2 z^{-2} + \beta_1 z^{-3} + \beta_0 z^{-4}}{1 + \alpha_3 z^{-1} + \alpha_2 z^{-2} + \alpha_1 z^{-3} + \alpha_0 z^{-4}}. \tag{6.31}$$

Note that the z^{-1} backward time shift operator acts on the discrete signal as

$$z^{-1} x(kT) = x[(k-1)T],$$

where T is the discrete sampling time interval and $x(kT)$ is the discrete value of x at sampling interval number k.

For any given discrete time input $X_{\mathrm{r}}(kT)$, the position response $X_{\mathrm{a}}(kT)$ can be found by rearranging the transfer function $G_{\mathrm{cl}}(z^{-1})$ as follows:

$$\begin{aligned} X_{\mathrm{a}}(k) = &-(\alpha_3 z^{-1} + \alpha_2 z^{-2} + \alpha_1 z^{-3} + \alpha_0 z^{-4}) X_{\mathrm{a}}(z^{-1}) \\ &+ K_{\mathrm{cl}} \cdot (z^{-1} + \beta_2 z^{-2} + \beta_1 z^{-3} + \beta_0 z^{-4}) X_{\mathrm{r}}(z^{-1}). \end{aligned} \tag{6.32}$$

The discrete time response of the position servo can be simulated by taking the inverse z^{-1} transform of Eq. (6.32). The resulting difference equation contains the discrete history of input and the past history of output position values.

Note that the position control loop can be analyzed in the continuous time (s) domain as an alternative to the discrete time (z) domain analysis. However, this approach requires approximating the digital filter ($D(z)$) in the s domain. One of the widely used approximations is $z = (1 + sT/2)/(1 - sT/2)$, which is called

Tustin's bilinear transformation. The corresponding digital filter becomes

$$D(s) = K_\mathrm{p}' \frac{s+a'}{s+b'}, \tag{6.33}$$

where

$$K_\mathrm{p}' = K_\mathrm{p} \frac{1-a}{1-b}, \quad a' = \frac{2}{T}\frac{1+a}{1-a}, \quad b' = \frac{2}{T}\frac{1+b}{1-b}$$

or

$$K_\mathrm{p} = K_\mathrm{p}' \frac{1-b'}{1-a'}, \quad a = \frac{Ta'-2}{Ta'+2}, \quad b = \frac{Tb'-2}{Tb'+2}.$$

In this case, the closed-loop transfer function of the position loop in Laplace domain becomes

$$G_\mathrm{cl}(s) = \frac{D(s)K_\mathrm{d}G_\mathrm{c}(s)}{1+D(s)K_\mathrm{d}G_\mathrm{c}(s)}. \tag{6.34}$$

Following Error in CNC Systems

There are two critical performance requirements from the feed drive servo. The first is to obtain a smooth transient response to avoid an oscillatory tool path during velocity changes; the second is to minimize the steady-state position error, the *following error*, to achieve precision multiaxis contouring. At steady-state contouring with a feeding velocity f_c (counts/s), the reference position is expressed as a ramp type input as follws:

$$X_\mathrm{r}(kT) = f_\mathrm{c} \cdot kT, \tag{6.35}$$

where k is the sampling counter. In the z domain, the ramp command is expressed as

$$X_\mathrm{r}(k) = f_\mathrm{c} \frac{Tz}{(z-1)^2}. \tag{6.36}$$

The corresponding following error is then derived as

$$e_\mathrm{ss} = \lim_{z\to 1} \frac{f_\mathrm{c}T}{(z-1)D(z)G_\mathrm{c}(z)}. \tag{6.37}$$

Substituting Eqs. (6.26), (6.29), and (6.35) into Eq. (6.37) gives the parametric expression for the following error in the feed drives as

$$e_\mathrm{ss} = \frac{f_\mathrm{c}K_3(1+b)}{K_1K_\mathrm{e}K_\mathrm{p}K_\mathrm{d}(1+a)}. \tag{6.38}$$

It is evident that the higher the open-loop transfer function gain is (i.e., $D(z)G_\mathrm{c}(z)$), the smaller the following error will be, which is desired for accurate multiaxis contouring operations. However, high open-loop gain is limited by the inertia of the mechanical drive system and the limits of motor torque and amplifier. The control engineer must tune the digital control (i.e., filter) parameters to achieve an optimum feed drive servoresponse without causing any oscillation or overshoot.

The steady-state error can also be estimated by the use of the continuous time domain transfer functions as

$$e_{ss} = \lim_{s \to 0} s \frac{f_c}{s^2[D(s)K_d G_c(s)]}, \tag{6.39}$$

where $G_c(s)$ and $D(s)$ are given in Eqs. (6.28) and (6.33).

6.4 STATE SPACE MODEL OF FEED DRIVE CONTROL SYSTEMS

A state space model for the feed drive servo is used to verify the derived model by using experimentally measured time domain response data. The servo is again divided into continuous and discrete sections for state space modeling [80].

The continuous part of the system consists of the velocity control loop (Eq. 6.28) and the up–down counter (Eq. 6.24). Three *states* – the armature current I_a, the angular velocity ω, and the actual position X_a – are derived as follows: From Figure 6.4, the armature voltage can be expressed as

$$V_a = K_v[K_I(S_g V_c - T_g H_g \omega) - K_a I_a]. \tag{6.40}$$

Substituting Eq. (6.12) into (6.40) yields

$$\frac{dI_a}{dt} = -\frac{K_v K_a + R_a}{L_a} I_a - \frac{K_b + K_v K_I T_g H_g}{L_a} \omega + \frac{K_v K_I S_g}{L_a} V_c. \tag{6.41}$$

Eliminating the motor torque T_m by substituting Eq. (6.13) into Eq. (6.14) gives

$$\frac{d\omega}{dt} = \frac{K_t}{J_e} I_a - \frac{B}{J_e} \omega - \frac{1}{J_e} T_s. \tag{6.42}$$

The inverse Laplace transform of the transfer function of the up–down counter and the encoder (Eq. 6.24) gives

$$\frac{dX_a(t)}{dt} = K_e \cdot \omega(t). \tag{6.43}$$

The state equations (6.41–6.43) can be organized in a standard state space form as follws:

$$\dot{x_c}(t) = A_c\, x_c(t) + B_c\, u_c(t), \tag{6.44}$$

where the *state vector* ($x_c(t)$) and the *input vector* ($u_c(t)$) are defined as

$$x_c(t) = \begin{bmatrix} I_a(t) \\ \omega(t) \\ X_a(t) \end{bmatrix}, \quad u_c(t) = \begin{bmatrix} V_c(t) \\ T_s(t) \end{bmatrix}$$

and A_c and B_c are constant matrices as follows:

$$A_c = \begin{bmatrix} -\dfrac{K_v K_a + R_a}{L_a} & -\dfrac{K_b + K_v K_I T_g H_g}{L_a} & 0 \\ \dfrac{K_t}{J_e} & -\dfrac{B}{J_e} & 0 \\ 0 & K_e & 0 \end{bmatrix},$$

$$B_c = \begin{bmatrix} \dfrac{K_v K_I S_g}{L_a} & 0 \\ 0 & -\dfrac{1}{J_e} \\ 0 & 0 \end{bmatrix}.$$

The state equation (6.44) represents the continuous part of the feed drive servosystem, which has the following discrete equivalent solution for an observation interval of T [80]:

$$x_c(k+1) = \Phi(T)\, x_c(k) + H(T)\, u_c(k), \tag{6.45}$$

where the state and input vectors at sampling interval k are defined as

$$x_c(k) = \begin{bmatrix} I_a(k) \\ \omega(k) \\ X_a(k) \end{bmatrix}, \quad u_c(k) = \begin{bmatrix} V_c(k) \\ T_s(k) \end{bmatrix},$$

$$\Phi(T) = e^{A_c T} = \begin{bmatrix} \phi_{11} & \phi_{12} & \phi_{13} \\ \phi_{21} & \phi_{22} & \phi_{23} \\ \phi_{31} & \phi_{32} & \phi_{33} \end{bmatrix},$$

$$H(T) = \int_0^T e^{A_c t} dt \cdot B_c = \begin{bmatrix} h_{11} & h_{12} \\ h_{21} & h_{22} \\ h_{31} & h_{32} \end{bmatrix}.$$

The matrix $\Phi(T)$ is computed from the eigenvalues of the A_c matrix or its Taylor series expansion for the discrete-time equivalent of the continuous-time system. Because the sampling interval (T) is small, the first three terms of the following Taylor series approximation may be sufficient for most applications:

$$\Phi(T) = e^{A_c T} = [I] + [A]T + [A]^2 \frac{T^2}{2!} + \cdots . \tag{6.46}$$

The discrete-time components of the position control loop consists of digital filter $D(z)$ and D/A converter gain K_d. The velocity command signal can be expressed in the z domain as

$$V_c(k) = K_p \frac{z+a}{z+b} \cdot K_d \cdot [X_r(k) - X_a(k)]. \tag{6.47}$$

The equation can be rearranged as

$$V_c(k) = K_p\, K_d\, [X_r(k) - X_a(k)] + V_d(k), \tag{6.48}$$

where

$$V_{\mathrm{d}}(k) = K_{\mathrm{p}}\, K_{\mathrm{d}}\, \frac{a-b}{z+b}\, [X_{\mathrm{r}}(k) - X_{\mathrm{a}}(k)]. \tag{6.49}$$

The new variables V_{c} and V_{d} can be treated as a fourth *state*. After rearranging Eqs. (6.49) and (6.47) and taking their inverse z transforms, we obtain the following discrete-time state equations:

$$\begin{aligned} V_{\mathrm{d}}(k+1) &= -b\, V_{\mathrm{d}}(k) + K_{\mathrm{p}}\, K_{\mathrm{d}}\, (a-b)\, [X_{\mathrm{r}}(k) - X_{\mathrm{a}}(k)], \\ V_{\mathrm{c}}(k) &= K_{\mathrm{p}}\, K_{\mathrm{d}}\, [X_{\mathrm{r}}(k) - X_{\mathrm{a}}(k)] + V_{\mathrm{d}}(k). \end{aligned} \tag{6.50}$$

The discrete-time state equation (6.50) can be combined with the state equation (6.45), which represents the discrete-time equivalent of the continuous part of the feed drive servo. An algebraic rearrangement yields the following complete state equations for the feed drive servo:

$$\begin{aligned} x(k+1) &= G(T) \cdot x(k) + \Gamma(T) \cdot u(k), \\ y(k) &= C_{\mathrm{s}} \cdot x(k) + D_{\mathrm{s}} \cdot u(k), \end{aligned} \tag{6.51}$$

where the state, input, and output vectors are defined, respectively, as

$$x(k) = \begin{bmatrix} V_{\mathrm{d}}(k) \\ I_{\mathrm{a}}(k) \\ \omega(k) \\ X_{\mathrm{a}}(k) \end{bmatrix}, \quad u(k) = \begin{bmatrix} X_{\mathrm{r}}(k) \\ T_{\mathrm{s}}(k) \end{bmatrix}, \quad y(k) = \begin{bmatrix} V_{\mathrm{c}}(k) \\ I_{\mathrm{a}}(k) \\ \omega(k) \\ X_{\mathrm{a}}(k) \end{bmatrix}.$$

The state matrix $G(T)$, input matrix $\Gamma(T)$, output matrix C_{s} and transmission matrix D_{s} are defined in order as follows:

$$G(T) = \begin{bmatrix} -b & 0 & 0 & -K_{\mathrm{p}}K_{\mathrm{d}}(a-b) \\ h_{11} & \phi_{11} & \phi_{12} & \phi_{13} - h_{11}K_{\mathrm{p}}K_{\mathrm{d}} \\ h_{21} & \phi_{21} & \phi_{22} & \phi_{23} - h_{21}K_{\mathrm{p}}K_{\mathrm{d}} \\ h_{31} & \phi_{31} & \phi_{32} & \phi_{33} - h_{31}K_{\mathrm{p}}K_{\mathrm{d}} \end{bmatrix},$$

$$\Gamma(T) = \begin{bmatrix} K_{\mathrm{p}}K_{\mathrm{d}}(a-b) & 0 \\ h_{11}K_{\mathrm{p}}K_{\mathrm{d}} & h_{12} \\ h_{21}K_{\mathrm{p}}K_{\mathrm{d}} & h_{22} \\ h_{31}K_{\mathrm{p}}K_{\mathrm{d}} & h_{32} \end{bmatrix},$$

$$C_{\mathrm{s}} = \begin{bmatrix} 1 & 0 & 0 & -K_{\mathrm{p}}K_{\mathrm{d}} \\ 0 & 1 & 0 & 0 \\ 0 & 0 & 1 & 0 \\ 0 & 0 & 0 & 1 \end{bmatrix}, \quad D_{\mathrm{s}} = \begin{bmatrix} K_{\mathrm{p}}K_{\mathrm{d}} & 0 \\ 0 & 0 \\ 0 & 0 \\ 0 & 0 \end{bmatrix}.$$

The output vector $y(k)$ gives access to three useful dynamic parameters in the feed drive servo; namely, the armature current, the angular velocity, and the position of the table for a given position command and applied cutting torque. Other states in the control system can be easily found by multiplying them with the appropriate gains according to the block diagram shown in Figure 6.4.

Example: Feed Drive Control System Design for a Vertical Milling Machine. A three-axis vertical research milling machine was retrofitted at the author's laboratory at the University of British Columbia. The retrofitted machine tool has a 5 kW ac motor connected to a spindle gear box. The three feeding axes (x, y, and z) of the machine have recirculating ballscrew drives with 60-, 40-, and 12-cm travel limits. All three axes are driven by permanent magnet dc servomotors powered by PWM amplifiers. The motors, which are directly connected to the leadscrew shafts, have an 11 ms mechanical (without table) time constant and a 5 ms electrical time constant, and the amplifiers have 30 A peak and 15 A continuous current delivery capacity. The amplifiers are the same type as explained in Section 6.2.4.

A personal computer (PC) is used as a master CNC. A 32-bit digital signal-processing board (DSP) with a three-axis motion control module is used to control positions of the three linear axes of the machine. Both cards reside in the PC bus. The motion control card has 24-bit programmable digital input/output (I/O) lines for logic control functions. Coolant, spindle, travel limit, emergency relay, etc. logic control signals are wired to the I/O lines to control auxiliary functions of the machine tool. The real-time linear, circular, and spline interpolation algorithms, and the position control of all axes, are executed in the DSP board. The discrete position commands are sent to the I/O board, which converts them into analog signal voltage via 16-bit dedicated D/A converters and sends them to the servoamplifiers. The I/O board and the real-time interpolation and control codes have been developed in the author's laboratory. Additional sensor-based intelligent machining process monitoring and control modules can be added to the CNC, which was designed with an open software and hardware architecture as explained in Chapter Seven. The machine tool can be used as a standard CNC machine by loading an in-house developed ISO NC language emulator into the PC. Feed, acceleration, deceleration, and digital filter parameters can be changed in real time by sending the desired values from the PC to the motion control unit [11]. This feature is particularly important and essential for adaptive control and machine tool monitoring tasks.

The block diagram of the control system is the same as shown in Figure 6.4. The constants of the dc motor, gains, and digital filter parameters are given in Table 6.1. The parameters of the velocity and position loop transfer functions [19] are calculated from Eqs. (6.19), (6.29), and (6.30) and are given in Table 6.2.

The calculated damping ratio and the natural frequencies of the velocity loop are found to be

$$\omega_{\rm n} = \sqrt{K_3} = 554 \ \text{(rad/s)} = 88\,\text{Hz},$$

$$\xi = \frac{K_2}{2\sqrt{K_3}} = 0.8384.$$

TABLE 6.1. Parameters of Feed Drive System

B	0.09 Nm/(rad/s)
H_g	0.08872 V/(rad/s)
J_e	0.0036 kg m^2
K_a	0.0643 A/A
K_b	0.3 V/(rad/s)
K_d	0.00488 V/count
K_e	636.62 counts/(rad/s)
K_I	25.5 A/V
K_p	9.2038
K_v	21.934 V/V
K_t	0.3 Nm/A
L_a	2 mH
R_a	0.4 Ω
S_g	0.0648 V/V
T_g	0.13183 V/V
a	−0.7161
b	−0.6681
p	5.08 mm
Continuous current supply = 15 A	
Peak current supply = 30 A	

The step response of the velocity loop is obtained by applying $V_c = 1.0$ V to the terminals of the amplifier and simultaneously measuring the velocity from the tachogenerator. The simulated step response (Eq. 6.22) of the velocity control servo is shown in Figure 6.7. The system seems to reach to the command velocity at about 10 ms without any overshoot. The frequency response of the velocity loop is analyzed to determine the bandwidth of the controller. The magnitude ratio ($M(\omega)$) and the phase angle ($\phi(\omega)$) of the closed velocity loop can be derived from Eq. (6.19) as follows:

$$M(\omega) = 20\log(K_1/K_3) - 20\log\left[(1-\omega^2/\omega_n^2)^2 + (2\xi\omega/\omega_n)^2\right]^{\frac{1}{2}} \tag{6.52}$$

$$\phi(\omega) = -\tan^{-1}\left[\left(2\xi\frac{\omega}{\omega_n}\right) / \left(1-\left(\frac{\omega}{\omega_n}\right)^2\right)\right]. \tag{6.53}$$

The frequency response of the velocity loop can be measured with a two-channel Fourier analyzer. The analyzer's random signal generator output is fed to the input terminal of the amplifier, and the velocity is measured from the tachogenerator output. Alternatively, harmonic (i.e., sine wave) signals at different frequencies can be given as an input, and the corresponding time delay (i.e., phase delay) and magnitude of the output can be measured. The simulated frequency response results are shown in Figure 6.7. The results indicate that the servo seems to be able to follow feeding velocity changes up to 40 Hz (250 rad/s), which is the bandwidth of the velocity loop.

TABLE 6.2. Calculated Parameters of the Feed Drive Transfer Function

$K_1 = 1.5102 \cdot 10^6$	$K_2 = 930.178$	$K_3 = 3.077 \cdot 10^5$	
$b_2 = 0.40691 \cdot 10^{-4}$	$b_1 = 1.2906 \cdot 10^{-4}$	$b_0 = 0.25539 \cdot 10^{-4}$	
$a_1 = -1.1992$	$a_0 = 0.3945$		
$\beta_2 = 2.4556$	$\beta_1 = -1.6438$	$\beta_0 = -0.4495$	
$\alpha_3 = -2.8616$	$\alpha_2 = 3.077$	$\alpha_1 = -1.4686$	$\alpha_0 = 0.261$
$K_{cl} = 0.0057$			

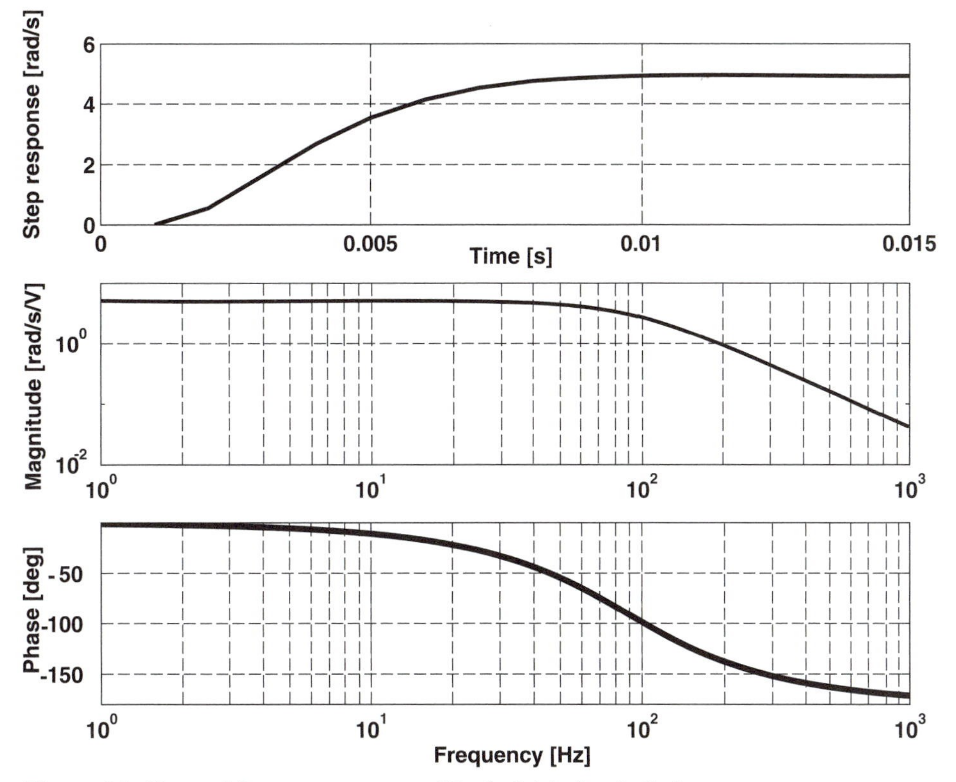

Figure 6.7: Step and frequency response of the feed drive's velocity loop.

In addition to the velocity loop, the position loop consists of an encoder, digital filter, and integrator (i.e., up–down counter). These are mostly digital and their transfer functions are known (see Fig. 6.4). The digital filter is tuned in such a way that the overall closed-loop position control system behaves like a second-order system with a rise time of $t_r = 30\,\text{ms}$ and an overshoot of $M_p = 0.1\%$. The following digital filter was identified:

$$D(z) = 9.2038\,\frac{z - 0.7161}{z - 0.6681}.$$

The calculated values of the state space parameters of the system are

$$\Phi(T) = e^{A_c T} = \begin{bmatrix} 0.3274 & -2.1160 & 0 \\ 0.0515 & 0.8718 & 0 \\ 0.0194 & 0.6049 & 1 \end{bmatrix},$$

$$H(T) = \int_0^T e^{A_c t} dt \cdot B_c = \begin{bmatrix} 11.3754 & 0.3481 \\ 0.5532 & -0.2639 \\ 0.1271 & -0.0859 \end{bmatrix},$$

$$G(T) = \begin{bmatrix} 0.6681 & 0 & 0 & 0.0022 \\ 11.3754 & 0.3274 & -2.1160 & -0.5112 \\ 0.5532 & 0.0515 & 0.8718 & -0.0249 \\ 0.1271 & 0.0194 & 0.6049 & 0.9943 \end{bmatrix},$$

$$\Gamma(T) = \begin{bmatrix} -0.0022 & 0 \\ 0.5112 & 0.3481 \\ 0.0249 & -0.2639 \\ 0.0057 & -0.0859 \end{bmatrix},$$

$$C = \begin{bmatrix} 1 & 0 & 0 & -0.0449 \\ 0 & 1 & 0 & 0 \\ 0 & 0 & 1 & 0 \\ 0 & 0 & 0 & 1 \end{bmatrix}, \quad D = \begin{bmatrix} 0.0449 & 0 \\ 0 & 0 \\ 0 & 0 \\ 0 & 0 \end{bmatrix}.$$

The steady-state position error can be calculated by substituting the servoparameter values in Eq. (6.37); thus,

$$e_{ss} = 0.0083\, f_c \quad \text{[counts]}.$$

The step, ramp, and frequency response of the digital position control loop are shown in Figure 6.8.

Measurements across a range of velocities showed that the steady-state error is not quite linear; this is mainly due to the effect of the static friction disturbance and unmodeled part of the dynamics. An average coefficient of friction is used for a general analysis and tuning of the positional control systems. However, for precision contouring, the influence of the static friction must be compensated by a more advanced control law than the simple digital filter presented in this text.

Example 9. A single-axis ballscrew-driven table is used in teaching identification and digital control, principles as shown in Figure 6.9. Develop the state space model of the system, and simulate the step and ramp input response.

Ballscrew Table Parameters			
Encoder gain	K_e(count/rad/s)	Current amplifier proportional gain	K_{vp} (V/V)
Equivalent inertia	J_e (kg m^2)	Current amplifier integral gain	K_{vi} (V/Vs)
Equivalent viscous damping	Nm/(rad/s)	Current command	i_a (V)
Disturbance torque	T_d (Nm)	Velocity amplifier proportional gain	K_{ip} (V/V)
Motor torque	T_m (Nm)	Velocity amplifier integral gain	K_{ii} (V/Vs)
Motor inductance	$L(H)$	Current sensor gain	K_a (V/A)
Motor resistance	$R(\Omega)$	Velocity amplifier gain	S_g (V/V)
Back electromotor constant	K_b (V/rad/s)	Motor torque constant	K_t (Nm/A)
Angular velocity of ballscrew	ω (rad/s)	Transfer function of controller	$G_c(s)$
Velocity feedback gain	T_g (V/count/s)	Position command	x_r (counts)
		Measured position	x_a (counts)

Table of Transfer Function Blocks	
Encoder	$\omega(s)\frac{K_e}{s} = x_a(s)$
Mechanical system	$(T_m - T_d)\frac{1}{J_e s + B} = \omega(s)$
Electrical winding of the motor	$(V_i - K_b\omega)\frac{1}{Ls+R} = i(s)$
Current PI amplifier	$(i_a - K_a i)\left(K_{vp} + \frac{K_{vi}}{s}\right) = V_i(s)$
Velocity PI amplifier	$(S_g V_c - \omega)\left(K_{ip} + \frac{K_{ii}}{s}\right) = i_a(s)$
Position error controller	$(x_r - x_a)G_c(s) = V_c(s)$

Motor electrical winding and mechanical system can be modeled as follows:

$$J\frac{d\omega}{dt} + B\omega = K_t i - T_d \rightarrow \frac{d\omega}{dt} = -\frac{B}{J}\omega + \frac{K_t}{J}i - \frac{1}{J}T_d \tag{6.54}$$

$$L\frac{di}{dt} + Ri = V_i - K_b\omega \rightarrow \frac{di}{dt} = -\frac{K_b}{L}\omega - \frac{R}{L}i + \frac{1}{L}V_i \tag{6.55}$$

Figure 6.8: Step, ramp, and frequency response of the feed drive's position loop.

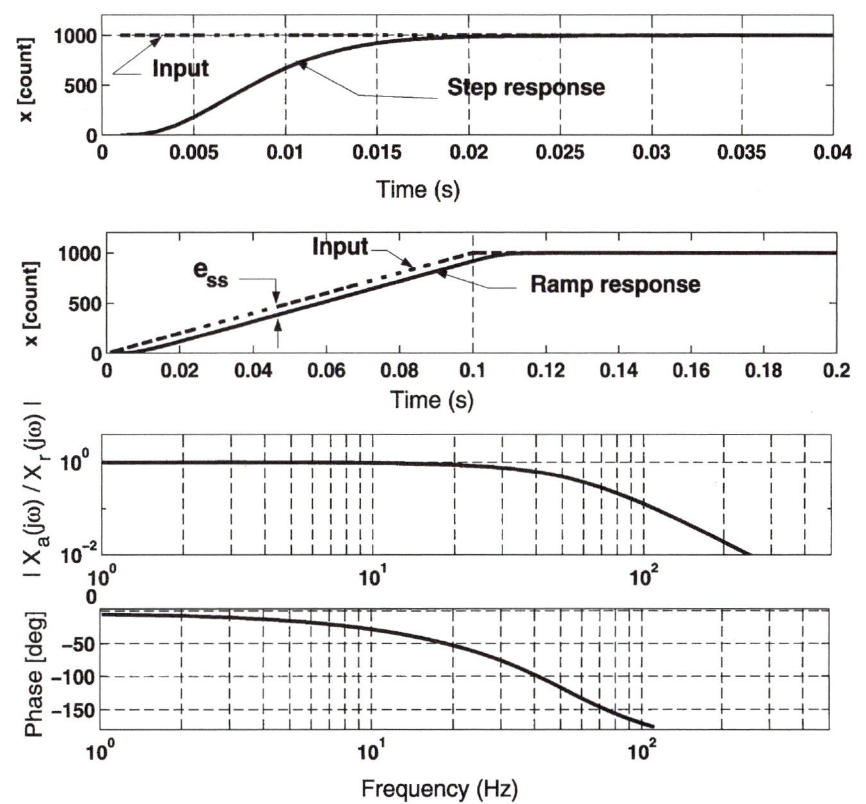

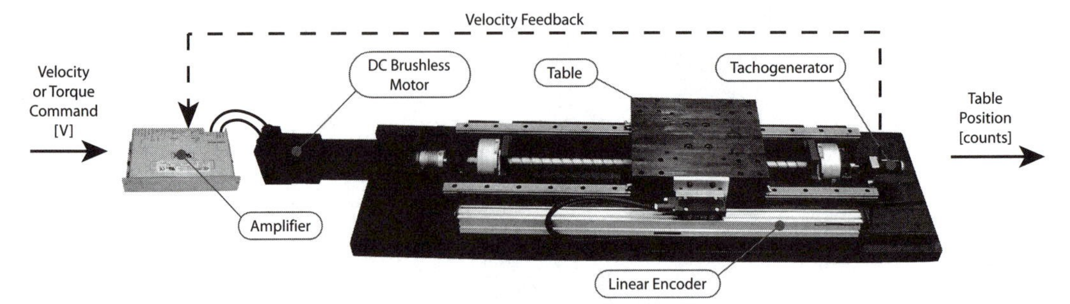

Figure 6.9: A ballscrew drive system used for teaching CNC design.

PI Current Loop

$$V_i = (i_a - K_a i)\left(K_{vp} + \frac{K_{vi}}{s}\right) = K_{vp}(i_a - K_a i) + K_{iv}\underbrace{\frac{1}{s}(i_a - K_a i)}_{z_1} \tag{6.56}$$

$$\text{Auxiliary state } z_1 = \frac{1}{s}(i_a - K_a i) \rightarrow \frac{dz_1}{dt} = i_a - K_a i \tag{6.57}$$

$$V_i = K_{pv}(i_a - K_a i) + K_{iv} z_1 \tag{6.58}$$

PI Velocity Loop

$$i_a = \left(S_g V_c - \omega\right)\left(K_{ip} + \frac{K_{ii}}{s}\right) = K_{ip}\left(S_g V_c - \omega\right) + K_{ii}\underbrace{\frac{1}{s}\left(S_g V_c - \omega\right)}_{z_2} \tag{6.59}$$

$$\text{Auxiliary state } z_2 = \frac{1}{s}\left(S_g V_c - \omega\right) \rightarrow \frac{dz_2}{dt} = S_g V_c - \omega \tag{6.60}$$

$$i_a = K_{ip}\left(S_g V_c - \omega\right) + K_{ii} z_2 \tag{6.61}$$

Rewriting the equation for $\frac{dz_1}{dt}$ and V_i by substituting i_a

$$\frac{dz_1}{dt} = i_a - K_a i = K_{ip}\left(S_g V_c - \omega\right) + K_{ii} z_2 - K_a i \tag{6.62}$$

$$V_i = K_{vp}(i_a - K_a i) + K_{vi} z_1 = K_{vp}\left(K_{ip}\left(S_g V_c - \omega\right) + K_{ii} z_2 - K_a i\right) + K_{vi} z_1 \tag{6.63}$$

$$V_i = K_{vp} K_{ip} S_g V_c - K_{vp} K_{ip} \omega + K_{vp} K_{ii} z_2 - K_{vp} K_a i + K_{vi} z_1 \tag{6.64}$$

By substituting V_i into $\frac{di}{dt}$

$$\frac{di}{dt} = -\frac{K_b}{L}\omega - \frac{R}{L}i + \frac{1}{L}V_i \tag{6.65}$$

$$= -\frac{K_b}{L}\omega - \frac{R}{L}i + \frac{K_{vp}K_{ip}S_g}{L}V_c - \frac{K_{vp}K_{ip}}{L}\omega + \frac{K_{vp}K_{ii}}{L}z_2 - \frac{K_{vp}K_a}{L}i + \frac{K_{vi}}{L}z_1 \tag{6.66}$$

$$= \left(-\frac{K_b + K_{vp}K_{ip}}{L}\right)\omega - \frac{R + K_{vp}K_a}{L}i + \frac{K_{vi}}{L}z_1 + \frac{K_{vp}K_{ii}}{L}z_2 + \frac{K_{vp}K_{ip}S_g}{L}V_c \tag{6.67}$$

The states are combined as follows:

$$\left\{\dot{x}\right\} = \begin{Bmatrix} \frac{d\omega}{dt} \\ \frac{di}{dt} \\ \frac{dz_1}{dt} \\ \frac{dz_2}{dt} \\ \frac{dx_a}{dt} \end{Bmatrix} = \begin{bmatrix} -\frac{B}{J} & \frac{K_t}{J} & 0 & 0 & 0 \\ -\frac{K_b+K_{vp}K_{ip}}{L} & -\frac{R+K_{vp}K_a}{L} & \frac{K_{vi}}{L} & \frac{K_{vp}K_{ii}}{L} & 0 \\ -K_{ip} & -K_a & 0 & K_{ii} & 0 \\ -1 & 0 & 0 & 0 & 0 \\ K_e & 0 & 0 & 0 & 0 \end{bmatrix} \begin{Bmatrix} \omega \\ i \\ z_1 \\ z_2 \\ x_a \end{Bmatrix} + \begin{bmatrix} 0 & -\frac{1}{J} \\ \frac{K_{pv}K_{pi}S_g}{L} & 0 \\ K_{ip}s_g & 0 \\ S_g & 0 \\ 0 & 0 \end{bmatrix} \begin{Bmatrix} V_c \\ T_d \end{Bmatrix} \tag{6.68}$$

Because we are interested in physical variables such as velocity (ω), current (i), equivalent current command (i_a), motor voltage (V_i)

$$\begin{Bmatrix} \omega \\ i \\ i_a \\ V_i \\ x_a \end{Bmatrix} = \begin{bmatrix} 1 & 0 & 0 & 0 & 0 \\ 0 & 1 & 0 & 0 & 0 \\ -K_{ip} & 0 & 0 & K_{ii} & 0 \\ -K_{vp}K_{ip} & -K_{vp}K_a & K_{vi} & K_{vp}K_{ii} & 0 \\ 0 & 0 & 0 & 0 & 1 \end{bmatrix} \begin{Bmatrix} \omega \\ i \\ z_1 \\ z_2 \\ x_a \end{Bmatrix} + \begin{bmatrix} 0 & 0 \\ 0 & 0 \\ K_{ip}S_g & 0 \\ K_{vp}K_{ip}S_g & 0 \\ 0 & 0 \end{bmatrix} \begin{Bmatrix} V_c \\ T_d \end{Bmatrix} \tag{6.69}$$

or in standard state space notation

$$\left.\begin{aligned} \left\{\dot{x}\right\} &= \left[A\right]\{x\} + \left[B\right]\{u\} \\ \{y\} &= \left[C\right]\{x\} + \left[D\right]\{u\} \end{aligned}\right\}. \tag{6.70}$$

The state space model of the machine indicated above represents the continuous part of the physical system. If the system needs to be controlled in discrete time domain, its ZOH equivalent must be considered which leads to state space model of the system in z domain. The digital controller then can be combined to achieve a full state space model of the closed-loop system in discrete time domain.

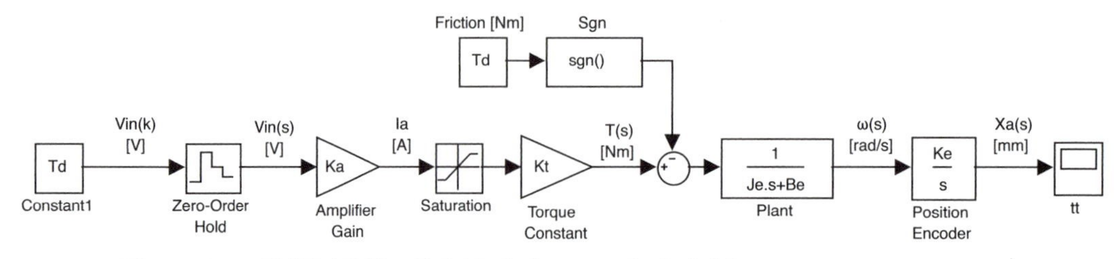

Figure 6.10: MATLAB Simulink block diagram of a feed drive system.

Example 10. A lead lag controller is designed to control a feed drive system whose MATLAB block diagram is given in Figure 6.10. The parameters of the drive are given as follows: $J_e = 7 \times 10^{-4}$ [kg m^2], $B_e = 0.00612$ [Nm/(rad/s)], $K_t = 0.72$ [Nm/A], $K_a = 0.887[A/V]$, $K_e = 20/(2\pi)$, $K_d = 1$ [V/mm] and sampling time of $T = 0.0002$ [s] used. The solution procedure is given as follows:

The open-loop transfer function of the drive is obtained as follows:

$$G_o(s) = \frac{K_a K_t K_e}{s(J_e s + B_e)} = \frac{K_0}{s(\tau_v s + 1)}, \tag{6.71}$$

where the gain (K_0) and time constant (τ_v) are $K_0 = \frac{K_a K_t K_e}{B_e} = 332.2$ [mm/V] $\tau_v = \frac{J_e}{B_e} = 0.1144$ [s]. The ZOH equivalent of the system is obtained to represent the drive dynamics in the discrete time domain.

$$G_o(z) = (1 - z^{-1})\mathbb{Z}\frac{G_o(s)}{s} = (1 - z^{-1})\mathbb{Z}\frac{K_0/\tau_v}{s^2(s + 1/\tau_v)}. \tag{6.72}$$

By applying partial fraction expansion rules, $G_o(s)/s$ can be transformed as follows:

$$\frac{1}{s^2(s+1/\tau_v)} = \frac{C_1}{s^2} + \frac{C_2}{s} + \frac{C_3}{s+1/\tau_v}$$

$$C_1 = \lim_{s\to 0} s^2 \frac{1}{s^2(s+1/\tau_v)} = \tau_v$$

$$C_2 = \lim_{s\to 0}\left\{\frac{1}{1!}\frac{d}{ds}\left(s^2\frac{1}{s^2(s+1/\tau_v)}\right)\right\} = \lim_{s\to 0}\frac{-1}{(s+1/\tau_v)^2} = -\tau_v^2$$

$$C_3 = \lim_{s\to -1/\tau_v}\left\{(s+1/\tau_v)\frac{1}{s^2(s+1/\tau_v)}\right\} = \tau_v^2$$

$$\frac{G_o(s)}{s} = \frac{K_0}{\tau_v}\tau_v\left(\frac{1}{s^2} - \frac{\tau_v}{s} + \frac{\tau_v}{s+1/\tau_v}\right) = K_0\left(\frac{1}{s^2} - \frac{\tau_v}{s} + \frac{\tau_v}{s+1/\tau_v}\right). \tag{6.73}$$

The ZOH equivalent of $G_o(s)$ is as follows:

$$G_0(z) = (1 - z^{-1})Z\frac{G_o(s)}{s}$$
$$= (1 - z^{-1})K_0\left(\frac{Tz^{-1}}{(1 - z^{-1})^2} - \frac{\tau_v}{1 - z^{-1}} + \frac{\tau_v}{1 - e^{-T/\tau_v}z^{-1}}\right)$$
$$G_0(z) = \frac{B(z)}{A(z)} = \frac{z^{-1}\left(b_1 z^{-1} + b_0\right)}{z^{-2}a_0 + z^{-1}a_1 + 1} = \frac{b_0 z + b_1}{z^2 + a_1 z + a_0}, \tag{6.74}$$

where

$$b_1 = K_0\left[\tau_v(1 - e^{-\frac{T}{\tau_v}}) - Te^{-\frac{T}{\tau_v}}\right] = 5.8014e - 005$$
$$b_0 = K_0\left[T - \tau_v(1 - e^{-\frac{T}{\tau_v}})\right] = 5.8048e - 005$$
$$a_0 = e^{-\frac{T}{\tau_v}} = 0.9983$$
$$a_1 = -(1 + e^{-\frac{T}{\tau_v}}) = -1.9983.$$

Note that one of the open-loop poles is on the unit circle , while the other is very close to the unit circle, i.e., $z^2 + a_1 z + a_0 = (z - 1)(z - 0.9983)$.

Assume that the position control loop is closed by a proportional controller with a gain K_p [V/mm]. The proportional controller is designed both in Laplace and discrete time domain as follow.

Laplace Domain Design

The ZOH is ignored and the closed-loop transfer function of the system is derived as follows:

$$G_{cl}(s) = \frac{K_p G_o(s)}{1 + G_o(s)} = \frac{K_p K_0}{\tau_v s^2 + s + K_p K_0}.$$

The roots of the characteristic equation ($p_1, p_2 = \left(-1 \pm \sqrt{1 - 4\tau_v K_p K_0}\right)/(2\tau_v)$), that is, the poles of the system, start from open-loop poles $K_p = 0 \rightarrow p_1 = 0, p_2 = -1/\tau_v = -1/0.1144 = -8.7413$; become a complex conjugate when $1 - 4\tau_v K_p K_0 = 0 \rightarrow K_p = 1/(4\tau_v K_0) = 1/(4 \times 0.1144 \times 332.2) = 0.0065783$, which gives identical poles at $p_1 = p_2 = -1/(2\tau_v) = -1/(2 \times 0.1144) = -4.3706$. If the proportional gain (K_p) is increased further, the system has complex poles that indicate an underdamped, oscillatory step response. If we wish to have a damping ratio of $\zeta = 0.8$, the proportional gain is selected by forcing the desired characteristic equation on the closed-loop system.

$$s^2 + \frac{1}{\tau_v}s + \frac{K_pK_0}{\tau_v} \equiv s^2 + 2\zeta\omega_n s + \omega_n^2 \tag{6.75}$$

$$s^2 + \frac{1}{0.1144}s + \frac{K_p 332.2}{0.1144} \equiv s^2 + 2 \times 0.8 \times \omega_n s + \omega_n^2$$

$$\omega_n = \frac{1}{2 \times 0.8 \times 0.1144} = 5.4633 \text{ [rad/s]}$$

$$K_p = \frac{\omega_n^2 \tau_v}{K_0} = \frac{5.4633^2 \times 0.1144}{332.2} = 0.0103.$$

Regardless of the magnitude of proportional gain K_p, the system remains stable despite the increased oscillatory behavior of second-order continuous systems.

Discrete Time Domain Design

The closed-loop transfer function of the system in z domain is

$$G_{cl}(z) = \frac{K_pG_o(z)}{1 + G_o(z)} = \frac{K_pK_0\left(b_0z + b_1\right)}{z^2 + (a_1 + K_pK_0b_0)z + a_0 + K_pK_0b_1}$$

$$p_{1,2} = \frac{-(a_1 + K_pK_0b_0) \pm \sqrt{(a_1 + K_pK_0b_0)^2 - 4(a_0 + K_pK_0b_1)}}{2},$$

with poles $(p_1, p_2 = \left(-(a_1+K_pK_0b_0) \pm \sqrt{(a_1+K_pK_0b_0)^2 - 4(a_0 + K_pK_0b_1)}\right)/2)$. If $K_p = 0$ the poles start from the open loop poles of the system at $p_1 = 1$ and $p_2 = 0.9983$. If $0 \leq K_p < 0.00658$, the poles are on the real axis and the system is damped. The poles will meet at $p_1 = p_2 = 0.9991$ when $K_p = 0.00658$. If $K_p > 30$ The poles will leave the unit circle and the system will be unstable in discrete time domain.

Lead–Lag Compensator Design in Laplace Domain

The controller structure is given as $C(s) = K(1 + \alpha sT)/(1 + sT)$, where gain K and α, T are the parameters of the compensator. A compensator will be designed to achieve 60 deg phase margin at crossover frequency of $\omega_g = 60$ Hz.

From the Bode plot of the plant in s domain, at frequency of 60 Hz (377 rad/s), the phase of the plant $G_o(s)$ is found to be -179[deg]. Additional phase lead of $\phi_l = +59\,\text{deg} = 1.03$ [rad] must be added by the lead compensator at $\omega_g = 60$ Hz $= 377$ [rad/s].

$$\alpha = \frac{1 + \sin\phi_l}{1 - \sin\phi_l} = \frac{1 + \sin 1.03}{1 - \sin 1.03} = 13.015$$

$$T = \frac{1}{\omega_g\sqrt{\alpha}} = \frac{1}{377\sqrt{13.015}} = 7.3525 \times 10^{-4}.$$

After the parameters are calculated, the gain of the controller needs to readjusted to ensure a unity gain at frequency 60 Hz. The gain of the system is −22.7dB at y 60 Hz; hence, the gain needs to be increased by +2.7 dB.

$$20\log K = +22.7 \rightarrow K = 10^{22.7/20} = 13.646.$$

Hence, the lead compensating controller becomes

$$C(s) = K\frac{1+\alpha sT}{1+sT} = 13.646\frac{1+9.5693\times 10^{-3}s}{1+7.3525\times 10^{-4}s}.$$

6.5 SLIDING MODE CONTROLLER

Sliding mode controller is a typical example of robust, nonlinear control system. An application and design of a sliding mode controller to a feed drive mechanism is given here as an example [10]. Open-loop block diagram of a feed drive system is sown in Figure 6.11.

The ballscrew drive system application considered here is controlled in a current mode, and its open loop transfer function between the table position (x [mm]) and amplifier command generated in the CNC (u[V]) is as follows:

$$\begin{aligned} x(s) &= \left[K_aK_tu(s) - T_c\right]\frac{r_g}{(Js+B)s} \\ &= \frac{K_aK_tr_g}{s(Js+B)}[u(s) - \frac{1}{K_aK_t}T_c(s)] \\ &= \frac{K}{s(s+p)}[u(s) - \frac{1}{K_aK_t}T_c(s)], \end{aligned} \tag{6.76}$$

where $K = K_aK_tr_g/J$ is the gain, $p = B/J$ is the velocity loop pole, and $T_c(s)$ is the torque disturbance caused by friction and the cutting process. In the linear motor drives, the inertia (J) is replaced by the table–workpiece mass, and the disturbance torque (T_c) is replaced by the cutting force and friction in the linear motor-driven systems. The differential equation of the drive system

Figure 6.11: Open-loop block diagram of a feed drive system powered under current mode of the amplifier.

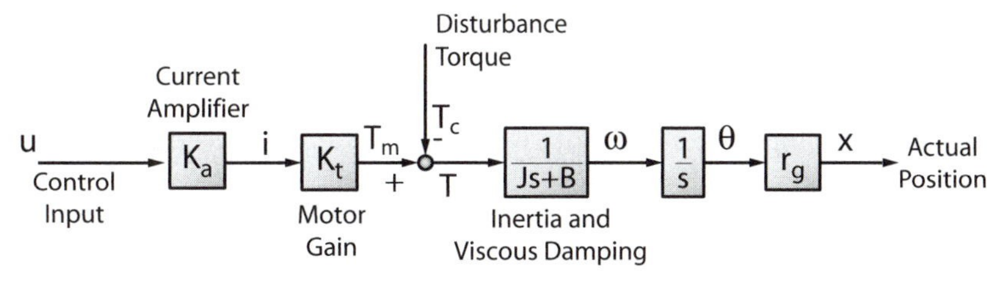

can be reorganized as follows:

$$\frac{1}{K}s^2x(s) + \frac{p}{K}sx(s) = u(s) - \frac{1}{K_aK_t}T_c(s) \tag{6.77}$$

$$\underbrace{\frac{J}{K_aK_tr_g}}_{J_e}\ddot{x}(t) + \underbrace{\frac{B}{K_aK_tr_g}}_{B_e}\dot{x}(t) = u(t) - \underbrace{\frac{1}{K_aK_t}T_c(t)}_{d(t)}$$

or normalizing the disturbance torque with respect to input,

$$J_e\ddot{x}(t) + B_e\dot{x}(t) = u(t) - d(t), \tag{6.78}$$

where the reflected disturbance at the input is $d(t) = T_c(t)/(K_aK_t)$. The acceleration of the drive can be separated from Eq. (6.78) as follows:

$$\ddot{x}(t) = \frac{1}{J_e}[-B_e\dot{x}(t) + u(t) - d(t)]. \tag{6.79}$$

The objective of the feed drive control system is that the controller must be capable of following the commanded trajectory and velocity with minimum error at high speeds, regardless of the slight variations in the inertia (J_e), viscous friction (B_e), and external disturbances (d) such as Coulomb friction and cutting forces reflected on the drive input. The conventional drive control systems, such as pole placement and feed-forward control techniques work quite well when the system is modeled accurately, and external disturbances are rejected by having a large bandwidth in linear drives. In addition, large transmission ratio in the ballscrew drives reduce the reflected torque on the rotating drive motor shaft significantly. However, when the friction is not modeled and compensated accurately, and when the external forces such as in linear drives are not determined before, conventional control techniques do not exhibit high tracking accuracy in high-speed machining, leading to inaccuracies in contour machining operations. The sliding mode controller belongs to a class of nonlinear control strategies, which are robust to such uncertainties and time variation in the drive dynamics.

There are two fundamental steps in designing a sliding mode controller: the selections of a sliding surface and a Lyapunov function. For accurate tracking of position and velocity, which are the key requirements from high-speed machine tools, the sliding surface (S) is selected as follows:

$$S = \lambda(x_r - x) + (\dot{x}_r - \dot{x}), \tag{6.80}$$

where $\lambda[1/s]$ is the desired but achievable bandwidth of the drive, x_r, x are the reference command and actual positions, and $\dot{x}_r, \dot{x}$ are the reference command and actual velocities of the drive, respectively. The control input (u) must be manipulated in such a way that, after a finite time, both the position and velocity of the drive approach reference command values ($x \to x_r, \dot{x} \to \dot{x}_r$), hence, forcing the value of position and velocity errors to be zero, that is, $S \to 0$.

Inertia (J_e) and viscous damping (B_e) on the machine drive are assumed to change slowly and insignificantly. The variation in the external disturbance caused by the cutting process and friction is considered to be strong, but with upper (d^+) and lower (d^-) limits measured on the machine. The external disturbance can be tracked by the following simple observer:

$$\dot{\hat{d}} = \rho\kappa S, \quad \hat{d}(k) = \hat{d}(k-1) + \rho\kappa S \cdot T, \tag{6.81}$$

where T[s] is the control period, k is the control interval counter in discrete time domain, ρ is the parameter adaptation gain ($\rho \approx 0.005$), and κ is used to limit the integral control of the disturbance as follows:

$$\kappa = \begin{Bmatrix} 0 & \text{if } \hat{d}(k) \leq d^- \text{ and } S \leq 0 \\ 0 & \text{if } \hat{d}(k) \geq d^+ \text{ and } S \geq 0 \\ 1 & \text{otherwise} \end{Bmatrix}. \tag{6.82}$$

Hence, the estimated disturbances are always kept within the predetermined bounds that is, $\hat{d}(k) \in [d^-, d^+]$.

The second step in sliding mode controller design is to select a Lyapunov function that is used to obtain a stable control law for a nonlinear system, that is, feed drive controlled by the nonlinear sliding mode control strategy. The following is a Lyapunov function:

$$\dot{V}(t) = \frac{1}{2}[J_e S^2 + \frac{(d-\hat{d})^2}{\rho}], \tag{6.83}$$

which resembles the summation of kinetic energy and the square of disturbance prediction error. Like the sliding surface, the selection of a specific Lyapunov function is based on experience and intuition. For asymptotic stability of nonlinear systems, the derivative of a Lyapunov function must be negative or zero, meaning that the rate of change in the energy and prediction error must decrease.

$$\dot{V}(t) = J_e S\dot{S} - \dot{\hat{d}}\frac{(d-\hat{d})}{\rho} < 0. \tag{6.84}$$

By substituting $\dot{S} = \lambda(\dot{x}_r - \dot{x}) + (\ddot{x}_r - \ddot{x})$ from Eq. (6.80), $\dot{\hat{d}} = \rho\kappa S$ from Eq. (6.81) and $\ddot{x}$ from Eq. (6.79),

$$\dot{V}(t) = J_e S\left[\lambda(\dot{x}_r - \dot{x}) + \ddot{x}_r\right] + SB_e\dot{x} - Su + Sd - S\kappa(d-\hat{d}) < 0. \tag{6.85}$$

Expressing $Sd - S\kappa(d-\hat{d}) = S\hat{d} + S(d-\hat{d})(1-\kappa)$, and noting that $S(d-\hat{d})(1-\kappa) < 0$ always due to imposed limit conditions ($\hat{d}(k) \in [d^-, d^+], \kappa = 0$ or 1, as well as the value of the sliding surface S) in Eq. (6.82), never being outside the following criteria will ensure the asymptotic stability condition at any condition ($\dot{V}(t) < 0$):

$$J_e S\left[\lambda(\dot{x}_r - \dot{x}) + \ddot{x}_r\right] + SB_e\dot{x} - Su + S\hat{d} = -K_s S^2, \tag{6.86}$$

where $K_s > 0$ is the control gain to be selected. The control law (u) is obtained from Eq. (6.86) as

$$u(k) = J_e\left[\lambda(\dot{x}_r(k) - \dot{x}(k)) + \ddot{x}_r(k)\right] + B_e\dot{x}(k) + \widehat{d}(k) + K_sS(k) \tag{6.87}$$

$$S(k) = \lambda[x_r(k) - x(k)] + [\dot{x}_r(k) - \dot{x}(k)],$$

where (k) is the control interval counter. The reference position, $x_r(k)$, velocity $\dot{x}_r(k)$, and acceleration $\ddot{x}_r(k)$ are obtained from the command generation algorithm running in the CNC system. The actual position $x(k)$ is measured from the encoder, and the actual velocity $\dot{x}(k)$ is estimated by taking the derivative of the measured position from the linear drive or measuring directly from a tachogenerator. However, evaluation of velocity and acceleration from discrete position commands and encoder readings may be noisy. The following simple low-pass filter can be used to smooth them:

$$\dot{x}_r(k) = \alpha\dot{x}_r(k-1) + \frac{1-\alpha}{T}[x_r(k) - x_r(k-1)]$$

$$\ddot{x}_r(k) = \alpha\ddot{x}_r(k-1) + \frac{1-\alpha}{T}[\dot{x}(k) - \dot{x}_r(k-1)]$$

$$\dot{x}(k) = \alpha\dot{x}(k-1) + \frac{1-\alpha}{T}\dot{x}_m(k),$$

where $\dot{x}_m(k)$ is the measured velocity from the tachogenerator. The filter gain is typically set to $\alpha \simeq 0.6$.

The disturbance compensation through on-line estimation is useful to minimize the effects of slowly varying cutting forces and friction. However, when the drive changes the direction of the velocity, especially at the corners and quadrants of the circular paths, the friction force reverses its direction and leaves glitches on the surface. If a major Coulomb friction force is known, it may be more advantageous to precompansate it at the feedforward command generation as follows:

$$u_{fc}(k) = \left\{\begin{array}{ccc} u_{fc}^{+} = T_f^{+}/(K_aK_t) & \rightarrow & \dot{x}_r(k) > 0 \\ 0 & \rightarrow & \dot{x}_r(k) = 0 \\ u_{fc}^{-} = T_f^{-}/(K_aK_t) & & \dot{x}_r(k) < 0 \end{array}\right\},$$

where T_f^{+}, T_f^{-} are the magnitudes of Coulomb friction measured in positive and negative directions of the motion. Hence, the overall control signal is obtained as follows:

$$u(k) = u_{smc}(k) + u_{fc}(k).$$

The sliding mode controller can attenuate the influence of external disturbances, while providing a good tracking performance with high bandwidth. The implementation of sliding mode controller is shown in Figure 6.12

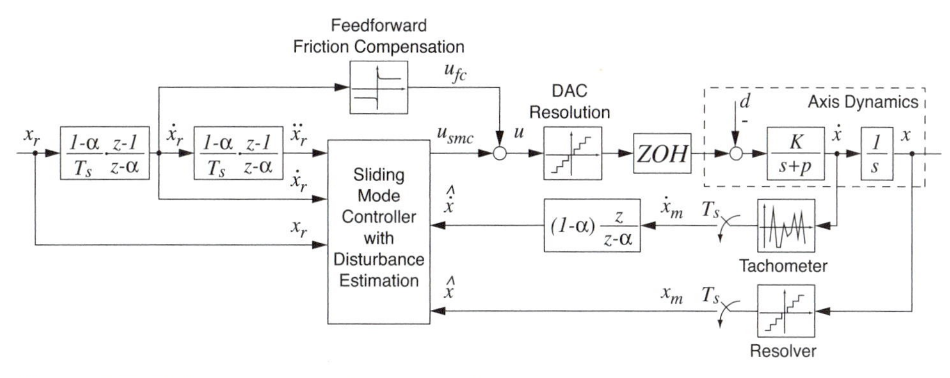

Figure 6.12: Sliding mode controller with feedforward friction compensation.

6.6 ACTIVE DAMPING OF FEED DRIVES

The ballscrew drives exhibit torsional flexibility at the motor shaft–screw coupling, screw itself, and nut as shown in Figure 6.13. The axial displacement of the screw is coupled with its torsional flexibility, and the screw may experience lateral flexibility which applies tension and compression loads on the table–guideway interface. The structural vibrations caused by the ballscrew assembly occur typically above the bandwidth frequency of the servodrive, i.e., more than 100 Hz. However they affect the surface finish quality and precision positioning accuracy during machining; hence, they need to be avoided. The linear cutting force (F_l) and table mass (m_t) are transmitted to motor as a reduced torque. The bandwidth and speed are increased by using two parallel ballscrew drives in the most recent, high-speed machine tools.

The mechanical drive system is represented by its rigid body motion when the structural dynamic flexibility is neglected, as presented in previous sections. However, it may be important to damp the structural dynamics of the machine that are excited by the cutting load and inertial forces during high-speed motions.

Transfer functions determine the relationship between the forces and positions of the table and motor because of the flexibility of the mechanical drives, and they replace the rigid body-based transfer function. If the machine is built, the transfer functions $G(s)$ are measured experimentally by applying white noise to the current amplifier to generate random torque (T_m), and the resulting angular position of the motor shaft (θ_2) and table position (x_2) are measured from rotary and linear encoders, respectively. The direct transfer function at the table (G_{tt}) is measured by applying impact load (F_l) and measuring the table vibrations (x_t) with the use of the accelerometer or displacement sensor. Alternatively, the transfer functions can be predicted from the finite element model of the drive structure by assuming a damping constant.

A simple structural dynamic model of the ball-screw drive system can be approximated by the reflected inertias at the motor (J_m) and leadscrew (J_l) connected by a torsional spring (k_t) and damping (c_t) elements as shown in Figure 6.13.

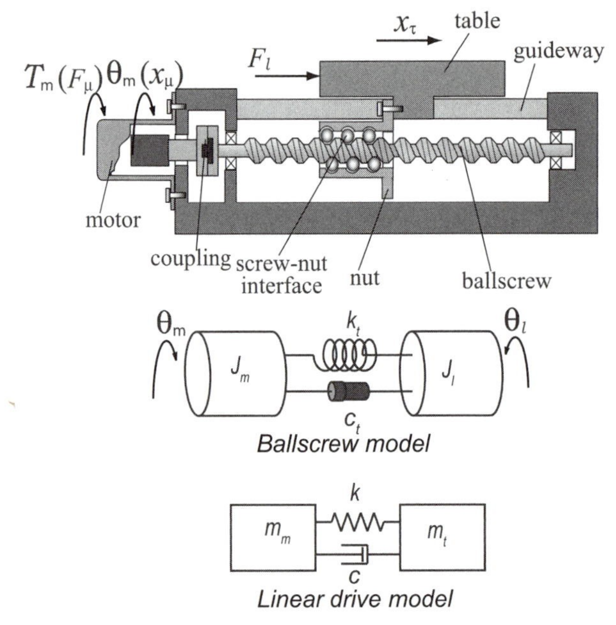

Figure 6.13: Torsional-axial vibrations of the ballscrew affect the positioning accuracy at the table (x_1).

The stiffness of the ballscrew drive varies as a function of table position as shown in Figure 6.14. The static stiffness is determined primarily by the equivalent axial stiffness of the ball screw-nut contact as outlined in [DIN 69051-6]. The ballscrew drive system, with bearings and the intermediate transmission or clutch, has a finite stiffness that assists in determining the static displacement of the table under load during the high-speed positioning of the table. The ballscrew is supported by thrust bearings at two ends. Bearings provide radial guidance to the screw and absorbs the feed forces in the axial direction.

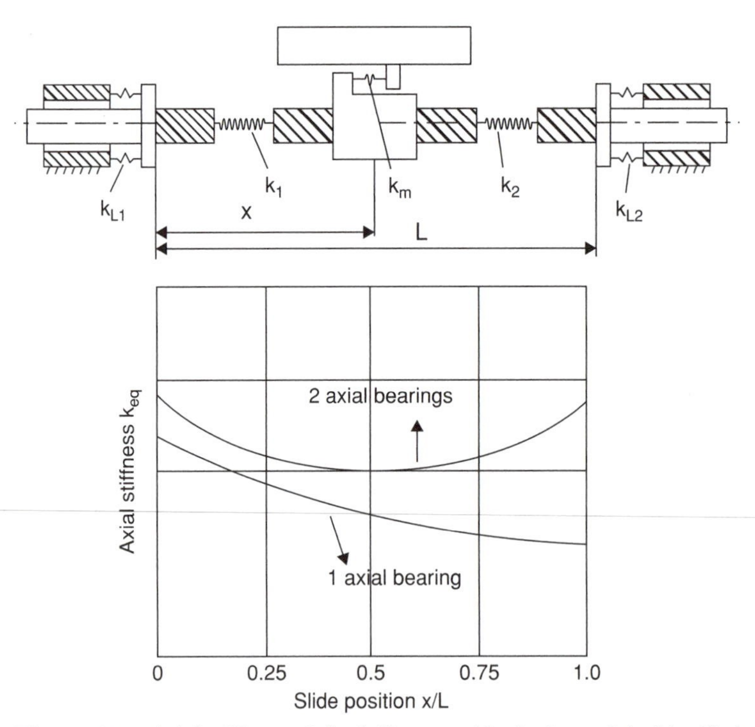

Figure 6.14: Axial stiffness of the ballscrew with single- and double-sided thrust bearings [DIN 69051-6].

If the bearings at both ends are fixed, the equivalent axial stiffness of the ballscrew system is given by the following:

$$k_t = \left[\frac{1}{k_i + k_{ii}} + \frac{1}{k_M} \right]^{-1}$$

where the stiffness terms contributed by the left (k_i) and right (k_{ii}) bearings are defined as follows:

$$k_i = \left[\frac{1}{k_1} + \frac{1}{k_{L1}} \right]^{-1}, \quad k_{ii} = \left[\frac{1}{k_2} + \frac{1}{k_{L2}} \right]^{-1}$$

The axial stiffness is reduced when the right bearing is preload free . As the table position changes, the axial stiffness of the drive varies, which leads to the time-varying dynamics of ballscrew drives. It must be noted that the largest flexibility is contributed by the screw, coupling, and nut in the system. For example, the torsional stiffness of the screw can be estimated as follows:

$$k_{ts} = \frac{GI}{L} \longleftarrow I = \frac{\pi d_p^4}{32},$$

where G [N/m^2] is the module of rigidity, which has a value of $G = 75 \times 10^9$ [N/m^2] for the steel. L and d_p are the length and the pitch diameter of the screw.

The connection can be considered at the motor–screw coupling or screw–nut coupling junctions for simplicity. By neglecting the viscous friction, the structural dynamics of the ballscrew system can be expressed by

$$\begin{aligned} J_m \frac{d^2\theta_m}{dt^2} + c_t(\frac{d\theta_m}{dt} - \frac{d\theta_l}{dt}) + k_t(\theta_m - \theta_l) = T_m \\ J_l \frac{d^2\theta_l}{dt^2} - c_t(\frac{d\theta_m}{dt} - \frac{d\theta_l}{dt}) - k_t(\theta_m - \theta_l) = T_L. \end{aligned} \tag{6.88}$$

If the equation of motion is transformed to the Laplace domain, then

$$\begin{bmatrix} J_m s^2 + c_t s + k_t & -(c_t s + k_t) \\ -(c_t s + k_t) & J_l s^2 + c_t s + k_t \end{bmatrix} \begin{Bmatrix} \theta_m \\ \theta_l \end{Bmatrix} = \begin{Bmatrix} T_m \\ T_l \end{Bmatrix} \tag{6.89}$$

which leads to transfer function matrix after the inversion of the transfer matrix as follows:

$$\begin{aligned} \begin{Bmatrix} \theta_m \\ \theta_l \end{Bmatrix} &= \frac{\begin{bmatrix} J_l s^2 + c_t s + k_t & c_t s + k_t \\ c_t s + k_t & J_m s^2 + c_t s + k_t \end{bmatrix}}{s^2 \left[s^2 J_l J_m + s c_t \left(J_l + J_m \right) + k_t \left(J_l + J_m \right) \right]} \begin{Bmatrix} T_m \\ T_l \end{Bmatrix} \\ &= \begin{bmatrix} G_{mm} & G_{mt} \\ G_{tm} & G_{tt} \end{bmatrix} \begin{Bmatrix} T_m \\ T_l \end{Bmatrix} \end{aligned} \tag{6.90}$$

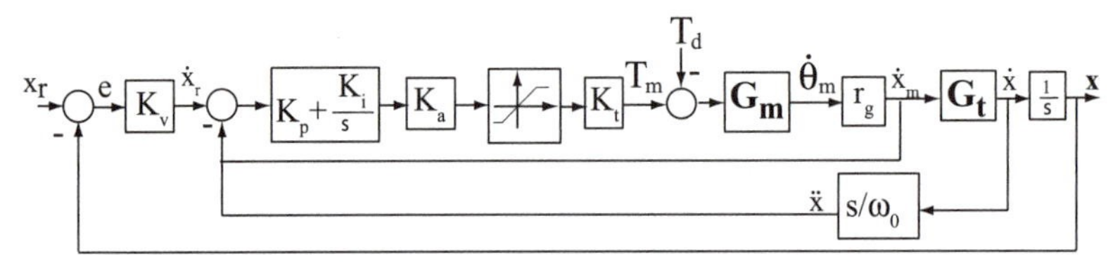

Figure 6.15: Active damping of a torsional-axial mode (ω_0) of ballscrew drive with an acceleration feedback (s^2/ω_0) in a cascaded control system.

The equation of the motion has the identical form if the linear mass, and translational motions are considered with the following transformations:

$$J_m = r_g^2 m_m, \; J_l = r_g^2 m_t$$

$$k_t = r_g^2 k_a, \;\; c_t = r_g^2 c_a$$

$$\theta_m = \frac{x_2}{r_g}, \quad \theta_l = \frac{x_1}{r_g}$$

$$T_m = \frac{h_p}{2\pi} F_m, \quad T_l = \frac{h}{2\pi} F_t$$

where transmission ratio $r_g = h_p/2\pi$ for a leadscrew with a pitch length of h_p. The transfer function of the system with linear masses (i.e., linear drives) becomes

$$\begin{Bmatrix} x_m \\ x_t \end{Bmatrix} = \frac{\begin{bmatrix} m_t s^2 + c_a s + k_a & c_a s + k_a \\ c_a s + k_a & m_m s^2 + c_a s + k_a \end{bmatrix}}{s^2 \left[s^2 m_t m_m + s c_a \left(m_t + m_m \right) + k_a \left(m_t + m_m \right) \right]} \begin{Bmatrix} F_m \\ F_t \end{Bmatrix} \tag{6.91}$$

The feed drive structure has a natural frequency as follows:

$$\omega_0 = \sqrt{\frac{k_t}{J_l J_m / \left(J_l + J_m \right)}} = \sqrt{\frac{k_a}{m_t m_m / \left(m_t + m_m \right)}} \text{[rad/s]},$$

which needs to be damped by the controller. Although several advanced control algorithms exist for active damping of modes, the application with a cascaded control architecture is illustrated here, because it is most commonly used in industrial CNCs. The cascaded control structure is shown in Figure 6.15. The controller has a current loop inside, surrounded by velocity and position control loops [21]. The current loop usually has about 1,000 Hz bandwidth with a PWM converter having more than 10 KHz modulator; hence, it is approximated by a gain K_a [A/V] here. The motor torque constant is K_t [Nm/A]. The cascaded controller uses a proportional gain (K_v) on position error (e), and a proportional and integral controller on the velocity error ($\dot{x}_r - \dot{x}_m$). The integral action is used to minimize the steady-state error caused by the disturbance (T_d) and lag caused by the transfer function of the system. The velocity at the motor shaft is usually indirectly measured from the rotary encoder mounted on the motor shaft. The encoder signals are digitally differentiated to obtain the velocity of the motor shaft ($s\dot{\theta}_m$). The angular velocity is scaled by the leadscrew transmission gain

(r_g) to obtain a linear velocity compatible with the velocity command generated by the CNC. The inertia and viscous damping forces are compensated by the feedforward and feedback terms, respectively, but they are not shown here for simplicity. G_m represents the transfer function between the angular velocity ($\dot{\theta}_m$) and torque at the motor shaft, whereas G_t represents the transfer function between the angular velocity of the ballscrew at the nut (i.e., table velocity) and the motor shaft's velocity. If the structural dynamics of the drive are neglected, G_m represents only the rigid body dynamics ($G_m = 1/(J_e s + B)$ and $G_t = 1$. The typical bandwidth of the velocity loop is about 100 Hz, and the position loop is approximately 30 Hz for linear drives with rigid body dynamics [87]. Ball screw drives have less bandwidth. The vibrations are damped by using acceleration feedback (s/ω_0) scaled at the resonance frequency ω_0. The acceleration can be measured either directly by a Ferraris sensor mounted between the stationary guide and leadscrew [88], or by taking a double derivative of position measurements obtained from the table-mounted linear encoder.

The transfer function between the motor torque and motor velocity can be expressed from Eq. (6.90) as follows:

$$G_m(s) = \frac{s\theta_m}{T_m} = sG_{mm}(s) = \frac{J_l s^2 + c_t s + k_t}{s\left[s^2 J_l J_m + s c_t (J_l + J_m) + k_t (J_l + J_m)\right]}. \tag{6.92}$$

The transfer function between the velocities at the table ($\dot{x}_t = \dot{x}$) and motor shaft is given as follows:

$$G_t(s) = \frac{\dot{x}}{\dot{x}_m} = \frac{s r_g \theta_l / T_m}{s r_g \theta_m / T_m} = \frac{sG_{mt}}{sG_{mm}} = \frac{c_t s + k_t}{J_l s^2 + c_t s + k_t}. \tag{6.93}$$

The closed velocity loop response is expressed as follows:

$$\begin{aligned} G_v(s) &= \frac{\dot{x}_m}{\dot{x}_r} = \frac{\left(\frac{K_p s + K_I}{s}\right) K_a K_t G_m(s) r_g}{1 + \left(\frac{K_p s + K_I}{s}\right) K_a K_t G_m(s) r_g} \\ &= \frac{K_a K_t r_g \left(K_p s + K_I\right)\left(J_l s^2 + c_t s + k_t\right)}{s^2\left[s^2 J_l J_m + s c_t (J_l + J_m) + k_t (J_l + J_m)\right] + K_a K_t r_g \left(K_p s + K_I\right)\left(J_l s^2 + c_t s + k_t\right)}. \end{aligned} \tag{6.94}$$

Velocity controller gains (K_p, K_i) are tuned to keep the system stable with a desired transient response. Although the indirect velocity feedback damps the system to a certain degree, it is not sufficient to damp the mode at frequency ω_0. The active damping is accomplished by adding the acceleration (s/ω_0) measured from the table to the velocity loop. The loop transfer function of the direct velocity loop with the acceleration feedback is

$$G_{vdo}(s) = G_v(s)G_t(s)\frac{s}{\omega_0}$$

$$= \frac{K_aK_tr_g(K_ps+K_I)(J_ls^2+c_ts+k_t)}{s^2\left[s^2J_lJ_m+sc_t(J_l+J_m)+k_t(J_l+J_m)\right]+K_aK_tr_g(K_ps+K_I)(J_ls^2+c_ts+k_t)}\frac{(c_ts+k_t)}{J_ls^2+c_ts+k_t}\frac{s}{\omega_0} \tag{6.95}$$

$$= \frac{K_aK_tr_g(K_ps+K_I)(c_ts+k_t)}{s^2\left[s^2J_lJ_m+sc_t(J_l+J_m)+k_t(J_l+J_m)\right]+K_aK_tr_g(K_ps+K_I)(J_ls^2+c_ts+k_t)}\frac{s}{\omega_0}. \tag{6.96}$$

A Bode diagram of the direct velocity loop with and without the acceleration feedback can be used to assess the strength of active damping. The final closed-loop transfer function of the direct velocity loop becomes

$$G_{vc}(s) = \frac{\dot{x}}{\dot{x}_r} = \frac{G_v(s)G_t(s)}{1+G_v(s)G_t(s)\frac{s}{\omega_0}}$$

$$= \frac{K_aK_tr_g(K_ps+K_I)(c_ts+k_t)\omega_0}{\omega_0\left\{s^2\left[s^2J_lJ_m+sc_t(J_l+J_m)+k_t(J_l+J_m)\right]+K_aK_tr_g(K_ps+K_I)(J_ls^2+c_ts+k_t)\right\}+ ++K_aK_tr_g(K_ps+K_I)(c_ts+k_t)s}. \tag{6.97}$$

Finally, the Bode diagram of position loop indicates the achievable gain K_v without making the system unstable.

$$G_{po}(s) = \frac{x}{e} = K_vG_{vc}(s)\frac{1}{s}. \tag{6.98}$$

The closed-loop transfer function of the complete position loop becomes

$$G_{po}(s) = \frac{G_{po}(s)}{1+G_{po}(s)} = \frac{K_vG_{vc}(s)}{s+K_vG_{vc}(s)}. \tag{6.99}$$

The performance of the active damping system can be illustrated with the following experimental setup created in the author's laboratory.

Example. The ballscrew setup shown in Figure 6.9 is used to demonstrate the active damping algorithm. The parameters of the drive are given as follows:

J_t [kg m^2]	2.6274×10^{-4}	h_p[m]	0.020
J_m [kg m^2]	5.0853×10^{-4}	d_p[m]	0.020
k_t [Nm/rad]	418.15	$r_g = h_p/2\pi$	0.0032
c_t [Nm/rad/s]	0.0121	L[m]	0.82
K_a [A/V]	0.887	G [GPa]	75
K_t [Nm/A]	0.72	$I[m^4] = \pi(0.8d_p)^4/32$	1.5708×10^{-8}
K_p	700	k_l [Nm/rad]=GI/L	1437
K_i	500	k_b [Nm/rad]	6000
K_v	?	k_{nut} [Nm/rad]	1300
		k_{coup} [Nm/rad]	6500

By serially connecting the bearing (k_b), leadscrew (k_l), coupling (k_{coup}), and nut (k_{nut}) stiffness, the equivalent torsional stiffness of the drive can be estimated as

$$\tilde{k}_t = \left[\frac{1}{k_l} + \frac{1}{k_b} + \frac{1}{k_{\text{nut}}} + \frac{1}{k_{\text{coup}}}\right]^{-1} = 568 \text{ [Nm/rad]}$$

whereas the measured torsional stiffness was $k_t = 418.15$ [Nm/rad]. The natural frequency of the drive structure is

$$\omega_0 = \sqrt{\frac{k_t}{J_l J_m / (J_l + J_m)}} = 1,553 \text{ [rad/s]} = 247 \text{ [Hz]}$$

The evaluated transfer functions related to the structure are

$$G_m(s) = \frac{0.0002627s^2 + 0.0121s + 418.2}{s\left(1.336 \times 10^{-7}s^2 + 9.33 \times 10^{-6}s + 0.3225\right)}$$

$$= \frac{1966.3\,(s + 23.03 + i1261.5)\,(s + 23.03 - i1261.5)}{s\,(s + 34.918 + i1553.3)\,(s + 34.918 - i1553.3)}$$

$$G_t(s) = \frac{c_t s + k_t}{J_l s^2 + c_t s + k_t} = \frac{0.0121s + 418.15}{2.6274 \times 10^{-4}s^2 + 0.0121s + 418.15}$$

$$= \frac{46.0531(s + 34558)}{(s + 23.03 + i1261.5)\,(s + 23.03 - i1261.5)}$$

The motor side has a natural frequency of 1,553 [rad/s] with 2.25% damping, whereas the table has 1,261 [rad/s] natural frequency with 1.8% damping.

The closed-loop transfer function of the velocity loop with an indirect feedback from the motor shaft is found as follows:

$$G_v(s) = \frac{0.0003739s^3 + 0.01748s^2 + 595s + 425}{1.336 \times 10^{-7}s^4 + 0.0003832s^3 + 0.34s^3 + 595s + 425}$$

$$= \frac{2798.2521(s + 0.7143)(s + 23.018 + i1261.3)(s + 23.018 - i1261.3)}{(s + 2554)(s + 0.7146)(s + 156.55 + i1310.9)(s + 156.55 + i1310.9)},$$

which is dominated by a damped natural frequency of 1310.9 [rad/s] with about 12 percent damping. The indirect velocity loop increased the damping from 2.25 percent to 12 percent. By cascading the table dynamics and adding the damping feedback (s/ω_0), the loop transfer function of the velocity loop is found as follows:

$$G_{vdo}(s) = G_v(s)G_t(s)\frac{s}{\omega_0}$$

$$= \frac{82.9339s(s + 3.456 \times 10^4)}{(s + 2554)\,(s + 156.55 + i1310.9)\,(s + 156.55 - i1310.9)}.$$

The frequency responses of indirectly closed velocity loop cascaded with the table dynamics $G_v(s)G_t(s)$ and with the added damping term $G_{vdo} = G_v(s)G_t(s)s/\omega_0$ are shown in Figure 6.16a. The damping term attenuates the loop transfer function at the lower frequency but does not change the magnitude at the resonance, but the phase is shifted upward by 90 deg.

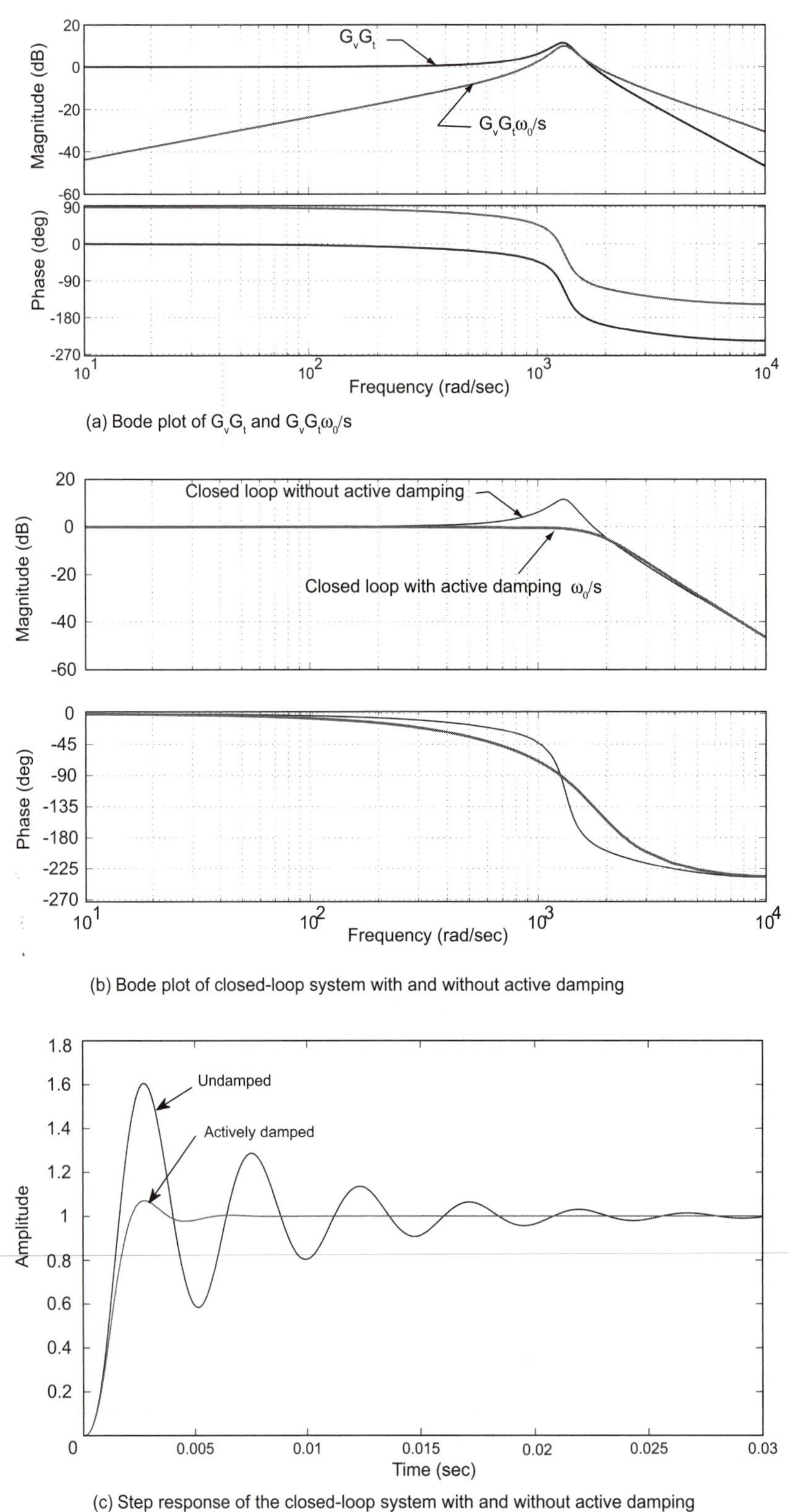

Figure 6.16: Active damping of velocity loop for a ballscrew drive system.

The closed-loop transfer function of the velocity loop with the acceleration feedback is evaluated as follows:

$$G_{vc}(s) = \frac{\dot{x}}{\dot{x}_r} = \frac{G_v(s)G_t(s)}{1 + G_v(s)G_t(s)\frac{s}{\omega_0}} = \frac{128851.252(s + 3.456 \times 10^4)}{(s + 1372)(s^2 + 1578s + 3.244 \times 10^6)}$$
$$= \frac{128851.252(s + 3.456 \times 10^4)}{(s + 1372)(s + 789 + i1619)(s + 789 - i1619)}, \quad (6.100)$$

which has a damped natural frequency of 1,619 [rad/s] and 43.8 percent damping ratio. The mode at 1,553 [rad/s] is now highly damped which can be seen from the frequency response function of the closed velocity loop with indirect velocity and direct acceleration feedback.

Frequency responses of the closed velocity loop with indirect velocity feedback ($G_v(s)G_t(s)$) and additional acceleration feedback (G_{vc}) are shown in Figure 6.16b. The magnitude at the resonance frequency zone is attenuated by 11.5 dB, and the mode is completely damped. The step response of the undamped and damped velocity loop clearly indicates the effect of damping as shown in Fig. 6.16c. The bandwidth of the velocity loop is therefore increased up to this frequency, which allows the designer to increase the position loop gain K_v to much higher values. High position gain means higher bandwidth with lower tracking and contouring errors in circular interpolation and contour machining. Furthermore, because the loop gain becomes higher, the disturbance stiffness of the controller against cutting and friction loads becomes higher.

6.7 DESIGN OF AN ELECTROHYDRAULIC CNC PRESS BRAKE

Sheet metal–forming machine tools, *press brakes*, are widely used in fabrication shops. A typical press brake consists of a moving ram, which holds a punch, and a die located on a bed frame (see Fig. 6.17). The motion of the ram is delivered by a pair of hydraulic actuators. Conventional presses are manually operated, and the end positions of the ram are manually set via limit switches or dead stops. Manual control leads to frequent tuning and adjustment of the machine and requires a number of trial bends until a satisfactory positioning and bending accuracy are achieved. In contrast, the CNC press provides flexibility in adapting the machine to different bending operations quickly, while delivering an accurate positioning. A simplified design of a computer control system for a press brake is presented in this section.

6.7.1 Hydraulic Press Brake System

Mechanical System

The press ram, which holds the punch at its free end, is held by two parallel hydraulic actuators (Fig. 6.17). The ram slides on lubricated guides, which

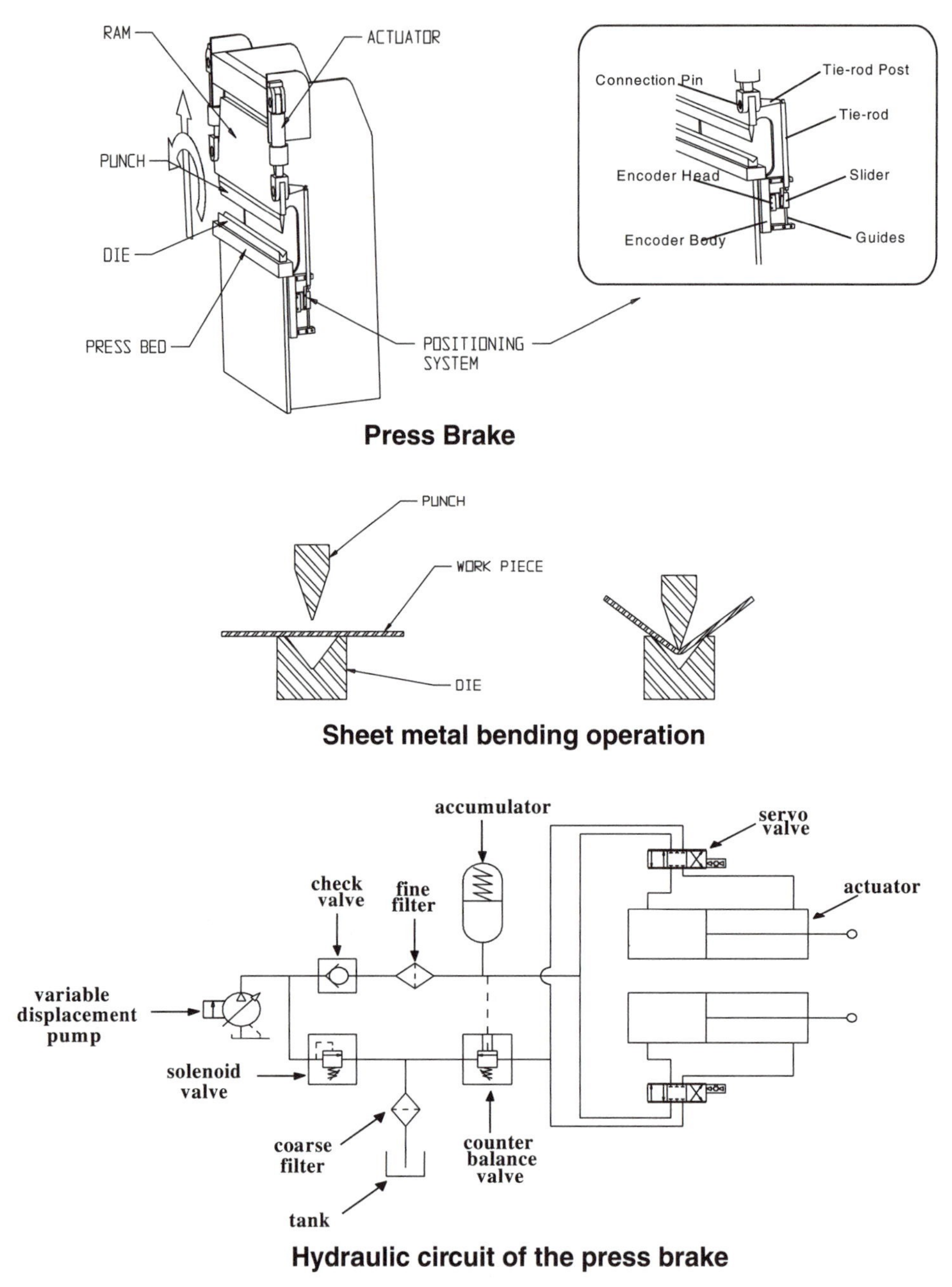

Figure 6.17: Hydraulic press brake system.

are modified to have cylindrical pads. The pads allow 3° tilt to the ram–punch assembly for press bending of sheet metals with a variable penetration depth at either side. The aim of the CNC system is to bring the punch rapidly toward the sheet metal laid on the V-shaped die and to form it at high pressure but with a slow penetration velocity to a desired depth. Depending on the offset between the two ends of the bend, the positions and velocities of the two hydraulic actuators that are holding the punch must be controlled by the CNC system.

The actual position of each side of the ram is monitored by linear encoders. To consider the relative static deformations between the press bed and ram, a sliding tie rod is designed as shown in Figure 6.17. The tie-rod posts are placed at the axes of the ram/actuator connections. High-precision ball joints are used to connect each tie rod to its respective post and slider. A linear encoder with 0.005 mm resolution was selected. The stationary encoder body was mounted on the press bed close to the die, and the moving read head was attached to the linear slider housed in a cylindrical guide with an adjustable stiffness. In place of the encoder, precision linear displacement transducers (LDTs), which use a tubular waveguide and a magnet, can be used as an integral part of the actuator rods.

Hydraulic System

The schematic diagram of the hydraulic system design is shown in Figure 6.17. A variable displacement pump with pressure feedback was chosen to deliver the hydraulic power. An accumulator, which keeps the system pressure within 10 percent of a set level and damps the pressure oscillations caused by the reciprocating pump pistons, was used. A check valve was placed between the accumulator and the pump to eliminate the possibility of the accumulator pressure driving the pump in reverse in the event of power outage. A counterbalance valve was used to lock the system whenever the system pressure was lost. A fine pressure filter was used in the supply line to keep any debris from reaching the sensitive pilot stage of the servovalves. A coarse low-pressure filter was added to the return line. The existing hydraulic actuators were modified to include servovalves and pressure transducers as shown in Figure 6.18, which would have been an integral part of the actuators on a production machine. The flow into and out of the actuator cylinder is controlled by a two-stage servovalve. It is rated as having a 20 gallons/min flow capacity with 85 Hz bandwidth by the manufacturer. The primary stage consists of an electronic torque motor driving the primary flapper valve, which provides a differential pressure across the ends of the secondary closed center spool valve with a zero overlap. The displacement of the secondary spool is fed back to the torque motor by way of a cantilever spring. This spool controls the flow into the cylinder. The hydraulic cylinder has two fluid chambers separated by a piston, which is connected to the ram from one side with a rigid rod. The inertia loads of the rod and ram assembly, Coulomb, and viscous friction loads in the ram guideways and piston cylinder interface, and the bending loads must be overcome by the pressurized fluid injected into the cylinder chamber.

CNC System

The hydraulic press was controlled by the in-house developed Open Architecture CNC System [11, 14].

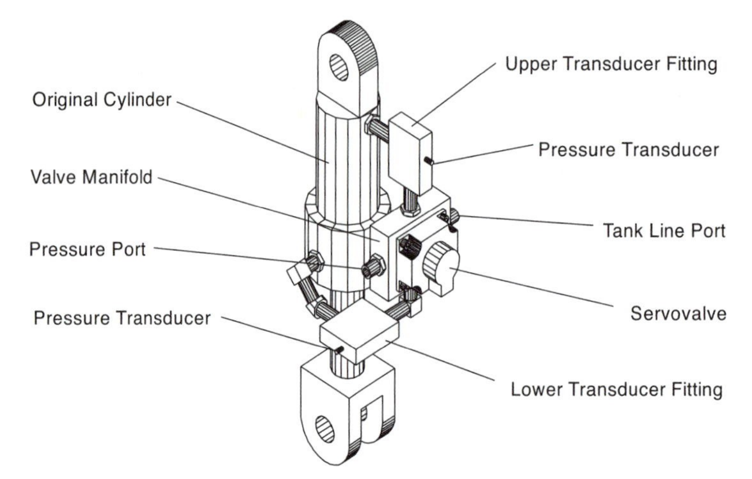

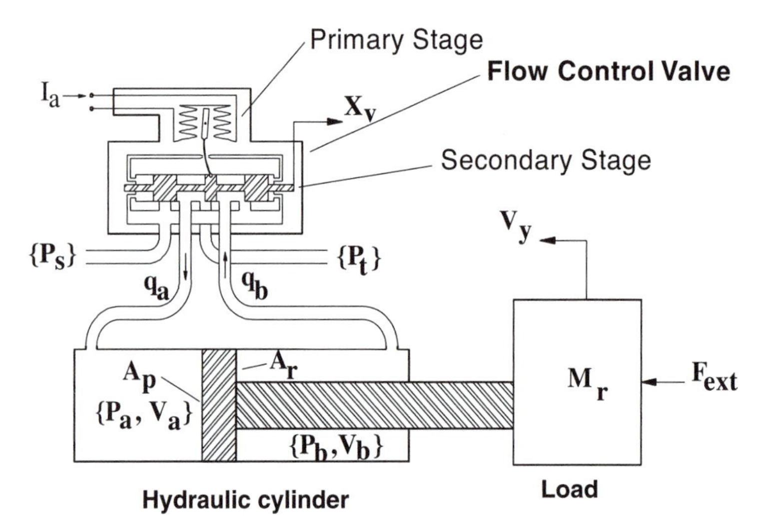

Figure 6.18: Electrohydraulic actuator assembly.

6.7.2 Dynamic Model of Hydraulic Actuator Module

An accurate dynamic model of the overall hydraulic system is rather difficult to obtain, because of the leakage in seals and connections, losses within flexible hoses, the compressibility of the fluid and changes in its viscosity, and the wear in the orifices and ports. In addition the flow-pressure expressions are nonlinear for a practical control algorithm design in a CNC system. However, even with some practical assumptions, a mathematical model of the hydraulic system is essential for the optimal design of the hydraulic circuit and for selecting components. Such an analysis has been made in designing the hydraulic circuit for the press [14]. Only a simplified model of the hydraulic system is presented here to provide a basic foundation for a digital press brake controller design suitable for a CNC.

The model is based on the valve–actuator assembly shown in Figure 6.18. The second-order dynamics of the spool is much faster than the electronic torque motor, which has first-order dynamics. The transfer function between the spool valve displacement and supplied current to the torque motor is approximated as

$$\frac{x_v(s)}{I_a(s)} = \frac{K_i}{\tau_i s + 1}, \tag{6.101}$$

where K_i and τ_i are the gain and time constant of the torque motor, respectively. Assume that the secondary spool is at null flow position, where both ports connected to either sides of the cylinder are closed by the zero-overlap spool. The flow across the valve port is proportional to the area of opening and to the square root of the pressure drop. Because the area of valve opening is proportional to the spool displacement, the following expressions can be used to represent the flow through the ports for the extending and retracting motion of the piston [78] as follows:

$$\left.\begin{aligned} q_a &= K_q x_v \sqrt{|P_s - P_a|}\ \mathrm{sign}(P_s - P_a), \\ q_b &= K_q x_v \sqrt{|P_b - P_t|}\ \mathrm{sign}(P_b - P_t), \end{aligned}\right\} \quad x_v \le 0,$$
$$\left.\begin{aligned} q_a &= K_q x_v \sqrt{|P_t - P_a|}\ \mathrm{sign}(P_t - P_a), \\ q_b &= K_q x_v \sqrt{|P_b - P_s|}\ \mathrm{sign}(P_b - P_s), \end{aligned}\right\} \quad x_v > 0, \tag{6.102}$$

where P_s, P_t, P_a, and P_b are the supply, tank, and port (A,B) pressures, respectively.

The flow coefficients for ports A and B are assumed to be equal to K_q. The flows into the cylinder move the piston according to the conservation of mass on each chamber as follows:

$$\begin{aligned} q_a &= -A_p v_y + \frac{V_a}{\beta}\frac{dP_a}{dt} + K_{lp}(P_a - P_b), \\ q_b &= -A_r v_y - \frac{V_b}{\beta}\frac{dP_b}{dt} + K_{lp}(P_a - P_b), \end{aligned} \tag{6.103}$$

where β and K_{lp} are the bulk modulus of the fluid and the coefficient of leakage past the piston, respectively. In the particular single-rod piston system, the piston areas (A_p, A_r) on each side of the chamber are different. The term with a differential pressure and fluid bulk modulus represents the change in the chamber volume due to the compressibility of the fluid.

The fundamental flow equations for the servovalve and actuator assembly given above (Eqs. 6.102–6.103) have several nonlinearities besides the square root term. The expressions are linearized here for simplicity in deriving the transfer function of the overall hydraulic actuator system. The following assumptions are made: Both sides of the piston area are equal (i.e., symmetric actuator with $A_p = A_r = A$); the tank is open to the atmosphere and its relative pressure is zero, $P_t = 0$; the flows into and out of the spool valve or cylinder chambers are equal. Defining the load flow as $q_l = q_a = q_b$, and the load pressure as $P_l = P_a - P_b$, we can obtain the following relationships from Eq. (6.102):

$$K_q x_v \sqrt{|P_s - P_a|}\,\text{sign}(P_s - P_a) = K_q x_v \sqrt{|P_b|}\,\text{sign}(P_b),$$

which leads to

$$P_s = P_a + P_b, \quad P_l = P_a - P_b, \quad \rightarrow \quad P_b = \frac{P_s - P_l}{2}.$$

Substituting $P_t = 0$ and P_b into q_b (Eq. 6.102) yields

$$q_l = (K_q\sqrt{P_s/2})x_v\sqrt{1 - P_l/P_s}. \tag{6.104}$$

The load flow is linearized by assuming small changes in the valve displacement (x_v) and load pressure (P_l). Linearizing q_l about the null flow position of the spool, we get

$$\begin{aligned} q_l &= \left.\frac{\partial q_l}{\partial x_v}\right|_{P_{lo},x_{v0}} x_v + \left.\frac{\partial q_l}{\partial P_l}\right|_{P_{lo},x_{vo}} P_l \\ &= K_{qo}x_v + K_{xo}P_l, \end{aligned} \tag{6.105}$$

where the normalized valve flow (K_{qo}) and pressure gains (K_{xo}) are

$$K_{qo} = K_q\sqrt{\frac{P_s - P_{lo}}{2}}, \quad K_{xo} = \frac{-K_q x_{v0}}{2\sqrt{2(P_s - P_{lo})}}.$$

The flow equations in the cylinder (Eq. 6.103) can also be linearized. If we let the total volume in the cylinder be V_t and the volume displaced in the chambers be ΔV, the volumes in the chambers become $V_a = V_t/2 + \Delta V$ and $V_b = V_t/2 - \Delta V$. If we add the flows in each chamber (i.e., $q_a + q_b = 2q_l$), and assume a constant supply pressure P_s and equal piston areas ($A_p = A_r = A$), and consider that

$$\frac{dP_l}{dt} = \frac{dP_a}{dt} - \frac{dP_b}{dt}, \quad \frac{dP_s}{dt} = \frac{dP_a}{dt} + \frac{dP_b}{dt} = 0,$$

the load flow from Eq. (6.103) becomes

$$q_l = -Av_y + \frac{V_t}{4\beta}\frac{dP_l}{dt} + K_{lp}P_l. \tag{6.106}$$

The equation of motion for the equivalent mass (M_r) at the piston rod with negligible viscous friction is

$$\sum F_{\leftarrow}^{+} = M_r\frac{dv_y}{dt} = A(P_b - P_a) + F_l, \tag{6.107}$$

where the load $F_L = F_{ext} - F_c\,\text{sign}(v_y)$. Here F_{ext} and F_c are the sheet metal bending and Coulomb friction forces, respectively. The transfer function of piston velocity can be derived by combining the Laplace transforms of Eqs. (6.105),

(6.106), and (6.107) as follows:

$$v_y(s) = \frac{K_x}{s^2 + 2\zeta_l \omega_l s + \omega_l^2} x_v + \frac{K_F(\tau_F s + 1)}{s^2 + 2\zeta_l \omega_l s + \omega_l^2} F_l, \tag{6.108}$$

where

$$\omega_l = 2A\sqrt{\frac{\beta}{V_t M_r}}, \quad \zeta_l = \frac{K_{lp} - K_{xo}}{A}\sqrt{\frac{\beta M_r}{V_t}},$$

$$K_x = \frac{-4\beta A K_{qo}}{V_t M_r}, \quad \tau_F = \frac{V_t}{4\beta(K_{lp} - K_{xo})}, \quad K_F = \frac{1}{\tau_F M_r}.$$

The open-loop transfer function indicates the influence of the external bending load and Coulomb friction on the ram velocity; the damping coefficient is proportional to the valve and piston leakage; and the stiffness of the actuator is proportional to the bulk modulus of the fluid and piston area whereas it is inversely proportional to the stroke. The linearized transfer function is useful for analyzing the general behavior of the actuator for a given set of initial design parameters. However, the nonlinearities and the unknown values of fluid viscosity, friction, and leakage in various seals and ports require the on-line identification of the overall system transfer function, so that a well-tuned digital control algorithm can be embedded into the CNC computer.

6.7.3 Identification of Electrohydraulic Drive Dynamics for Computer Control

Although mathematical modeling is a useful tool for selecting hydraulic components, and determining the general order of the assembled system, the effects of valve deadband, hysteresis, stiction, leakage, and transport delay make the accurate modeling of the dynamics of hydraulic system for computer control difficult. A series of system identification tests are usually conducted to evaluate the discrete transfer function of the system for computer control.

Several characteristics of the hydraulic system must be considered in designing the identification experiments. Although the gain of the closed-center spool valve is largest at the null-flow position, the deadband and hysteresis are also dominant. In addition, the amplitude of the input signal must be large enough so that the steady-state flow gain identified is not biased by the valve-spool stiction. Because a significant Coulomb friction exists between the ram and guideways, the input excitation signal chosen must not reverse the motion and thus the static friction load. Furthermore, because we are using a single-rod asymmetrical actuator, the dynamics of the system in extending and retracting will not be the same. Finally, overall flow-pressure relationships are nonlinear owing to square root terms, and their linearization is acceptable only at small regions of actuator movements. Identification tests areconducted by considering the above-mentioned characteristics of the hydraulic system. First, a series of step response tests are conducted in extending and retracting pistons separately. An approximate order and time delay of the system was estimated, which was later enhanced by pseudo random binary sequence (PRBS) excitation and least-squares identification techniques. In both excitations, the minimum input signal amplitude was at least 3 percent of the maximum amplitude required to overcome the effects of nonlinear valve gain and spool friction. A dc offset is added to the input signal to prevent any reversal of piston velocity (and Coulomb friction load), which constrains the identification to either extending or retracting.

The input signals, and the corresponding actuator position, velocity, and pressure outputs, are collected synchronously with a multichannel digital scope. The digital data are later transferred to a computer and processed by MATLAB's Systems Identification Tool Box [1]. The digital position feedback, the encoder feedback to the CNC, is disabled, and input signals are sent to servovalve amplifiers as analog voltages via D/A converters, which have a gain of $K_{\mathrm{d}} = 20/2^9$ [V/count/s]. The velocities of both actuators are measured by magnetic field–based velocity transducers instrumented to the tie-rod slider mechanism. The data-sampling frequency was fixed at 1 kHz during identification tests.

The open-loop discrete transfer function of the system was identified from the sampled values of PRBS inputs to the servovalve amplifier and outputs of the velocity transducer at 1-ms time intervals, which are equivalent to loop-closure periods. The system's open-loop discrete transfer function can be represented in the form of

$$G_{\mathrm{o}}(z^{-1}) = \frac{v_y(z^{-1})}{u(z^{-1})} = \frac{z^{-d}B(z^{-1})}{A(z^{-1})}, \tag{6.109}$$

where d is the system's deadtime and

$$B(z^{-1}) = b_0 + b_1 z^{-1} + b_1 z^{-2} + \cdots + b_{n_{\mathrm{b}}} z^{-n_{\mathrm{b}}},$$
$$A(z^{-1}) = 1 + a_1 z^{-1} + a_2 z^{-2} + \cdots + a_{n_{\mathrm{a}}} z^{-n_{\mathrm{a}}}.$$

Alternatively, the system's response – the piston velocity (v_y) at time interval k – can be expressed in an auto regressive moving average (ARMA) form as follows:

$$v_y(k) = \{\phi(k)\}^T \{\theta\}, \tag{6.110}$$

where the measurement vector is

$$\{\phi(k)\}^T = [u(k-d)\; u(k-d-1)\; \cdots\; u(k-d-n_{\mathrm{b}}), -v_y(k-1)\; \cdots\; -v_y(k-n_{\mathrm{a}})] \tag{6.111}$$

and the transfer function parameter vector is

$$\{\theta\} = [b_0\, b_1 \cdots b_{n_\mathrm{b}} \;\; a_1\, a_2 \cdots a_{n_\mathrm{a}}]^T. \tag{6.112}$$

The values of the transfer function, or the components of the time-invariant parameter vector $\{\theta\}$, are estimated from N sets of measurements using an off-line standard least-squares (LS) method, which is summarized in Appendix B. The MATLAB Identification Tool Box [1] has the LS identification routines used here. The input signal amplitude was chosen to be as 600 mV, or 3% of the maximum valve input to overcome valve stiction. In addition, this value is well within the range of valve input signals during the forming operation.

It was found that a first-order model represents the system dynamics at a low-frequency band, while a third-order model captures the dynamic mode contributed by the load and actuator at frequencies more than 100 Hz. The delays in the system vary between 4 and 5 (ms) for each model. The left and right actuators have different dynamic characteristics, which reflect the machining and friction characteristics of each actuator–guideway assembly. In addition, each actuator exhibits different dynamics in extending and retracting motions owing to its asymmetrical piston. Although high-order models represent the system dynamics over a wide frequency range, a first-order model with five time delays is satisfactory for the extending motion of the actuators over the low-frequency band where the forming operation takes place. The identified open-loop transfer function between the actuator velocity (v_y) and valve amplifier input (u) is expressed as

$$\frac{v_y}{u} = \frac{z^{-5}(b_0 + b_1 z^{-1})}{1 + a_1 z^{-1}}. \tag{6.113}$$

The excitation of high-order modes is avoided by applying smooth velocity changes using the trapezoidal velocity profile as shown in Figure 6.19.

The reader is reminded of the following for the open-loop transfer function parameters of the press shown in Table 6.3:

- $a_1 = a_1' + 1$, since a_1' belongs to position (see next section).
- The values b_0 and b_1 should be multiplied by the D/A converter gain of the CNC to obtain the correct transfer function units between position input and output in *counts*. A nine-bit two-parallel pulse width modulated circuit with ± 10 V is used as an analog to digital converter. The gain of the digital to analog converter circuit is $K_\mathrm{d} = 20/2^9 = 0.0391$ [V/(count/s)].

6.7.4 Digital Position Control System Design

We need to design a digital control algorithm with a minimum computation time and high control loop closure frequency while providing a low positioning error. The performance of the positioning system during the bend and dwell operations determines the accuracy of the press. In these cases, the actuators are either extending or holding their position against a forming load.

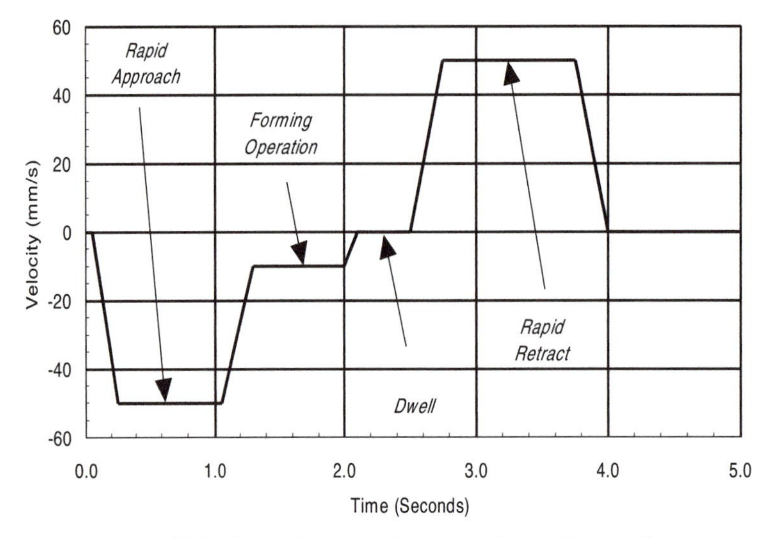

Velocity profile for a sheet metal bending cycle.

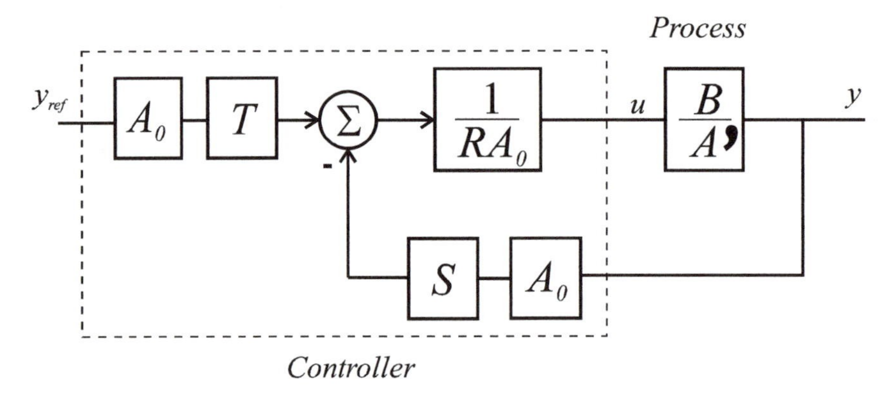

Block diagram of a pole-placement control system.

Figure 6.19: Press brake control system.

Therefore, the first-order models that represent the velocity dynamics during actuator extensions are used for the digital control of the press ram position. The position of the ram is measured with a linear encoder attached to the ram–tie-rod assembly. The linear encoder and position counter in the CNC act as an integrator ($1/(1-z^{-1})$) and provide the actual position of the actuator end point at discrete control intervals $T_s = 1$ ms. The resulting open-loop transfer function between the amplifier input (u [V]) and actuator position (y) becomes

$$\frac{B(z^{-1})}{A'(z^{-1})} = \frac{y}{u} = \frac{v_y}{u}\left(\frac{1}{1-z^{-1}}\right) = \frac{z^{-5}(b_0 + b_1 z^{-1})}{1 + a'_1 z^{-1} + a'_2 z^{-2}}, \tag{6.114}$$

TABLE 6.3. Pole-Placement Axis Control Law Parameters for the Hydraulic Press

$[A_m] = 1 + a_{m1}z^{-1} + a_{m2}z^{-2}$	$1 - 1.429z^{-1} + 0.4724z^{-2}$ **Desired Model**
Left Axis – y_1	**Pole-Placement Design Parameters**
$[B] = b_0 + b_1z^{-1}$	$(1{,}573 - 702\,z^{-1})$ [count/V] $\cdot k_d$
$[A] = 1 + a'_1z^{-1} + a'_2z^{-2}$	$1 - 1.871z^{-1} + 0.871z^{-2}$
$[T] = t_0$	4.8837×10^{-4}
$[S] = s_0 + s_1z^{-1}$	$0.00444 - 0.00395z^{-1}$
$[R] = r_0 + r_1z^{-1} + \cdots + r_5z^{-5}$	[1 0.442 0.4284 0.4165 0.4062 0.1244]
Right Axis – y_2	**Pole-Placement Design Parameters**
$[B] = b_0 + b_1z^{-1}$	$(982 - 505z^{-1})$ [count/V] $\cdot k_d$
$[A] = 1 + a'_1z^{-1} + a'_2z^{-2}$	$1 - 1.861z^{-1} + 0.861z^{-2}$
$[T] = t_0$	7.4717×10^{-4}
$[S] = s_0 + s_1z^{-1}$	$0.00645 - 0.0057z^{-1}$
$[R] = r_0 + r_1z^{-1} + \cdots + r_5z^{-5}$	[1 0.432 0.4154 0.401 0.3887 0.1306]

Note that the D/A converter gain is $k_d = 20/2^9$ [V/counts/s].

where $a'_1 = a_1 - 1$ and $a'_2 = -a_1$. Here, (u) is in volts and y unit is in counts, where 1 count = 0.005 mm. The open-position loop transfer function can be expressed in an alternate form using a forward shift operator (z) for convenience as follows:

$$\frac{y(k)}{u(k)} = \frac{B(z^{-1})}{A'(z^{-1})} = \frac{b_0z + b_1}{z^4(z^2 + a'_1z + a'_2)} \quad \left[\frac{\text{counts}}{V}\right]. \tag{6.115}$$

The above transfer function has a first-order numerator (i.e., $\deg(B(z)) = 1$) and a sixth-order denominator (i.e., $\deg(A(z)) = 4 + 2 = 6$). A pole-placement control scheme presented by Astrom and Wittenmark [29] allows performance-based design, where the performance criteria are 1) to have a high gain to reject bending and friction load disturbances, 2) to compensate delays in the open-loop transfer function, and 3) to have low steady-state positioning errors. The block diagram of the pole-placement controller is shown in Figure 6.19, where the input is the command and the output is the actual position of the actuator, respectively. B/A is the open-loop transfer function of the system given in Eq. (6.113). The digital controller consists of a feed forward filter $T(z)$, a feedback filter $S(z)$, controller poles $R(z)$, and an observer $A_0(z)$. The transfer function of the closed-loop control system can be derived from the block diagram (Fig. 6.19) as follows:

$$\frac{y(k)}{y_{\text{ref}}(k)} = \frac{(BT)/(A_0A'R)}{1 + (BA_0S)/(RA_0A')} = \frac{BT}{A'R + BS}. \tag{6.116}$$

The control polynomials R, S, T, and A_0 are designed in such a way that the closed-loop response of the overall system behaves like a desired model transfer function as follows:

$$\frac{y(k)}{y_{\text{ref}}(k)} = \frac{BT}{A'R + BS} = \frac{A_0 B_m}{A_0 A_m}, \tag{6.117}$$

where polynomials A_m and B_m contain the desired poles and zeros of the system. The desired system response was chosen to have second-order dynamics, with a settling time of $t_s \approx 4/(\xi\omega_n)$ for underdamped systems and $t_s \approx -\ln 0.05/(\omega_n(\xi - \sqrt{\xi^2 - 1}))$ for overdamped systems. The desired characteristic equation is selected as

$$A_{m0} = s^2 + 2\xi\omega_n s + \omega_n^2 = s^2 + 750s + 62{,}500,$$

with a natural frequency of 250 rad/s. The roots of the overdamped ($\xi > 1$) characteristic equation are

$$s_{i,ii} = \omega_n(-\xi \pm \sqrt{\xi^2 - 1}) = (-95.4915, -654.5085).$$

The corresponding discrete time domain roots for a $T = 1$ ms sampling interval become

$$z_i = e^{s_i T} = 0.9089, \quad z_{ii} = e^{s_{ii} T} = 0.5197,$$

with the desired discrete transfer function of

$$A_{m0} = (z - z_i)(z - z_{ii}) = z^2 - 1.429z + 0.4724 = z^2 + a_{m1}z + a_{m2}.$$

The resulting desired model transfer function is selected to have the same numerator (i.e., zeros) and the same order of the denominator of the open-loop process dynamics as follows:

$$\frac{B_m}{A_m} = \frac{y(k)}{y_{\text{ref}}(k)} = \frac{b_{m0}(b_0 z + b_1)}{z^4(z^2 + a_{m1}z + a_{m2})} \left[\frac{\text{count}}{\text{count}}\right], \tag{6.118}$$

where y_{ref} is the commanded reference position and b_{m0} is a scale factor to ensure a unity gain in the overall closed-loop system as follows:

$$\left|\frac{B_m(z)}{A_m(z)}\right|_{z=1} = 1, \quad \rightarrow \quad b_{m0} = \frac{1 + a_{m1} + a_{m2}}{b_0 + b_1}.$$

Based on the design criterion of causality rules (i.e., the present input must not depend on the future outputs of the system) proven by Astrom and Wittenmark [29], the control polynomials must have the following orders:

$$\begin{aligned}
&\deg(A_0(z)) \geq 2\ \deg(A'(z)) - \deg(A_m(z)) - 1 = 5, \quad \text{let } \deg(A_0(z)) = 5,\\
&\deg(R(z)) = \deg(A_0(z)) + \deg(A_m(z)) - \deg(A'(z)) = 5 + 6 - 5 = 5,\\
&\deg(S(z)) \leq \deg(R(z)), \quad \text{let } \deg(S(z)) = 5,\\
&\deg(T(z)) = \deg(A_0(z)) = 5.
\end{aligned}$$

The corresponding control polynomials are as follows:

$$\begin{aligned}
A_0 &= z^5,\\
R(z) &= z^5 + r_1 z^4 + r_2 z^3 + r_3 z^2 + r_4 z + r_5,\\
S(z) &= s_0 z^5 + s_1 z^4 + s_2 z^3 + s_3 z^2 + s_4 z + s_5.
\end{aligned} \tag{6.119}$$

By considering the equivalence of the actual closed-loop and desired transfer function models (Eq. 6.117), we can identify the parameters of the control polynomials from the solution of the following Diophantine equations [29]:

$$\begin{aligned} &B(z)T(z) \equiv A_0(z)B_m(z), \\ &A'(z)R(z) + B(z)S(z) \equiv A_0(z)A_m(z). \end{aligned} \tag{6.120}$$

By considering the numerator terms, we have

$$(b_0 z + b_1)T(z) \equiv z^5 b_{m0}(b_0 z + b_1) \rightarrow T(z) = b_{m0} z^5. \tag{6.121}$$

When $A'(z)R(z) + B(z)S(z) \equiv A_0(z)A_m(z)$ is considered, we get

$$\begin{aligned} &z^4(z^2 + a_1' z + a_2)(z^5 + r_1 z^4 + r_2 z^3 + r_3 z^2 + r_4 z + r_5) \\ &\quad + (b_0 z + b_1)(s_0 z^5 + s_1 z^4 + s_2 z^3 + s_3 z^2 + s_4 z + s_5) \\ &\quad \equiv z^5 z^4 (z^2 + a_{m1} z + a_{m2}). \end{aligned}$$

The resulting expressions for the parameters are

$$\begin{aligned} &r_1 = a_{m1} - a_1', \quad r_2 = a_{m2} - a_2' - a_1' r_1, \\ &2 < j < (d = 5) \rightarrow r_j = -(a_1' r_{j-1} + a_2' r_{j-2}), \quad j = 3, 4, \\ &r_{\mathrm{d}} = r_5 = \left[\left(a_1' \frac{b_1}{b_0} - a_2'\right) r_{\mathrm{d}-1} + a_2' \frac{b_1}{b_0} r_{\mathrm{d}-2}\right] \Big/ \left[a_1' - a_2' \frac{b_0}{b_1} - \frac{b_1}{b_0}\right], \\ &s_0 = -\frac{r_{\mathrm{d}} + a_1' r_{\mathrm{d}-1} + a_2' r_{\mathrm{d}-2}}{b_0}, \quad s_1 = -\frac{a_2' r_{\mathrm{d}}}{b_1}, \quad s_j = 0, \quad j = 2, \ldots, d, \end{aligned}$$

where delay is $d = 5$. Because the parameters of the open-loop transfer function are assumed to be time invariant, the Diophantine equations are solved once during the design. The computed numerical values used for the right- and left-actuator position control are presented in Table 6.3.

From the block diagram of the control system, the position command input is generated according to the following expression:

$$R(z)u(k) = T(z)y_{\mathrm{ref}}(k) - S(z)y(k). \tag{6.122}$$

Substituting the control polynomial parameters (R, T, S) leads to the following control law being executed at each loop closure time T_{s}:

$$u(k) = t_0 y_{\mathrm{ref}}(k) - \sum_{j=0}^{1} s_j y(k-j) - \sum_{j=1}^{5} r_j u(k-j). \tag{6.123}$$

A series of position jumps are commanded to both actuators to test the step response behavior of the controlled press (see Fig. 6.20). It is observed that both actuators have approximately matched dynamic response, which is essential in coordinated precision bending, and the system has a total rise time of about 12 ms. Although the same control law was used in both extending (negative direction) and retracting motions, the actuators do not exhibit overshoots, which is a basic requirement for a machine tool control system. To test the

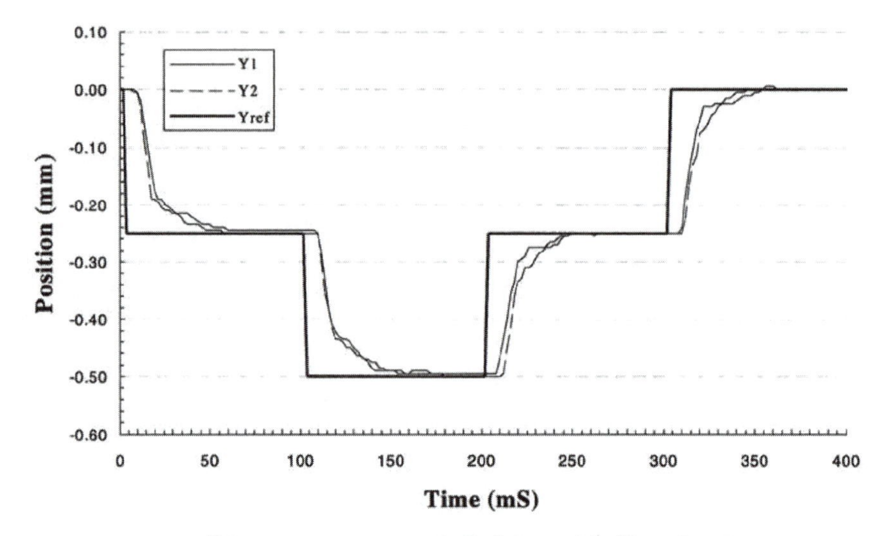

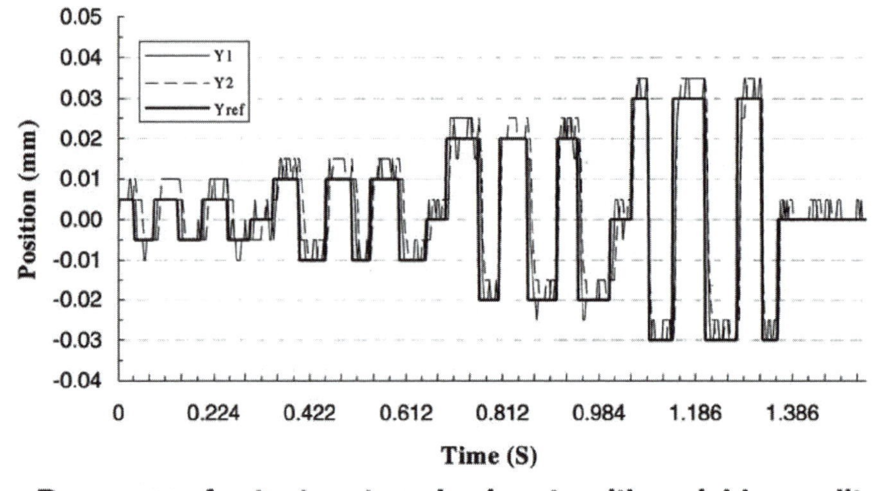

Figure 6.20: The response of the right- and left-axis positioning systems to step changes in the position command.

influence of deadband and hysteresis, which are caused by the valve-spool stiction and Coulomb friction in the guides, a series of pulse position commands with an increasing amplitude were demanded from the CNC system. The actuator positions were monitored at 2 ms intervals and are shown in Figure 6.20. The results indicate that the deadband of the controlled motion is within 1 encoder unit (0.005 mm), which is quite satisfactory.

The computer control strategy presented here is applicable to many other motion or process control applications found in practice.

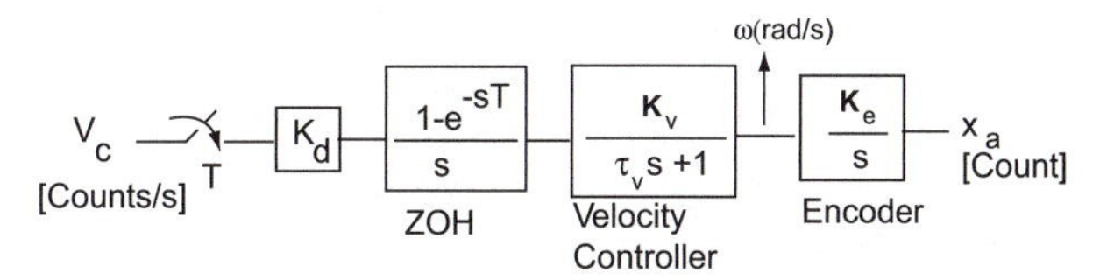

Figure 6.21: Open-loop transfer function of axis drive system.

6.8 PROBLEMS

1. An open-loop block diagram of a feed drive system is given in Figure 6.21. The given feed drive system is to be controlled with a proportional+derivative (PD) type controller that has a transfer function of $\frac{u(s)}{e(s)} = K_p + K_d s$, where the position-tracking error in [count] is $e(s) = x_r(s) - x(s)$, and $u(s)$ is the control signal [V].($K_g = 8.956$ Nm/V, $K_m = 50$ rad/s/Nm, $\tau_m = 1.565$ s, $K_e =$ 3,183 counts/rad.)
   a. Draw the closed-loop system's block diagram, considering the commanded position $x_r'(s)$ [count] and the disturbance torque $T_d(s)$[Nm] as the inputs, and the measured table position $x(s)$[count] as the output.
   b. Obtain the closed-loop expression for the measured table position as a function of the commanded position and the disturbance torque in the form $x(s) = G_x(s)x_r(s) + G_d(s)T_d(s)$.
   c. Design the PD controller gains (i.e., K_p and K_d) so that the closed-loop response to a step position command has a rise time of $t_r = 0.025[s]$ and a maximum overshoot of $M_p = 1$ [percent]. Design the controller in the Laplace domain.
   d. Assuming that the position control loop is to be closed digitally at a sampling period of $T = 0.001$[s], obtain the control law (i.e., difference equation) that needs to be implemented in the control computer by transforming the continuous-time PD into the z domain using Euler (backward difference) approximation ($s \cong \frac{z-1}{Tz}$).
   e. Sketch the frequency response graph (only magnitude at 3 points, $\omega = 0, \omega_n, 10\omega_n$) of the closed-loop disturbance transfer function (i.e., $G_d(s) = x(s)/T_d(s)$).
   f. Compute the steady-state tracking error for a machining operation where a feed motion of $f = 10^5$ [counts/s] is commanded against a constant cutting torque of $T_d = 10$ Nm in the opposing direction.
2. CNC Design and Analysis Project: A simplified block diagram of a CNC milling machine's feed drive system is shown in Figure 6.22. The parameters of the system are given below:

   $H_g = 0.0913$ V/(rad/s); tachogenerator constant.
   $K_d = 0.0781$ V/counts; D/A converter gain.
   $K_e = 636.6$ counts/rad; encoder gain.
   $K_p =?$; digital filter gain.
   $a =?$; digital filter's zero.
   $b =?$; digital filter's pole.
   $T = 0.001$ s ; sampling period.

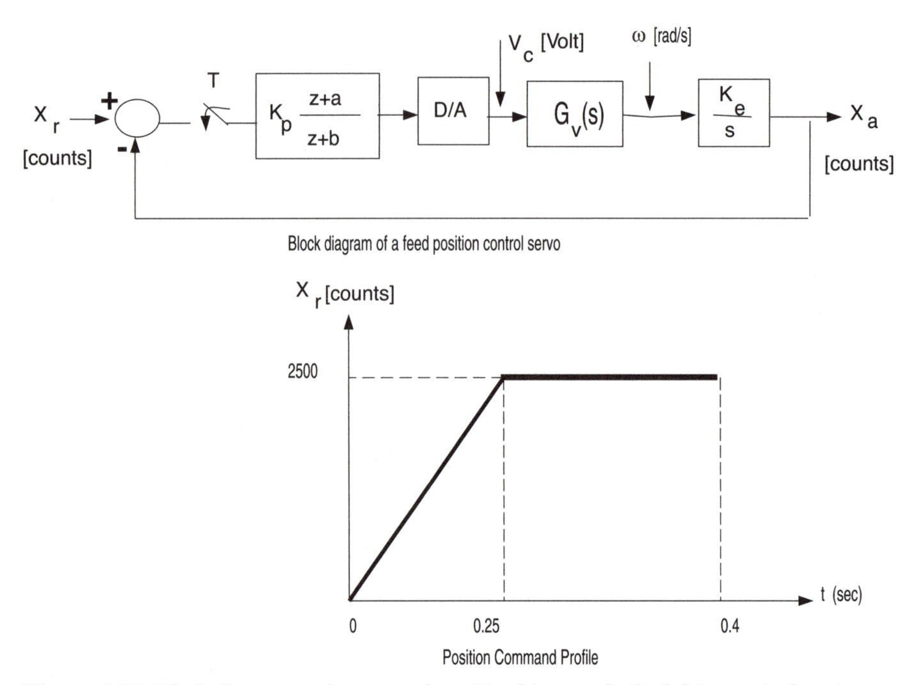

Figure 6.22: Block diagram and command position history of a feed drive control system.

a. The transfer function of the velocity loop ($G_\mathrm{v}(s)$) is not known. Measure the transient response of the velocity loop by applying a rectangular waveform with $V_\mathrm{c} = 1\,\mathrm{V}$ amplitude to the servoamplifier. The actual velocity of the feed drive is measured from the tachogenerator output, which has a gain of H_g [V/(rad/s)]. Identify the transfer function of the velocity loop ($G_\mathrm{v}(s) = \omega(s)/V_\mathrm{c}(s) = ?$) by assuming that it can be approximated as a first-order system. Simulate the step response of the velocity loop in the continuous time domain using Laplace transforms, and compare the response with the experimental measurements by plotting them on the same graph. (Note: If the experimental setup is not available, assume $G_\mathrm{v}(s) = 11/(0.008s + 1)$.)
b. Measure the frequency response of the velocity loop. Plot the magnitude ratio and phase difference up to 100 Hz. Determine the bandwidth of the velocity loop.
c. Design the digital filter of the position control system in such a way that the overshoot is negligibly small (i.e., $M_\mathrm{p} = 1$ percent) and the rise time is about 20 ms. You can design the control law either in the s or the z domain.
d. Assuming that the CNC computer is sending velocity signals (V_c) at $T = 1$-ms intervals to the servoamplifier via a D/A converter, identify the transfer function of the entire position loop in the z or the s domain. Simulate the response of the position loop for the input shown in Figure 6.22 using the discrete transfer function of the system. Identify the steady-state error theoretically and compare it with the error indicated by the simulation.

e. Derive the state-space response of the position loop. Simulate the position and velocity responses of the position control system for the input profile given in Figure 6.22. Compare the position response profile with the transfer function simulation results obtained in the previous question.
f. Design a real-time interpolation algorithm for a combination of linear and circular tool paths. Simulate the given tool paths on a PC. If possible, try the algorithms on a real machine tool in a laboratory.
g. Combine the real-time interpolation and the time domain feed drive servosimulation algorithms. Analyze the contouring errors on a circular tool path by increasing the feed velocities.

3. A High-Speed Linear Motor Driven XY Table Design: A high speed, two-axis XY table is driven by linear (i.e., flat bed) dc motors and linear guides. The assembly drawing and the block diagram of the continuos velocity control loop are provided in Figure 6.23. The actual position (x_a [m]) of the head is measured with a linear encoder. The position loop is closed in the digital computer, and the position error is passed from a digital filter $D(z) = K_p\frac{z+a}{z+b}$ at T discrete time intervals. The digitally filtered error is sent to a linear motor amplifier via a 12-bit D/A converter with a voltage range of ± 10 V. The x axis of the table has the following parameters: $K_i = 9{,}700$ A/A, $K_p = 31.8$ V/v, $K_a = 0.25$ V/A, $K_b = 13.78$ V/(m/s), $L_a = 0.75$ mH, $R_a = 2.8\,\Omega$, $T_g = 0.068$ V/V, $H_g = (1\text{ V})/(0.03\text{ m/s})$, $H_i = 5.246$, $K_t = 15.568$ V/V, $K_e = 1\,\mu$m/count. The saturation limits of the PWM voltage amplifier are ± 159 V, and the control interval is $T = 0.001$ s.

a. Derive the transfer function of the velocity loop ($G_v(s)$).
b. The inner velocity loop has a high bandwidth; thus, it can be approximated as a gain $E = 4$ A/V as shown in Figure 6.23. The frequency response of the velocity loop is measured by using the experimental setup shown. The magnitude of the system at frequencies $\omega = 0$, 1,000 rad/s were measured to be $|G_v(0)| = 0.1076$ m/s/V and $|G_v(1{,}000)| = 0.0139$ m/s/V, respectively. Estimate the table mass (M_a (kg)) and reference signal gain (S_g). What are the gain and the time constant of the velocity loop?
c. Design the digital filter ($D(z)$) in such a way that the overall closed-position loop behaves like a second-order *overdamped* system with two negative real poles, $s_1 = -100$ rad/s, $s_2 = -150$ rad/s. (K_p, a, b =?)
d. Find the steady-state error of the position control loop for a steady-state feed velocity of $f = 0.1$ m/s.
e. Remove the velocity feedback from the system ($T_g = H_g = 0$), and derive the closed-loop transfer function of the complete position loop (in the s or the z domain). Consider the inner loop in derivations.
f. Simulate current, velocity, position, and velocity command (V_c) for a unit step and ramp inputs with a constant Coulomb friction of 10 N. Control software tools can be used for the simulation.

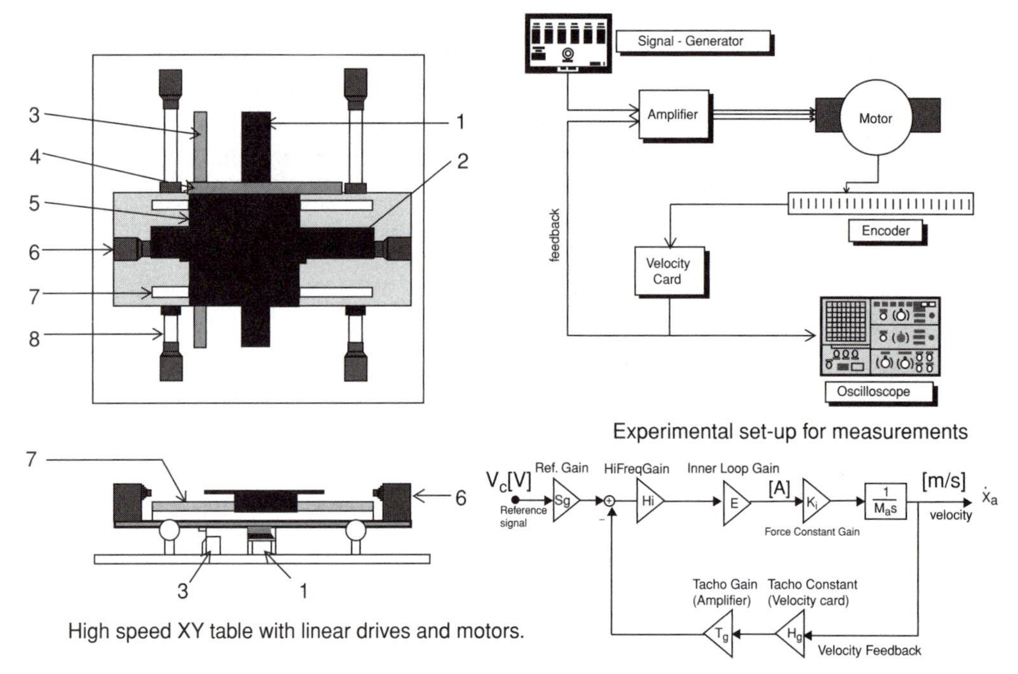

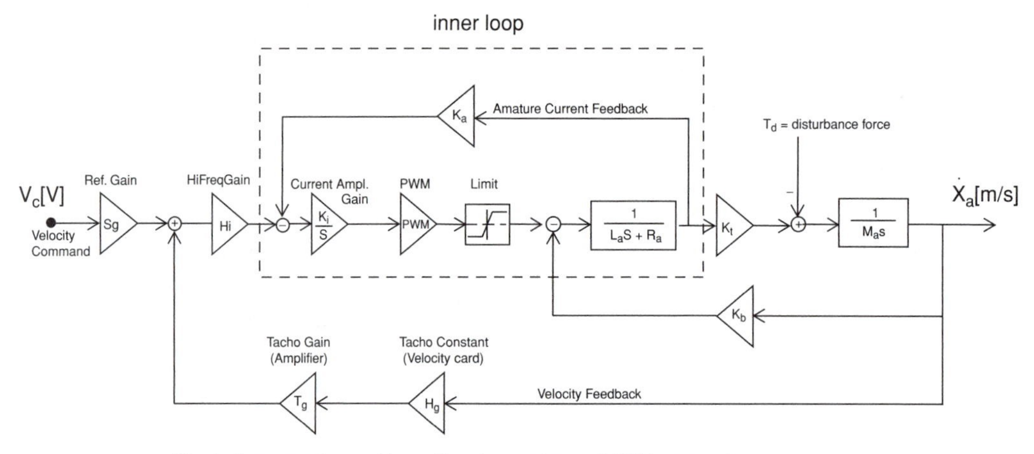

Block diagram of one drive without encoder and CNC computer.

Figure 6.23: A high-speed linear XY table system. Parts: 1,2 = dc linear motors for x and y axis; 3,4 = linear encoders for x, y drives; 5 = top positioning plate; 6 = spring-loaded bumper; 7,8 = linear support rails for x, y drives.

4. A model of a leadscrew feed drive system with a direct table position feedback encoder is shown in Figure 6.24. The motor is powered by the current amplifier and drives the leadscrew system that delivers motion to the machine table, which is supported by lubricated guideways. The position of the table is measured by using a linear encoder, and this measurement is used by the servocontroller for generating the necessary control signals, so that the table

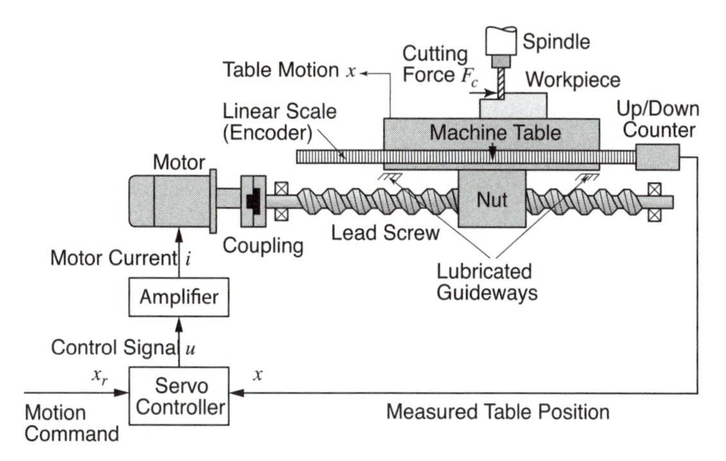

Figure 6.24: Machine tool feed drive control system.

closely follows the commanded motion profile. (Linear Encoder: a linear optic scale attached to the table, and it replaces rotary encoder.)

No rotary velocity or position feedback is used on the motor. It is assumed that current regulation loop between the amplifier and the motor armature is much faster than the rest of the system dynamics; hence, the amplifier is regarded as a static gain that produces motor current proportional to the applied control signal. The electrical winding of the motor can be neglected. The effect of disturbances such as cutting force are to be modeled as equivalent disturbance torque (T_d) acting on the motor shaft.

a. Draw the block diagram of the open-loop system (i.e., servocontroller not connected to the physical system). Express the transfer function of each part in the system in the Laplace domain. Use symbolic representation of the system on the block diagram. The input is a signal voltage [V] to the amplifier, and the output is in [counts]. The motor disturbance torque is in [Nm]. Substitute the values of each parameter in the open-loop transfer function block diagram using the following parameters of the machine:

Current amplifier gain	K_a	4 [A/V]
Motor torque gain	K_t	2.239 [Nm/A]
Leadscrew pitch length	h_p	20 [mm/rev]
Rotational inertia of rotor, coupling, and leadscrew	J	0.027[kg m^2]
Mass of Table and Nut	m_t	225 [kg]
Mass of workpiece	m_w	200 [kg]
Reflected viscous friction at the motor shaft	B_e	0.020 [Nm/(rad/s)]
Linear encoder resolution of quadrature decoding		1 [count/μ m]
Digital to analog converter gain (**neglect**)		1

b. Express the open-loop transfer function of the system both symbolically and numerically.

c. Design a lead compensator to achieve 60 degrees phase margin at a crossover frequency of $\omega = 60$ [Hz]. Plot the frequency response of the open-loop system by including the compensator, and show that the design criteria are met.

5. CNC Hydraulic System Design Project: You are assigned to design a digital control system for a hydraulic actuator driving a lumber bandsaw head. The physical control system consists of an amplifier driving a servovalve–piston/cylinder actuator carrying the bandsaw head. The actual position (x_a [μm]) of the head is measured with a linear encoder circuit, which has a gain of K_e [μm/mm], and the linear velocity of the head motion is monitored in [mm/s] by a velocity transducer. The position loop is closed in the digital computer, and the pole placement control law explained in Section 6.7.4 is used for the position control of the actuators. The resulting control variable is sent to a servovalve amplifier via a D/A converter, which has a gain of K_d. The input voltage to the velocity loop (V_c [V]) is received from the D/A converter, and the output velocity is measured from a velocity transducer in u [mm/s]. The velocity loop is found to behave like a first-order continuous system with a gain of K_v [(mm/s)/V] and time constant of τ_v [s].

a. Draw the block diagram of the complete control system.
b. Derive the state-space equations for the velocity u [mm/s] and head position x_a [μm]. Neglect the disturbance load.
c. The gain and the time constant of the velocity loop were estimated from a simple frequency response test by sending V_c [V] $= \sin 100t$ to the servovalve amplifier and measuring the resulting sinusoidal velocity response. The measured velocity was observed to have a maximum amplitude of $u = 70.71$ mm/s and a time delay of $T_d = 0.0078$ s with respect to the sinusoidal input V_c. What are the gain and the time constant of the velocity loop?
d. The control interval is $T = 0.001$ s, the encoder circuit has a resolution of 1,000 μm/mm (i.e., 1 count $=$ 1 μm), and the D/A circuit has a 10-bit resolution with a voltage range of ± 5 V. Design the pole-placement control law in such a way that the overall closed-position loop behaves like a second-order underdamped system with a natural frequency of $\omega_n = 250$ rad/s and a damping ratio of $\zeta = 1.5$.
e. Express the states (u [mm/s], x_a [μm]) as difference equations (from item b) with numerical values. Assume the input to be series of $x_r(k)$.

CHAPTER SEVEN

SENSOR-ASSISTED MACHINING

7.1 INTRODUCTION

The first step in automating machining systems was the introduction of computer numerically controlled (CNC) machine tools. The primary function of CNC is to automatically execute a sequence of multiaxis motions according to a part geometry. However, safe, optimal, and accurate machining processes are generally planned by manufacturing engineers based on their experience and understanding of the process. It is difficult to predict vibration, tool wear and breakage, thermal deformation of the machine tools, and similar process-based events by using off-line theoretical models. In addition to engineering the process plans before actual machining, the machine tools are instrumented with vibration, temperature, displacement, force, vision, and laser sensors to improve the productivity and reliability of the cutting operations on-line. The sensors must have reliable frequency bandwidth, have a good signal-to-noise ratio, and provide signals with reliable correlation to the state of the process. They must also be practical for installation on machine tools. The measured sensor signals are processed by real-time monitoring and control algorithms, and the corrective actions are taken by the CNC accordingly. The corrective actions may be manipulation of spindle speed, feed, tool offsets, compensation of machine tool positions, feed stop, and tool change depending on the process monitoring and control application. Such a sensor-assisted cutting is called *intelligent machining* in the literature [16, 17]. The architecture of CNC must be organized in such a way that it allows real-time manipulation of the machine tool's operating conditions. In other words, CNC must be *open* to allow integration of user-developed real-time application programs. The following *Intelligent Machining Module* has been developed at the author's laboratory [11, 18] for modular integration of cutting process monitoring and control tasks to CNC machine tools.

7.2 INTELLIGENT MACHINING MODULE

An Intelligent Machining Module (IMM) has been designed to run on existing commercial CNC systems that allow limited manipulation of cutting conditions by the end users. The IMM runs on a digital signal-processing (DSP) board with analog sensor signal-processing capability. Various intelligent machining

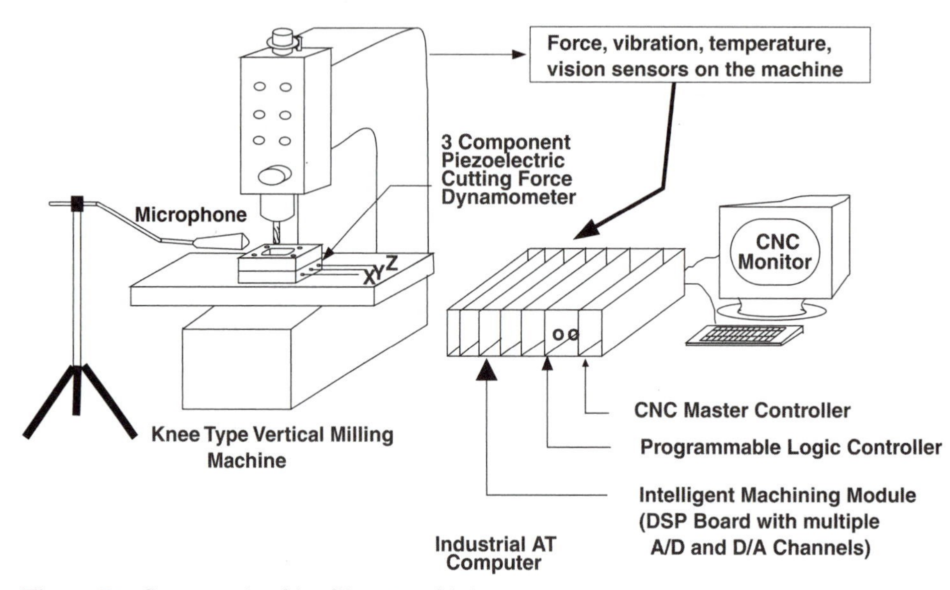

Figure 7.1: Sensor-assisted intelligent machining system.

tasks, such as adaptive control, tool condition monitoring, and process control, can simultaneously run on the system. The user can reconfigure the system by using script commands from the supplied signal-processing and data collection library. Each function is called a Plug In Module (PIM), and the IMM has a mechanism of integrating new, user-developed PIMs to the script command library. The IMM is configured to communicate with commercial semiopen CNCs through the PC–CNC communication links and software. Several IMM systems have been installed in industry. Some examples include adaptive control, chatter detection, and tool failure–monitoring applications on a five-axis machining center controlled by a FANUC™ CNC with a PC interface. The IMM sends feed and spindle speed change, machining halt, tool change, tool offset, and other numerically controlled (NC) commands accepted by the CNC controllers.

7.2.1 Hardware Architecture

Real-time IMM algorithms can run on any DSP board that has a shared memory with the host PC and analog channels to collect sensor signals. Analog outputs of the sensors, such as cutting force, vibration, temperature, and pressure sensors, can be connected to the DSP board's data acquisition module (see Fig. 7.1). The DSP board is controlled by PC software running under the Windows NT operating system. Some commercial CNC systems either allow high-speed communication with an external PC or they have an internal PC. Machine tool commands, such as *feed change, spindle speed change, tool offset, tool change*, and similar commands can be sent from C programs running on the

PC to the executive of the commercial CNC. Such CNC systems, which allow limited interface with user applications, are called "semiopen CNC systems" in this book. The IMM has been designed to run on semiopen CNC systems.

7.2.2 Software Architecture

The IMM is a combination of a PC system manager running on the host PC and a DSP system manager running on each DSP board in the PC bus. The DSP system manager is configured to act as a server, whereas the PC manager acts as a client. All real-time sensor data collection and signal-processing activities are handled in the DSP. The PC system manager initializes the user-requested functions, sets up signal-processing networks, and exchanges information among the DSP, user, and CNC. To eliminate any hardware dependence, a generic DSP interface was developed as a software module on the host PC. Each DSP driver supplied by the DSP manufacturer communicates with the rest of the software via the generic DSP interface, which is hardware independent. The generic DSP interface provides high-level services, such as command buffers, immediate data transfer protocol, and circular and double buffers for data transfer. The IMM also features a generic I/O interface in the DSP software, for high-level access to analog I/O, interrupts, and timers, and it has a low level I/O driver for DSP-dependent operations. Different I/O and DSP boards can easily be supported by simply changing hardware-dependent drivers in the PC and the DSP software without modifying any user-developed functions.

Signal-Processing Networks (SPNs)

The SPN is a collection of application-specific tasks, which are called plug in modules (PIMs) in the IMM as shown in Figure 7.2a. Each SPN runs on a single DSP board. The PIMs in an SPN are executed sequentially, which eliminates the need for explicit synchronization, thereby enabling one to use a very fast communication scheme without endangering the integrity of the transferred data. An SPN can have multiple levels running at different frequencies, so that data collection and processing can be done at one frequency, and additional processing and decision making can be done at another frequency. The PIMs and the links between the PIMs can be configured at runtime using a high-level script language [11].

Plug in Module

The PIM interface allows application-specific tasks to be implemented at an abstract level without knowing the hardware details of the underlying system. Functions such as filters, control algorithms, transfer function identifiers, and fast Fourier transforms can be implemented as PIMs so that they can be integrated with the rest of the system. The inputs and outputs of a PIM are shown in Figure 7.2b.

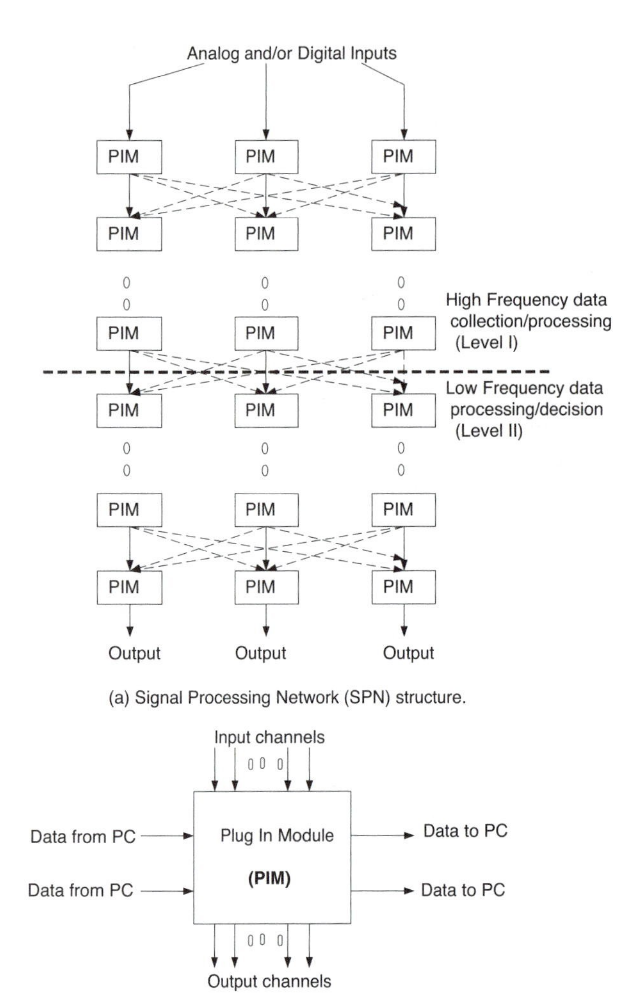

(a) Signal Processing Network (SPN) structure.

(b) Plug In Module (PIM) structure.

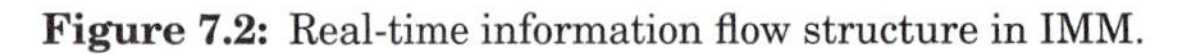

Figure 7.2: Real-time information flow structure in IMM.

7.2.3 Intelligent Machining Application

The following script commands are set by a user for adaptive cutting force control and tool breakage detection in a five-axis milling application. Cutting force sensors, with a calibration factor of 1,000 N/V, are connected to analog channels 0, 1, and 2 of the DSP board. Cutting forces are sampled at 2,000 Hz, they are low pass filtered at 300 Hz, and the peak and average resultant forces are identified. There are four flutes on the end mill, and the spindle speed is 600 rev/min (10 Hz). In the second level, which corresponds to tooth-passing frequency (40 Hz), the peak cutting force is used by the adaptive force control algorithm and the average force is used by the tool breakage detection function. The adaptive control PIM calculates the required feed to keep the force constant, and the tool failure algorithm sends a feed hold signal if chipping is detected. Both algorithms pass the feed and feed hold commands to the CNC via a PC handler.

The system is set up via the following script commands:

```
Sensors are connected to analog channels 0,1, and 2 of DSP board.
INPUT SCALE 0, 1, 2; 1000, 1000, 1000 // Calibration: 200 N/V
BEGIN 0,1,2; 2000 // Sample input channels 0, 1, and 2 at 2,000 Hz.
LOWPASS FILTER(0; 16; 300) // Apply low-pass filter of length 16.
LOWPASS FILTER(1; 16; 300) // with a cutting frequency of 300 Hz
LOWPASS FILTER(2; 16; 300) // to channels 0, 1, and 2.
PEAK DETECT(0,1,2 -> 5 ; ; )  // Write peak and average of force vector
AVERAGE DETECT(0,1,2 -> 7; ; ) // available on channels 0, 1, and 2 to
                               // software channels 5 and 7.
SECOND LEVEL Timer 40 // Start 2nd algorithm level to run at 40 Hz
// Run adaptive control using peak force available on channel 5.
```

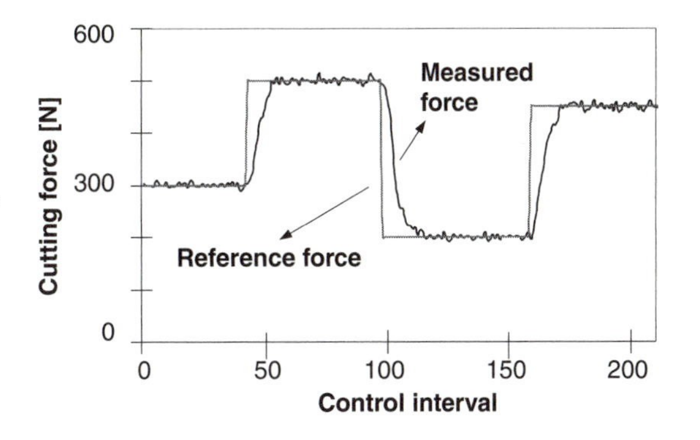

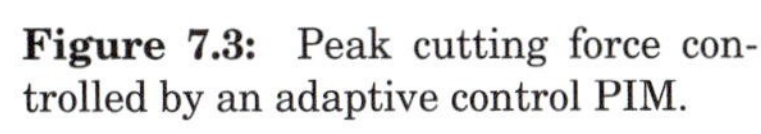
Figure 7.3: Peak cutting force controlled by an adaptive control PIM.

```
    ADAPTIVE CONTROL ( 5, 6 ; ; 200, 200, 4, 0.2, 0.05 )
  // Run tool breakage detection using average force on channel 7.
    TOOL BREAKAGE DETECTION ( 7, 8 ; ; )
  IMMEDIATE OUTPUT TO PC ( 6, 8 ; ; ) // Send the calculated feed and
                                       // feed hold commands to the PC.
  DOUBLE BUFFERED OUTPUT TO PC( 5, 6 ; ; ) // Store peak force and
                                           // identified feed in a buffer.
  END                                      // End second loop.
  // Send feed override value to semi-open CNC.
  IMMEDIATE HANDLER = Write feed override and emergency stop (0.05)
  // Save all data in a disk file.
  DOUBLE HANDLER = Write to disk (c:results.dat)
```

The graphical user interface of the IMM system records any signal or parameter indicated within the script command file. A sample window of commanded and measured cutting forces is shown in Figure 7.3. The fundamental philosophy behind the IMM system is that it allows integration of new, user-developed functions (i.e., PIMs) to the script library, and a new application can rapidly be configured by calling script commands in the desired sequence. The mathematical details of sample intelligent machining tasks – *adaptive cutting force control, tool breakage detection*, and *chatter avoidance* – are provided in the following sections.

7.3 ADAPTIVE CONTROL OF PEAK FORCES IN MILLING

7.3.1 Introduction

There are several physical constraints in the machining of metals. Some of these constraints are machine tool dependent, such as maximum torque and power available from the spindle drives. Other constraints are tool and workpiece dependent. Overloading of a carbide insert must be avoided by keeping the maximum chip thickness removed at a level that would not increase the principal tensile stress in the cutting wedge beyond the ultimate tensile

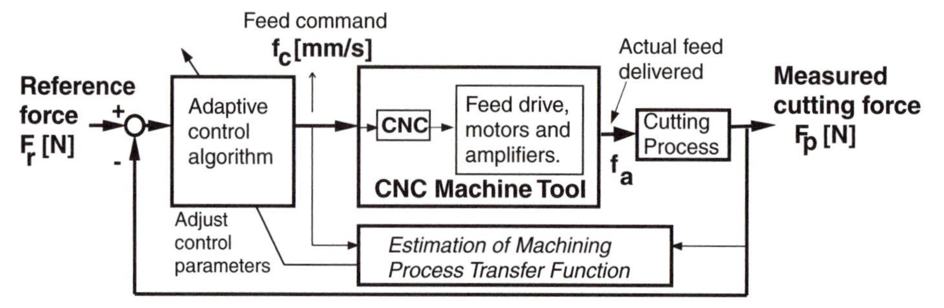

Figure 7.4: Block diagram of a general adaptive control system in machining.

strength of the tool material. Maximum static deflections left on the finished surface must be within the tolerance of the workpiece. Maximum resultant cutting force on a slender end mill must be kept safely below a limit value to prevent fracture of the shank. Here, we present two different adaptive control strategies that constrain the maximum resultant force at a safe level by adaptively manipulating the chip load or the feed rate.

A general block diagram of a typical adaptive control system is shown in Figure 7.4. The input to the system is the reference or desired level of the maximum cutting force. The actual cutting forces are measured via sensors mounted on the table or on the spindle and are collected at 3 to 5 degrees of angular rotation intervals to capture the peak resultant force. The peak force at each tooth or spindle period is evaluated and passed to the adaptive control law. When there are multiple teeth on the cutter, it is recommended to use peak forces per spindle revolution. Otherwise runouts on the cutter would give fluctuating peak forces at each tooth period. The adaptive control would send similarly fluctuating feed rates to the machine tool drives, which have a low bandwidth. As a result, the adaptive control system would produce an oscillatory response, which is not desirable. Hence, the peak force ($F_{\rm p}(k)$) is evaluated at each spindle period (k) and subtracted from the set or desired force level ($F_{\rm r}(k)$). The adaptive control algorithm determines a new feed command to minimize the force error. The feed command ($f_{\rm c}(k)$) is sent to the CNC unit, which has its own digital position control law executed at comparatively small time intervals (e.g., 0.100 ms). The CNC unit sends voltage to the feed drive motors, which move the table at an actual feed velocity of $f_{\rm a}$ [mm/s]. Because machine tool drive control servos are tuned to be overdamped with zero overshoot, they can be approximated to have first-order dynamics with an average time constant of 0.1 ms. Note that this value may depend on the type of the machine tool and can be quite small for high-speed machine tool drives. The cutting process feels the change in the chip load or the peak cutting force ($F_{\rm p}(k)$) at least after one tooth period. The chatter vibration–free cutting process can be approximated as a first-order system with a time constant equal to one or more tooth periods but less than the spindle period. As the width or depth cut changes on the workpiece, the peak cutting force varies along the tool path. Hence, the first-order dynamics of the process has time-varying parameters. The combined CNC, machine tool feed drive, and cutting process can have an approximate second-order dynamics ($G_c(s)$). The peak force is evaluated from measurements and sent to an on-line *process identification algorithm*, which estimates the digital parameters of the combined machine tool, CNC, and time-varying cutting process as the cutter travels through holes, slots, and other features of the machined part. The time-varying coefficients of the process are estimated from the feed command input and peak force output of the process. Estimated parameters of the machine tool and cutting process are used to update the parameters of the adaptive control law at each control interval. Because the control law parameters are adjusted according to the changes in the cutting process parameters at each sampling interval, the control system is *adaptive* to the varying workpiece geometry. An adaptive control law, running at the control computer, computes new feed rate command value at each sampling interval. The feed command moves the machine tool table according to the servodynamics of the feed drive. The actual feed velocity of the table changes the chip thickness and therefore the cutting force produced during machining. The adaptive control loop ensures that the actual force felt by the milling cutter is always equal to the reference constant force value which is safely below the breakage limit. In the following sections we present step-by-step the designs of a pole-placement controller and a generalized predictive adaptive controller.

7.3.2 Discrete Transfer Function of the Milling Process System

As explained above, the machine tool control and drive system can be approximated by a first-order continuous system [100] as follows:

$$G_m(s) = \frac{f_a(s)}{f_c(s)} = \frac{1}{\tau_m s + 1}, \tag{7.1}$$

where the f_a and f_c are the actual output and command input values of the feed speed in [mm/s]. The feed or chip load per revolution can be found by $h[\text{mm/rev}] = f_c/(Nn)$, where N is the number of teeth on the milling cutter and n [rev/s] is the spindle speed. A turning process can be modeled by assigning $N = 1$. The cutting force does not change instantly with the feed, and the process can be approximated to have first-order dynamics as follows:

$$G_p(s) = \frac{F_p(s)}{f_a(s)} = \frac{K_c a b(\phi_{st}, \phi_{ex}, N)}{Nn} \frac{1}{\tau_c s + 1}, \tag{7.2}$$

where K_c [N/mm^2], a [mm], and $b(\phi_{st}, \phi_{ex}, N)$ are the cutting constant, the depth of cut, and immersion function, respectively. The immersion function ($b(\phi_{st}, \phi_{ex}, N)$) does not have any units and may change between 0 and $\sim N$, depending on the immersion angle and the number of teeth in the cut. Both the axial depth of cut (a) and the radial depth of cut (b) may change along the tool path depending on the workpiece geometry; hence, a and b are time varying. Note that when chatter vibrations are present, the process is very unstable, producing large-amplitude, oscillating cutting forces. The process has complex, high-order nonlinear dynamics and can not be controlled by manipulating the feed with an adaptive control system. Chatter control must be treated separately, and the adaptive control must be deactivated when chatter vibrations are present.

Although the machine tool dynamics can be considered as time invariant, it is more convenient to treat a combined machine tool and cutting process as time varying for the practical application of adaptive control on multiaxis machine tools. The combined transfer function of the system is

$$G_c(s) = \frac{1}{(\tau_m s + 1)} \frac{K_c ab}{Nn(\tau_c s + 1)} = \frac{K_p}{(\tau_m s + 1)(\tau_c s + 1)}, \tag{7.3}$$

where the process gain is K_p[N/mm/s] $= K_c ab/(Nn)$. In reality, τ_c may change owing to static deflections of the tool–workpiece structure, which in turn affect the chip load. However, the above approximation is valid for practical simulation studies of the adaptive control algorithms. Because the machining process is controlled at spindle period T, the zero-order hold equivalent of $G_c(s)$ is considered

$$G_c(z) = \frac{F_p(k)}{f_c(k)} = (1 - z^{-1})\mathcal{Z}\frac{G_c(s)}{s} = \frac{b_0 z + b_1}{z^2 + a_1 z + a_2}, \tag{7.4}$$

where k is the spindle rotation counter, z is the forward shift operator, and

$$b_0 = K_p \frac{\tau_m(1 - e^{-T/\tau_m}) - \tau_c(1 - e^{-T/\tau_c})}{\tau_m - \tau_c},$$

$$b_1 = K_p \frac{\tau_c(1 - e^{-T/\tau_c})e^{-T/\tau_m} - \tau_m(1 - e^{-T/\tau_m})e^{-T/\tau_c}}{\tau_m - \tau_c},$$

$$a_1 = -(e^{-T/\tau_m} + e^{-T/\tau_c}),$$

$$a_2 = e^{-T(1/\tau_c + 1/\tau_m)}.$$

The discrete process parameters (a_1, a_2, b_0, b_1) depend on the workpiece geometry, and they may change during machining. For example, during the machining of engine blocks or aircraft wings, both the width of cut (i.e., immersion) and the axial depth of cut may change according to the workpiece and selected tool path geometry.

To design an adaptive control system for regulating the maximum cutting forces, the unknown time-varying parameters (a_1, a_2, b_0, b_1) must be estimated at each spindle period. A recursive least-square (RLS) algorithm is used to estimate the time-varying parameters [48]. Henceforth, $\hat{a}_1, \hat{a}_2, \hat{b}_0$, and $\hat{b}_1$ will be used for estimates of the cutting process parameters. The following RLS algorithm is executed in a sequential manner at each spindle period (k):

$$\hat{\theta}(k) = \hat{\theta}(k-1) + K(k)[F_p(k) - \varphi(k)^T \hat{\theta}(k-1)], \tag{7.5}$$

$$K(k) = \frac{P(k-1)\varphi(k)}{\lambda + \varphi(k)^T P(k-1)\varphi(k)}, \tag{7.6}$$

$$P(k) = \frac{P(k-1)}{\lambda}[I - K(k)\varphi(k)^T], \tag{7.7}$$

where the parameter vector is

$$\hat{\theta}^T = [\hat{a}_1 \; \hat{a}_2 \; \hat{b}_0 \; \hat{b}_1]$$

and the regressor vector is

$$\hat{\varphi}(k-1)^T = [-F_p(k-1) \quad -F_p(k-2) \quad f_c(k-1) \quad f_c(k-2)].$$

$K(k)$ is called the estimation gain with a user-selected forgetting factor of $0.5 < \lambda < 1$. Typically λ is selected between 0.8 and 0.95 to discount the influence of previous measurements on the estimation. The covariance matrix $P(k)$ is a square matrix with dimensions of $N_p \times N_p$, where N_p is the number of parameters to be estimated (i.e., $N_p = 4$ in this application). The algorithm is started by assigning guessed initial values for the parameters (i.e., $\theta(0) = [0.1 \; 0.1 \; 0.1 \; 0.1]$) and a large initial value for the covariance matrix ($P(0) = 10^5$). Care must be taken in using the RLS algorithm. When the adaptive control algorithm runs for a long period and the process does not change significantly, the covariance matrix becomes either too small or too large, which creates numerical instability. The problem can be avoided by monitoring the trace of the covariance matrix (i.e., $\mathrm{tr}(P(k)) = \sum_{i=1}^{4} P_{ii}$). Whenever the trace is too small or too large, the covariance matrix can be reset to an initial value.

7.3.3 Pole-Placement Control Algorithm

Once the process parameters are estimated on-line, a regular pole-placement control algorithm [29] can be designed to constrain the cutting forces (see Fig. 7.5). The closed-loop transfer function of the control system is given by

$$\frac{F_p(k)}{F_r(k)} = \frac{T(z)B(z)}{A(z)R(z) + B(z)S(z)}. \tag{7.8}$$

The polynomials $S(z^{-1})$, $T(z^{-1})$, and $R(z^{-1})$ represent feedback, feedforward, and error regulators, respectively, which have to be determined adaptively at each spindle period. The control algorithm regulates the feeding velocity $f_c(k)$ as

$$R\, f_c(k) = TF_r(k) - S\, F_p(k). \tag{7.9}$$

The aim of the adaptive controller based on pole placement design is to make the closed-loop transfer function between the reference force F_r and the actual

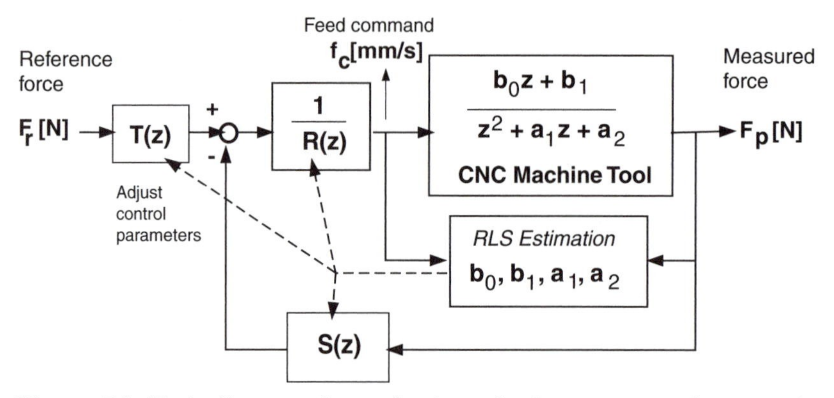

Figure 7.5: Block diagram of an adaptive pole-placement control system in machining.

force F_p obey the following desired model dynamics:

$$\frac{F_\mathrm{p}(k)}{F_\mathrm{r}(k)} = \frac{B_m(z)}{A_m(z)}. \tag{7.10}$$

Here, $A_m(z)$ is designed to satisfy the transient response characteristics of the controller. A second-order dynamics is selected to represent the desired response as follows:

$$A_m(z) = z^2 - 2e^{-\zeta\omega_\mathrm{n}T}\cos(\omega_\mathrm{n}\sqrt{1-\zeta^2}T)z + e^{-2\zeta\omega_\mathrm{n}T}. \tag{7.11}$$

This corresponds to a second-order continuous system with a desired damping ratio of ζ and natural frequency of ω_n (i.e., $s^2 + 2\zeta\omega_\mathrm{n}s + \omega_\mathrm{n}^2$) with a discrete control interval of T, which is equal to the spindle period. A damping ratio of $\zeta = 0.8$ and a rise time (T_r) equal to the larger of three spindle periods or feed drive servo's rise time can be selected for practical applications. The natural frequency corresponds to $\omega_\mathrm{n} = 2.5/(T_\mathrm{r})$ rad/s. With these criteria, the discrete closed-loop model transfer function with a unit gain is

$$\frac{F_\mathrm{p}(k)}{F_\mathrm{r}(k)} = \frac{B_m}{A_m} = \frac{1 + m_1 + m_2}{z^2 + m_1z + m_2}, \tag{7.12}$$

where $m_1 = -2e^{-\zeta\omega_\mathrm{n}T}\cos(\omega_\mathrm{n}\sqrt{1-\zeta^2}T)$ and $m_2 = +e^{-2\zeta\omega_\mathrm{n}T}$. For the system to behave like the desired model, the following equivalence must hold:

$$\frac{BT}{AR + BS} = \frac{B_m}{A_m}. \tag{7.13}$$

The design problem is to find the polynomials R, S, and T by using the machining plant's and desired model's transfer functions. The numerator of the machining process transfer function can be expressed as follows:

$$B(z) = B^-(z)B^+(z) = b_0(z + b_1/b_0). \tag{7.14}$$

In reality, the zero of the machine tool plant is always stable and within the unit circle (i.e., $|b_1/b_0| < 1.0$). However, RLS estimation does not necessarily predict the parameters b_0 and b_1 accurately. To keep the control system robust, separate control systems are designed depending on whether the process has a stable or an unstable zero.

Case i: Process Zero Is Stable ($||b_1//b_0|| < 1.0$). Let

$$B^-(z) = b_0, \quad B^+(z) = z + b_1/b_0.$$

The control parameters can be designed in such a way that the stable zero can be canceled by one of the system's closed-loop poles. To cancel B^+ with one of the closed-loop transfer function poles, as A and B are coprime, R must be designed to have a factor of B^+ as follows:

$$R = B^+ \cdot R' = (z + b_1/b_0) \cdot R'. \tag{7.15}$$

The closed-loop transfer function (Eq. 7.13) becomes

$$\frac{B^+B^- T}{B^+(AR' + B^-S)} = \frac{B^- T}{AR' + B^-S} = \frac{A_0 \cdot B_m(z)}{A_0 \cdot A_m(z)}, \tag{7.16}$$

where the observer A_0 is included to ensure good tracking and causality in the design. (Note: Causality means that the control input does not depend on the future inputs and outputs of the system.) The control polynomials R', S, and T can be solved from the following identities:

$$\begin{aligned} B^-T &\equiv A_0B_m, \\ AR' + B^-S &\equiv A_0A_m, \end{aligned} \tag{7.17}$$

where the second identity is called the *Diophantine equation*. The degrees of machining process transfer functions (i.e., the order of polynomials B and A) are

$$\deg(B) = \deg(B^+) + \deg(B^-) = 1 + 0, \quad \deg(A) = 2 \tag{7.18}$$

and $\deg(A_m) = 2$. The *degree* of the polynomials are determined from the causality conditions and the order of process transfer functions is summarized as follows:

$$\begin{aligned} \deg(A_0) &\geq 2 \cdot \deg(A) - \deg(A_m) - \deg(B^+) - 1 = 2 \times 2 - 2 - 1 - 1 = 0, \\ \deg(R) &= \deg(A_0) + \deg(A_m) - \deg(A) = 1 + 2 - 2 + 1 = 2, \\ \deg(S) &= \deg(A) - 1 = 2 - 1 = 1, \\ \deg(T) &= \deg(A_0) + \deg(B_m) - \deg(B^-) = 1 + 0 - 0 = 1. \end{aligned}$$

The resulting control polynomials are

$$\begin{aligned} A_0(z) &= z, \\ R'(z) &= z + r_1, \\ S(z) &= s_0z + s_1, \\ T(z) &= t_0z. \end{aligned} \tag{7.19}$$

The Diophantine equation ($AR' + B^-S = A_0A_m$) becomes

$$(z^2 + a_1z + a_2)(z + r_1) + b_0(s_0z + s_1) \equiv z(z^2 + m_1z + m_2)$$

and

$$B^{-}T \equiv A_0 B_m.$$

From the two equivalence, the control polynomial parameters are derived as follows:

$$\begin{aligned}
&r_1 = m_1 - a_1, \quad \rightarrow && R(z) = z^2 + (r_1 + b_1/b_0)z + (b_1/b_0)r_1, \\
&s_0 = \frac{m_2 - a_1 r_1 - a_2}{b_0}, \quad s_1 = -\frac{a_2 r_1}{b_0}, \quad \rightarrow && S(z) = s_0 z + s_1, \\
&t_0 = \frac{1 + m_1 + m_2}{b_0}, \quad \rightarrow && T(z) = t_0 z,
\end{aligned} \tag{7.20}$$

where $R(z) = B^{+}R' = (z + b_1/b_0)(z + r_1)$.

The feed rate is calculated at each spindle period from the control law as follows:

$$f_c(k) = \frac{T(z)}{R(z)} F_r(k) - \frac{S(z)}{R(z)} F_p(k).$$

After substituting parameters and scaling polynomials by z^{-2}, we obtain

$$\begin{aligned}
f_c(k) = t_0 F_r(k-1) - \left(r_1 + \frac{b_1}{b_0}\right) f_c(k-1) - r_1 \frac{b_1}{b_0} f_c(k-2) \\
- s_0 F_p(k-1) - s_1 F_p(k-2).
\end{aligned} \tag{7.21}$$

Here, the process parameters are replaced by the on-line estimated values $(\hat{b}_0, \hat{b}_1, \hat{a}_1, \hat{a}_1)$ so that the control system can adapt itself to the time-varying machining process. The real-time application of the adaptive control algorithm can be summarized as follows:

- Collect the maximum cutting force at spindle revolution k.
- Estimate the process parameters (b_0, b_1, a_1, a_2) by using the RLS algorithm.
- Solve for the control polynomials R, S, and T (Eq. 7.20).
- Calculate the feed rate (Eq. 7.21) and send it to the CNC.

Case ii: Process Zero Is not Stable $(||b_1//b_0|| > 1.0)$. Let

$$B^{+}(z) = 1, \quad B^{-}(z) = b_0 z + b_1.$$

The procedure is identical to that of case i except that the process zero is not canceled and it must be included in the model transfer function as follows:

$$t_0 = \frac{1 + m_1 + m_2}{b_0 + b_1},$$

and the Diophantine equation becomes

$$(z^2 + a_1 z + a_2)(z + r_1) + (b_0 z + b_1)(s_0 z + s_1) \equiv z(z^2 + m_1 z + m_2).$$

The remaining solution procedure is identical to the previous case.

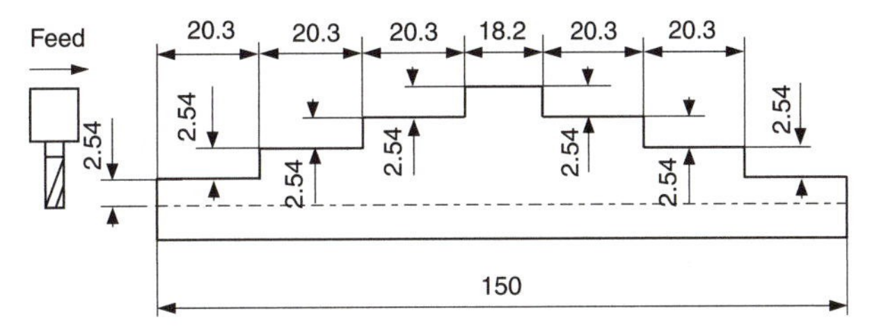

Figure 7.6: Workpiece geometry to test step response of adaptive milling force control algorithms. Workpiece material: Aluminum 6067. Cutter: 4-fluted carbide end mill. Spindle speed = 715 rev/min.

Machining Test Results

The adaptive pole-placement control algorithm is integrated to the IMM as a PIM, and tested on a milling machine controlled by an open architecture CNC. The workpiece geometry and cutting conditions are given in Figure 7.6. The workpiece was prepared to have step changes in the axial depth of cut to test the transient response and robustness of the control algorithm. Whenever an unstable zero is estimated by the RLS algorithm, the pole-placement control law is switched to the second case as outlined above. The reference peak force was set to $F_r = 1{,}200$ N. The experimental results are given in Figure 7.7. The peak resultant force is always kept at the desired 1,200 N level, and the algorithm gives some overshoots when the axial depth of cut increases stepwise. However, the control algorithm is stable and rapidly converges. The corresponding feeds identified by the control law are also given, along with the estimated process parameters (b_0, b_1, a_1, a_2) by the RLS PIM. The control interval corresponds to the spindle rotation period.

7.3.4 Adaptive Generalized Predictive Control of Milling Process

In general, most adaptive control algorithms work fairly well when they are tuned for a specific machine tool at a constant speed, such as the adaptive pole-placement control presented above. The pole-placement algorithm depends on the spindle speed, which is the sampling period (T). A robust and modular adaptive control algorithm, which is spindle speed independent, free of covariance drifting in parameter estimation, and robust to variable time delays between the feed command and actual feed delivered by the CNC is required so that it can be integrated to an open-architecture CNC system with a minimal amount of modifications. Based on the extensive experience of machining process control strategies in the author's laboratory [5, 100], an adaptive generalized predictive control (GPC) method was found to satisfy the requirements most [16]. GPC is more robust to future transient changes in the cutting forces, that is, the geometry that can be obtained from a computer-aided design (CAD) system [100], and the varying time delay between the feed command generated by the adaptive control module and its actual execution by the CNC system. The fundamental control law was presented by Clarke et al. [40], and its specific application to machining control is briefly summarized as follows.

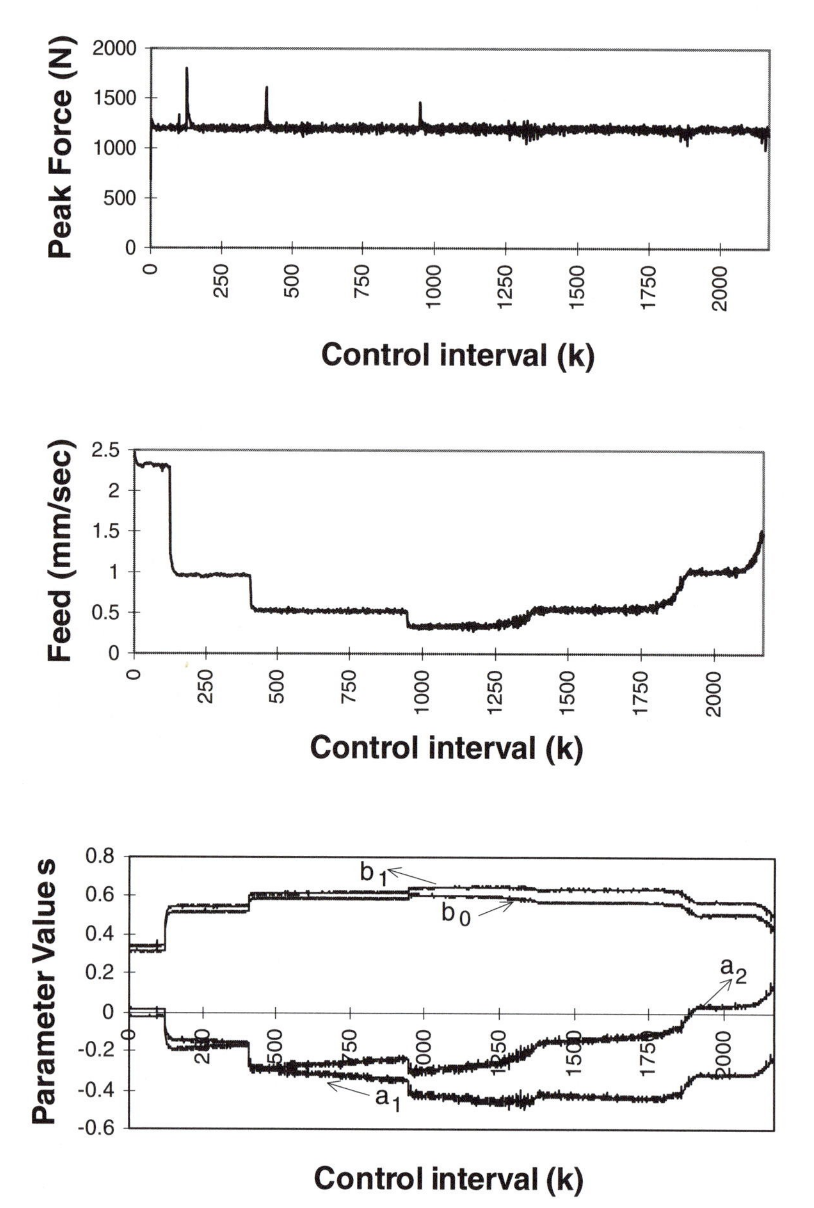

Figure 7.7: Milling test results of adaptive pole-placement control system. See Figure 7.6 for the part geometry and cutting conditions.

The combined transfer function of the feed drive dynamics and chatter-free cutting process can be expressed as [16] follows:

$$G_c(z^{-1}) = \frac{F_p(k)}{f_c(k)} = \frac{z^{-1}B(z^{-1})}{A(z^{-1})}$$

$$\text{[3pt]} = \frac{z^{-1}(b_0 + b_1 z^{-1} + b_2 z^{-2})}{1 + a_1 z^{-1} + a_2 z^{-2}}, \tag{7.22}$$

where $f_c(k)$ is the command feed sent to the CNC system by the adaptive control module at spindle period k. With respect to the previous transfer function (Eq. 7.4), the order of the machining process is increased to account for the nonlinear relationship between the feed rate (i.e., chip thickness) and the cutting force. The parameters of polynomials B and A may vary with time depending on the changes in the radial and axial depths of cut and the slight changes in the feed drive dynamics. They are estimated recursively at each spindle or adaptive control interval k from measured peak force F_p and commanded feed f_c vectors. A modified RLS algorithm, which avoids parameter drifting caused by the absence of steady excitations (i.e., changes in the workpiece geometry), was used. The covariance matrix drifting toward instability is avoided by tracking the trace of the covariance matrix, and updating it only when there is a change in the process. The recursive parameter identification algorithm with covariance tracking is given in Appendix B. The GPC method is devised for an ARIMAX model. Invoking this model naturally leads to the adoption of an integrator to the controller, thus eliminating steady-state offsets. The machining process model is therefore rearranged in the ARIMAX form as follows:

$$F_p(k) = \frac{B(z^{-1})}{A(z^{-1})} f_c(k-1) + \frac{\zeta(k)}{\Delta A(z^{-1})}, \tag{7.23}$$

where $\Delta = (1 - z^{-1})$ and $\zeta(k)$ is assumed to be an uncorrelated random noise sequence. Then a j step ahead prediction of $F_p(k)$ based on Eq. (7.23) can be obtained by expressing the noise term by its partial fraction expansion as follows:

$$\frac{z^j \zeta(k)}{\Delta A(z^{-1})} = z^j E_j \zeta(k) + \frac{F_j(k)}{\Delta A(z^{-1})} \zeta(k),$$

which leads to the Diophantine equation as follows:

$$1 = E_j(z^{-1}) A(z^{-1}) \Delta + z^{-j} F_j(z^{-1}), \tag{7.24}$$

$$\deg(E_j(z^{-1})) = j - 1, \ \deg(F_j(z^{-1})) = \deg(A(z^{-1})) = 2.$$

Here, E_j and F_j are polynomials uniquely defined, given $A(z^{-1})$ and the prediction interval j. Once they are calculated for one value of j, simple iterations can be used to calculate them for all other j; see the next section. All the noise components are in the future; therefore, provided that the output data up to time k and $f_c(k + j - 1)$ are available, the j step ahead prediction of peak cutting force F_p at spindle interval k can be given by

$$\hat{F}_p(k + j) = G_j(z^{-1}) \Delta f_c(k + j - 1) + F_j F_p(k), \tag{7.25}$$

where $G_j(z^{-1}) = E_j B(z^{-1})$. The delay in the process is assumed to be one here, which may change slightly depending on the spindle speed.

A general case is considered by selecting the minimum and maximum prediction output horizons as $N_1 = 1$ and $N_2 = 4$, respectively. The GPC control law takes predicted output values within the next four spindle revolutions when manipulating the control parameter (i.e., the feed). Equation (7.25) contains present (i.e., k), past (i.e., $k-i$), and future (i.e., $k+i$) values of feed and forces. GPC considers that future control inputs (i.e., feed rates $f_c(k+j)$) do not change beyond the control horizon NU, which is selected to be $NU=1$ here. Thus, $\Delta f_c(k+1) = \Delta f_c(k+2) = \Delta f_c(k+3) = 0$. Equation (7.25) can be partitioned as

$$\{\hat{F}_p\} = \{G_I\}\Delta f_c(k) + \{f\}, \tag{7.26}$$

where vector dimensions are $[4 \times 1]$, and their derived contents are given in the following section. Note that vector $\{f\}$ contains present and past measured peak forces $F_p(k-i), i = 0, 1, 2$ and past feed commands $f_c(k-i), i = 1, 2$, and recursively computed polynomials G_j and F_j, as well. According to GPC strategy, the feed rate is calculated by minimizing the expected value of a quadratic cost function containing future predicted errors within the output horizon (i.e., the next four spindle revolutions) between the actual and reference peak forces over the control horizon (i.e., the next spindle revolution) as follows:

$$J(f_c, k) = E\sum_{j=N_1}^{N_2} [\hat{F}_p(k+j) - F_r(k+j)]^2 + \lambda[\Delta f_c(k)]^2, \tag{7.27}$$

where λ, the weighting factor on the control input increment $f_c(k) - f_c(k-1)$, is selected as $\lambda = 0.2$ to soften the impact of sudden changes in the geometry on the adaptive control law. The minimization of the cost function ($\partial J/\partial f_c = 0$) leads to the input feed command at spindle interval k as follows:

$$f_c(k) = f_c(k-1) + \frac{1}{\{G_I\}^T\{G_I\} + \lambda}\{G_I\}^T(\{F_r\} - \{f\}). \tag{7.28}$$

The parameters of polynomials E_j, F_j, and G_j are recursively calculated followed by the control law given above (Eq. 7.28) at each spindle period. The computed vector feed $f_c(k)$ is sent to the CNC master controller, which constrains the actual machining force at a desired reference level. The feed is bounded between the user-defined minimum and maximum limits for safety.

Recursive Computation of Polynomials E_j, F_j, and G_j

The recursive algorithm of GPC is briefly summarized here. For $j=1$ the Diophantine equation becomes

$$1 = \Delta A(z^{-1})E_1(z^{-1}) + z^{-1}F_1(z^{-1}), \tag{7.29}$$

which yields $e_{10} = 1,\ f_{10} = -(a_1 - 1),\ f_{11} = -(a_2 - a_1)$, and $f_{12} = a_2$.

For $j > 1$, consider the difference of Diophantine equations at j and $j+1$ as follows:

$$\Delta A[E_{j+1} - E_j] + z^{-j}[z^{-1}F_{j+1} - F_j] = 0,$$

TABLE 7.1. Polynomials E_j, F_j, and G_j

j	E_j	F_j	G_j
1	e_0	$f_{10}+f_{11}z^{-1}+f_{12}z^{-2}$	$g_{10}+g_{11}z^{-1}+g_{12}z^{-2}$
2	$e_0+e_1z^{-1}$	$f_{20}+f_{21}z^{-1}+f_{22}z^{-2}$	$g_{20}+g_{21}z^{-1}+g_{22}z^{-2}+g_{23}z^{-3}$
3	$e_0+e_1z^{-1}+e_2z^{-2}$	$f_{30}+f_{31}z^{-1}+f_{32}z^{-2}$	$g_{30}+g_{31}z^{-1}+g_{32}z^{-2}+g_{33}z^{-3}$ $+g_{34}z^{-4}$
4	$e_0+e_1z^{-1}+e_2z^{-2}+e_3z^{-3}$	$f_{40}+f_{41}z^{-1}+f_{42}z^{-2}$	$g_{40}+g_{41}z^{-1}+g_{42}z^{-2}+g_{43}z^{-3}$ $+g_{44}z^{-4}+g_{45}z^{-5}$

which leads to

$$e_{10}=e_{20}=e_{30}=e_{40}=e_0=1;\; e_{21}=e_{31}=e_{41}=e_1;\; e_{32}=e_{42}=e_2;$$
$$e_{43}=e_3;$$

and

$$e_{j-1}=f_{j-1,0};\; f_{j,i}=f_{j-1,i+1}-e_{j-1}(a_{i+1}-a_i),\quad i=0,1,2.$$

The polynomial G_j is then calculated as follows:

$$G_j=E_jB=E_j(b_0+b_1z^{-1}+b_2z^{-2}). \tag{7.30}$$

The equations are solved recursively at each prediction step j. The results are given in Tables 7.1 and 7.2.

The contents of the vectors are given as

$$\{\hat{F}_\mathrm{p}\}=\begin{Bmatrix}\hat{F}_\mathrm{p}(k+1)\\ \hat{F}_\mathrm{p}(k+2)\\ \hat{F}_\mathrm{p}(k+3)\\ \hat{F}_\mathrm{p}(k+4)\end{Bmatrix},\quad \{F_\mathrm{r}\}=\begin{Bmatrix}F_\mathrm{r}(k+1)\\ F_\mathrm{r}(k+2)\\ F_\mathrm{r}(k+3)\\ F_\mathrm{r}(k+4)\end{Bmatrix},\quad \{G_\mathrm{I}\}=\begin{Bmatrix}g_0\\ g_1\\ g_2\\ g_3\end{Bmatrix}, \tag{7.31}$$

TABLE 7.2. Recursively Calculated Parameters of Polynomials E_j, F_j, and G_j

j	Parameters
1	$e_0=1;\; f_{10}=-(a_1-1),\; f_{11}=-(a_2-a_1),\; f_{12}=a_2;$ $g_0=g_{10}=b_0,\; g_{11}=b_1,\; g_{12}=b_2$
2	$e_1=f_{10};\; f_{20}=f_{11}-e_1(a_1-1),\; f_{21}=f_{12}-e_1(a_2-a_1),\; f_{22}=e_1a_2;$ $g_{20}=b_0,\; g_1=g_{21}=b_1+e_1b_0,\; g_{22}=b_2+e_1b_1,\; g_{23}=e_1b_2$
3	$e_2=f_{20};\; f_{30}=f_{21}-e_2(a_1-1),\; f_{31}=f_{22}-e_2(a_2-a_1),\; f_{32}=e_2a_2;$ $g_{30}=g_0,\; g_{31}=g_1,\; g_2=g_{32}=b_2+e_1b_1+e_2b_0,\; g_{33}=e_1b_2+e_2b_1,\; g_{34}=e_2b_2$
4	$e_3=f_{30};\; f_{40}=f_{31}-e_3(a_1-1),\; f_{41}=f_{32}-e_3(a_2-a_1),\; f_{42}=e_3a_2;$ $g_{40}=g_0,\; g_{41}=g_1,\; g_{42}=g_2,\; g_3=g_{43}=e_1b_2+e_2b_1+e_3b_0,$ $g_{44}=e_2b_2+e_3b_1,\; g_{45}=e_3b_2$

$$\{f\} = \begin{Bmatrix} g_{11} & g_{12} & f_{10} & f_{11} & f_{12} \\ g_{22} & g_{23} & f_{20} & f_{21} & f_{22} \\ g_{33} & g_{34} & f_{30} & f_{31} & f_{32} \\ g_{44} & g_{45} & f_{40} & f_{41} & f_{42} \end{Bmatrix} \begin{Bmatrix} \Delta f_c(k-1) \\ \Delta f_c(k-2) \\ F_p(k) \\ F_p(k-1) \\ F_p(k-2) \end{Bmatrix}. \tag{7.32}$$

Machining Test Results

The adaptive GPC algorithm is also integrated to the IMM as a PIM and tested on a milling machine controlled by an open-architecture CNC. The workpiece geometry and cutting conditions are given in Figure 7.6, which is the same as the pole-placement control tests shown in Figure 7.7. The reference peak force was set to F_r =1,200 N. The experimental results are given in Figure 7.8. The peak resultant force is always kept at the desired 1,200 N level, and the algorithm gives some overshoots when the axial depth of cut increases stepwise. However, the control algorithm is stable and rapidly converges. The corresponding feeds identified by the control law are also given, along with the estimated process parameters $(b_0, b_1, b_2, a_1, a_2)$ by the RLS PIM. The control interval corresponds to the spindle rotation period.

7.3.5 In-Process Detection of Tool Breakage

The tool breakage detection algorithm uses the average resultant cutting forces per tooth period (m), which can be written as follows:

$$F_a(m) = \frac{\sum_{i=1}^{I} \sqrt{F_x(i)^2 + F_y(i)^2}}{I}, \tag{7.33}$$

where I is the number of force samples collected at tooth period m. When runout and tooth breakage are not present, and the cutter is not in a transient geometric zone (i.e., the part–cutter intersection geometry is not changing), all teeth on the milling cutter produce equal average cutting forces, and the first differences of the average cutting forces,

$$\Delta F_a(m) = F_a(m) - F_a(m-1) = (1 - z^{-1})F_a(m), \tag{7.34}$$

will be zero. Otherwise, the average cutting forces will reflect the changes in the chip load and be nonzero. If the cutter runs into a transient geometry (i.e., varying entry and exit angles due to holes, slots, and voids), the differenced forces will reflect the trend [22]. Because the feed per tooth period is considerably smaller than the diameter of the cutter and the transient geometry changes, a first-order adaptive time series filter can remove the slow varying dc trend caused by the changes in the workpiece geometry as follows:

$$\epsilon_1(m) = (1 - \hat{\phi}_1 z^{-1})\Delta F_a(m), \tag{7.35}$$

where $\hat{\phi}_1$ is estimated from measurements $\Delta F_a(m)$ using the standard RLS technique at each tooth period. However, such a filter may still produce

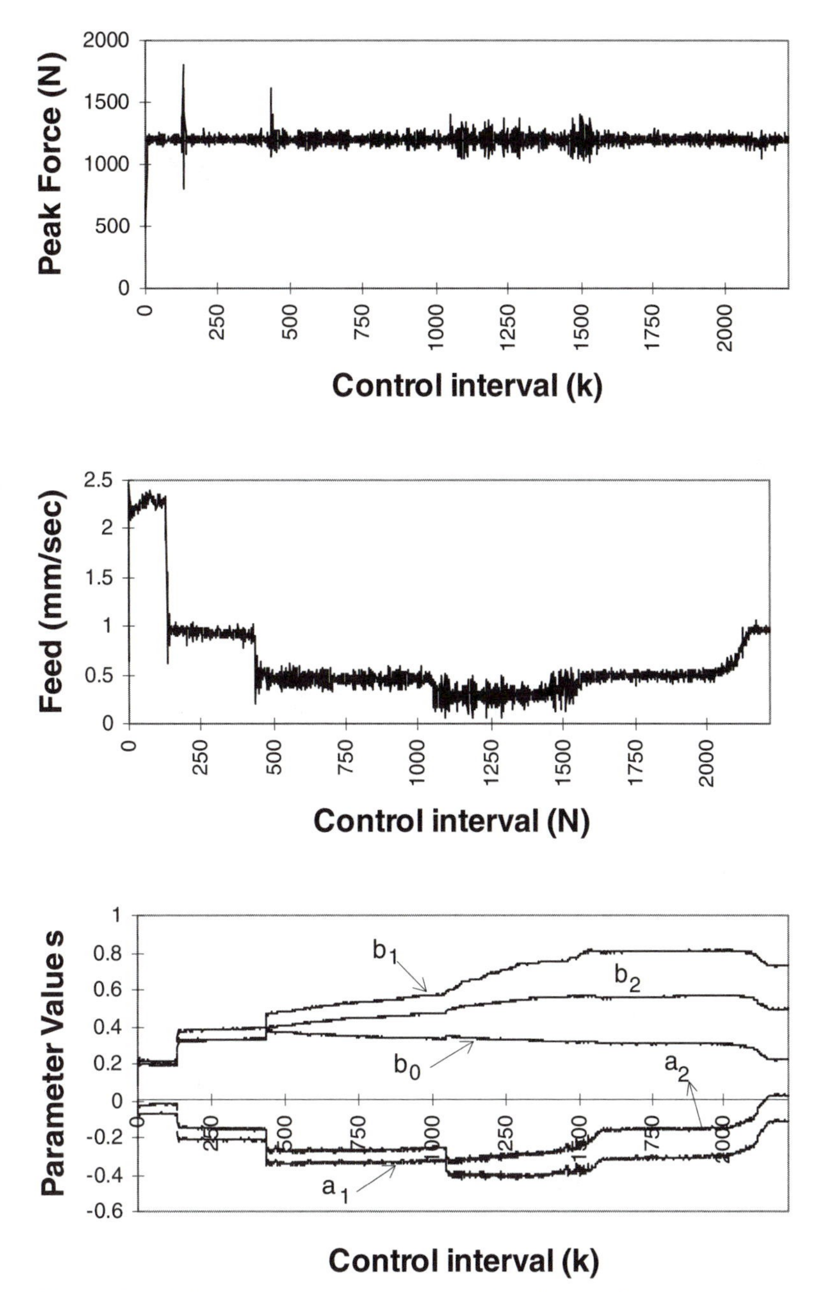

Figure 7.8: Milling test results of a generalized predictive adaptive control system. See Figure 7.6 for the part geometry and cutting conditions.

high-amplitude residuals [3] at each tooth period if the runouts on the cutter teeth are not the same. The runout of each tooth can be removed by comparing the tooth's performance by itself one revolution before as follows:

$$\Delta^N F_a(m) = F_a(m) - F_a(m-N). \tag{7.36}$$

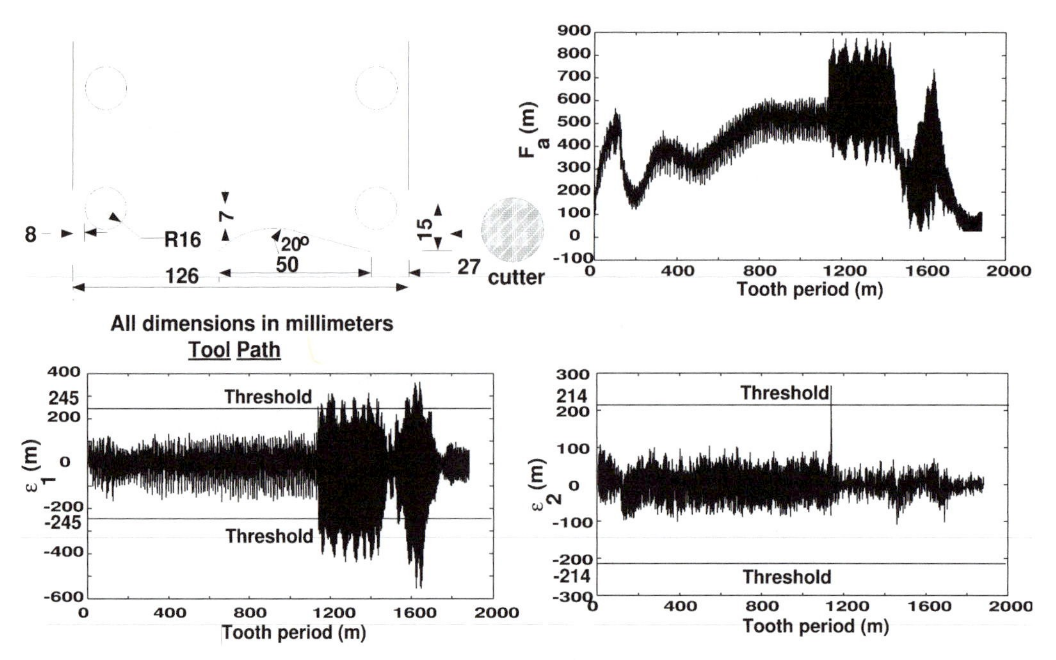

Figure 7.9: Typical tool breakage event. One flute has a chipped edge at tooth period 1,138. Cutting conditions: 25.4 mm diameter high-speed steel cutter with 4 flutes, spindle speed = 566 rev/min, feed rate = 0.1 mm/tooth, depth of cut = 7 .62 mm; radial width of cut is varying.

The resulting differences are again passed from a first-order adaptive time series filter to remove the possible dc trend left by the changes in the geometry as follows:

$$\epsilon_2(m) = (1 - \hat{\phi}_2 z^{-1})\Delta^N F_a(m). \tag{7.37}$$

The two adaptive time series filters are run recursively in parallel at every tooth period. When the cutting forces increase from a level of air machining at the beginning of cut, the maximum residuals of both filters are measured during the first five revolutions of spindle. It is assumed that the cutter is not broken during this period. The residuals contain the influence of runout on the cutter used and the noise, but not the geometric transients, which are filtered by the algorithm. The breakage thresholds are selected by scaling the maximum residuals by user-defined factors α_1 and α_2,

$$\text{LIMIT}_1 = \alpha_1 \cdot \max\{\epsilon_1\}, \quad \text{LIMIT}_2 = \alpha_2 \cdot \max\{\epsilon_2\}, \tag{7.38}$$

and they are used during the rest of the machining with the cutter. A tool breakage event is assumed whenever both residuals exceed their thresholds, $\epsilon_1(m) > \text{LIMIT}_1$ and $\epsilon_2(m) > \text{LIMIT}_2$ followed by an additional transient assurance check in the following revolution $\epsilon_1 > \text{LIMIT}_1$.

Figure 7.9 illustrates a sample testing of the algorithm for peripheral milling of a workpiece with transient voids and holes. The cutting test was repeated twice under identical cutting conditions, but one with a good cutter and the other with a chipped end mill. Later, the collected average force data are assembled at various transient geometry locations to test the robustness of the algorithm. The measured force and processed residuals are also shown in Figure 7.9. The intuitively selected threshold factors were $\alpha_1 = \alpha_2 = 2.0$, which correspond to $\text{LIMIT}_1 = 245$ N and $\text{LIMIT}_2 = 214$ N. The tool failure is detected because both residuals exceed their thresholds at the tool breakage event ($m = 1{,}138$), and ϵ_1 continuously exceeds its threshold thereafter without forgetting the breakage event.

The same algorithm can be applied to any sensor measurements provided that the signals can be correlated to cutting forces at the operating frequency (i.e., bandwidth) of the machine tool. For example, the feed drive motor current has been used at low cutting speeds for tool condition monitoring [4].

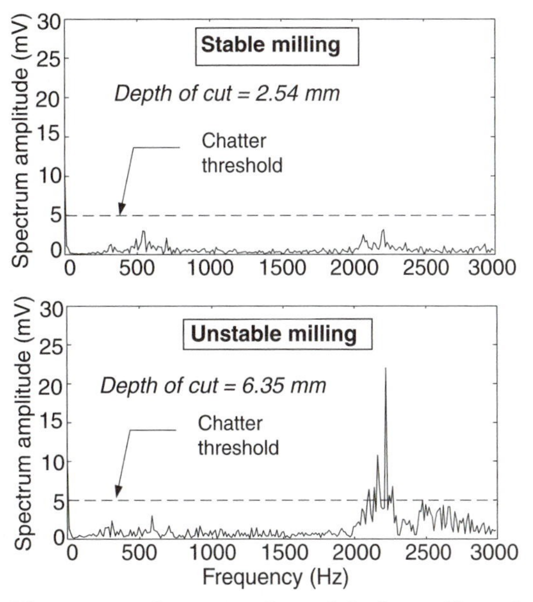

Figure 7.10: Spectrum of sound during stable and unstable machining. Cutting conditions: 19.05 mm diameter high-speed steel cutter with four flutes, spindle speed = 885 rev/min, feed rate = 0.01 mm/tooth, helix angle = 0, tool overhang = 74.0 mm, half-immersion up-milling, workpiece = Al 7075-T6 alloy.

7.3.6 Chatter Detection and Suppression

The chatter is detected by continuously monitoring the amplitude of the sound spectrum measured by a microphone. It was shown previously that the variation of spindle speed slightly improves the stability in milling, but limited spindle torque/power and frequency bandwidth prevent practical application of the method [9]. Smith and Delio [97] suppress the chatter by matching the tooth-passing frequency with the on-line detected chatter frequency. This results in the operation of milling at the highest stability lobe, which is the most favorable speed zone for chatter-free machining. Alternatively, Weck et al. [117] proposed a method of automatically reducing the axial depth of cut for chatter suppression. The reduction in the axial or radial depth of cut decreases the gain of the dynamic milling operation, which has a closed-loop behavior. Although the later strategy lowers the productivity, it always works and is especially applicable in the low spindle speed range. Because chatter occurs close to one of the structural frequencies, which are generally higher than 500 Hz, high spindle speed is required to avoid chatter by matching the tooth and chatter frequencies. Here, the power spectrum of the sound is calculated, and the maximum amplitude that occurs at the chatter frequency is searched every 250 ms. A sample spectrum of milling operations with and without chatter vibrations are shown in Figure 7.10. The magnitude of spectrum at the chatter frequency of 2,300 Hz is about six times larger than the one in chatter-free milling. The cutting conditions and the microphone location were identical, except that the depth of cut was increased from 2.54 mm to 6.35 mm in the unstable milling with chatter vibrations. It was observed that once the microphone is set for a particular machine, the sound spectrum does not change significantly with the changes in the feed rate and axial and radial depths of cut when there are no chatter vibrations. The chatter threshold is then selected as 5 mV, which is well above the maximum amplitude observed in stable, chatter-free machining tests. The CNC system assumes that chatter is present whenever the measured sound spectrum amplitude exceeds the threshold (5 mV) during milling on the particular machine tool setup [16].

7.4 INTELLIGENT POCKETING WITH THE IMM SYSTEM

The generalized predictive control methods of cutting forces, tool breakage detection, and chatter avoidance algorithms explained in the previous sections were integrated to our open-architecture CNC/IMM system. All three algorithms run in parallel and are able to manipulate the CNC system. The complete system was tested on a pocket milling application (see Fig. 7.11). The full axial depth of cut was programmed in the NC code sent to the CNC system. Default minimum and maximum feed rates are automatically assumed whenever the adaptive force control job is envoked by the user. The adaptive control manipulates the feed within the feed rate range during machining. The cutter fully penetrates to the bottom of the pocket at path position 1 and starts finishing the wall as it moves along the x axis toward 2. The wall's dimensional tolerance is 0.1 mm, and the cutter stiffness at the tip was 2,700 N/mm. The reference force to be controlled is selected to be the maximum force normal to the wall (F_y) and is set to 270 N to constrain the static deflection of the end mill within the tolerance. The system detects the chatter, stops the feed, decreases the axial depth of cut by 1 mm, and continues to cut with the feed regulated by the adaptive controller. The system automatically reduced the depth of cut from 6.35 mm to 3.35 mm where the chatter diminished (see Fig. 7.11). The adaptive control system adjusted the feed to 74.3 mm/min. When the tool moves from 2 to 3, the normal force to the wall is F_x, which is kept at 270 N. The same algorithm is repeated. If the chatter had occurred, the depth of cut would have been further reduced from 3.35 mm. The procedure is repeated at paths 3–4 and 4–5. The minimum depth of cut was assumed to be removed during the first layer, the NC program was updated, and the remaining wall was finished in one more pass, automatically. The cutter was positioned to path point 6, and the remaining island was removed with a reference resultant force set to 600 N to avoid shank failure. The tool breakage monitoring was always active during machining, and the algorithm never indicated a breakage event (see Fig. 7.12). Figure 7.13 shows the sound spectrum for machining the pocket. After the machining of the pocket, the walls had a 0.095 mm maximum form error, which was within the tolerance specified.

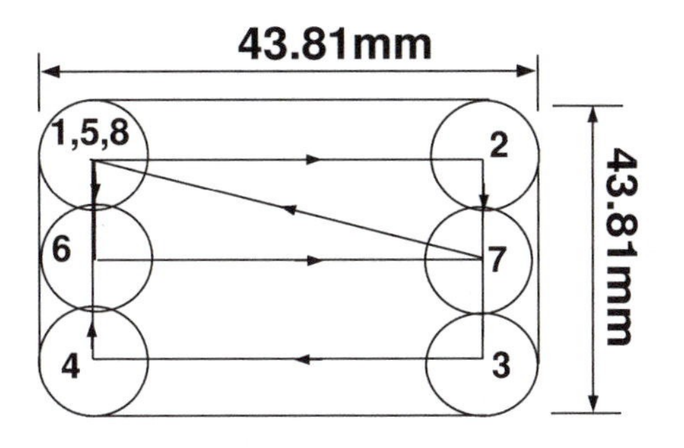

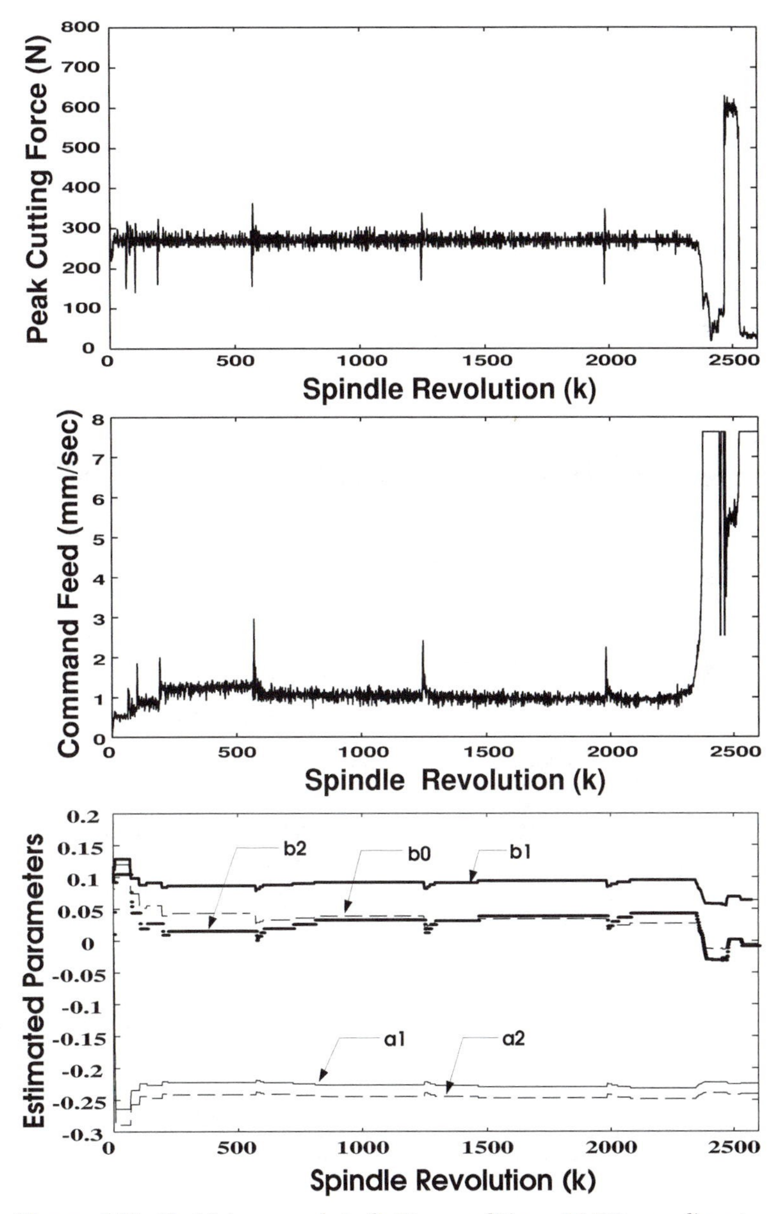

Figure 7.11: Machining a pocket. Cutting conditions: 15.875 mm diameter high-speed steel cutter with two flutes, spindle speed = 1,500 rev/min, varying feed rate, helix angle = 30 degrees, tool overhang = 76.2 mm, workpiece = Al 7075-T6 alloy.

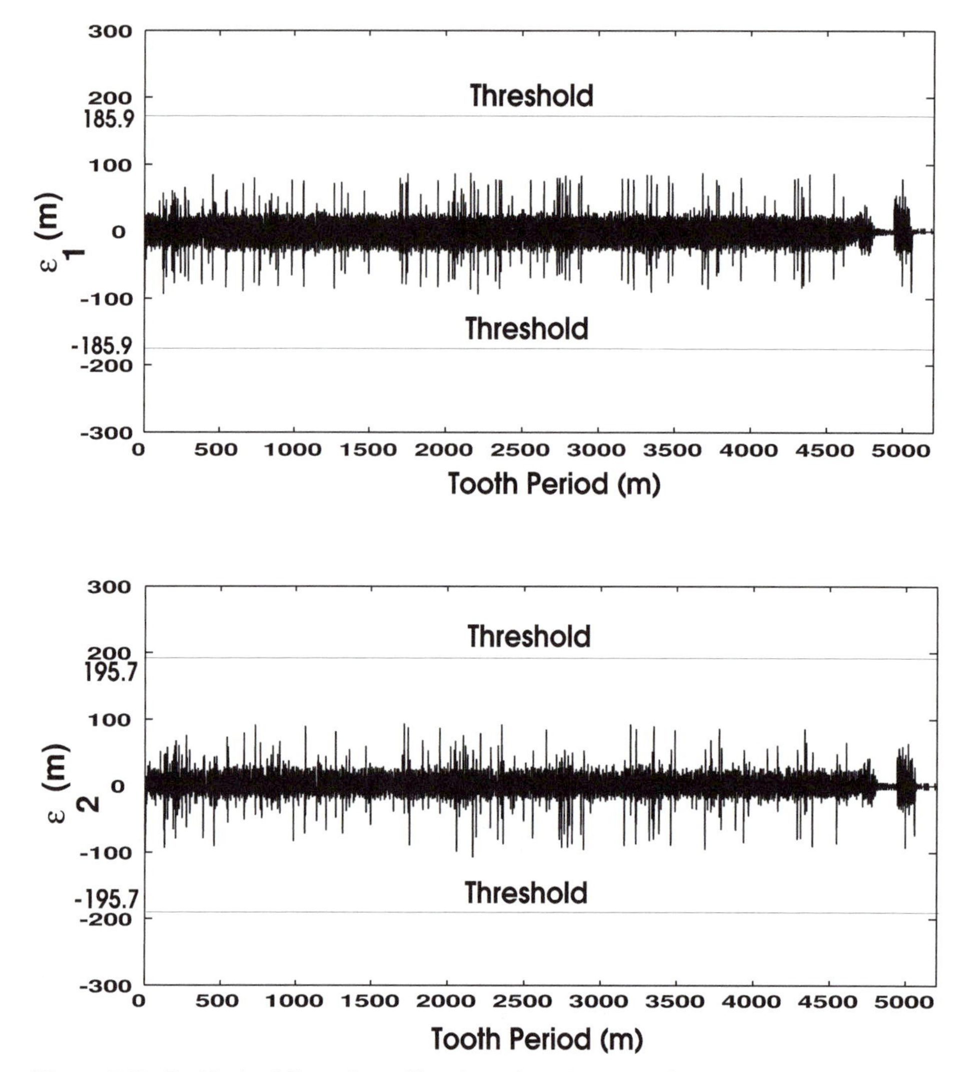

Figure 7.12: Residuals of filters for tool breakage detection in machining the pocket.

Other process control algorithms, such as thermal deformation compensation, collision detection, and in-process probing, can be added to the system by simply registering the new PIMs to the IMM library.

7.5 PROBLEMS

1. Design an adaptive pole-placement control system to keep cutting forces at a desired reference level in a turning process. The spindle speed is kept constant at 1,200 rev/min, and the resultant cutting force constant is given as $K_c = 1{,}500$ N/mm^2. The axial depth of cut is increased by 1.0 mm every 50 revolutions of the spindle, starting with 1.0 mm initial depth. When the spindle completes 250 revolutions, the axial depth of cut is decreased at the same rate. The machine tool's feed control servodynamics can be

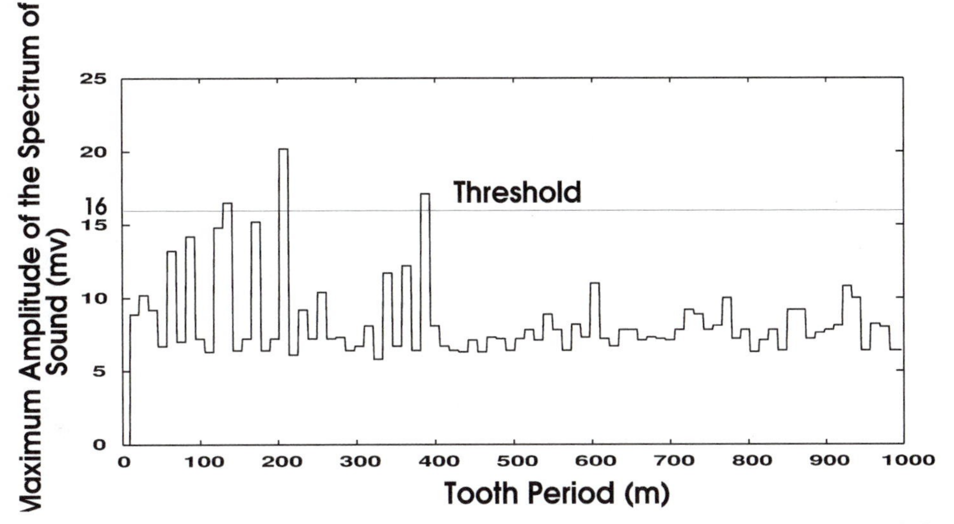

Figure 7.13: Maximum amplitude of sound spectrum during machining the pocket. (Note: Only a part of the first pass of machining is shown in the figure.)

approximated as a first-order continuous system with a time constant of $\tau_m = 0.1$ s. The vibration-free cutting process has an approximate time constant of $\tau_c = 0.65\ T$, where T is the spindle period. The adaptive control system must behave like a second-order system with a damping ratio of 0.95 and rise time of four spindle revolutions.

2. Design an adaptive Proportional Integral Darivative (PID) controller for the same process given in Problem 1.
3. Design an adaptive GPC for the same process given in Problem 1.
4. How can you modify the adaptive force controller so that it keeps the cutter deflections within 0.2 mm? Discuss the practical aspects of the application in minimizing the surface errors caused by the static deflections of the cutter.
5. Write a tool breakage detection program in the C language. Test the algorithm by simulating a tool breakage event within the milling force prediction algorithm.
6. How can you detect tool breakage by monitoring the power spectrum of cutting forces?
7. A piezo actuator–based tool holder is designed for precision positioning of cutting tools during turning operations. The picture of the piezo actuator–based tool holder mounted on the turret of a CNC turning center and its mechanical and control diagram are shown in Figure 7.14. One end of the piezo element rests on a solid, rigid wall, whereas the other end is bolted to a flexure mechanism carrying a tool holder. A signal voltage (u(V)) is supplied by the computer to the piezo electric high-voltage amplifier, which delivers a very high negative voltage (v(V)). When a high voltage is applied to the piezo element, it pushes or pulls the flexure carrying the tool holder. The displacement of the tool holder is measured by a laser. With the use of a

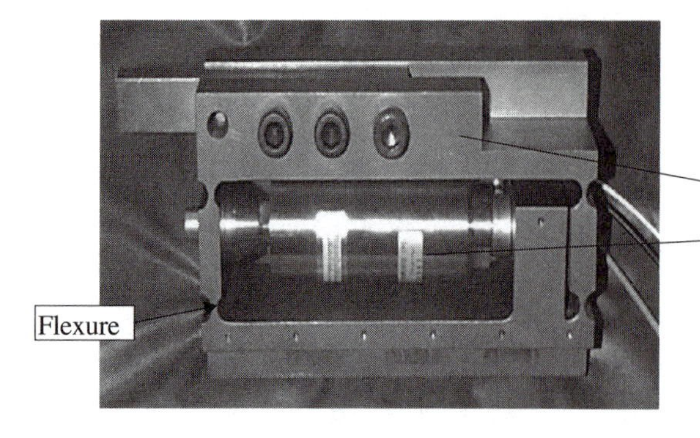

Photograph of the piezo actuator assembly.

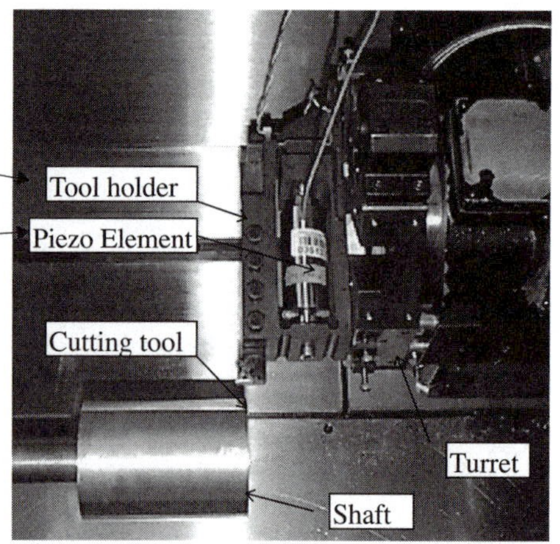

Piezo actuator mounted on CNC turning center turret.

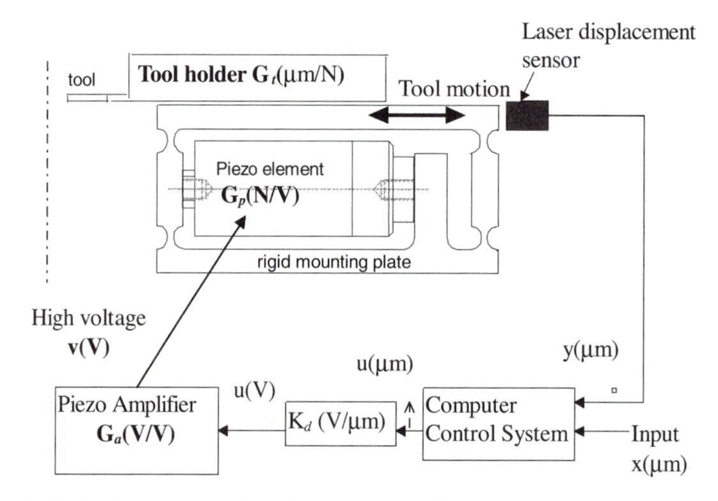

Figure 7.14: A piezo actuator–based precision tool holder system mounted on a CNC turning center turret. See Problem 7 for details.

digital control system, it is possible to control the position of the tool tip within 0.1 m accuracy. The actuator is used within a frequency bandwidth of 700 Hz and thus any dynamics higher than 1,000 Hz can be neglected. The frequency response behavior of each component in the system is shown in Figure 7.15. The tool holder can be modeled as an equivalent mass (m), spring (k), and damping (c) elements as shown in the figure. Design a control system by strictly following the order of instructions outlined below:

a. Draw the block diagram of a computer control system for the precision positioning of the tool tip.
b. Identify the transfer function of each physical component in the block diagram. Use symbols, but summarize the numerical values of each symbol at the end.

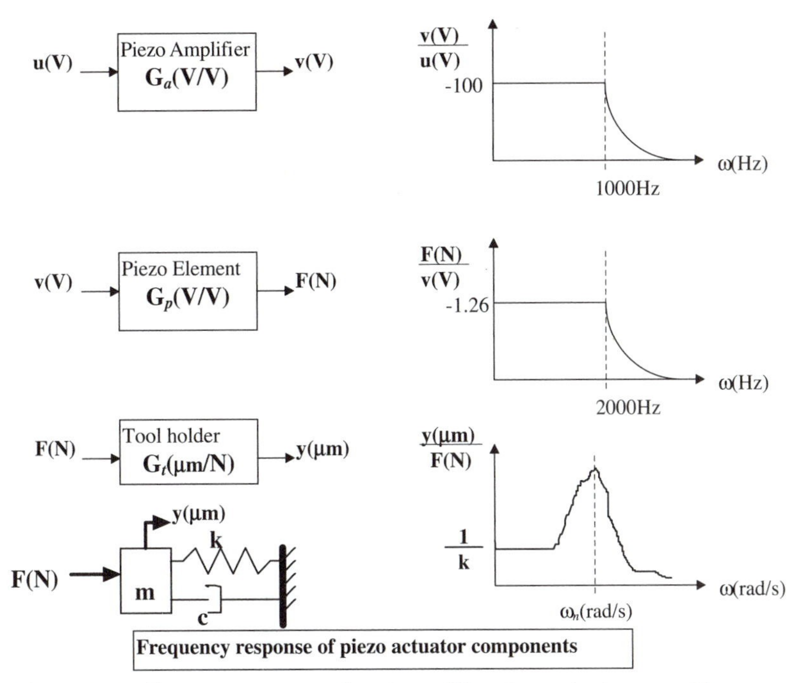

Figure 7.15: Frequency response functions of the piezo actuator assembly.

c. Assuming that the control/sampling interval is T seconds, express open-loop continuous and discrete (zero-order hold equivalent) transfer functions of the entire physical system algebraically. (Do not use numbers.)

d. Draw the block diagram of a pole-placement control system for ultra-positioning of the tool tip.

e. Design a pole-placement control law in such a way that the settling time and overshoot are t_s, M_p, respectively. (Find the corresponding desired natural frequency (w_m) and damping ratio (η) first, and use these two symbols in your algebraic derivations.) Solve the design problem analytically using only symbolic algebra.

f. The following numerical values are provided: $T = 0.0004$ s, $t_s = 0.005$ s, $M_p = 1$ percent. Tool holder: $m = 2.1$ kg, $c = 0.0006324$ N/(m/s), $k = 33.55$ N/m. D/A converter (K_d): 16 bit, -3V to $+3$V range. Summarize the numerical values of all unknowns in the following order:

- G_a, G_p, G_t: Transfer functions of piezo amplifier, piezo element, and tool holder.
- $G_o(s), G_o(z)$: Open-loop transfer functions of the entire physical piezo actuator assembly.
- B_m/A_m: Desired transfer function in z domain.
- R, S, T: Control law parameters.
- $u(k) = f(x, u, y)$: Expression for control input.

LAPLACE AND *z* TRANSFORMS

A.1 INTRODUCTION

The Laplace and z transform methods are useful for simplifying the analysis and design of linear time-invariant systems in continuous time and in discrete time, respectively. For example, considering a set of differential equations representing a system in continuous time, Laplace transform theory simplifies the problem by replacing the given set of differential equations by a set of algebraic equations.

In this appendix we will consider only linear time-invariant dynamic systems. For a more detailed study of the Laplace, z, and related transforms such as the Fourier and starred Laplace transforms and their interrelationships, refer to established control texts (e.g., Ogata [80, 81] and Kuo [65]).

The use of Laplace (s) and z transforms are sometimes confused by engineers who have limited experience in control theory. The following simple example is introduced first to illustrate the physical meaning and the use of both transformations in engineering.

Consider a simple first-order system (i.e., a servomotor) whose input is x (i.e., voltage given to the amplifier) and output is y (i.e., motor velocity). The system transfer function in the Laplace (i.e., continuous) domain is

$$G(s) = \frac{Y(s)}{X(s)} = \frac{K}{\tau s + 1} = K\frac{1/\tau}{s + 1/\tau}, \tag{A.1}$$

where K is the gain (at $s = 0$ or $t = +\infty$) and τ is the time constant of the motor. If a step input with an amplitude of A (i.e., physically A volts) is given to the system, we want to know what happens to the output y (i.e., motor velocity [rad/s]). Considering that the Laplace transform of step input is $L[x(t)] = X(s) = A/s$, we have

$$Y(s) = X(s) \cdot K\frac{1/\tau}{s + 1/\tau} = KA\frac{1/\tau}{s(s + 1/\tau)}. \tag{A.2}$$

The inverse Laplace transform of Eq. (A.2) provides the response (i.e., velocity) of the system (motor) for a step input (i.e., step change in the supply voltage) with an amplitude of A [in volts] in the continuous time domain as follws:

$$y(t) = L^{-1}Y(s) = KA(1 - e^{-t/\tau}). \tag{A.3}$$

It is evident from (A.3) that after some time the output will settle down at a steady state value (velocity) KA (i.e., rad/s). The settling time depends on the time constant of the system (τ). The smaller τ is, the faster the system will reach its steady-state target (final velocity) KA.

Now, let us consider that the command to the system is sent from a computer. A numerical value for input x is sent at discrete time intervals T from the computer to the physical system (i.e., motor amplifier). The number is converted to a voltage by a digital to analog (D/A) converter circuit. However, the computer sends the values only at T [s] intervals, and the value given to the physical device (i.e., amplifier) remains constant during the time interval T. The command generation is now completely in the discrete time domain; hence, the input must be converted to the z domain. The z transform of step input $x = A$ is $Z[x(kT)] = X(z) = Z(A) = A/(1 - z^{-1})$. Although the amplifier–servomotor system receives the input commands in discrete time intervals T, it works in the continuous time domain because it is a physical system. To combine the discrete (digital computer) and continuous (i.e., servomotor) parts under one domain, we have to obtain a discrete equivalent transfer function of the physical system. The D/A converter circuit is the physical bridge between the two domains, and it has to be considered. It is most practical to model the D/A circuit by a sample and hold with a zero-order approximation, that is, $ZOH = (1 - e^{-sT})/s$. Because $z^{-1} = e^{-sT}$, a ZOH equivalent of the continuous physical system is converted into the discrete time domain as follows:

$$\begin{aligned} G(z) &= (1 - z^{-1}) Z \frac{G(s)}{s} = (1 - z^{-1}) \left[K \frac{1/\tau}{s(s + 1/\tau)} \right] \\ &= (1 - z^{-1}) \frac{K z^{-1} (1 - e^{-T/\tau})}{(1 - z^{-1})(1 - e^{-T/\tau} z^{-1})} \\ &= \frac{z^{-1} [K(1 - e^{-T/\tau})]}{1 - e^{-T/\tau} z^{-1}}. \end{aligned} \tag{A.4}$$

Now that the system is completely transferred to the z domain, we can analyze its response for a step input generated by the computer. The steady-state response of the system for a step input with amplitude A can be found as follows:

$$y_{ss} = \lim_{z=1} X(z) \cdot G(z) = \lim_{z=1} \frac{A}{1 - z^{-1}} \cdot \frac{z^{-1}[K(1 - e^{-T/\tau})]}{1 - e^{-T/\tau} z^{-1}} = KA, \tag{A.5}$$

which is identical to the value found from (A.3) for the continuous system. We can look at the system response at discrete time intervals $k = 0, 1, 2, 3, \ldots, k$ that are equally spaced with T [s] periods, called the *sampling* or *control interval*. If the computer sends the step change in the following order: $x(0) = 0, x(1) = A,\ x(2) = A,\ x(3) = A, \ldots, x(k) = A$, then noting that $z^{-1}x(k) = x(k - 1)$ and $z^{-2}x(k) = x(k - 2)$, the system's response in discrete time intervals can be found from (A.4) as follows:

$$\begin{aligned} & y(k) = G(z) x(k), \\ & [(1 - e^{-T/\tau}) z^{-1}] y(k) = z^{-1}[K(1 - e^{-T/\tau})] x(k), \\ & y(k) = +e^{-T/\tau} y(k - 1) + [K(1 - e^{-T/\tau})] x(k - 1). \end{aligned} \tag{A.6}$$

The reader can simulate both systems (Eqs. A.3 and A.6) by assigning numerical values to the system and comparing the results (e.g., $A = 10,\ K = 2, T = 1,\ \tau = 0.2$).

The handling of Laplace (continuous) and z domain transformations must be clearly understood before designing machine tool or machining process control algorithms.

In this appendix, besides introducing the derivation of the (inverse) Laplace and z transforms we will also present (inverse) Laplace and z transforms using the MATLAB© Symbolic Math Toolbox [1].

A.2 BASIC DEFINITIONS

Let $x(t)$ be a continuous function of time t with sampled values represented by the sequence

$$\{x(kT)\} = \{x(0), x(T), x(2T), \ldots\}$$

or

$$\{x(k)\} = \{x(0), x(1), x(2), \ldots\}, \tag{A.7}$$

where T is the sampling period, which is sometimes not explicitly indicated. In Eq. (A.7), it is assumed that

$$x(t) = 0 \quad \text{for } t < 0 \quad \text{and} \quad x(kT) = 0 \quad \text{for } k < 0, \tag{A.8}$$

where k is a nonnegative integer ($k = 0, 1, 2, \ldots$).

The Laplace transform of $x(t)$ is defined as

$$\mathcal{L}[x(t)] = X(s) = \int x(t)e^{-st}dt. \tag{A.9}$$

The z transform of $x(t)$ or of the corresponding sequence $\{x(kT)\}$ (see Eq. A.7) is defined as

$$Z[x(t)] = X\{x(kT)\} = \sum_{k=0}^{\infty} x(kT)z^{-k},$$

or dropping the sampling period T, we have

$$Z[x(t)] = X\{x(k)\} = \sum_{k=0}^{\infty} x(k)z^{-k} \tag{A.10}$$

and $x(kT) = Z^{-1}[X(z)]$ is the inverse z transform. Both Laplace and z transforms given by (A.9) and (A.10) are one-sided transforms (i.e., the integral and summation in (A.8) and (A.10) start from zero), which are justified following (A.8). These one-sided transforms are the ones that are widely used in controls.

Conceptually, the z transform is easier to understand than the Laplace transform because the z transform involves summation, whereas the Laplace transform involves integration. For example, consider the ramp function as follows:

$$x(t) = \begin{Bmatrix} At & t \geq 0, \\ 0 & \text{otherwise.} \end{Bmatrix} \tag{A.11}$$

or the corresponding unit ramp sequence

$$x(t)|_{t=kT} = x(kT) = \begin{cases} AkT & k = 0, 1, 2, \ldots, \\ 0 & \text{otherwise.} \end{cases} \tag{A.12}$$

Following (A.9) and (A.11) and, for example, using integration by parts, the Laplace transform of $x(t)$ is obtained as

$$X(s) = \mathcal{L}[At] = \int_0^{\infty} Ate^{-st}dt = \frac{A}{s^2}. \tag{A.13}$$

However, the z transform of (A.12) is a straightforward summation as follows:

$$X(z) = \sum_{k=0}^{\infty} AkT \cdot z^{-k} = AT(z^{-1} + 2z^{-2} + 3z^{-3} + \cdots), \tag{A.14}$$

which converges for $|z| > 1$. However, (A.14) is not in a very useful form because it is an open-ended series (i.e., the summation has infinitely many terms). By the use of the power series, the z transform (A.14) can be written in closed form as follows:

$$X(z) = AT\frac{z^{-1}}{(1 - z^{-1})^2} = AT\frac{z}{(z - 1)^2}. \tag{A.15}$$

Although, as mentioned, the z transform is conceptually easier to understand than the Laplace transform, in textbooks on controls, the Laplace transform is considered before the z transform. The reason is that, although in the analysis and design of continuous-time control systems the Laplace transform suffices, in the case of digital control systems both the Laplace and z transforms are necessary. This is because the digital control systems have both continuous-time (e.g., plant) and discrete-time (e.g., controller) components as illustrated in the previous section.

As shown in (A.15), the z transform of a function or a sequence can be written both using the negative or positive powers of z. One word of caution is that when the negative powers of z are used, one has to be careful about the zeros at the origin of the z plane [80]. To this end, consider

$$X(z) = \frac{z}{z - 3} = \frac{1}{1 - 3z^{-1}}, \tag{A.16}$$

which, when written in terms of the positive powers of z, clearly has a pole at $z = 3$ and a zero at the origin. However, when $X(z)$ is written in terms of the negative powers of z, although the pole at $z = 3$ is clearly seen, the zero at the origin may not be that obvious. In this appendix the positive powers of z will be used to obtain the residues. However, whenever the z transform table has to be used, the expressions obtained will be written in terms of the negative powers of z.

Given a transfer function in the s domain, the corresponding transfer function in the z domain can be obtained by using several different methods such as: 1) backward difference method; 2) forward difference method; 3) matched pole-zero mapping method; 4) bilinear transformation (or Tustin) method (with frequency prewarping); 5) impulse invariance method; and 6) step invariance

TABLE A.1. Table of Laplace and z Transforms [80]

$x(t)$	$X(s)$	$x(kT)$	$X(z)$
Impulse	—	$\delta(k)=\left\{\begin{smallmatrix}1 & k=0\\ 0 & k\neq 1\end{smallmatrix}\right\}$	1
Delay	—	$\delta(n-k)=\left\{\begin{smallmatrix}1 & k=n\\ 0 & k\neq n\end{smallmatrix}\right\}$	z^{-k}
1	$\frac{1}{s}$	1	$\frac{1}{1-z^{-1}}$
e^{-at}	$\frac{1}{s+a}$	e^{-akT}	$\frac{1}{1-e^{-aT}z^{-1}}$
t	$\frac{1}{s^2}$	kT	$\frac{Tz^{-1}}{(1-z^{-1})^2}$
t^2	$\frac{2}{s^3}$	$(kT)^2$	$\frac{T^2z^{-1}(1+z^{-1})}{(1-z^{-1})^3}$
t^3	$\frac{6}{s^4}$	$(kT)^3$	$\frac{T^3z^{-1}(1+4z^{-1}+z^{-2})}{(1-z^{-1})^4}$
$1-e^{-at}$	$\frac{a}{s(s+a)}$	$1-e^{-akT}$	$\frac{z^{-1}(1-e^{-aT})}{(1-z^{-1})(1-e^{-aT}z^{-1})}$
$e^{-at}-e^{-bt}$	$\frac{b-a}{(s+a)(s+b)}$	$e^{-akT}-e^{-bkT}$	$\frac{z^{-1}(e^{-aT}-e^{-bT})}{(1-e^{-aT}z^{-1})(1-e^{-bT}z^{-1})}$
te^{-at}	$\frac{1}{(s+a)^2}$	kTe^{-akT}	$\frac{Te^{-aT}z^{-1}}{(1-e^{-aT}z^{-1})^2}$
$(1-at)e^{-at}$	$\frac{s}{(s+a)^2}$	$(1-akT)e^{-akT}$	$\frac{1-(1+aT)e^{-aT}z^{-1}}{(1-e^{-aT}z^{-1})^2}$
t^2e^{-at}	$\frac{2}{(s+a)^3}$	$(kT)^2e^{-akT}$	$\frac{T^2e^{-aT}(1+e^{-aT}z^{-1})z^{-1}}{(1-e^{-aT}z^{-1})^3}$
$\sin\omega t$	$\frac{\omega}{s^2+\omega^2}$	$\sin\omega kT$	$\frac{z^{-1}\sin\omega T}{1-2z^{-1}\cos\omega T+z^{-2}}$
$\cos\omega t$	$\frac{s}{s^2+\omega^2}$	$\cos\omega kT$	$\frac{1-z^{-1}\cos\omega T}{1-2z^{-1}\cos\omega T+z^{-2}}$
$e^{-at}\sin\omega t$	$\frac{\omega}{(s+a)^2+\omega^2}$	$e^{-akT}\sin\omega kT$	$\frac{e^{-aT}z^{-1}\sin\omega T}{1-2e^{-aT}z^{-1}\cos\omega T+e^{-2aT}z^{-2}}$
$e^{-at}\cos\omega t$	$\frac{s+a}{(s+a)^2+\omega^2}$	$e^{-akT}\cos\omega kT$	$\frac{1-e^{-aT}z^{-1}\cos\omega T}{1-2e^{-aT}z^{-1}\cos\omega T+e^{-2aT}z^{-2}}$
		a^k	$\frac{1}{1-az^{-1}}$
		$a^{k-1}, k=1,2,3\ldots$	$\frac{z^{-1}}{1-az^{-1}}$
		ka^{k-1}	$\frac{z^{-1}}{(1-az^{-1})^2}$
		$a^k\cos k\pi$	$\frac{1}{1+az^{-1}}$

method [80]. Among these methods the impulse invariance uses the z transform directly.

Note that the examples given by Eqs. (A.11) and (A.12) are elementary cases and, therefore, rather than using the transformation formulas as in (A.13) and (A.14), we could directly use Table A.1 to obtain the Laplace and z transforms. To simplify the problem in nonelementary cases, it is perhaps easier to use Laplace and z transform theorems given in Tables A.2 and A.3 whenever possible. In fact, in Eqs. (A.13) and (A.15) we already made use of the first property of Laplace and z transform theorems.

TABLE A.2. Properties of Laplace Transforms [80]

$\mathcal{L}[Af(t)] = AF(s)$
$\mathcal{L}[f_1(t) \pm f_2(t)] = F_1(s) \pm F_2(s)$
$\mathcal{L}\pm\left[\frac{d}{dt}f(t)\right] = sF(s) - f(0\pm)$
$\mathcal{L}\pm\left[\frac{d^2}{dt^2}f(t)\right] = s^2F(s) - sf(0\pm) - \dot{f}(0\pm)$
$\mathcal{L}[e^{-at}f(t)] = F(s+a)$
$\mathcal{L}[f(t-a)1(t-a)] = e^{-as}F(s)$
$\mathcal{L}[tf(t)] = -\frac{dF(s)}{ds}$
$\mathcal{L}\left[\frac{1}{t}f(t)\right] = \int_s^\infty F(s)ds$

Next, we will consider another example by which we will demonstrate using the table of Laplace transform properties (Table A.2):

$$\mathcal{L}\{Ae^{-\alpha t}\cos\omega t\} = A \cdot \mathcal{L}\{e^{-\alpha t}\cos\omega t\} \tag{A.17}$$

where A, a, and ω are real. From the Laplace transform table (A.1) we have

$$\mathcal{L}\{\cos\omega t\} = \frac{s}{s^2+\omega^2}, \tag{A.18}$$

and, using the Laplace transform property of real translation in the s domain (see Table A.2), we obtain

$$\mathcal{L}\{Ae^{-\alpha t}\cos\omega t\} = A\frac{s+\alpha}{(s+\alpha)^2+\omega^2}. \tag{A.19}$$

Note that the function considered "$e^{-\alpha t}\cos\omega t$" represents a damped sinusoidal signal and, owing to its importance in engineering systems, it is given in the Laplace transform Table A.1 as well.

In the rest of this appendix we will consider inverse Laplace and z transforms. In the case of inverse Laplace transforms, attention will be focused on the partial fraction expansion method. In the case of the inverse z transform, besides the partial fraction expansion method, we will also consider the residue method. The residue method, which is also known as the inversion residue method as well as the inversion formula method, leads to the inverse z transforms directly without resorting to the z transform table.

TABLE A.3. Properties of z Transforms [80]

$x(k)$ or $x(k)$	$Z[x(t)]$ or $Z[x(k)]$
$ax(t)$	$aX(z)$
$ax_1(t)+bx_2(t)$	$aX_1(z)+bX_2(z)$
$x(t+T)$ or $x(k+1)$	$zX(z)z - x(0)$
$x(t+2T)$	$z^2X(z) - z^2x(0) - zx(T)$
$x(t+kT)$	$z^kX(z) - z^kx(0) - z^{k-1}x(T) - \cdots - zx(kT-T)$
$x(t-kT)$ or $x(n-k)$	$z^{-k}X(z)$
$tx(t)$	$-Tz\frac{d}{dz}X(z)$
$e^{-at}x(t)$	$X(ze^{aT})$
$\nabla x(k) = x(k) - x(k-1)$	$(1-z^{-1})X(z)$
$\sum_{k=0}^{n} x(k)$	$\frac{1}{1-z^{-1}}X(z)$

A.3 PARTIAL FRACTION EXPANSION METHOD

The partial fraction expansion method is a mathematical tool similar to the continued fraction method. The partial fraction expansion method is also utilized to obtain inverse Laplace, z, and Fourier transforms among which we will consider the first two.

In the case of inverse transforms, the partial fraction expansion method enables us to write a given a rational function in s or z in such a way that we can use the Laplace or z transform tables directly or after some manipulations.

Consider the transfer function

$$X(p) = \frac{N(p)}{(p-p_1)\cdots(p-p_\mathrm{r})\cdots(p-p_\mathrm{n})}, \tag{A.20}$$

which is assumed to have distinct poles and no pole-zero cancellation. In (A.20) the degree of the numerator polynomial $N(p)$ is less than the degree of the denominator polynomial, which is n. In other words $X(p)$ is a strictly proper rational function.

Note that in the case of the Laplace transform we will set $p = s$ and in the case of the z transform we will set $p = z$. In the rest of this appendix we will consider the negative powers of z only when we use the z transform table.

If the degree of the numerator polynomial $N(p)$ of (A.20) is equal to or greater than n, long division is used yielding

$$X(p) = F(p) + \tilde{X}(p), \tag{A.21}$$

where $F(p)$ is a polynomial of positive powers of p and is a strictly proper rational function. If $X(p)$ represents a transfer function of an engineering system, following the causality condition, $F(p)$ will be a constant.

In the case of distinct poles, partial fraction expansion of (A.20) is written as

$$X(p) = \frac{A_1}{p-p_1} + \cdots + \frac{A_\mathrm{r}}{p-p_\mathrm{r}} + \cdots + \frac{A_\mathrm{n}}{p-p_\mathrm{n}}, \tag{A.22}$$

where $A_i (i = 1, \ldots, n)$ is the residue associated with pole p_i. Noting that all poles in (A.22) are assumed to be distinct, the residues are obtained according to

$$A_i = [(p-p_i)X(p)]|_{p=p_i}, \tag{A.23}$$

where (A.23) is obtained by multiplying (A.22) by $p - p_i$ and evaluating at $p = p_i$. Note that (A.22) and (A.23) are valid when poles are real or complex conjugate ($i = 1, 2, \ldots, n$). In the later case, for example, if $p_2 = p_1^*$ then $A_2 = A_1^*$ where the asterisk indicates complex conjugate.

In the case of multiple poles,

$$X(p) = \frac{A_1}{p-p_1} + \frac{A_{21}}{p-p_2} + \frac{A_{22}}{(p-p_2)^2} + \cdots + \frac{A_{2j}}{(p-p_2)^j} + \cdots \frac{A_{2\mathrm{r}}}{(p-p_2)^r}, \tag{A.24}$$

where the multiplicity of pole p_2 is assumed to be r; the residues associated with the multiple pole are derived from

$$A_{2j} = \frac{1}{(r-j)!}\left\{\frac{d^{r-j}}{dp^{r-j}}[(p-p_2)^r X(p)]\right\}\Bigg|_{p=p_r} \quad (j = 1, \ldots, r). \tag{A.25}$$

Note that the residues can alternatively be obtained by solving a set of algebraic equations, as will be demonstrated in Example 1.

The following examples demonstrate obtaining the residues by using the partial fraction method.

Example 1. To demonstrate the partial fraction expansion for single poles, consider the following rational function with distinct poles:

$$X(p) = \frac{1}{(p+2)(p+4)} = \frac{A_1}{p+2} + \frac{A_2}{p+4}. \tag{A.26}$$

to demonstrate the partial fraction expansion for single poles.

Following (A.23) and (A.26) the partial fraction can be written as

$$X(p) = \frac{0.5}{p+2} + \frac{-0.5}{p+4}. \tag{A.27}$$

Note that the residues $A_1 = 0.5$ and $A_2 = -0.5$ alternatively could also be obtained from a set of linear equations as follows:

$$\begin{bmatrix} 1 & 1 \\ 4 & 2 \end{bmatrix} \cdot \begin{bmatrix} A_1 \\ A_2 \end{bmatrix} = \begin{bmatrix} 0 \\ 1 \end{bmatrix}, \tag{A.28}$$

which obtained from

$$\frac{A_1(p+4) + A_2(p+2)}{(p+2)(p+4)} = \frac{1}{(p+2)(p+4)}, \tag{A.29}$$

where we equate the equal powers of p in the numerator.

Example 2. Next consider the partial fraction expansion in the case of multiple poles as follows:

$$X(p) = \frac{p+0.4}{(p+0.1)^2(p+0.2)}, \tag{A.30}$$

which, using (A.25), yields

$$X(p) = \frac{-20}{p-0.1} + \frac{3}{(p-0.1)^2} + \frac{20}{p-0.2}, \tag{A.31}$$

where each term of (A.31) appears in the Laplace transform table (see Example 5).

Example 3. In the case of inverse z transforms, the poles at the origin have to be handled carefully. Therefore, in this example, we choose a transfer function with a double pole at the origin as follows:

$$X(p) = \frac{1}{(p-0.8)^2 p^2}. \tag{A.32}$$

With the use of the residue formulas (A.23) and (A.25) the partial fraction expansion of (A.32) can be written as

$$X(p) = \frac{3.9}{p - 0.8} + \frac{1.56}{(p - 0.8)^2} + \frac{3.9}{p} + \frac{1.56}{p^2}. \quad \text{(A.33)}$$

A.4 PARTIAL FRACTION EXPANSION METHOD TO DETERMINE INVERSE LAPLACE AND z TRANSFORMS

In this section, we will use the partial fraction expansion method to obtain the inverse Laplace and z transforms. Note that, in the case of the z transform, if the transfer function $X(z)$ is written using the positive powers of z and has at least one zero at the origin, then instead of $X(z)$ we can also consider $X(z)/z$ [80].

Example 4. To demonstrate the inverse Laplace transform in the case of single poles, consider the transfer function (A.26) in the s domain by substituting $p = s$ as follows:

$$X(s) = \frac{1}{(s + 2)(s + 4)}. \quad \text{(A.34)}$$

Following (A.27), we obtain $x(t)$ using the Laplace transform table as

$$x(t) = \mathcal{L}^{-1}[X(s)] = 0.5\mathcal{L}^{-1}\frac{1}{s + 2} - 0.5\mathcal{L}^{-1}\frac{1}{s + 4} = 0.5e^{-2t}u(t) - e^{-4t}u(t), \quad \text{(A.35)}$$

where $u(t)$ is the unit step.

Example 5. Next, we consider the inverse Laplace transform of transfer functions (A.30) with multiple poles as follows:

$$X(s) = \frac{s + 0.4}{(s + 0.1)^2(s + 0.2)}. \quad \text{(A.36)}$$

Following (A.31) and using the Laplace transform table, we have

$$\begin{aligned} x(t) = \mathcal{L}^{-1}[X(s)] &= -20\mathcal{L}^{-1}\frac{1}{s + 0.1} + 3\mathcal{L}^{-1}\frac{1}{(s + 0.1)^2} + 20\mathcal{L}^{-1}\frac{1}{s + 0.2} \\ &= (-20e^{-0.1t} + 3te^{-0.1t} + 20e^{-0.2t})u(t). \end{aligned} \quad \text{(A.37)}$$

Example 6. To demonstrate the inverse z transform we start by considering the transfer functions with single poles as in the case of the inverse Laplace transform (see Example 4). Letting $p = z$ in (A.26) we get

$$X(z) = \frac{1}{(z + 2)(z + 4)} = \frac{1}{z^2 + 6z + 8}. \quad \text{(A.38)}$$

Equation (A.27) can be written in Z^{-1} domain by multiplying the numerator and denominator by z^{-2} as follows:

$$X(z) = \frac{z^{-2}}{1 + 6z^{-1} + 8z^{-2}}. \quad \text{(A.39)}$$

For discrete input $u(k)$, the output $x(k)$ can be evaluated at each step k as a recursive function of past input and output values, that is,

$$x(k) = u(k-2) - 6x(k-1) - 8x(k-2). \tag{A.40}$$

However, the contributions of each root-pole on both transfer function as well as on the value of each response for any given input state can be explained better by looking at the partial fraction of the transfer function. Equation (A.27) has the following partial fraction expansion:

$$X(z) = \frac{0.5}{z+2} + \frac{-0.5}{z+4}. \tag{A.41}$$

Note that if the transfer function is given in terms of the negative powers of z, we will write the transfer function in terms of the positive powers of z before deriving the partial fractions.

From (A.41) we have

$$\begin{aligned} x(k) = Z^{-1}[X(z)] &= 0.5Z^{-1}\frac{1}{z+2} - 0.5Z^{-1}\frac{1}{z+4} \\ &= 0.5Z^{-1}\frac{z^{-1}}{1+2z^{-1}} - 0.5Z^{-1}\frac{z^{-1}}{1+4z^{-1}}, \end{aligned} \tag{A.42}$$

which, with the use of the tables and simplifying the results obtained, yields

$$x(k) = (-1)^{k-1}\{2^{k-2} - 2^{2k-3}\}u(k-1). \tag{A.43}$$

Here, $u(k-1)$ is the one-sample delayed unit step sequence.

Note that, even though partial fractions are obtained by using positive powers of z, the inverse z transform is obtained from the table that is given in terms of the negative powers of z.

Example 7. Consider the rational function (A.30) with multiple poles, to demonstrate the inverse z transform. Substituting $p = z$ we get

$$X(z) = \frac{z+0.4}{(z+0.1)^2(z+0.2)} \tag{A.44}$$

and from (A.31) we have

$$\begin{aligned} x(k) = Z^{-1}[X(z)] &= -20Z^{-1}\frac{1}{z+0.1} + 3Z^{-1}\frac{1}{(z+0.1)^2} + 20Z^{-1}\frac{1}{z+0.2} \\ &= -20Z^{-1}\frac{z^{-1}}{1+0.1z^{-1}} + 3Z^{-1}\frac{z^{-2}}{(1+0.1z^{-1})^2} + 20Z^{-1}\frac{z^{-1}}{1+0.2z^{-1}}. \end{aligned} \tag{A.45}$$

Considering the second term on the right-hand side of (A.45) and using the z transform and the shifting property we have

$$Z^{-1}\left[z^{-1}\left(\frac{z^{-1}}{(1+0.1z^{-1})^2}\right)\right] = (k-1)(-0.1)^{k-2}u(k-1), \tag{A.46}$$

from which, using (A.45), we obtain

$$x(k) = \{1 + 2^k - 3k\}(-1)^{k-1}(0.1)^{k-2}u(k-1). \tag{A.47}$$

Example 8. As a last example to demonstrate the use of the partial fraction expansion in obtaining the inverse z transform of transfer functions with multiple poles at the origin, consider the rational function (A.32) with $p = z$ as follows:

$$X(z) = \frac{1}{(z-0.8)^2 z^2}. \tag{A.48}$$

Following (A.33), we have

$$X(z) = \frac{3.9}{z-0.8} + \frac{1.56}{(z-0.8)^2} + \frac{3.9}{z} + \frac{1.56}{z^2}, \tag{A.49}$$

$$\begin{aligned} x(k) = Z^{-1}[X(z)] &= 3.9Z^{-1}\frac{z^{-1}}{1-0.8z^{-1}} \\ &\quad + 1.56Z^{-1}\frac{z^{-2}}{(1-0.8z^{-1})^2} + 3.9Z^{-1}[z^{-1}] + 1.56Z^{-1}[z^{-2}], \end{aligned} \tag{A.50}$$

and using

$$Z^{-1}\left[z^{-1}\frac{z^{-1}}{(1-0.8z^{-1})^2}\right] = (k-1)(0.8)^{k-2}u(k-1),$$

$$Z^{-1}\left[\frac{z^{-1}}{1-0.8z^{-1}}\right] = (0.8)^{k-1}u(k-1),$$

$$Z^{-1}[z^{-1}] = \delta(k-1) = \begin{Bmatrix} 1 & k = 1, \\ 0 & k \neq 1, \end{Bmatrix}$$

$$Z^{-1}[z^{-2}] = \delta(k-2) = \begin{Bmatrix} 1 & k = 2, \\ 0 & k \neq 2, \end{Bmatrix}$$

we obtain

$$x(k) = (k-3)(0.8)^{k-4}u(k-3). \tag{A.51}$$

APPENDIX B

OFF-LINE AND ON-LINE PARAMETER ESTIMATION WITH LEAST SQUARES

B.1 OFF-LINE LEAST-SQUARES ESTIMATION

The transfer function of systems can be identified either by using impulse, step, and frequency response techniques in the continuous or discrete time domain or using regression techniques in the discrete time domain. When the process is to be controlled by a computer, it may be easier to identify the transfer function parameters of the system by collecting input and output data at uniform discrete time intervals that are equal to the computer control period.

Consider a general transfer function with d samples delay, with orders nb and na in the numerator and denominator, respectively (such a system is called ARIMA(na, d, nb) in the literature), as follows:

$$G_o(z^{-1}) = \frac{z^{-d}(b_0 + b_1 z^{-1} + b_2 z^{-2} + \cdots + b_{nb} z^{-nb})}{+ a_1 z^{-1} + a_2 z^{-2} + \cdots + a_{na} z^{-na}}. \tag{B.1}$$

The dead time can be easily observed by applying step, pulse, or harmonic input to the system and then measuring the time delay between the input and output on a digital scope. It is usually difficult to guess the order of a transfer function (i.e., d, na, nb) if we do not have any engineering feeling about the process. In such cases, one has to resort to stochastic order identification techniques, which are not covered here. In machine tool engineering, however, the orders of drives are usually low, up to about second or third order if we neglect the dynamic characteristics of the structure. Furthermore, a good engineer has to have the basic understanding of the drive design and control process provided in Chapter Six. It is customary to start with a low-order transfer function assumption and increase the order if the identification does not yield good output estimates with the measurements.

Let us assume a simple process with the following discrete transfer function:

$$G(z^{-1}) = \frac{y(k)}{u(k)} = \frac{z^{-2}(b_0 + b_1 z^{-1})}{1 + a_1 z^{-1}}, \tag{B.2}$$

which is a first-order process whose output (y) has a two-sample period delay in responding to an input (u). By expanding the transfer function, we get

$$(1 + a_1 z^{-1}) y(k) = [z^{-2}(b_0 + b_1 z^{-1})] u(k),$$

$$y(k) = -a_1 y(k-1) + b_0 u(k-2) + b_1 u(k-3), \tag{B.3}$$

or

$$y(k) = \{-y(k-1) \quad u(k-2) \quad u(k-3)\}\{a_1 \quad b_0 \quad b_1\}^T \phi(k)^T \theta, \tag{B.4}$$

where k is the sampling counter, $\phi(k)^T = \{-y(k-1) \; u(k-2) \; u(k-3)\}$ is the regression or measurement vector, and $\theta = \{a_1 \; b_0 \; b_1\}^T$ is the unknown parameter vector. The three unknown parameters are *estimated* from the N measurements, which are greater than the number of unknowns (i.e., $N \gg 3$). If $\hat{\theta} = \{\hat{a}_1 \; \hat{b}_0 \; \hat{b}_1\}$ are the estimated parameters, the error between the estimated and actual measurement at interval k is

$$\epsilon(k) = y(k) - \hat{y}(k) = y(k) - \phi(k)^T \hat{\theta}. \tag{B.5}$$

Assume that a pulse or random input containing a wide range of frequencies is given to the system, and N number of inputs and outputs are collected at fixed time intervals. The following set of expressions can be written for each measurement:

$$\begin{Bmatrix} \epsilon(3) \\ \epsilon(4) \\ \vdots \\ \epsilon(k) \\ \vdots \\ \epsilon(N) \end{Bmatrix} = \begin{Bmatrix} y(3) \\ y(4) \\ \vdots \\ y(k) \\ \vdots \\ y(N) \end{Bmatrix} - \begin{bmatrix} \{-y(2) & u(1) & u(0)\} \\ \{-y(3) & u(2) & u(1)\} \\ \vdots & \vdots & \vdots \\ \{-y(k-1) & u(k-2) & u(k-3)\} \\ \vdots & \vdots & \vdots \\ \{-y(N-1) & u(N-2) & u(N-3)\} \end{bmatrix} \begin{Bmatrix} a_1 \\ b_0 \\ b_1 \end{Bmatrix},$$

or, in matrix form,

$$Y = \Phi\theta, \tag{B.6}$$

where

$$Y = \{y(3) \quad y(4) \quad \cdots y(k) \quad \cdots y(N)\}^T,$$
$$\Phi = \{\phi(3) \quad \phi(4) \quad \cdots \phi(k) \quad \cdots \phi(N)\}^T.$$

The least-squares identification method is based on minimizing the sum of the squares of errors evaluated from all measurements [48]. The sum of the squares of errors, or *cost function*, to be minimized is

$$J(\theta, N) = \sum_{k=3}^{N} \epsilon(k)^2 = (Y - \Phi\theta)^T (Y - \Phi theta) = (Y^T - \theta^T \Phi^T)(Y - \Phi\theta). \tag{B.7}$$

By expanding the expression further, we have

$$J(\theta, N) = Y^T Y - Y^T \Phi\theta - \theta^T \Phi^T Y - \theta^T \Phi^T \Phi\theta.$$

The parameters are obtained by minimizing the error or cost function, that is,

$$\frac{\partial J(\theta, N)}{\partial \theta} = \begin{Bmatrix} \dfrac{\partial J(\theta, N)}{\partial \theta_1} \\ \dfrac{\partial J(\theta, N)}{\partial \theta_2} \end{Bmatrix} = -2\Phi^T Y + 2\Phi^T \Phi\theta = 0,$$

which leads to the estimates of parameters from N measurements as follows:

$$\theta = [\Phi^T \Phi]^{-1} \Phi^T Y. \tag{B.8}$$

The least-squares technique presented here is a useful tool for identifying the parameters of a discrete transfer function with any order. The matrix multiplication and inversion can be evaluated by either writing a short computer code or by using popular software tools.

B.2 RECURSIVE PARAMETER ESTIMATION ALGORITHM

An on-line, recursive, regularized constant-trace algorithm is given below [48]:

$$\hat{\theta}(t) = \hat{\theta}(t-1) + a(t)k(t)(F_\mathrm{p}(t) - \phi^T(t)\hat{\theta}(t-1)), \tag{B.9}$$

$$k(t) = P(t-1)\phi(t)(1 + \phi^T(t)P(t-1)\phi(t))^{-1}, \tag{B.10}$$

$$\bar{P}(t) = \bar{P}(t-1) - a(t)k(t)\phi^T(t)P(t-1), \tag{B.11}$$

$$P(t) = c_1 \frac{\bar{P}(t)}{\mathrm{tr}(\bar{P}(t))} + c_2 I, \tag{B.12}$$

$$a(t) = \begin{cases} \bar{a} & \text{if } |F_\mathrm{p}(t) - \phi^T(t)\hat{\theta}(t-1)| > 2\delta, \\ 0 & \text{otherwise}, \end{cases} \tag{B.13}$$

where $\hat{\theta}(t) = [\hat{a}_1\ \hat{a}_2\ \hat{b}_0\ \hat{b}_1\ \hat{b}_2]^T$ is the estimated parameter vector, $k(t)$ is the estimation gain, $\phi(t) = [-F_\mathrm{p}(k-1)\ \ -F_\mathrm{p}(k-2)\ \ f_\mathrm{c}(k-1)\ \ f_\mathrm{c}(k-2)\ \ f_\mathrm{c}(k-3)]^T$ is the regression vector or observation vector, $\bar{P}(t)$ is the auxiliary covariance matrix, $\mathrm{tr}(\bar{P}(t)) = \sum_{i=1}^{5} \bar{P}_{ii}$ is the trace of the auxiliary covariance matrix, and $P(t)$ is the covariance matrix. Here, $c_1 > 0$, $c_2 \geq 0$, and δ is an estimate of the magnitude of the tolerable fluctuation of the output of the process or noise.

BIBLIOGRAPHY

[1] *MATLAB Users Guide*. MathWorks, Inc., Natick, MA, 1992.
[2] R.J. Allemang and D.L. Brown. Multiple input experimental modal analysis – a survey. *International Journal of Analytical and Experimental Modal Analysis*, 44:37–44, 1986.
[3] Y. Altintas. In-process detection of tool breakages using time series monitoring of cutting Forces. *International Journal of Machine Tool and Manufacturing*, 28(2):157–172, 1988.
[4] Y. Altintas. Prediction of cutting forces and tool breakage in milling from feed drive current measurements. *ASME Journal of Engineering for Industry*, 114(4):386–392, 1992.
[5] Y. Altintas. Direct adaptive control of end milling process. *International Journal of Machine Tools and Manufacture*, 34(4):461–472, 1994.
[6] Y. Altintas. Analytical prediction of three dimensional chatter stability in milling. *JSME International Journal, Series C: Mechanical Systems, Machine Elements and Manufacturing*, 44(3):717–723, 2001.
[7] Y. Altintas, C. Brecher, M. Weck, and S. Witt. Virtual machine tool. *CIRP Annals*, 54(2):651–674, 2005.
[8] Y. Altintas and E. Budak. Analytical prediction of stability lobes in milling. *CIRP Annals*, 44(1):357–362, 1995.
[9] Y. Altintas and P. Chan. In-process detection and supression of chatter in milling. *International Journal of Machine Tools and Manufacture*, 32:329–347, 1992.
[10] Y. Altintas, K. Erkorkmaz, and W. Zhu. Sliding mode controller design for high speed feed drives. *CIRP Annals*, 49(1):265–270, 2000.
[11] Y. Altintas and N.A. Erol. Open architecture modular tool kit for motion and machining process control. *CIRP Annals*, 47(1):295–300, 1998.
[12] Y. Altintas, M. Eyniyan, and M. Onozuka, Identification of dynamic cutting force coefficients and chatter stability with process damping. *CIRP Journal of Manufacturing Science and Technology*, 57(1):371–374, 2008.
[13] Y. Altintas and J.H. Ko. Chatter stability of plunge milling. *CIRP Annals*, 55(1):361–364, 2006.
[14] Y. Altintas and A.J. Lane. Design of an electro-hydraulic CNC press brake. *International Journal of Machine Tools and Manufacture*, 37:45–59, 1997.
[15] Y. Altintas and P. Lee. A general mechanics and dynamics model for helical end mills. *CIRP Annals*, 45(1):59–64, 1996.
[16] Y. Altintas and K. Munasinghe. Modular CNC system design for intelligent machining. Part II: Modular integration of sensor based milling process monitoring and control tasks. *ASME Journal of Manufacturing Science and Engineering*, 118:514–521, 1996.

[17] Y. Altintas and W.K. Munasinghe. A hierarchical open architecture CNC system for machine tools. *CIRP Annals*, 43(1):349–354, 1994.

[18] Y. Altintas, N. Newell, and M. Ito. Modular CNC design for intelligent machining. Part I: Design of a hierarchical motion control module for CNC machine tools. *ASME Journal of Manufacturing Science and Engineering*, 118:506–513, 1996.

[19] Y. Altintas and J. Peng. Design and analysis of a modular CNC system. *Computers in Industry*, 13(4):305–316, 1990.

[20] Y. Altintas, G. Stepan, D. Merdol, and Z. Dombovari. Chatter stability of milling in frequency and discrete time domain. *CIRP Journal of Manufacturing Science and Technology*, (1):35–44, 2008.

[21] Y. Altintas, A. Verl, C. Brecher, L. Uriarte, and G. Pritschow. Machine tool feed drives. *CIRP Annals of Manufacturing Technology*, 60(2):779–796, 2011.

[22] Y. Altintas, I. Yellowley, and J. Tlusty. The detection of tool breakage in milling operations. *Journal of Engineering for Industry*, 110:271–277, 1988.

[23] Y. Altintas, and Spence, A. End milling force algorithms for CAD systems. *CIRP Annals*, 40(1):31–34, 1991.

[24] E.J.A. Armarego. *Material Removal Processes – An Intermediate Course*. The University of Melbourne, 1993.

[25] E.J.A. Armarego and R.H. Brown. *The Machining of Metals*. Prentice-Hall, 1969.

[26] E.J.A. Armarego, D. Pramanik, A.J.R. Smith, and R.C. Whitfield. Forces and power in drilling-computer aided predictions. *Journal of Engineering Production*, 6:149–174, 1983.

[27] E.J.A. Armarego and M. Uthaichaya. A mechanics of cutting approach for force prediction in turning operations. *Journal of Engineering Production*, 1(1):1–18, 1977.

[28] E.J.A. Armarego and R.C. Whitfield. Computer based modelling of popular machining operations for force and power predictions. *CIRP Annals*, 34:65–69, 1985.

[29] K.J. Astrom and B. Wittenmark. *Computer-Controlled Systems*. Prentice Hall Inc., 1984.

[30] G. Boothroyd. Temperatures in orthogonal metal cutting. *Proceedings of the Institution of Mechanical Engineers*, 177:789–802, 1963.

[31] G. Boothroyd. *Fundamentals of Machining and Machine Tools*. McGraw-Hill, 1975.

[32] E. Budak and Y. Altintas. Flexible milling force model for improved surface error predictions. In *Proceedings of the 1992 Engineering System Design and Analysis, Istanbul, Turkey*, pages 89–94. ASME, 1992. PD-Vol. 47-1.

[33] E. Budak and Y. Altintas. Peripheral milling conditions for improved dimensional accuracy. *International Journal of Machine Tools and Manufacture*, 34(7):907–918, 1994.

[34] E. Budak and Y. Altintas. Analytical prediction of chatter stability in milling. Part I: General formulation; Part II: Application of the general formulation to common milling systems. *ASME Journal of Dynamic Systems, Measurement and Control*, 120:22–36, 1998.

[35] E. Budak, Y. Altintas, and E.J.A. Armarego. Prediction of milling force coefficients from orthogonal cutting data. *ASME Journal of Manufacturing Science and Engineering*, 118:216–224, 1996.

[36] CAMI. *Encylopedia of the APT Programming Language*. Computer Aided Manufacturing International, Arlington,Texas, 1973.

[37] L.H. Chen and S.M. Wu. Further investigation of multifaceted drills. *Trans. ASME J. Engineering for Industry*, 106:313–324, 1984.

[38] R.Y. Chiou and S.Y Liang. Chatter stability of a slender cutting tool in turning with wear effect. *International Journal of Machine Tools and Manufacture*, 38(4):315–327, 1998.

[39] B.E. Clancy and Y.C. Shin. A comprehensive chatter prediction model for face turning operation including the tool wear effect. *International Journal of Machine Tools and Manufacture*, 42(9):1035–1044, 2002.

[40] D.W. Clarke, C. Mohtadi, and P.S Tuffs. Generalized predictive control. Part I: The basic algorithm. *Automatica*, 23(2):137–148, 1987.

[41] R.C. Colwell. Predicting the angle of chip flow for single point cutting tools. *Transactions of the ASME*, 76:199–204, 1954.

[42] M.K. Das and S.A. Tobias. The relation between static and dynamic cutting of metals. *International Journal of Machine Tools and Manufacture*, Vol. 7:63–89, 1967.

[43] B.R. Dewey. *Computer Graphics for Engineers*. Harper & Row Publishers, New York, 1988.

[44] D.J. Ewins. *Modal Testing Theory and Practice*. Research Studies Press, Baldock, United Kingdom, 1984.

[45] M. Eyniyan and Y. Altintas. Chatter stability of general turning operations with process damping. *Journal of Manufacturing Science and Engineering, Transactions of the ASME*, 131(4):1–10, 2009.

[46] H.J. Fu, R.E. DeVor, and S.G. Kapoor. A mechanistic model for the prediction of the force system in face milling operations. *ASME Journal of Engineering for Industry*, 106:81–88, 1984.

[47] D.J. Galloway. Some experiments on the influence of various factors on drill performance. *Transactions of the ASME*, 79:139, 1957.

[48] C.G. Goodwin and K.S. Sin. *Adaptive Filtering Prediction and Control*. Prentice Hall, Inc., 1984.

[49] E. Govekar, J. Gradišek, M. Kalveram, T. Insperger, K. Weinert, G. Stépàn, and I. Grabec. On stability and dynamics of milling at small radial immersion. *CIRP Annals-Manufacturing Technology*, 54(1):357–362, 2005.

[50] W.G. Halvorsen and D.L. Brown. Impulse technique for structural frequency response. *Sound & Vibration*, vol. 11/11:8–21, 1977.

[51] R.E. Hohn, R. Sridhar, and G.W. Long. A stability algorithm for a special case of the milling process. *ASME Journal of Engineering for Industry*, vol. 90:325–329, 1968.

[52] T. Insperger and G. Stépán. Stability of the milling process. *Periodica Polytechnica*, 44:47–57, 2000.

[53] T. Insperger and G. Stépán. Updated semi-discretization method for periodic delay-differentilal quations with discrete delay. *International Journal for Numerical Methods in Engineering*, 61:117–141, 2004.

[54] I.S. Jawahir and C.A. van Luttervalt. Recent developments in chip control research and applications. *CIRP Annals*, 42/2:49–54, 1993.

[55] S. Kaldor and E. Lenz. Drill point geometry and optimization. *ASME Journal of Engineering for Industry*, 104:84–90, 1982.

[56] S. Kalpakjian. *Manufacturing Engineering and Technology*. Addison-Wesley Publishing Company, 1995.

[57] R.I. King. *Handbook of High Speed Machining Technology*. Chapman and Hall, 1985.

[58] W.A. Kline, R.E. DeVor, and I.A. Shareef. The prediction of surface accuracy in end milling. *ASME Journal of Engineering for Industry*, 104:272–278, 1982.

[59] W.A. Kline, R.E. DeVor, and W.J. Zdeblick. A mechanistic model for the force system in end milling with application to machining airframe structures. In: *North American Manufacturing Research Conference Proceedings, Dearborn, MI*, page 297. Society of Manufacturing Engineers, Vol. XVIII, 1980.

[60] W. Kluft, W. Konig, C.A. Lutterwelt, K. Nagayama, and A.J. Pekelharing. Present knowledge of chip form. *CIRP Annals*, 28/2:441–455, 1979.

[61] F. Koenigsberger and J. Tlusty. *Machine Tool Structures. Vol. I: Stability Against Chatter*. Pergamon Press, 1967.

[62] W. Konig and K. Essel. New tool materials – wear mechanism and application. *CIRP Annals*, 24/1:1–5, 1975.

[63] Y. Koren. *Computer Control of Manufacturing Systems*. McGraw Hill, 1983.

[64] J. Krystof. *Berichte uber Betriebswissenschaftliche Arbeiten, Bd., 12*. VDI Verlag, 1939.

[65] C.K. Kuo. *Digital Control Systems*. 2nd Edition. Saunders College Publishing. R. Worth, 1992.

[66] UBC Manufacturing Automation Laboratory. *CUTPRO ©Advanced Machining Process Measurement and Simulation Software*. MAL Manufacturing Automation Lab. Inc., Vancouver, BC, Canada, 2000.

[67] E.H. Lee and B.W. Shaffer. Theory of plasticity applied to the problem of machining. *Journal of Applied Mechanics*, 18:405–413, 1951.

[68] G.C.I. Lin, P. Mathew, P.L.B. Oxley, and A.R. Watson. Predicting cutting force for oblique machining conditions. *Proceedings of the Institution of Mechanical Engineers*, 196:141–148, 1982.

[69] T.N. Loladze. Nature of brittle failure of cutting tool. *CIRP Annals.*, 24/1:13–16, 1975.

[70] W.K. Luk. The direction of chip flow in oblique cutting. *International of Journal of Production Research*, 10:67–76, 1972.

[71] M.E. Martellotti. An analysis of the milling process. *Transactions of the ASME*, 63:677–700, 1941.

[72] M.E. Martellotti. An analysis of the milling process. Part II: Down milling. *Transactions of the ASME*, 67:233–251, 1945.

[73] M.E. Merchant. Basic mechanics of the metal cutting process, *ASME Journal of Applied Mechanics*, vol. 11:A168–A175, 1944.

[74] M.E. Merchant. Mechanics of the metal cutting process. *Journal of Applied Physics*, vol. 16:267–279, 1945.

[75] M.E. Merchant. Mechanics of the metal cutting process. II: Plasticity conditions in orthogonal cutting. *Journal of Applied Physics*, 16:318–324, 1945.

[76] S.D. Merdol and Y. Altintas. Multi frequency solution of chatter stability for low immersion milling. *Journal of Manufacturing Science and Engineering, Transactions of the ASME*, 126(3):459–466, 2004.

[77] H.E. Merrit. Theory of self-excited machine tool chatter. *ASME Journal of Engineering for Industry*, 87:447–454, 1965.

[78] H.E. Merrit. *Hydraulic Control Systems*. John Wiley & Sons, New York, 1967.

[79] I. Minis, T. Yanushevsky, R. Tembo, and R. Hocken. Analysis of linear and nonlinear chatter in milling. *CIRP Annals*, 39:459–462, 1990.

[80] K. Ogata. *Modern Control Engineering*. Prentice Hall, Inc., 1970.

[81] K. Ogata. *Discrete Time Control Systems*. Prentice Hall, Inc., 1987.

[82] H. Opitz and F. Bernardi. Investigation and calculation of the chatter behavior of lathes and milling machines. *CIRP Annals*, 18:335–343, 1970.

[83] P.L.B. Oxley. *The Mechanics of Machining*. Ellis Horwood Limited, 1989.

[84] W.B. Palmer and P.L.B. Oxley. Mechanics of orthogonal machining. *Proceedings of the Institution of Mechanical Engineers*, 173(24):623–654, 1959.

[85] J. Peters and P. Vanherck. Machine tool stability and incremental stiffness. *CIRP Annals*, 17/1, 1967.

[86] W.H. Press, B.P. Flannery, S.A. Tenkolsky, and W.T. Vetterling. *Numerical Recipes in C*. Cambridge University Press, 1988.

[87] G. Pritschow. A comparison of linear and conventional electromechanical drives. *CIRP Annals*, 47/2:541-547, 1998.

[88] G. Pritschow, C. Eppler and W-D. Lehner. Ferraris sensor: the key for advanced dynamic drives. *CIRP Annals*, 52/1:289-292, 2003.

[89] S.S Rao. *Mechanical Vibrations*. Addison-Wesley Publishing Company, 1990.

[90] H. Ren. Mechanics of machining with chamfered tools. Master's thesis, The University of British Columbia, 1998.

[91] D.M. Ricon, Ulsoy, A.G. Effects of drill vibrations on cutting forfces and torque. *CIRP Annals*, 43(1):59–62, 1994.

[92] J.C. Roukema and Y. Altintas. Generalized modeling of drilling vibrations. Part I: Time domain model of drilling kinematics, dynamics and hole formation. *International Journal of Machine Tools and Manufacture*, 47(9):1455–1473, 2007.

[93] J.C. Roukema and Y. Altintas. Generalized modeling of drilling vibrations. Part II: Chatter stability in frequency domain. *International Journal of Machine Tools and Manufacture*, 47(9):1474–1485, 2007.

[94] T.L. Schmitz and K.S. Smith. *Machining Dynamics – Frequency Response to Improved Productivity*. Springer, 2009.

[95] E. Shamoto and Y. Altintas. Prediction of shear angle in oblique cutting with maximum shear stress and minimum energy principle. *ASME Journal of Manufacturing Science and Engineering*, 121:399–407, 1999.

[96] M.C. Shaw. *Metal Cutting Principles*. Oxford University Press, 1984.

[97] S. Smith and T. Delio. Sensor-based control for chatter-free milling by spindle speed selection. In *Proceedings of ASME 1989 WAM: Winter Annual Meeting*, 18:107–114, 1989.

[98] S. Smith and J. Tlusty. An overview of modelling and simulation of the milling process. *ASME Journal of Engineering for Industry*, 113(2):169–175, 1991.

[99] A. Spence and Y. Altintas. A solid modeller based milling process simulation and planning system. *ASME Journal of Engineering for Industry*, 116(1):61–69, 1994.

[100] A.D. Spence and Y. Altintas. CAD assisted adaptive control for milling. *Trans. ASME J. Dynamic Systems, Measurement and Control*, 116:61–69, 1991.

[101] L.N. Srinivasan and Q.J. Ge. Parameteric continuous and smooth motion interpolation. *Journal of Mechanical Design*, 118(4):494–498, 1996.

[102] G.V. Stabler. Fundamental geometry of cutting tools. *Proceedings of the Institution of Mechanical Engineers*, Vol. 165:14–26, 1951.

[103] G.V. Stabler. The chip flow law and its consequences. *Proceedings of 5th Machine Tool Design and Research Conference*, 243–251, 1964.

[104] D.A. Stephenson. Material characterization for metal-cutting force modeling. *Journal of Engineering Materials and Technology*, 111:210–219, 1989.

[105] G. Stute. *Electrical Feed Drives for Machine Tools*. Edited by H. Gross, John Wiley Inc., 1983.

[106] J.W. Sutherland and R.E. DeVor. An improved method for cutting force and surface error prediction in flexible end milling systems. *ASME Journal of Engineering for Industry*, 108:269–279, 1986.

[107] A.O. Tay, M.G. Stevenson, G. De Vahl Davis, and P.L.B. Oxley. A numerical method for calculating temperature distributions in machining, from force and shear angle measurements. *International Journal of Machine Tool Design and Research*, 16:335–349, 1976.

[108] F.W. Taylor. On the art of cutting metals. *Transactions of the ASME*, vol. 28:31–350, 1907.

[109] THK. *THK LM System Ball Screws, Catalog No. 75-IBE-2*. T H K Co., Japan, 1996.

[110] J. Tlusty and F. Ismail. Basic nonlinearity in machining chatter. *CIRP Annals*, 30:21–25, 1981.

[111] J. Tlusty and T Moriwaki. Experimental and computational identification of dynamic structural methods. *CIRP Annals*, 25:497–503, 1976.

[112] J. Tlusty and M. Polacek. The stability of machine tools against self-excited vibrations in machining. *International Research in Production Engineering, ASME*, vol. 1:465–474, 1963.

[113] S.A Tobias. *Machine Tool Vibration*. Blackie and Sons Ltd., 1965.

[114] S.A. Tobias and W. Fishwick. A theory of regenerative chatter. *The Engineer – London*, vol. 205:139–239, 1958.

[115] E.M. Trent. *Metal Cutting*. Butterworths, 1977.

[116] F.C. Wang, S. Schofield, and P. Wright. Open architecture controllers for machine tools. Part II: A real time quintic spline interpolator. *Journal of Manufacturing Science and Engineering*, 120(2):425–432, 1998.

[117] M. Weck, E. Verhagg, and M. Gather. Adaptive control of face milling operations with strategies for avoiding chatter-vibrations and for automatic cut distribution. *CIRP Annals*, 24(1):405–409, 1975.

[118] I. Zeid. *CAD/CAM, Theory and Practice*. McGraw Hil, New York, 1991.

[119] N.N. Zorev. Interrelationship between shear processes occurring along tool face and on shear plane in metal cutting. *ASME Proceedings of International Research in Production Engineering Research Conference, New York*, pages 42–49, New York. September 1963.

[120] N.N. Zorev. *Metal Cutting Mechanics*. Pergamon Press, 1966.

INDEX